Mathematics of Epidemics on Networks

Interdisciplinary Applied Mathematics

Volume 46

More information about this series at http://www.springer.com/series/1390

István Z. Kiss • Joel C. Miller • Péter L. Simon

Mathematics of Epidemics on Networks

From Exact to Approximate Models

 Springer

István Z. Kiss
Department of Mathematics
University of Sussex
Falmer, Brighton, UK

Joel C. Miller
Applied Mathematics
Institute for Disease Modeling
Bellevue, WA, USA

Péter L. Simon
Institute of Mathematics
Eötvös Loránd University
Budapest, Hungary

ISSN 0939-6047   ISSN 2196-9973   (electronic)
Interdisciplinary Applied Mathematics
ISBN 978-3-319-84494-7   ISBN 978-3-319-50806-1   (eBook)
DOI 10.1007/978-3-319-50806-1

Mathematics Subject Classification (2010): 00A71, 00A72, 05C82, 34C11, 34C23, 34D20, 35Q84, 37G10, 37N25, 47D06, 47N40, 60J27, 60J28, 60J75, 60J80, 60J85, 60K20, 60K35, 82B43, 91D30, 92C42, 92D30

Printed on acid-free paper

This Springer imprint is published by Springer Nature
The registered company is Springer International Publishing AG
The registered company address is: Gewerbestrasse 11, 6330 Cham, Switzerland

In author order, we dedicate this book to our wives Katalin, Anja, and Nadinka and our children Louisa and ε_2; Maaike and Jojanneke; and Nadinka, Zsófia, Viktória, Bernadett, and Brúnó.

Preface

Over the past decade, the use of networks has led to a new modelling paradigm combining several branches of science, including physics, mathematics, biology and social sciences. The spread of infectious diseases between nodes in a network has been a central topic of this growing field. The fundamental questions are easily stated, but answering them draws on observations and techniques of many fields.

There is a long successful history of mathematical modelling informing policies to mitigate the impact of infectious disease. Typically, models divide the population into compartments based on infection status and use simple assumptions about mixing and movements between these compartments. Over time, these models have grown more sophisticated to more accurately incorporate the contact structure of the population and to take advantage of increased computational resources. For example, sexually transmitted diseases have been investigated using high-dimensional compartmental models separating individuals by contact rates, socio-economic status and many other factors. However, when we make the additional observation that partnerships may be long-lasting, a new paradigm is needed, leading naturally to a network representation of the population structure.

Progress in model development has been extremely fast and has attracted interest from a diverse set of researchers. The fundamental objective is to combine the underlying population contact structure and the properties of the infectious agent to yield an understanding of the resulting spectrum of epidemic behaviours. To do this, researchers translate observed population and disease properties into a well-defined model. In many cases, the model sits at the interface of graph/network theory, stochastic processes and probability theory, dynamical systems, and statistical physics. The diversity of researcher backgrounds and the variety of applications considered have led to the development of many different modelling approaches. As the field matures, there is a need to increase understanding of how these different models fit together, how they relate to the underlying assumptions and how to develop an appropriate mathematical framework to unify different approaches.

This book sets out to make a contribution to modelling epidemics on networks by synthesising a large pool of models, ranging from exact and stochastic to approximate differential equation models, so that we may:

1. recognise underlying model assumptions and the resulting model complexity;
2. provide a mathematical framework with which we can describe observed phenomena and predict future scenarios;
3. permit direct comparison of the main models and provide their hierarchy; and
4. identify research gaps and opportunities for further rigorous mathematical exploration.

Chapter 1 introduces the reader to the fundamentals of disease transmission models and the underlying networks. Chapter 2 takes a rigorous probabilistic view and frames disease transmission on a network as a continuous-time Markov chain. In contrast, Chapter 3 builds a hierarchy of models starting at the node level which depend on the node–neighbour pairs, which in turn depend on triples formed by considering the next-nearest neighbours. Chapter 4 focuses on mean-field and pairwise models and their analysis on homogeneous networks. Chapter 5 extends approaches of Chapter 4 to heterogeneous networks and introduces effective degree models. In Chapter 6, the focus is primarily on SIR epidemics, and percolation theory methods are used to derive the low-dimensional edge-based compartmental model. Chapter 7 brings the different SIR models together, showing that under reasonable assumptions, the high-dimensional models of earlier chapters reduce to the low-dimensional model of Chapter 6. Chapter 8 extends the earlier models to account for the simultaneous spread of the disease and change in the network, considering several scenarios for how networks vary in time. Chapter 9 generalises the pairwise and edge-based compartmental models to non-Markovian epidemics, leading to integro-differential and delay differential equations. Chapter 10 starts from a Markov chain to derive the Fokker–Planck equation for the distribution of the number of infected individuals as a function of time and uses the resulting partial differential equation (PDE) to investigate epidemic processes. Finally, Chapter 11 shows that our models can perform surprisingly well even in networks, including empirically observed networks, for which the assumptions they are based on do not appear to be satisfied. The Appendix gives efficient simulation algorithms and discusses issues encountered in simulating epidemics on networks.

With more space, we would have liked to make a stronger emphasis on probabilistic models. Moreover, we would have examined epidemic control measures such as vaccination and contact tracing, as well as household models. Many other topics, for example, multilayer networks (networks with multiple types of connections), are left out, although many of the techniques we discuss apply to them. An additional topic, deserving of a book on its own, would be the use of real-world data to parametrise network models.

This book contains a number of rigorous mathematical arguments and proofs. However, a guiding principle throughout is to appeal to and be useful for audiences in fields outside of mathematics. Some quantitative sophistication will be necessary; in particular, previous exposure to linear algebra, calculus, differential equations, dynamical systems and basics of probability and stochastic processes would be useful. We do not assume knowledge of graph theory.

Advanced undergraduate and graduate students can use the book as a foundation for learning the main modelling and analysis techniques. There are many exercises designed to develop a deeper understanding of the topic. Models and results of immediate applicability are signposted through the use of grey boxes.

We use this format to highlight readily implementable models or to summarise model outcomes, such as steady states, final epidemic size, basic reproductive ratio R_0, probability of an epidemic, etc.

Doctoral students, researchers and experts in this area can use the book not only as a reference guide or synthesis of the major modelling frameworks and model analysis tools but also to (i) confirm the validity and optimal range of applicability of models, (ii) understand how mathematical tools have been and are used in network modelling and (iii) identify further synergies between mainstream mathematical methods and problems arising in network modelling.

To enhance the flow of the presentation, citations to previous research are concentrated either at the beginning or end of chapters. This allows us to (a) build up models from the ground up by unifying different approaches leading to synthesised models and (b) cite further new developments that we could not cover.

Pseudocode for efficient epidemic simulation algorithms is given in the Appendix, and ready-to-run source code is available at the following website:

https://springer-math.github.io/Mathematics-of-Epidemics-on-Networks/

These include stochastic simulation of SIS and SIR on networks and numerical solutions of many differential equation models we present in the book. An extensive Python package using NetworkX [130] is provided, and many of these are also available in Matlab. We hope to add additional languages. These will help readers to complete many of the simulation-based exercises proposed in the book and may assist other researchers with their own projects. Other resources are available; for example, a useful package in C++ is EpiFire [143]. Solutions to exercises will be made available for instructors who use the book. Inevitably, small errors creep into any book. Please contact us directly for solutions or to report errors.

Acknowledgements: The authors wish to thank their former and current co-workers and collaborators, research students and their current and former institutions. IZ Kiss thanks the University of Leeds, the University of Oxford and the University of Sussex; JC Miller thanks Pennsylvania State University (Penn State), Monash University (in particular MAXIMA) and the Institute for Disease Modeling and PL Simon thanks the Department of Applied Analysis and Computational Mathematics and the whole Institute of Mathematics at the Eötvös Loránd University in Budapest, Hungary. The community at tex.stackexchange. com has helped with intricacies of LaTeX. The authors thank Dr. John Haigh from the University of Sussex for reading early drafts with great care and attention to detail. Finally, the authors thank their families for providing constant support and encouragement.

Final thoughts: We would like to end with a memorable summary of our book about epidemics on networks. We hope this epidemic sonnet works:

> *When partnerships endure so long that to*
> *Disease they are like frozen ties that bind,*
> *Mass action fails us till new paradigms*
> *Emerge; and networks then are useful tools.*
>
> *Equation counts are exponential till*
> *Reduced — through automorphic symmetries*
> *Or caref'ly cutting out some vertices.*
> *But yet complexity is too high still.*
>
> *And so our mod'ler must approximate*
> *And close equations — but not too simply.*
> *For she must doubly count a high degree.*
> *Or, she may watch diseases percolate.*
>
> *With these techniques our mod'ler has new keys*
> *To learn how partnerships affect disease.*

Brighton, UK   István Z. Kiss
Bellevue, WA, USA   Joel C. Miller
Budapest, Hungary   Péter L. Simon
November 2016

About the Authors

Dr I.Z. Kiss is a Reader in the Department of Mathematics at the University of Sussex with his research at the interface of network science, stochastic processes and dynamical systems. His work focuses on the modelling and analysis of stochastic epidemic processes on static and dynamic networks. His current interests include the identification of rigorous links between approximate models and their rigorous mathematical counterparts and the formulation of new models for more complex spreading processes or structured networks.

Dr J.C. Miller is a Senior Research Scientist at the Institute for Disease Modeling in Seattle. He is also a Senior Lecturer at Monash University in Melbourne with a joint appointment in Mathematics and Biology. His research interests include dynamics of infectious diseases, stochastic processes on networks and fluid flow in porous media. The majority of his work is at the intersection of infectious disease dynamics and stochastic processes on networks.

Prof P.L. Simon is a Professor at the Institute of Mathematics, Eötvös Loránd University, Budapest. He is a member of the Numerical Analysis and Large Networks research group and the Head of Department of Applied Analysis and Computational Mathematics. His research interests include dynamical systems, partial differential equations and their applications in chemistry and biology. In particular, his work focuses on the modelling and analysis of network processes using differential equations.

Contents

Chapter 1
Introduction to networks and diseases

Mathematical models are caricatures of real systems that aim to capture the fundamental mechanisms of some process in order to explain observations or predict outcomes. No model — no matter how complicated — is perfect, or in the words of George Box [46]: "All models are wrong; some models are useful". A useful model can provide valuable insights which improve our understanding of a system, and ultimately informs our decision-making.

Infectious diseases are responsible for a significant health and economic burden on society. New diseases appear frequently, and old diseases persist. Mathematical models have been used recently to guide policy makers responding to the emergence of diseases including foot and mouth, SARS, H1N1 influenza and Ebola; for controlling established diseases such as HIV, cholera and seasonal influenza; and for guiding strategies to eliminate diseases such as polio and Guinea worm.

A variety of modelling approaches have been used, ranging from simple compartmental models that assume a well-mixed population divided into a few compartments to large-scale agent-based models that simulate individuals' movements within cities or between countries. The simple models allow us to analytically explain how the primary mechanisms influence important characteristics such as whether an epidemic will happen, how long it will last, and how large it will be. To obtain these simple explanations, we must sacrifice accuracy by neglecting secondary effects. At the other end of the spectrum, one can develop models that incorporate much more detail about both the disease and individual-level interactions. These models may have more predictive power, but their complexity can mask the underlying causal relations. The way in which model realism, complexity, and the insight a model offers are related is represented in Fig. 1.1. This figure shows where, in our opinion, classical simple compartmental models, complex but tractable network models and data-driven agent-based models sit. As the figure suggests, models provide maximum insight when the right balance of realism and simplicity is met. Where this balance lies may depend on the specific question under investigation.

© Springer International Publishing AG 2017
I.Z. Kiss et al., *Mathematics of Epidemics on Networks*, Interdisciplinary Applied Mathematics 46, DOI 10.1007/978-3-319-50806-1_1

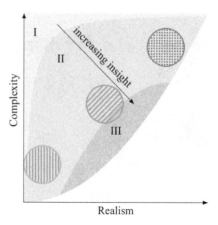

Fig. 1.1: A caricature of the relation between model realism, complexity and the insight the model provides. The lower right region in yellow is excluded — a minimum amount of complexity is needed to capture a given level of realism. Insight is limited by model complexity (which reduces our ability to understand causal mechanisms) and realism (we cannot learn about an effect that a model does not include). Moving from lighter shaded regions (I) to darker regions (III) increases the insight we gain. We expect the most insight in some intermediate region where the model contains the most significant effects while still being understandable. The circular areas with vertical lines, oblique lines and dots are the areas where simple deterministic models, more complex but tractable network models and data-driven agent-based models based on simulations sit. In truth, the three paradigms blend together, and the distinctions between them are less clear.

Network science has become a well-established and productive research area starting from the late 1990s and early 2000s. Initial studies focused on empirical networks and aimed to understand how networks emerge and what are their fundamental properties. It turned out that degree heterogeneity (variation in the number of neighbours different nodes have) is a ubiquitous feature of many real-world networks, ranging from biological to socio-technical networks. Subsequently, a better understanding of how real-world networks emerge and evolve has led to the development of a myriad of theoretical/synthetic network models. These models in turn lead to an improved understanding of the impact of network properties on the unfolding of processes on networks, such as disease transmission, flow of information and memes, as well as the flow of goods, travel or financial transactions.

The overarching theme of this book is the analysis of stochastic processes unfolding on networks in order to understand the final state and the intermediate dynamics. The main tools we use are mean-field and related deterministic models, as well as some stochastic approaches such as percolation theory. The central challenge is the development of simple models which capture sufficient detail to give meaningful results. Figure 1.2 shows that three networks with the same average degree (number of neighbours per node) but different degree distributions lead to epidemics that be-

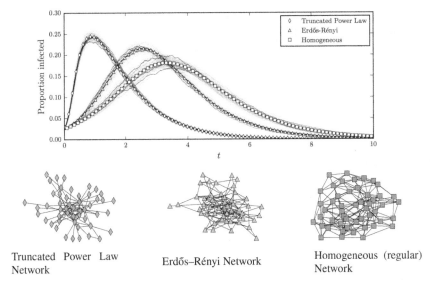

Fig. 1.2: A comparison of three networks with the same number of nodes and edges, but different distribution of degrees. It is prohibitively difficult to study an exact model of the stochastic dynamics, so we rely on stochastic simulation. (Top) Simulated SIR epidemics in networks of $N = 10^5$ nodes, each with an average degree (number of edges per node) of 6, taking the per-edge transmission rate $\tau = 0.5$ and recovery rate $\gamma = 1$. Light curves show a cloud of 50 simulations per network. Thin black curves highlight a few simulations. The thick black curve gives the average of the simulations. Symbols show predictions from analytical techniques developed in Chapters 4–6. The simulations begin with 1% infected, and $t = 0$ is set when prevalence reaches 2.5% in order to reduce stochastic noise in the average. (Bottom) Sample small networks exhibiting the degree distributions of the networks used in the simulations, in order of decreasing heterogeneity. (Left) The truncated power law network: a node has degree k with probability proportional to $k^{-1.5}$ for k between 1 and 61, and 0 otherwise. (Middle) An Erdős–Rényi network in which each pair of nodes has an edge independently of any other with probability $6/(N-1)$. (Right) A homogeneous network in which all nodes have exactly 6 neighbours.

have differently. The figure demonstrates that (a) the properties of the network have a significant impact on the epidemic and that (b) it is possible to approximate the average behaviour of the stochastic process using mean-field models, which then allow us to derive analytical results that shed light on the impact of network properties on the likelihood of an epidemic outbreak and its size. By comparing the simulations in the top panel, it is evident that the epidemic grows faster as the variance or diversity in the number of links that nodes have increases. Moreover, epidemics also peak at lower values as we move from left to right, from power law to uniform networks.

Due to the ubiquitous nature of networks and the modelling flexibility that they offer, the study of networks and modelling stochastic processes on networks have been developed in parallel in many different areas, including physics, mathematics, biology and social sciences, and a large number of network models with or without dynamics have been developed. Disease spread is perhaps the most widely studied dynamic process on networks. The theoretical analysis of epidemic dynamics has led to a variety of stochastic and deterministic mean-field models which have great practical and theoretical relevance and are closely related. Progress in model development has been fast, and diverse models have been developed heuristically. However, rigorous mathematical analysis has started to catch up with modelling. These include:

1. derivation of pairwise models directly from the exact probabilistic model;
2. derivation of models and exactness proofs for SIR epidemics on tree networks;
3. comparison theorems between exact stochastic and mean-field models;
4. exactness of the edge-based compartmental model for Configuration Model networks in the limit of network size going to infinity;
5. model hierarchy results which help us understand model assumptions, complexity and interrelatedness; and
6. usage of mathematically rigorous tools, such as partial differential equations (PDEs), to analyse mean-field models and link them directly to the exact probabilistic counterpart.

Some of these results are presented or developed for the first time here. Since this book aims to provide a synthesis of both model development and their analysis, in this introduction we give the main ingredients for this journey. We cover aspects including the disease itself, networks and the mathematical techniques needed to connect the two.

1.1 Mathematical modelling of epidemics: the basics

A feature that distinguishes an infectious disease such as influenza from diseases such as cancer or arthritis is that the presence of infected individuals directly causes additional infections. This immediately creates a feedback loop: infected individuals cause more infections, leading to exponential growth, which reduces the susceptible pool and limits future spread. Mathematical modelling of epidemic propagation goes back nearly a century. One of the earliest models was developed and analysed extensively by Kermack and McKendrick [174] for diseases which provide immunity. Several textbooks have been published on mathematical epidemiology in the last decades: without aiming to be exhaustive, we mention here the books by Anderson and May [4], Mollison [223], Isham and Medley [153], Diekmann and Heesterbeek [78], Daley and Gani [74], Brauer, van den Driessche and Wu [48], Keeling and Rohani [168], Brauer and Castillo-Chavez [47], and Diekmann, Heesterbeek and Britton [81].

Many of these provide an overarching treatment of the topic, ranging from characterisation of diseases, model development, model analysis and epidemic control to predictions. The "need to know" components of epidemic models are presented below and should be sufficient to follow our account. It is also worth noting that the present book has a strong network focus and simple compartmental and stochastic models are presented only to set the scene.

Developing a basic model of how diseases progress within an individual and how they spread between individuals is the first important step in modelling. For example, a simplistic view of treatable/curable sexually transmitted infections is that individuals can either be susceptible (S) or infectious (I) and that upon being infected an individual can get treatment and then return to the susceptible cohort. This is shown in Fig. 1.3a, where the arrows represent paths along which individuals can move between compartments and the associated fluxes are determined by details of the transmission mechanism and recovery process.

Compartmental models can be refined and made disease and application specific by adding extra classes or setting up the compartments differently to reflect observable features of the infection process. For example, diseases which confer immunity after the first infection are better represented by a model shown in Fig. 1.3b. If the disease is seasonal or it confers short-lived immunity, the model can be set up as in Fig. 1.3c. Finally, if the latency period plays an important role and the disease confers immunity, then the model in Fig. 1.3d may be appropriate.

It is possible to further refine such models to account for behavioural traits or for other individual differences. For example, for sexually transmitted diseases it is possible to further divide the S and I compartments to account for individuals that engage in high-risk behaviour. This increases model realism at the cost of including more compartments. However, there is a limit to how far such models can be refined, and eventually other methods, such as network models, are needed.

The main task of disease modelling is to establish and formalise the flow or transitions between compartments so that one can track the number of individuals in the different compartments. As we show below, for the deterministic case this exercise leads to systems of ordinary differential equations (ODEs), where the variables represent the sizes of the different classes or compartments and their evolution equations encode the transmission and the physiology of the disease. For stochastic models, the sizes of the compartments will be random variables on an integer lattice with rates of moving between the elements of the lattice encoding the information about the disease. In such cases, the theory of Markov chains and branching processes helps us to make sense of, and analyse, the resulting models. Examples for both deterministic and stochastic models are given below.

1.1.1 Deterministic epidemic models: compartmental SIS & SIR

We will use the SIS and SIR disease models. The SIS model has two compartments: susceptible (*S*) and infected (*I*), and the SIR model has an additional recovered (*R*)

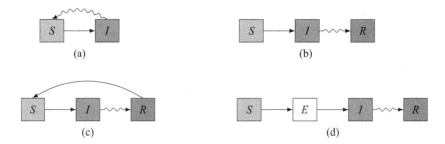

Fig. 1.3: Flow diagrams showing the flux between compartments consisting of individuals with the same status from the viewpoint of disease evolution, e.g. (*S*) - susceptible, (*E*) - exposed, infected but not infectious, (*I*) - infected and transmitting and (*R*) - recovered, removed or immune. The figure shows flow diagrams for (a) SIS, (b) SIR, (c) SIRS and (d) SEIR epidemic models.

compartment (see Fig. 1.3a and Fig. 1.3b). The model is written in the form of ODEs for the sizes of these compartments and is based on the idea that two transitions may happen between the compartments: infection (transition from S to I) and recovery (transition from I to S or to R). The rate of infection is assumed to be proportional to the sizes of the S and I compartments, while the rate of recovery is assumed to be proportional to the size of the infected compartment. Taking a dot to denote the time derivative, the SIS compartmental model is written in the form:

$$\dot{S}(t) = \gamma I(t) - \beta I(t)\frac{S(t)}{N}, \tag{1.1a}$$

$$\dot{I}(t) = \beta I(t)\frac{S(t)}{N} - \gamma I(t), \tag{1.1b}$$

where $S(t)$ and $I(t)$ denote the size of the susceptible and infected compartments at time t, and β and γ are positive constants called infection and recovery rates, respectively. It may be helpful to think of β as the rate at which infected individuals make infection-transmitting contacts. Then, the total rate of infectious contacts is βI, but only a fraction S/N of these are to susceptible individuals and thus lead to a new infection. It can be immediately seen that one of the equations can be omitted from the system, since $S(t) + I(t) = N$ is constant. Hence, the single differential equation for $I(t)$ is $\dot{I}(t) = \beta(1 - I(t)/N)I(t) - \gamma I(t)$, which can be solved by separation of variables. Instead of presenting the formula of the solution, we characterise its qualitative behaviour. The equation may have two steady states (or equilibria): the disease-free steady state $I_{df} = 0$ and the endemic steady state $I_e = N\left(1 - \frac{\gamma}{\beta}\right)$, which exists only if the "basic reproductive ratio" $R_0 = \beta/\gamma > 1$. Moreover, if $R_0 < 1$, the disease-free steady state is stable, while for $R_0 > 1$, the disease-free steady state is unstable and the endemic steady state $I_e = N\left(1 - \frac{1}{R_0}\right)$ is stable. The time dependence of

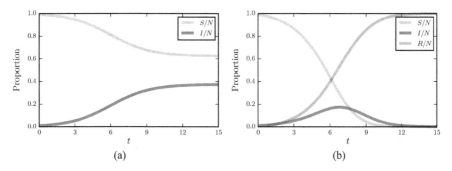

Fig. 1.4: The time dependence of the relative sizes of the compartments to the whole population for (a) an SIS epidemic and (b) an SIR epidemic. The recovery rate is $\gamma = 1$ and the infection rate is $\beta = 1.6$.

the number of infected and susceptible individuals is shown in Fig. 1.4 (a) for the case $R_0 > 1$. (The curves are the solutions of the ODEs divided by the population size N.)

For the SIR epidemic, the transition out of the infected compartment leads to the recovered one; hence, the SIR compartmental model can be written in the form:

$$\dot{S}(t) = -\beta I(t)\frac{S(t)}{N}, \tag{1.2a}$$

$$\dot{I}(t) = \beta I(t)\frac{S(t)}{N} - \gamma I(t), \tag{1.2b}$$

$$\dot{R}(t) = \gamma I(t), \tag{1.2c}$$

where the same notations are used as above and $R(t)$ denotes the size of the recovered compartment. A quick observation is that if $R_0 = \beta/\gamma < 1$, then I decreases, while if $R_0 > 1$, I will increase if the initial value of S is close enough to N. As for the SIS model, we can eliminate one equation using $S + I + R = N$. Fig. 1.4 (b) shows typical time dependence of compartment sizes.

Both S and R are bounded between 0 and N, and the signs of the right-hand sides show that they are monotonic functions; hence, they approach a limit. Because $\dot{R} \to 0$, we conclude I approaches 0. Thus, the system tends to a disease-free steady state as time goes to infinity; however, this steady state depends on the initial condition. One of the most important characteristics of the epidemic is the final epidemic size, R_∞, which is the limit of $R(t)$ as $t \to \infty$. This can be determined from the ODEs by simple integration as follows. Writing the first equation in the form $\dot{S}(t) + \frac{\beta}{N}S(t)I(t) = 0$ and multiplying it by the integrating factor $\exp\left(\frac{\beta}{N\gamma}R(t)\right)$, we get that $S(t)\exp\left(\frac{\beta}{N\gamma}R(t)\right)$ is constant in time. The value is determined by the initial condition. Assuming that $R(0) = 0$ and using that $S(\infty) + R(\infty) = N$, we have

$$N - R_\infty = S(0) \exp\left(-\frac{\beta}{N\gamma} R_\infty\right). \tag{1.3}$$

This implicit equation can be solved to arbitrary accuracy by iteration.

Compartmental models assume that links are switching at an infinitely fast rate, which guarantees that each transmission attempt is to a new individual. This makes the transmission more efficient compared to a case where contacts are fixed and the infection attempts play out over the same link or links. Specifically, the compartmental models lack the capability to fully account for the local depletion of susceptible individuals, which in a realistic contact network setting is an important factor, for example, in a household, school or the networks induced by sexually transmitted diseases. We note that compartmental models can also be formulated in discrete time, where the ODEs are replaced by difference equations.

1.1.2 Stochastic epidemic models: compartmental SIS & SIR

As alluded to previously, the starting point of many epidemic models is a stochastic formulation, which even for the simplest models involves an implicit network of connectivity between individuals. For the SIS model, $S(t)$ and $I(t) = N - S(t)$ are now random variables taking values from the set $\{0, 1, \ldots, N\}$. The focus of the model is then on determining evolution equations for the probabilities of observing a particular number of susceptible and infected individuals at a given time t, e.g. $P(I(t) = j) = p_j(t)$, where $j \in \{0, 1, \ldots, N\}$. Assume further that:

1. each individual can contact or is in contact with any other individual (the network of contacts is in fact a fully connected network);
2. infection is transmitted across a link between a susceptible and an infected individual at rate τ (so τN corresponds to β);
3. each infected individual recovers at rate γ independently of all others and of the network; and
4. both processes are Markovian (the infection and the recovery processes take place at points of Poisson processes with corresponding rates).

Treating this process as a continuous-time Markov chain and given the state of the system at time t, $(S,I)(t)$, the following two transitions are possible:

$$(S,I) \xrightarrow{\tau SI} (S-1, I+1), \tag{1.4a}$$

$$(S,I) \xrightarrow{\gamma I} (S+1, I-1), \tag{1.4b}$$

where the rates encode the transmission and recovery processes. Whether the next event is an infection or recovery is simply determined at random but relative to the magnitude of the two rates. Such a process can be given in terms of Kolmogorov equations (also referred to as master equations)

$$\dot{p}_j(t) = \tau(j-1)(N-j+1)p_{j-1} - (\tau j(N-j) + \gamma j)p_j + \gamma(j+1)p_{j+1}, \quad (1.5)$$

where $p_j(t)$ is the probability that we observe j infectious individuals at time t.

For the SIR model, the random variables are $S(t)$, $I(t)$, and $R(t) = N - S(t) - I(t)$. These take values from the set $\{0,1,\ldots,N\}$. Using the same notational convenience, $P(S(t) = i, I(t) = j) = p_{i,j}(t)$, and treating the epidemic as a continuous-time Markov chain, the transitions are given by

$$(S,I,R) \xrightarrow{\tau SI} (S-1,I+1,R), \qquad (1.6a)$$

$$(S,I,R) \xrightarrow{\gamma I} (S,I-1,R+1). \qquad (1.6b)$$

The master equation in this case is

$$\dot{p}_{i,j}(t) = \tau(i+1)(j-1)p_{i+1,j-1} - (\tau ij + \gamma j)p_{i,j} + \gamma(j+1)p_{i,j+1}, \qquad (1.7)$$

where $p_{i,j}(t)$ is the probability there are i susceptible, j infected and $N - (i+j)$ recovered individuals at time t, with $0 \leq i + j \leq N$.

How is the model above different from the deterministic one and what similarities are there? We begin by focusing on the initial stage of an epidemic. It turns out that this can be well approximated by a birth-and-death process, where a birth equates to a susceptible becoming infected while a death represents the recovery of an infected individual. Hence, the birth-and-death process can be written as

$$(S,I,R) \xrightarrow{\tau NI} (S-1,I+1,R), \qquad (1.8a)$$

$$(S,I,R) \xrightarrow{\gamma I} (S,I-1,R), \qquad (1.8b)$$

where we have assumed that initially the number of susceptible individuals is approximated well by N. Now from the theory of birth-and-death processes, we have that the population is certain to become extinct if the birth rate is less than or equal to the death rate,

$$\tau NI \leq \gamma I \Leftrightarrow R_0 = \frac{\tau N}{\gamma} \leq 1.$$

However, if $R_0 > 1$ and we start with i_0 infected individuals, two outcomes are possible: either no epidemic or an epidemic, with probabilities

$$\mathbb{P}(\text{no epidemic}) \simeq \left(\frac{1}{R_0}\right)^{i_0}, \qquad (1.9a)$$

$$\mathbb{P}(\text{epidemic}) \simeq 1 - \left(\frac{1}{R_0}\right)^{i_0}. \qquad (1.9b)$$

These results follow from the "gambler's ruin" problem. Fig. 1.5 plots the distribution of the final epidemic size from simulations. The outcome shows excellent agreement with the birth-and-death process approximation. A proportion of 0.3316 of the simulations lead to an epidemic, while equation (1.9b) predicts the epidemic probability to be $\mathbb{P}(\text{epidemic}) \simeq 1 - 2/3 = 1/3$. The main differences of stochastic compared to the deterministic models are:

1. $R_0 > 1$ does not necessarily lead to an outbreak;
2. the final size is described by a distribution and not by a single value; and
3. the stochastic SIS process is eventually absorbed by the disease-free state.

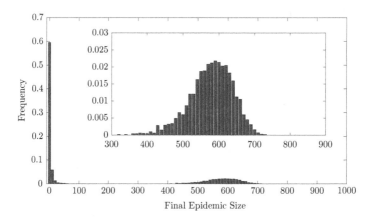

Fig. 1.5: Final epidemic size distribution for a fully connected network and SIR epidemic, with the inset showing an enlarged part of the main plot. The parameter values are $N = 1000$, number of initial infecteds $i_0 = 1$, recovery rate $\gamma = 1$ and infection rate $\tau = 1.5/N$ with 10000 simulations.

The all-to-all connectivity between individuals is a crude simplification of reality. In many cases contacts are more accurately represented as long-lasting links for a small subset of pairs of individuals or nodes in a network. Multiple infection attempts across the same link are possible, in which case the simplifying assumptions break down. The use of networks to model contact patterns opens up the possibility to account for features observed in realistic contact patterns, such as

- high heterogeneity or variation in the number of contacts that individuals have;
- the propensity of individuals to preferentially connect to other individuals of similar or dissimilar properties; or
- the tendency of already connected individuals to share further common friends,

and real-world data can be used to parametrise network models that include all of these features.

As with the deterministic models, it is possible to consider stochastic models in discrete time. Significantly, stochastic models are much more flexible in terms of the choice of inter-event time distributions. We can easily eliminate the assumption of exponential inter-event times, implicitly used above where all processes were Markovian. For deterministic models this is more complicated, and we consider non-Markovian deterministic models in Chapter 9. Efficient simulation algorithms are outlined in Appendix A.1.

1.1.3 Linking stochastic and deterministic compartmental models

Based on Fig. 1.2, which shows excellent agreement between the average of many individual stochastic simulations and a corresponding deterministic mean-field model, it is obvious that a rigorous mathematical link between the two exists. Even though the rigorous link is not discussed in detail here, we use the simple example of a fully connected network to help relate the two models.

We consider the SIS epidemic model in a fully connected network. With the deterministic model in mind, we consider the expected number of infectious individuals at time t rather than its distribution. Hence, the focus is on $[I](t) = \sum_{j=0}^{N} j p_j(t)$. The evolution equation for $[I](t)$, i.e. $[\dot{I}](t) = \sum_{j=0}^{N} j \dot{p}_j(t)$, can be obtained using Eq. (1.5). Simple algebra leads to

$$[\dot{I}](t) = \tau N[I](t) - \tau E(I^2) - \gamma[I](t).$$

This can be rewritten as

$$[\dot{I}](t) = \tau[I](t)(N - [I](t)) - \gamma[I](t) + \tau \text{Cov}(I, N - I), \qquad (1.10)$$

where we have used that $\text{Cov}(I, N - I) = E((I - [I])(N - I - (N - [I]))) = ([I]^2 - E(I^2))$. Now to follow a rigorous mathematical approach, one notes that the evolution equation for $[I](t)$ is not closed and depends on a new variable, namely $\text{Cov}(I, N - I) = ([I]^2 - E(I^2))$. There are at least two different ways to tackle the problem:

1. derive an evolution equation for the second moment $E(I^2)$ and thus implicitly for $\text{Cov}(I, N - I)$; or
2. break the dependency on higher order moments by employing some form of closure or approximation.

Neither approach is perfect. In general, the first leads to yet more new variables, so equations for higher moments are needed. This eventually forces us to go up to full system size. This is not feasible for any system of realistic size due to the large number of equations. The second approach seems more attractive if a "good" closure can be found. The availability of a closure can be situation dependent and could heavily depend on the model and its parameters.

As it turns out, the second approach is viable. Before we attempt a closure, we note that $\text{Cov}(I, N - I) \leq 0$, since the number of infected and susceptible individuals are negatively correlated. It is now informative to compare our equations with the deterministic model (1.1b) with $\beta \simeq \tau N$, taking $\tilde{I}(t)$ to be the deterministic solution (and using $N - \tilde{I}$ to be the number susceptible)

Deterministic: $\quad \dot{\tilde{I}}(t) = \tau \tilde{I}(t)(N - \tilde{I}(t)) - \gamma \tilde{I}(t) \,,$ (1.11)

Stochastic: $\quad [\dot{I}](t) = \tau [I](t)(N - [I](t)) - \gamma [I](t) + \tau \mathrm{Cov}(I, N - I)$ (1.12)

$$\leq \tau [I](t)(N - [I](t)) - \gamma [I](t) \,.$$

It is evident that the two equations differ only in the negative covariance term. The agreement should be good if $\mathrm{Cov}(I, N - I)$ is small. Focusing on closing the system, the most straightforward closure is the approximation $E(I^2) = [I]^2$, which implies that $\mathrm{Cov}(I, N - I) = 0$. Applying this to Eq. (1.12) leads to the deterministic version of the SIS model. For systems of reasonable size, this provides a good approximation (see Fig. 4.5). We will show rigorously in Chapter 3 that deterministic models can capture average behaviour well and, as a general rule of thumb, they tend to over-estimate the true stochastic process.

Unfortunately, more often than not, direct analysis of the explicit stochastic process is difficult especially when formulated on arbitrary networks, and gaining information about the temporal dynamics is often also not feasible. However, it turns out that mean-field models, which rely on "clever" averaging at different levels, provide a strong alternative. Mean-field models, almost without exception, rely on closures or approximations. The aim of these is mainly to curtail the number of equations in order to obtain tractable systems which are close enough to the true stochastic process. As we shall show, such methods have proven extremely useful as they provide the means to interpret and analyse stochastic processes.

1.2 Networks

Real contact network data, see Fig. 1.6, and the careful and objective analysis of the transmission process provide a strong case for the use of networks in modelling disease processes.

We highlight the role of partnership duration in order to explain why considering network structure can be essential to understanding disease transmission. Duration of partnerships can have a significant impact on the spread of an infectious disease, and it is particularly relevant for models focusing on households, schools or other small communities or for diseases which require close contact to transmit. Classical "well-mixed" models treat susceptible depletion as occurring at a global scale as the disease spreads, and so the immediate neighbourhood of an individual remains fully susceptible early on in an epidemic. This tends to overestimate early spread. Further, in an SIR epidemic the neighbourhood of individuals infected later will tend to be less depleted of susceptibles than for a node in a region the disease has already visited, and so the equations may be inaccurate at later times as well. For an SIS epidemic, an individual with many contacts may recover only to find many infected individuals around him/her, and thus become reinfected. This can cause "high-degree" individuals to form the centre of persistent infectious islands and may be particularly important for sexually transmitted diseases. The mathematical model will underestimate the high degree nodes' probability of reinfection, perhaps with significant consequences.

Many systems of interacting units are naturally represented and visualised as a network with nodes representing the units, and edges representing the interactions. This representation is conceptually simple and intuitive, leading to the use of the network paradigm across fields as diverse as social sciences, neuroscience, biology, linguistics, finance, engineering, computer science, archaeology and marketing. In Fig. 1.6, we give a few examples of empirical networks.

The well-established mathematical field of graph theory, which is concerned with the rigorous analysis of structural properties of graphs, can be rightly considered to be a precursor of network science as it is known today, and there is significant cross-pollination between the two. The textbook by Diestel [82] gives a good overview of graph theory, while the highly relevant branch of random graphs can be found in the monograph by Bollobás [43] and a recent book by Frieze and Karoński [103]. The description of large networks by using graph limits is presented in the book by Lovász [201].

However, graph theory and network science have different emphases, even though the mathematical object studied is the same. As a crude distinction, graph theory is typically interested in determining which graphs have a particular property or which properties a particular graph has, while network theory is typically focused more on taking a network underlying some process, determining the network's characteristics and using those characteristics to understand some emergent property of the original process. Graph theory dates back to Euler [95]. Network science goes back only a few decades. Introductions to it can be found in the book by Newman [236], the monograph by Caldarelli [55], the textbook by Estrada [93], the handbook by Bollobás, Kozma, and Miklós [44] and the book by Cohen and Havlin [68].

In the context of this book, the nodes of a network are individuals who may become infected. The edges represent contacts along which the disease can spread. Thus, network structure provides an important, powerful constraint on disease spread.

1.2.1 Basic tools for representing networks

Real-world networks are commonly represented by graphs. Depending on the system being modelled, the graph may be undirected, directed, weighted, time-dependent or various combinations of these. We start from a basic mathematical definition of a graph, according to which a graph G is given by a pair (V,E), where V is a set, the elements of which are called vertices (or nodes in the network context), and $E \subset V \times V$ is a set of pairs of vertices, called edges (or links in the network context). For example, a line graph with three vertices is given by $V = \{1,2,3\}$ and $E = \{(1,2),(2,1),(2,3),(3,2)\}$. The graph is represented visually by points (vertices) connected by segments (edges). In the case of the above line graph, the pairs $(1,2)$ and $(2,1)$ are represented by a single segment connecting vertices 1 and 2. If each edge appears in both directions in the set E, then the graph is called undirected, otherwise the graph is called directed. For example, the line graph given by $V = \{1,2,3\}$ and $E = \{(1,2),(2,3)\}$ is a directed graph.

For a finite graph with N vertices, the vertex set is taken as $V = \{1,2,\ldots,N\}$. The set E can be encoded in the *adjacency matrix* whose entry g_{ij} is 1 if $(i,j) \in E$, otherwise it is zero. A finite graph is determined uniquely by its adjacency matrix.

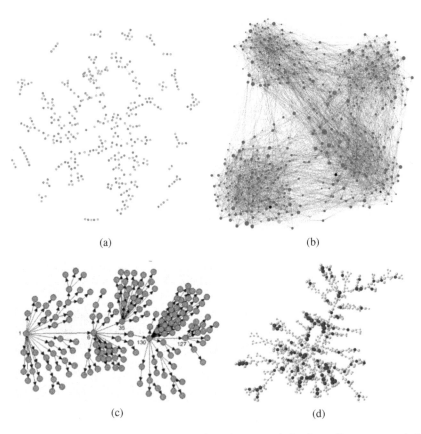

(a)   (b)

(c)   (d)

Fig. 1.6: An illustration of some networks relevant to infectious disease spread. (a) A network of romantic relationships from a high school "Jefferson high" [33] (some component shapes such as an isolated pair were observed multiple times, but shown only once here). The network demonstrates heterogeneity in the number of partners, and it shows that there are relatively few short cycles. (b) A measured network of individual–individual interactions within a school, showing that the network has substructures [69]. (c) Observed transmission chain of SARS in Singapore showing that some individuals infected far more than others (largely due to heterogeneity in infectiousness) [62]. (d) Partial observations of a network of university students. The red nodes were observed to have symptoms of H1N1 influenza. They had more observed contacts than other nodes [64].

We will occasionally use G to denote the adjacency matrix of the graph G, with the context making it clear whether we are thinking of the graph or its adjacency matrix. The adjacency matrix of the toast network, Fig. 1.7(a), is given in Fig. 1.7(b). The matrix is symmetric if and only if the graph is undirected. In certain cases, it is useful to put weights on the edges of the graph. Then, the entries of the adjacency matrix are arbitrary non-negative numbers. In this case, the graph is called a weighted graph. From a computational viewpoint, a much more efficient way to store networks is by using linked lists, see Fig. 1.7(c), or sparse matrices.

From a practical viewpoint, networks can arise in three different ways:

- based on real-world data;
- graph-theoretical or combinatorial graphs, e.g. Petersen graph, Cayley tree; and
- theoretical/synthetic models where networks are sampled from a set of networks that observe certain prescribed properties.

One of the most well-known and widely studied random graph models is called the Erdős–Rényi graph model with N nodes and m edges. It consists of all N-node graphs having m edges: $\{G_1, G_2, \ldots, G_n\}$, where $n = \binom{M}{m}$ with $M = N(N-1)/2$, and the probability of picking each graph is the same, namely $1/n$. A closely related model which is also usually referred to as the Erdős–Rényi graph model places each edge independently with probability p. Every graph on N nodes is possible, and the probability of a particular graph is $p^e(1-p)^{M-e}$, where e is the number of edges. For "k-regular" random graphs (random graphs in which all nodes have k neighbours), the set of graphs $\{G_1, G_2, \ldots, G_l\}$ consists of all k-regular (labelled) graphs and the probability of picking each graph is $1/l$. More generally, a random graph model is a probability distribution over a certain set of graphs. Formally, a random graph is given by a set of graphs $\{G_1, G_2, \ldots, G_n\}$ and a set of non-negative numbers $\{p_1, p_2, \ldots, p_n\}$, for which $p_1 + p_2 + \cdots + p_n = 1$. A realisation of this random graph model is G_i with probability p_i.

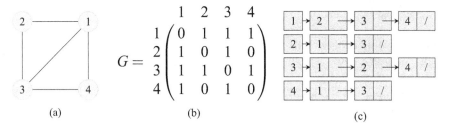

(a)   (b)   (c)

Fig. 1.7: (a) The toast network represented by (b) an adjacency matrix and (c) a linked list, where the head of the list represents a node with all its neighbours enumerated in its list.

1.2.2 Characterising networks

In studying disease spread in a city, we are unlikely to be able to find the correct adjacency matrix, and indeed a similar city would have a different adjacency matrix, but very similar disease dynamics. Our expectation and hope is that the disease spread is governed by a few quantitative network-specific measures such as degree distribution, preferential mixing, clustering, etc. which can be found by sampling the population. Below, we provide a succinct overview of the main network measures with a focus on what we use in the book.

Degree and degree distribution: The most important characteristic is the degree distribution of the graph. The out-degree of a vertex is the number of edges starting from it. Expressed in terms of the adjacency matrix, $k_i^{\text{out}} = \sum_{j=1}^N g_{ij}$. The in-degree of a vertex is the number of edges going to it, i.e. $k_i^{\text{in}} = \sum_{j=1}^N g_{ji}$. For an undirected graph, these are equal; this number is called the degree of the node and is denoted by k_i. The average degree of a graph is

$$\langle K \rangle = \frac{1}{N} \sum_{i=1}^N k_i = \frac{1}{N} \sum_{i=1}^N \sum_{j=1}^N g_{ij}.$$

In some places we will use n or $\langle k \rangle$ for the average degree in order to maintain consistency with the standard literature on a given model. If every vertex has the same degree, then the graph is called regular.

If the degrees occurring in the graph are denoted by $d_1, d_2, \ldots, d_L$ and N_ℓ is the number of vertices with degree d_ℓ, then the degree distribution is $p_\ell = N_\ell/N$. For a regular graph, $L = 1$ and $N_1 = N$, that is all nodes have the same degree d_1. For bimodal graphs with two different degrees, we have d_1 and d_2, i.e. $L = 2$. For an Erdős–Rényi graph, every degree from 0 to $N - 1$ may occur, that is $L = N$, $d_\ell = \ell - 1$, and it can be shown that the distribution p_ℓ is binomial. We note that in many cases each degree in the range 1 to M may occur as a degree in the graph. In that case, $d_\ell = \ell$ for all $\ell = 1, 2, \ldots, M$, in which case the notation d_ℓ is not needed. Instead, the degree of a general node is denoted by k, the number of nodes of degree k is denoted by N_k and the degree distribution is given by $p_k = N_k/N$, where $N = N_1 + N_2 + \cdots + N_M$ is the total number of nodes. Often, we use $P(k)$ instead of p_k.

Assortative or disassortative mixing: The next level of characterising the graph is to determine how the nodes with different degrees are connected to each other. This can be specified by an $L \times L$ matrix with entries $n_{\ell j}$ yielding the total number of edges connecting nodes of degree ℓ with nodes of degree j. These numbers must satisfy the compatibility conditions

$$\sum_{j=1}^L n_{\ell j} = d_\ell N_\ell, \quad \text{for all} \quad \ell = 1, 2, \ldots, L.$$

For random mixing, when the edges of the nodes are connected randomly, we have

$$n_{\ell j} = \frac{d_\ell d_j N_\ell N_j}{\displaystyle\sum_{i=1}^M d_i N_i}.$$

If this relation does not hold, then the mixing is called preferential.

Specifying the degree distribution p_ℓ and the mixing coefficients $n_{\ell j}$ does not determine the graph structure. This can be seen even in small graphs. For example, there are two distinct 3-regular graphs on $N = 6$ nodes. Both can be drawn as hexagons with three diagonals. In the first one, denoted by G_1, the diagonals are $(1,4), (2,6), (3,5)$. In the second one, denoted by G_2, the diagonals are $(1,4), (2,5), (3,6)$ (see Fig. 1.8). By counting the number of triangles, we can see these are different (not isomorphic) graphs. Any other 3-regular 6-node graph will look like one of these.

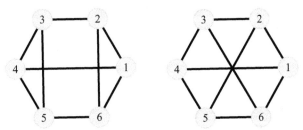

Fig. 1.8: Two 3-regular graphs over six nodes that are indistinguishable by degree and assortativity, but not isomorphic.

Clustering and higher order structure: The example above shows that characterising networks may require graph properties beyond just p_k and $n_{\ell j}$. A frequently used property is clustering, which measures the probability that two neighbours of a randomly chosen node share an edge to form a triangle (see Fig. 1.9). We define the clustering coefficient more formally in Section 4.2.3. Clustering can have a significant impact on epidemic dynamics, as it leads to a quicker depletion of susceptible nodes around infected ones. This occurs because edges closing a triangle can be "wasted" if infection crosses the other edges first. It turns out that other higher order structures, over more than three nodes, may also be relevant. Hence, instead of triangles further subgraphs may be taken into consideration (see Fig. 1.9).

Fig. 1.9: A few examples of ways in which clustering can arise in networks. Subgraphs over three and four nodes with clustering.

As we will show in the book, by studying stochastic processes on networks, one can uncover direct relationships between (a) conditions that ensure analytical tractability or allow for the validation of the performance of closures and (b) classical graph theoretical properties such as strongly connectedness or node- and edge-level properties, e.g. cut-vertices and bridges. In some cases, such links extend to specific classes of graphs, such as tree graphs. Some of these properties are defined below.

Shortest path length and strong connectedness: A directed path connecting two vertices $u, v \in V$ is a sequence of vertices $v_0, v_1, \ldots, v_n \in V$ such that $(v_i, v_{i+1}) \in E$ for all i and $v_0 = v$, $v_n = u$. A graph is called strongly connected if there is a directed path from any node to any other. The distance between two vertices is the number of edges in the shortest directed path connecting the two vertices. (This might be infinite if the graph is not strongly connected.) The greatest distance between any two vertices is called the diameter of the graph. In the case of undirected graphs, we simply say that the graph is connected.

Cycles and tree graphs: A path is called a cycle if its endpoints coincide. A connected graph without cycles is called a tree.

1.2.3 Network-generating algorithms

In the case of large networks, the adjacency matrix is typically not available. However if some properties of the network are known, then for simulation purposes a random graph with the known properties is created. The aim of this section is to introduce some basic, widely used network-generating algorithms.

We often use the "Configuration Model", or "Molloy-Reed" network class [224, 225], which is used to fit an observed degree distribution. Let $d_1, d_2, \ldots, d_L$ again denote the degrees of the graph and $N_1, N_2, \ldots, N_L$ the number of nodes of each degree. The average degree is

$$\langle K \rangle = \frac{1}{N} \sum_{\ell=1}^{L} N_\ell d_\ell.$$

Our formula is equivalent to $\langle K \rangle = \sum k p_k$, where the probability that a randomly chosen node has degree k is $p_k = N_k / N$. The degree heterogeneity of the graph can be characterised by the variance of the degree distribution:

$$\frac{1}{N} \sum_{\ell=1}^{L} N_\ell (d_\ell - \langle K \rangle)^2 = \langle K^2 \rangle - \langle K \rangle^2,$$

where

$$\langle K^2 \rangle = \frac{1}{N} \sum_{\ell=1}^{L} N_\ell d_\ell^2$$

denotes the second moment of the degree distribution.

To generate a Configuration Model network, each node is assigned a degree at random and independently of all others but chosen according to the degree distribution p_k. Each node gets a number of stubs equal to its degree and stubs in the network are joined into pairs at random to form edges (see Fig. 1.10). When we generate a network according to this rule, nodes have no individual preference for low-degree

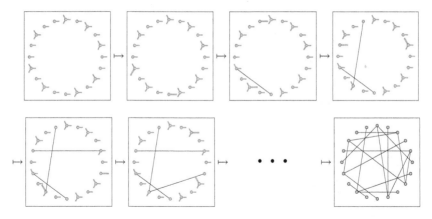

Fig. 1.10: The steps towards generating a Configuration Model network with $P(1) = P(3) = 1/2$. All nodes are independently assigned a degree and then assigned the appropriate number of stubs. Pairs of stubs are randomly chosen and joined together to form an edge until no unpaired stubs remain.

or high-degree neighbours; however, the rule induces an apparent bias towards high-degree neighbours. When a node's stub is joined to a neighbour, the probability it connects to a node of degree k is proportional to the total number of stubs of degree k nodes, and so the probability of joining to a degree k node is $P_n(k) = kp_k/\langle K \rangle$, and thus higher degree nodes are more likely to appear as neighbours than lower degree nodes [96].

This process may create self-edges or repeated edges. For most degree distributions, the probability a random node has a self-edge or repeated edge scales like $1/N$ as $N \to \infty$. The expected number of such nodes approaches a constant for large N. The *proportion* affected thus goes to zero and we can simply discard self-edges or repeated edges without affecting dynamics. An alternate perspective of disease spread in the Configuration Model [239] yields an equivalent result by assigning the stubs, only joining them when the disease attempts to transmit.

In this book, we will frequently consider the following networks: (a) regular random, (b) bimodal random, (c) Erdős–Rényi random and (d) power law (also referred to as scale-free) random networks. We briefly describe these networks:

a. The simplest Configuration Model network is the regular random network, for which $L = 1$ and where all nodes have the same degree d_1 and $N_1 = N$. The average degree of this network is obviously $\langle K \rangle = d_1$.

b. In order to study the effect of graph heterogeneity, we will use a bimodal random network, for which $L = 2$ and nodes can have either degree d_1 or d_2. The total number of nodes in the network is $N = N_1 + N_2$ and $\langle K \rangle = (d_1 N_1 + d_2 N_2)/N$.

c. We will also consider Erdős–Rényi random graphs created by linking each possible pair of nodes with probability p. The degree distribution of this graph is binomial, with the average degree $\langle K \rangle = (N-1)p$. The Erdős–Rényi random graph can be considered as a Configuration Model random graph based on a binomial degree distribution.

d. Scale-free random networks having nodes with high degree can be given by power law degree distributions. In the limit of large k, these are given by $P(k) \sim k^{-\alpha}$, with some power $\alpha > 1$. The numbers L, d_ℓ and N_ℓ are chosen as follows. For each node, i, with $i = 1, 2, \ldots, N$, a degree is chosen from the infinite distribution $Ck^{-\alpha}$, where C is a normalisation constant. Once the list of degrees is determined, one can deduce the degree sequence $d_1, d_2, \ldots, d_L$ and the degree frequencies $N_1, N_2, \ldots, N_L$. Then, a Configuration Model random graph is created with N_ℓ nodes of degree d_ℓ, where $\ell = 1, 2, \ldots, L$. There are different ways of choosing a random integer from the infinite distribution $Ck^{-\alpha}$. We suggest the following simple algorithm. Choose a uniform random number $r \in [0, 1]$, then take the integer part of $r^{\frac{1}{1-\alpha}}$. The distribution of these integers will be $Ck^{-\alpha}$. The random graph constructed in this way can contain nodes with very high degree, but there will be degrees that are missing from the distribution. An important property of power law networks is that if α lies between 2 and 3, then $\langle K^2 \rangle$ is infinite, but $\langle K \rangle$ is finite in the large-network limit.

In our simulations, we use Configuration Model random networks with a cut-off power law degree distribution, as discussed in [245]. These random networks are given by a minimal degree $k_{\min}$, a maximal degree $k_{\max}$ and a power α. The degree distribution of the graph is $P(k) = Ck^{-\alpha}$ for $k = k_{\min}, k_{\min} + 1, \ldots, k_{\max}$ with the normalisation constant C given by

$$\frac{1}{C} = \sum_{k=k_{\min}}^{k_{\max}} k^{-\alpha}.$$

This means that integers from $k_{\min}$ to $k_{\max}$ occur but there are no degrees outside the pre-specified range.

Many other network models exist, most notably algorithms for generating scale-free networks by preferential attachment [23, 246] and algorithms for creating clustered networks [21, 122, 149, 213, 237, 264–266, 315]. An additional class of random networks is the Exponential Random Graph Model (ERGM) [203]; it has seen many applications in social sciences.

1.3 Disease spread on networks: the main topic of the book

The main topic of this book is the study of continuous-time stochastic epidemic models on networks. A number of review articles have explored related topics, such as Boccaletti et al. [36], Danon et al. [75], Pastor-Satorras et al. [247], Rock et al. [267] and Funk et al. [107], and a good introduction to modelling dynamic processes on networks is the tutorial by Porter and Gleeson [253]. The first, and probably the simplest, approach of investigating these processes is stochastic simulation.

However, our focus is on the mathematical modelling and analysis of stochastic processes, besides and beyond stochastic simulation, and the topics covered in the book are described in detail in Section 1.3.3.

When we model epidemic spread in networks, individuals are represented by nodes and the contact pattern amongst them is encoded by the edges of the network. For most cases in the book, we consider static, unweighted networks with an SIS or SIR disease spreading across edges. Nodes can have the following statuses: susceptible S, infected and infectious I, and recovered/immune/removed R. At any instant, the network's state is given by the statuses of all N nodes. Thus, there are 2^N or 3^N states in the SIS and SIR case, respectively. That is, the state space consists of N-tuples of symbols from the set $\{S,I\}$ or $\{S,I,R\}$. The state changes in time and the infection and recovery processes give the rates of change of the nodes' statuses.

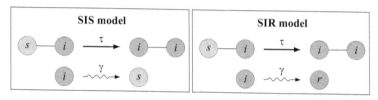

Fig. 1.11: An infected node infects its susceptible neighbour at rate τ. An infected node recovers with rate γ, independently of the status or number of his/her contacts. After recovering, for the SIS model, infected nodes become susceptible, but for the SIR model, recovery results in immunity. The nodes play no further role.

Both epidemic types are driven by two processes: (a) infection across edges and (b) recovery, which is network independent (see Fig. 1.11). Depending on the choice of modelling paradigm, the models can be classified as discrete or continuous-time models and deterministic or stochastic models. The usual assumptions are:

- transmission from an infected node to a susceptible node occurs across an edge as a Poisson process with rate τ, the "transmission rate"; and
- an infected node recovers as a Poisson process with rate γ, the "recovery rate".

All these events are considered to be independent. Thus, a susceptible node with k infectious contacts becomes infected according to a Poisson process at rate $k\tau$, i.e. a pooled Poisson process. It is important to note that once a node is infected, the time until it recovers is independent of the status of any other node in the network.

As a result of assuming Poisson processes, it follows that time to infection and time to recovery are exponentially distributed. This is a reasonable starting assumption but can be relaxed, as shown in Chapter 9.

1.3.1 Stochastic simulation

A first step towards studying the stochastic process of epidemic spread on a network is individual-based stochastic simulation. Such a stochastic process can be rigorously simulated by keeping track of all possible events in the network and the rate at which these happen.

One way to perform this simulation is to step forward in time in small intervals Δt. At each step, we can calculate the probability that a node will change status within the next interval. Then, by generating random numbers we can choose which changes occur and then move to the next time step. A significant limitation of this approach is that when the time step is large, multiple events may happen that would affect one another [97]. For example, a node might both recover and transmit in the same interval, but we cannot tell what happens first. If a shorter interval were chosen, we might see that the recovery happens first, and so the transmission should not occur. However, the simulation may become prohibitively slow.

Some alternative approaches exist. Briefly, we can use Gillespie simulations or event-driven simulations. In a Gillespie simulation, we calculate the time of the next event by knowing the combined rate of all possible events at the current time and choosing a time from an exponential distribution with that rate. We then use another random number to determine which of the possible events occurs. This provides an exact stochastic simulation. It is difficult to immediately adapt this to the non-Markovian case where inter-event times are not exponential random variables. However, recent research in this area provides extensions of the Gillespie or Gillespie-like algorithm to the general case when infection, recovery or other events are characterised by general inter-event times [42].

In an event-driven simulation, we keep a priority queue of upcoming events. At each step, we take the next event from the queue and process it. When a node becomes infected, we are able to calculate in advance which nodes it will transmit to, when it will transmit and when it will recover. We place each of these events in the queue and then draw the next event. This process is efficient and can be adapted to non-Markovian processes. We provide pseudocode for the Gillespie and event-driven simulations and discuss them in more detail in Appendix A.1.

1.3.2 Mathematical modelling

Our assumptions imply that the most basic mathematical model is a Markov chain with discrete state space. The full state space of the Markov chain contains all 2^N or 3^N possible states of the whole network. Assuming that infection and recovery events occur as Poisson processes leads to a continuous-time Markov chain. Usually our goal in creating a mathematical model is to be able to predict the probability the network is in a given state or aggregation of states or to calculate the expected number of nodes being susceptible, infected, or recovered as a function of time.

Previously for the compartmental models, we aggregated the states based on only the number of individuals in each status (if there were j individuals infected, we did not concern ourselves with which j individuals they were). In the network situation, this is not enough: knowing exactly which nodes are infected or susceptible is important. We will see that how (and even whether) we can exactly aggregate some network states depends on the network's structure. Often exact aggregation is not feasible, so we must turn to approximate models. In many cases, the spread in a large population approaches an apparently deterministic behaviour, and so much of our focus is on finding deterministic equations that approximate this behaviour.

1.3.3 Topics covered and not covered in the book

The book covers many aspects, including model development based on continuous-time Markov chains, the detailed introduction, treatment and analysis of a large number of models used in the network community, as well as clearly describing model assumptions, interdependencies and hierarchy. The book also dedicates some chapters to the treatment of some actively studied areas in network sciences, such as dynamic or adaptive networks, non-Markovian epidemics on networks and PDE approximations. Below, we give a more detailed chapter-by-chapter description, allowing the reader to more efficiently navigate the book. Finally, we summarise some topics that we were not able to cover but are significant for network-based stochastic models.

The most general Markov chain model is presented and investigated in Chapter 2. Because it consists of 2^N or 3^N states, the system of *master equations* contains this many equations. Chapter 2 shows how to reduce this system of equations into a much smaller system, without making any approximations. This method requires the network to have particular special structures. We refer to these as *top-down* models.

Generally, for large networks an exact reduction is not possible and thus approximations are used. Approximate models can be classified as individual- and population-level. In the first case, the differential equations are formulated in terms of the probabilities of individual nodes having a given status at time t with $P(X_i(t) = A)$ being the probability of node i having status A, $A \in \{S, I, R\}$. We refer to these models as *bottom-up models*. The terms *individual-based mean-field* and *individual-based pairwise* models are also used in the literature. These models are introduced and analysed in Chapter 3.

An alternative approach is to write the differential equations in terms of population-level quantities, such as the expected (or average) number of susceptible, infected and recovered nodes, or the average number of edges connecting different types of nodes. These models are called *mean-field* and *pairwise models* according to the level of closures. Mean-field models are written at the node level and the closure is applied to pairs. Pairwise models are formulated in terms of singles (nodes) and pairs (edges), and the closure is applied at the level of triples.

These two types of models are introduced and studied in detail in Chapter 4. This type of coarse-graining is not satisfactory when the network is strongly heterogeneous, i.e. there are nodes with low and high degree. Then, instead of using $[S](t)$ the average number of susceptible nodes in the network, the differential equations are formulated in terms of new variables, such as $[S_k](t)$, which denotes the average number of susceptible nodes of degree k. These models, called *heterogeneous mean-field* and *heterogeneous pairwise models*, are introduced and analysed in Chapter 5. The heterogeneous pairwise model yields excellent approximation for Configuration Model random graphs; however, its size is still too large for analytical investigations. Hence, reduced models, the *compact pairwise* and *super compact pairwise models*, are developed and analysed. The effect of degree heterogeneity can be captured also by *effective degree models*, which are dealt with in Chapter 5.

Originating in statistical physics, *percolation theory* has been used as a basis for studying propagation on networks. These ideas have given insights into modelling SIR epidemics on networks and led to the development of *edge-based compartmental* models. This approach is investigated in Chapter 6. Some percolation-like methods for SIS disease are briefly mentioned as well.

The models developed so far use different variables and different modelling approaches to describe the same spreading process on the network; hence, it is natural to investigate their relation to each other. Chapter 7 deals with this question for SIR diseases, showing strong relations between the three main approaches, derived in completely separate ways, namely the pairwise, the effective degree and the edge-based compartmental models. In particular, *the compact pairwise, compact effective degree* and *edge-based compartmental* models are equivalent for SIR epidemics, in the sense that they can be derived from each other.

Up to and including Chapter 7, we consider only static networks. However, the network structure may also vary as the process is unfolding. This leads to the theory and modelling of *dynamic or adaptive networks*, where the dynamics on the network and of the network are considered concurrently. Depending on the *rewiring mechanism*, several dynamic network models emerge: link-number-conserving rewiring, degree-preserving rewiring, random link activation deletion, link-status-dependent activation deletion, and link deactivation and activation on a fixed network. Chapter 8 studies these rewiring mechanisms using the pairwise, effective degree and edge-based compartmental models, with particular focus on *bifurcation* analysis, with more exotic behaviours, including *bistability* and *oscillations*.

Chapter 9 focuses on *non-Markovian epidemics on networks* and culminates with a generalisation of pairwise and edge-based compartmental models to non-Markovian epidemics. More specifically, the pairwise model is generalised to account for Markovian transmission with an arbitrary recovery process on homogenous networks, with the edge-based model fairing better with an extension to arbitrary transmission and recovery processes on heterogenous networks. These extensions provide the framework for a more systematic understanding of non-Markovian dynamics.

All the above approximating models can be compared to individual-based stochastic simulations to study their performance. However, we also want a more rigorous theory to investigate their accuracy by comparing them to the exact Markov chain model. This is carried out for one-step processes in Chapter 10, where we show how to introduce mean-field equations for the moments, a *partial differential equation* for the *probability generating function* and the *Fokker–Planck equation* of the probability distribution given by the process. These are then used to derive rigorous results about the accuracy of approximations.

We derive our models through closure assumptions. In many networks, however, these assumptions break down. Chapter 11 compares our model predictions with simulations in an empirical network as well as in Watts–Strogatz and Barabási–Albert networks. We find surprisingly good agreement. Although presented at the end of this book, this material would be appropriate anywhere after Chapter 7. We end the book with a short appendix which focuses on techniques for efficient stochastic simulation.

The mathematical investigation of stochastic epidemic models on networks is a huge research field. As alluded to before, its treatment or even an overarching review of the subject is too large for a single book. As the content of the chapters suggests, our approach is dominated by differential equations and analysis tools from dynamical systems. While the percolation model discussed in the book uses techniques from branching processes, we would have liked to present more results, with a stronger emphasis on probabilistic models and tools. Most notably, we mention research by Ball and co-workers [13–18], Britton and co-workers [3, 49–51, 81], Janson et al. [156] and Decreusefond et al. [76].

Closely related to the content of the book, and a natural extension to the models presented in the book, would have been the consideration of epidemic control measures such as vaccination [10, 11, 118, 148, 150] and contact tracing [91, 149], as well as the consideration of household models [10, 11, 17, 118, 148]. Equally, we would have liked to dedicate a chapter to recent progress in network control [71, 198], with special focus on controlling network-based epidemic models [67, 129, 242, 277, 329, 331] and linkages between classical control theory and network models. We would have also liked to have had the chance to parametrise some of our models with real data to perform some parameter inference tests.

While the focus of the book is on epidemic propagation on networks, other processes, e.g. voter model, neuronal networks, rumour spreading, etc., can be studied with similar tools. Extensive literature on topics similar to those presented in our book, but with a different focus or in a different context, can be found in books by Newman, Barabási, and Watts [238], Bornholdt and Schuster [45], Barrat, Barthelemy and Vespignani [26], Jackson [154], Easley and Kleinberg [92], Draief and Massoulié [86] and Fu, Small, and Chen [104].

Last, but not least, we finish by noting that while the book presents a snapshot of the results and developments in the field to date, we envisage that huge progress will be made over a short period of time in existing research areas involving networks and that many new research directions will emerge. The most likely trends for further developments, highlighted as challenges in [20, 139, 248], are towards:

- increasing model realism further and developing models that are more readily parametrised with real data;
- deriving network inference from partially observed data;

- incorporating social and behavioural aspects into models; and
- linking control theory and network-based epidemic models, thus designing network-centred interventions.

From a mathematical viewpoint, the main challenge remains to find more synergies in how existing mathematical tools, e.g. delay-differential equations, stochastic differential equations, martingale theory, PDEs, topology, etc., can be employed to give a more rigorous basis to the fast-developing field of network models and to grow it into a stronger mathematical research area.

Chapter 2
Exact propagation models on networks: top down

Chapter 1 introduced SIS and SIR diseases and some weaknesses of compartmental models that can be remedied by considering networks. In this chapter, we begin our network-based investigation by setting up the problem we will study. We introduce the full stochastic model and show how to derive exact equations to calculate the probability of the entire network being in a given state. We refer to such models as "top down". The resulting system will generally have too many equations for practical analysis, so we discuss techniques to derive exact equations at a coarser level.

The diseases occur at the nodes, with the status of a node changing over time. At any given time, we assume the only factors that can affect the probability of a node changing its status are its current status and the statuses of its immediate neighbours. For example, the rate at which a susceptible node becomes infected depends only on how many of its neighbours are infected. Although neighbours can influence a node's transitions, the rate can also be independent of its neighbours. In particular, an infected node recovers at a rate that does not depend on any neighbour's status.

If each of the N nodes can have one of m different statuses, then the total number of distinct states of the network is m^N, so for SIS there are 2^N states and for SIR there are 3^N. We refer to the collection of all states as the state space, denoted $\{S_1, S_2, \ldots, S_n\}$. Ideally, we would like to predict the state of the network at any time. However, for a stochastic dynamic process, at best we might hope to know each state's probability. Even this may be impractical, and so we may settle for simply knowing the probability that a given number of individuals have each status.

In this chapter, we will develop exact equations for these probabilities. The full system of equations is generally quite large. Through careful aggregation of states we may find it is easier to calculate the probability that the system is in one of a collection of states. The methods we develop are not specific to SIS or SIR disease, but apply generally to a wide range of dynamic processes spreading on networks. It is simpler to consider a generic process rather than focusing on disease. So we will

© Springer International Publishing AG 2017

I.Z. Kiss et al., *Mathematics of Epidemics on Networks*, Interdisciplinary Applied Mathematics 46, DOI 10.1007/978-3-319-50806-1_2

develop the mathematical theory for the more general problem, with an emphasis on the aspects relevant to disease spread. To aid the construction of our mathematical models, we make a few explicit assumptions:

1. time is considered to be continuous,
2. a node has only a finite number of possible statuses,
3. the process is stochastic,
4. the inter-event times are exponentially distributed,
5. the parameter of this exponential distribution may depend on the status of the node and on the number of neighbouring nodes of each status,
6. transitions at different nodes are independent.

Ideally, we would like to know the probability the system has a given state at a given time, but this is likely to be impractical. Thus, the aim of the investigation is to determine

- the probability that a node has a given status at a given time,
- the expected number of nodes having a given status at a given time.

These assumptions lead to a continuous-time Markov chain with finite state space. This is a widely used and intensively studied field of mathematics with an extensive literature both from the theoretical and applied points of view. This book is not aimed at introducing Markov chains in general. For this, any of the following provides a good point of reference [124, 157, 161, 162, 171, 259]. Our topic is restricted to a special class of Markov chains which originate from exact propagation models on networks.

We note that relaxing the first requirement above leads to discrete-time models, which are partially investigated in Chapter 6. Further discussion of discrete-time models in the context of network epidemics can be found in [97, 102]. Relaxing the fourth requirement leads to non-Markovian models, where the transition times can be non-exponential. These will be studied in Chapter 9. If the fifth assumption is relaxed, then network models will turn into hypergraph models, where the processes evolve via hyperedges instead of edges (see, for example, [38, 111, 193]).

The methods we develop in this Chapter allow us to significantly simplify the mathematical models we study, without using any approximations. However, we will see that this process is not always easy, and the simplified models may still be quite complex. Later in the book we will explore ways to develop approximate models which are more tractable.

2.1 An introductory example

To motivate our approach and illustrate the processes we consider, we will analyse the state space of an SIS disease propagating on a triangle. We will derive equations governing the probabilities that the system is in each state. Then, we will derive a reduced system that captures the important details of the model. The remainder of the chapter develops techniques for more general cases.

Example 2.1. The state space of the Markov chain for an SIS disease in a triangle is

$$\{SSS, SSI, SIS, ISS, SII, ISI, IIS, III\},$$

where, for example, SSI represents the state where nodes 1 and 2 are susceptible and node 3 is infected. Figure 2.1 shows the possible states and the possible transitions between those states, highlighting the SII state. At most, one transition can happen at a time (though the time between transitions can be arbitrarily small).

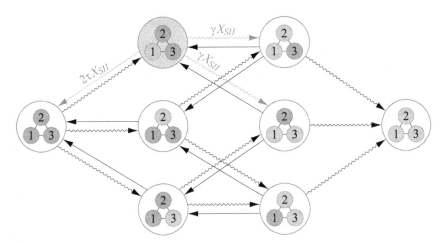

Fig. 2.1: Illustration of all eight network states and associated transitions for an SIS disease spreading in the fully connected network with 3 nodes. Susceptible (◯) and infected (◉) nodes are denoted by filled circles of different colours. Note that the SSS state is an absorbing state: there is no path out. The SII state and the fluxes of probability going out are highlighted (but not the fluxes in).

We define X_{ABC} to be the probability that the state of the network is ABC, where A, B and C may be S or I. If the transition corresponds to recovery of a single node, it occurs at the *recovery rate* γ, while if it corresponds to infection of a single node, it occurs at the *per-contact infection* or *transmission rate* τ times the number of infected neighbours. For example, if the system is in state SII, the rate at which it moves to SIS is γ. In contrast, the rate at which the system moves to III is 2τ. From this and following Fig. 2.1, we note that the flux of probability from state ABC to another is given by the rate at which that transition happens times X_{ABC}. Summarising the above yields

$$\dot{X}_{SSS} = \gamma(X_{SSI} + X_{SIS} + X_{ISS}),$$
$$\dot{X}_{SSI} = \gamma(X_{SII} + X_{ISI}) - (2\tau + \gamma)X_{SSI},$$
$$\dot{X}_{SIS} = \gamma(X_{SII} + X_{IIS}) - (2\tau + \gamma)X_{SIS},$$
$$\dot{X}_{ISS} = \gamma(X_{ISI} + X_{IIS}) - (2\tau + \gamma)X_{ISS},$$
$$\dot{X}_{SII} = \gamma X_{III} + \tau(X_{SSI} + X_{SIS}) - 2(\tau + \gamma)X_{SII},$$

$$\dot{X}_{ISI} = \gamma X_{III} + \tau(X_{SSI} + X_{ISS}) - 2(\tau + \gamma)X_{ISI},$$
$$\dot{X}_{IIS} = \gamma X_{III} + \tau(X_{SIS} + X_{ISS}) - 2(\tau + \gamma)X_{IIS},$$
$$\dot{X}_{III} = -3\gamma X_{III} + 2\tau(X_{SII} + X_{ISI} + X_{IIS}).$$

These equations are the *forward Kolmogorov equations* of the system and are also called *master equations*. A solution to the master equations gives the probability that the system is in each state at each given time.

The equations become simpler if we are willing to accept knowing just the probability a given number of nodes have each status. Rather than calculating a given state's probability, we calculate the probability the system is in one of several states collected or grouped based on how many nodes are infected. Define Y_i to be the probability exactly i nodes are infected; that is, $Y_0 = X_{SSS}$, $Y_1 = X_{ISS} + X_{SIS} + X_{SSI}$, $Y_2 = X_{IIS} + X_{ISI} + X_{SII}$ and $Y_3 = X_{III}$; the equations become

$$\dot{Y}_0 = \gamma Y_1, \tag{2.1a}$$
$$\dot{Y}_1 = 2\gamma Y_2 - \gamma Y_1 - 2\tau Y_1, \tag{2.1b}$$
$$\dot{Y}_2 = 2\tau Y_1 - 2\gamma Y_2 - 2\tau Y_2 + 3\gamma Y_3, \tag{2.1c}$$
$$\dot{Y}_3 = 2\tau Y_2 - 3\gamma Y_3. \tag{2.1d}$$

This simplification works for a triangle, but it often fails for more general networks.

Exercise 2.1. Starting from $Y_1 = X_{ISS} + X_{SIS} + X_{SSI}$ for SIS disease in a triangle, find an equation in terms of the X_{ABC} variables for $\dot{Y}_1$. Then, simplify this by using Y_1 and Y_2 to derive equation (2.1b). Continue using this approach to complete the derivation of system (2.1).

Exercise 2.2. Using the same variables Y_0, Y_1, Y_2 and Y_3 as for the triangle, attempt to repeat Exercise 2.1 to derive a system like system (2.1) if the 1–3 edge does not exist. What goes wrong? [You will have to modify the equations for X_{ABC} to account for the fact that 1 and 3 cannot transmit to each other.]

In this chapter, we give a more rigorous basis for what we have just done. This will give us the mathematical tools needed to derive master equations for more complicated systems and to identify when it is possible to simplify these, without approximation, by collecting or *lumping* states together. We will see that for large networks lumping often will not reduce the system of equations enough. Even for small networks lumping often leaves us with a large system of equations. The process will generally only be practical for special networks. So later chapters show how to systematically simplify the dynamics to arrive at tractable approximate equations.

2.2 Continuous-time Markov chains

We briefly summarise the basics of continuous-time Markov chains, which we will use to derive master equations. Let the state space of the Markov chain be the set $\{S_1, S_2, \ldots, S_n\}$. For the 3-node SIS system of Fig. 2.1, $n = 2^3 = 8$.

Definition 2.1 *If the system is in state S_i, the rate at which it transitions to state $S_j \neq S_i$ is defined to be $h(S_i, S_j)$. We define $h(S_i, S_i) = -\sum_{j \neq i} h(S_i, S_j)$.*

So $-h(S_i, S_i)$ is the total rate at which the system would transition from S_i to any other state. Thus for each i

$$\sum_j h(S_i, S_j) = 0. \tag{2.2}$$

As a shorthand, we often use $a_{ij} = h(S_i, S_j)$.

If δt is the length of a short time interval, then the probability of transition from S_i to S_j in the interval $(t, t + \delta t)$ is $h(S_i, S_j) \delta t + o(\delta t)$, where $o(\delta t)$ represents an error such that $o(\delta t)/\delta t \to 0$ as δt decreases to 0. Similarly, with probability $1 - h(S_i, S_j) \delta t + o(\delta t)$, that event will not take place. Hence, if the system state at time t is denoted by $S(t)$, then

$$P\big(S(t + \delta t) = S_j | S(t) = S_i\big) = h(S_i, S_j) \delta t + o(\delta t).$$

This relation enables us to formulate master equations for the probabilities of each state. Let $X_i(t) = P\big(S(t) = S_i\big)$ denote the probability that the system is in state i at time t. Then, the law of total probability leads to

$$X_j(t + \delta t) = P\big(S(t + \delta t) = S_j\big) = \sum_{i=1}^n P\big(S(t + \delta t) = S_j | S(t) = S_i\big) P\big(S(t) = S_i\big)$$

$$= \left(\sum_{i \neq j} h(S_i, S_j) \delta t X_i(t) \right) + (1 - h(S_j, S_j) \delta t) X_j(t) + o(\delta t),$$

Subtracting $X_j(t)$, replacing $h(S_i, S_j)$ with a_{ij}, dividing by δt and taking $\delta t \to 0$ yields the master equation

$$\dot{X}_j(t) = \sum_{i=1}^n a_{ij} X_i(t).$$

This is a linear system of ordinary differential equations of the form

$$\dot{X} = PX \tag{2.3}$$

where the matrix P is the transpose of the matrix of transition rates, that is $P_{ji} = a_{ij}$ for $j \neq i$ and $P_{jj} = -\sum_{k \neq j} a_{jk}$. The entries in each column sum to zero, representing the fact that if the system leaves one state, it enters another, so every increase in one state's probability is balanced by a decrease in the probability of another. We note that often the transpose of P is used, and then X is a row vector, not a column. However, we use this formulation since it is more convenient from a dynamical system point of view. The master equations of a network process are formulated in the next section.

2.3 Master equations for arbitrary networks

The aim of this section is to show how to formulate master equations for a process spreading on an arbitrary network, assuming transitions of a node depend only on the statuses of the node and its neighbours.

2.3.1 State space and transition rates for arbitrary dynamics

The network is given by the adjacency matrix of the corresponding undirected graph with N nodes: $G = (g_{ij})_{i,j=1,2,\ldots,N}$. Here, $g_{ij} = 1$ if nodes i and j are connected, or $g_{ij} = 0$ otherwise. Plainly, $g_{ij} = g_{ji}$ and we take $g_{ii} = 0$. Let $\{Q_1, Q_2, \ldots, Q_m\}$ be the possible statuses nodes can take. Then, a state of the network can be specified by an N-tuple $(q_1, q_2, \ldots, q_N)$, where $q_i \in \{Q_1, Q_2, \ldots, Q_m\}$ is the status of node i. The matrix in the master equation Eq. (2.3) is of size $m^N \times m^N$. This is a high-dimensional matrix, but with many zero entries because many transitions are not possible.

Because events happen independently and intervals are exponentially distributed, at most one event happens at any given time. Thus, if the states $\mathcal{S}_\alpha = (q_1, q_2, \ldots, q_N)$ and $\mathcal{S}_\beta = (s_1, s_2, \ldots, s_N)$ differ at more than one node, then the transition from $\mathcal{S}_\alpha$ to $\mathcal{S}_\beta$ is not possible and $h(\mathcal{S}_\alpha, \mathcal{S}_\beta) = 0$. If they differ in exactly one node, i, that is $q_i \neq s_i$ but $q_l = s_l$ for all $l \neq i$, then the transition rate depends on the statuses of node i, its neighbours and the process we study. For example, for the specific case of SIS or SIR dynamics, if $q_i = S$, $s_i = I$ and node i has n infected neighbours, then the transition rate is $n\tau$. In other words, node i is infected at rate $n\tau$. In the case of an arbitrary dynamic, let $q_i = Q$ and $s_i = T$, i.e. node i changes from status Q to status T. We take $n_{Q_1}, n_{Q_2}, \ldots, n_{Q_m}$ to be the number of neighbours node i has of each status. We define $f_{QT}(n_{Q_1}, n_{Q_2}, \ldots, n_{Q_m})$ to be the rate at which node i transitions from status Q to T if it has n_{Q_l} neighbours of status Q_l. Thus, summarising, the rate of transition from state $\mathcal{S}_\alpha$ to state $\mathcal{S}_\beta$ can be given as:

$$h(\mathcal{S}_\alpha, \mathcal{S}_\beta) = \begin{cases} 0 & \mathcal{S}_\alpha \text{ and } \mathcal{S}_\beta \text{ differ for at least two nodes,} \\ f_{QT}(n_{Q_1}, n_{Q_2}, \ldots, n_{Q_m}) & \mathcal{S}_\alpha \text{ and } \mathcal{S}_\beta \text{ differ in exactly one} \\ & \text{node } i \\ -\sum_{l \neq \alpha} h(\mathcal{S}_\alpha, \mathcal{S}_l) & \mathcal{S}_\alpha = \mathcal{S}_\beta. \end{cases}$$

$$(2.4)$$

2.3.2 State space and transition rates for binary dynamics

To be more specific, we take binary dynamics $m = 2$ as in SIS disease. Nodes can have one of two statuses, say Q and T. In deriving the master equations, we follow ideas developed independently in [288], and [312]. Considering $m = 2$ has many advantages. First, many fundamental processes can be described by two statuses, including the SIS model in epidemiology [4], the QAQ (i.e. quiescent–active–quiescent) model in neuroscience [70], the voter model in social sciences [58, 291, 314] and the Ising model in statistical physics [77, 85, 115, 194], to name just a few. Second, model formulation is clearer and more transparent in this case, and understanding the binary dynamics model provides a direct and simple way towards a generalisation to more than two statuses. The following section will extend this to SIR disease.

We group the 2^N states into $N+1$ subsets $\mathcal{C}^k$ for $k = 0, 1, \ldots, N$. Here, the superscript k is used for indexing and not as an exponent. We take $\mathcal{C}^k$ to be the set of states with k nodes of status T. There are two notable special cases: $\mathcal{C}^0$ is a subset with a single element, namely the state with all nodes having status Q: $\mathcal{C}^0 = \{QQ\cdots Q\}$, and $\mathcal{C}^N$ is also a subset with a single element, namely the state with all nodes having status T: $\mathcal{C}^N = \{TT\cdots T\}$.

More generally, the states in $\mathcal{C}^k$ are denoted by $\mathcal{S}_1^k, \mathcal{S}_2^k, \ldots, \mathcal{S}_{c_k}^k$, where $c_k = \binom{N}{k}$ is the number of different ways in which k nodes of status T can be placed on the network. The status of the lth node of state $\mathcal{S}_j^k$ will be denoted by $\mathcal{S}_j^k(l)$; thus, $\mathcal{S}_j^k(l) = Q$ or $\mathcal{S}_j^k(l) = T$. For example, when an SIS epidemic is considered on a triangle network, then $\mathcal{C}^0 = \{SSS\}$, $\mathcal{C}^1 = \{SSI, SIS, ISS\}$, $\mathcal{C}^2 = \{SII, ISI, IIS\}$ and $\mathcal{C}^3 = \{III\}$.

The state of the system can change as follows:

Transition of a node from status Q to T: A node of status Q transitions to status T, that is an $\mathcal{S}_j^k \to \mathcal{S}_i^{k+1}$ type transition, where j and i are chosen such that there exists l for which $\mathcal{S}_j^k(l) = Q$, $\mathcal{S}_i^{k+1}(l) = T$ and $\mathcal{S}_j^k(m) = \mathcal{S}_i^{k+1}(m)$ for all $m \neq l$. The rate of this transition is given by $f_{QT}(n)$, where n denotes the number of neighbours of node l of status T when the system is in state $\mathcal{S}_j^k$. We note that in the general case, as given in Subsection 2.3.1, the transition rate f_{QT} may depend on the numbers of all types of neighbours. For simplicity, we assume that f_{QT} depends only on the number of neighbours of status T and not on the number of neighbours of status Q.

Transition of a node from status T to Q: A node of status T transitions to status Q, that is an $\mathcal{S}_j^k \to \mathcal{S}_i^{k-1}$ type transition, where j and i are chosen such that there exists l for which $\mathcal{S}_j^k(l) = T$, $\mathcal{S}_i^{k-1}(l) = Q$ and $\mathcal{S}_j^k(m) = \mathcal{S}_i^{k-1}(m)$ for all $m \neq l$. This means that states $\mathcal{S}_j^k$ and $\mathcal{S}_i^{k-1}$ differ only at the l'th position. We assume that the T to Q transition depends only on the number of neighbours of status Q. Thus, the transition rate is $f_{TQ}(n)$, where n denotes the number of neighbours of node l of status Q.

For illustration, we consider three different processes: SIS, QAQ with a hyperbolic tangent transition rate and a voter-like model with two statuses, where both transitions depend on the status of the neighbours.

Example 2.2. Consider an SIS disease propagating on a network. A node can be susceptible S or infected I. Using the above notation, let Q be S and T be I. There are two transitions: infection and recovery. In order to specify the dynamics, the transition rates f_{SI} and f_{IS} have to be specified. In this case, $f_{SI}(n) = \tau n$ with τ the per-contact infection rate, $f_{IS}(n) = \gamma$, with γ, the recovery rate. That is, the infection rate is proportional to the number of infected neighbours, while the recovery rate is independent of the statuses of the neighbours.

Example 2.3. A network of neurones is considered with purely excitatory connections. Within the network, neurones are considered to be either quiescent (Q) or active (A) with the following two types of transitions. A quiescent neurone with n

active neighbours becomes active with rate $f_{QA}(n) = \omega \tanh(n)$, with some parameter ω, which is called synaptic weight, while an active neurone becomes quiescent with rate $f_{AQ}(n) = \alpha$ with some parameter α, which is called the deactivation rate. That is, the activation rate depends in a non-linear way on the number of active neighbours, while the deactivation rate is independent of the statuses of the neighbours.

Example 2.4. Consider a voter-like model, where the nodes of the network represent the voters, which can be of status A or B, and the neighbours of a node can change their status as follows. A node of status A becomes B at rate $f_{AB}(n) = an$ with some parameter a, where n denotes the number of B neighbours of the node. Similarly, a node of status B becomes A at rate $f_{BA}(n) = bn$ with some parameter b, where n denotes the number of A neighbours of the node. Thus, in this case both transitions depend on the statuses of the neighbours.

We note that an alternative rate function is widely used: $f_{AB}(n_A, n_B) = a\frac{n_B}{n_A + n_B}$, where n_A and n_B denote the number of neighbours of status A and B, respectively. The corresponding $B \to A$ transition rate is $f_{BA}(n_A, n_B) = b\frac{n_A}{n_A + n_B}$, expressing the fact that the rate depends on the ratio of the number of different neighbours.

2.3.3 Master equations for binary dynamics

We begin our derivation of master equations in the special case of SIS disease spreading on a triangle. We then show how this generalises for the derivation of an arbitrary binary process spreading on a network.

Example 2.5. In Fig. 2.1, we have $N = 3$ nodes. There are $2^3 = 8$ distinct states the system can take. We write X_{ABC} to be the probability that the system state is $S = ABC$. We have already seen that the master equations are

$$\dot{X}_{SSS} = \gamma(X_{SSI} + X_{SIS} + X_{ISS}),$$
$$\dot{X}_{SSI} = \gamma(X_{SII} + X_{ISI}) - (2\tau + \gamma)X_{SSI},$$
$$\dot{X}_{SIS} = \gamma(X_{SII} + X_{IIS}) - (2\tau + \gamma)X_{SIS},$$
$$\dot{X}_{ISS} = \gamma(X_{ISI} + X_{IIS}) - (2\tau + \gamma)X_{ISS},$$
$$\dot{X}_{SII} = \gamma X_{III} + \tau(X_{SSI} + X_{SIS}) - 2(\tau + \gamma)X_{SII},$$
$$\dot{X}_{ISI} = \gamma X_{III} + \tau(X_{SSI} + X_{ISS}) - 2(\tau + \gamma)X_{ISI},$$
$$\dot{X}_{IIS} = \gamma X_{III} + \tau(X_{SIS} + X_{ISS}) - 2(\tau + \gamma)X_{IIS},$$
$$\dot{X}_{III} = -3\gamma X_{III} + 2\tau(X_{SII} + X_{ISI} + X_{IIS}).$$

These can be rewritten as $\dot{X} = PX$, where $X = (X_{SSS}, X_{SSI}, X_{SIS}, X_{ISS}, X_{SII}, X_{ISI}, X_{IIS}, X_{III})$ is a vector and

$$P = \begin{pmatrix} 0 & \gamma & \gamma & \gamma & 0 & 0 & 0 & 0 \\ 0 & -2\tau-\gamma & 0 & 0 & \gamma & \gamma & 0 & 0 \\ 0 & 0 & -2\tau-\gamma & 0 & \gamma & 0 & \gamma & 0 \\ 0 & 0 & 0 & -2\tau-\gamma & 0 & \gamma & \gamma & 0 \\ 0 & \tau & \tau & 0 & -2\tau-2\gamma & 0 & 0 & \gamma \\ 0 & \tau & 0 & \tau & 0 & -2\tau-2\gamma & 0 & \gamma \\ 0 & 0 & \tau & \tau & 0 & 0 & -2\tau-2\gamma & \gamma \\ 0 & 0 & 0 & 0 & 2\tau & 2\tau & 2\tau & -3\gamma \end{pmatrix}.$$

As noted following equation (2.3), the columns sum to 0. The vector X can be divided into four parts, $X = ((X_{SSS}),(X_{SSI},X_{SIS},X_{ISS}),(X_{SII},X_{ISI},X_{ISS}),(X_{III}))$, based on the number of nodes infected. Thus, $X = (X^0,X^1,X^2,X^3)$, where X^k is a subvector whose entries all correspond to k infections. Performing the corresponding division of the rows and columns of P leads to the block-tridiagonal form

$$P = \begin{pmatrix} B^0 & C^0 & 0 & 0 \\ A^1 & B^1 & C^1 & 0 \\ 0 & A^2 & B^2 & C^2 \\ 0 & 0 & A^3 & B^3 \end{pmatrix},$$

where

$$B^0 = (0), \qquad C^0 = (\gamma,\ \gamma,\ \gamma),$$

$$A^1 = \begin{pmatrix} 0 \\ 0 \\ 0 \end{pmatrix}, \qquad B^1 = \begin{pmatrix} -2\tau-\gamma & 0 & 0 \\ 0 & -2\tau-\gamma & 0 \\ 0 & 0 & -2\tau-\gamma \end{pmatrix}, \qquad C^1 = \begin{pmatrix} \gamma & \gamma & 0 \\ \gamma & 0 & \gamma \\ 0 & \gamma & \gamma \end{pmatrix},$$

$$A^2 = \begin{pmatrix} \tau & \tau & 0 \\ \tau & 0 & \tau \\ 0 & \tau & \tau \end{pmatrix}, \qquad B^2 = \begin{pmatrix} -2\tau-2\gamma & 0 & 0 \\ 0 & -2\tau-2\gamma & 0 \\ 0 & 0 & -2\tau-2\gamma \end{pmatrix}, \qquad C^2 = \begin{pmatrix} \gamma \\ \gamma \\ \gamma \end{pmatrix},$$

$$A^3 = (2\tau,\ 2\tau,\ 2\tau), \qquad B^3 = (-3\gamma).$$

We have

$$\dot{X}^k = A^k X^{k-1} + B^k X^k + C^k X^{k+1}, \qquad k = 0,1,2,3.$$

We return now to a general binary dynamic process and formulate the master equations with Q replacing S and T replacing I. Let $X_j^k(t)$ be the probability the system is in state S_j^k at time t. Let

$$X^k(t) = (X_1^k(t),X_2^k(t),\dots,X_{c_k}^k(t))$$

be a c_k-dimensional vector for $k = 0,1,\dots,N$. The above transitions determine the master equations in the form of (2.3) for the probability functions $X_j^k(t)$. However, as before, the matrix P is block-tridiagonal

$$P = \begin{pmatrix} B^0 & C^0 & 0 & 0 & \cdots & 0 \\ A^1 & B^1 & C^1 & 0 & \vdots & \vdots \\ 0 & A^2 & B^2 & C^2 & \vdots & \vdots \\ 0 & 0 & A^3 & B^3 & \ddots & 0 \\ \vdots & \cdots \cdots & \ddots & \ddots & C^{N-1} \\ 0 & \cdots \cdots & 0 & A^N & B^N \end{pmatrix}. \tag{2.5}$$

and

$$\dot{X}^k = A^k X^{k-1} + B^k X^k + C^k X^{k+1}, \qquad k = 0, 1, \ldots, N, \tag{2.6}$$

where A^0 and C^N are zero matrices. Thus, equation (2.6) is in the form (2.3) with X a column vector made up of the entries of X^0, followed by $X^1, X^2, \ldots, X^N$.

The A^k matrices capture the transition from Q to T, while the C^k matrices describe the transition from T to Q. These matrices depend on the structure of the network and the transition rates f_{QT} and f_{TQ}. We now investigate the structure of these matrices.

Exercise 2.3. Show that if $S_i^{k-1} \in \mathcal{C}^{k-1}$ differs from $S_j^k \in \mathcal{C}^k$ at only a single node, l, then $S_i^{k-1}(l) = Q$ and $S_j^k(l) = T$.

In class $\mathcal{C}^{k-1}$, there are c_{k-1} elements, and in class $\mathcal{C}^k$, there are c_k elements; hence, matrix A^k has c_k rows and c_{k-1} columns. The entry in the ith row and jth column of the matrix A^k is denoted by $A_{i,j}^k$. It gives the transition rate from S_j^{k-1} to S_i^k, which is non-zero only in the case where the states differ at only a single node l with $S_j^{k-1}(l) = Q$ and $S_i^k(l) = T$. The rate depends on the number of status T neighbours of node l, that is $A_{i,j}^k = h(S_j^{k-1}, S_i^k)$, yielding

$$A_{i,j}^k = \begin{cases} 0 & S_j^{k-1} \text{ and } S_i^k \text{ differ in more than one node,} \\ f_{QT}(n_T(l, S_j^{k-1})) & S_j^{k-1} \text{ and } S_i^k \text{ differ only for node } l, \end{cases} \tag{2.7}$$

where $n_T(l, S)$ is the number of status T neighbours of node l in state S.

In order to better understand the role of the A^k matrix, consider its jth column, which corresponds to the system leaving S_j^{k-1}. Let $\Omega_Q(S_j^{k-1})$ denote the set of nodes l which have status Q, that is, $\Omega_Q(S_j^{k-1}) = \{l : S_j^{k-1}(l) = Q\}$. For each $l \in \Omega_Q(S_j^{k-1})$, define $i(l)$ so that $S_{i(l)}^k$ is the (unique) state that would result from changing only node l's status to T. Then, equation (2.7) becomes

$$A_{i,j}^k = \begin{cases} 0 & i \neq i(l) \text{ for any } l \in \Omega_Q(S_j^{k-1}), \\ f_{QT}(n_T(l, S_j^{k-1})) & i = i(l) \text{ for (a unique) } l \in \Omega_Q(S_j^{k-1}). \end{cases}$$

We conclude that the sum of the elements in the jth column of matrix A^k is

$$\sum_{i=1}^{c_k} A_{i,j}^k = \sum_{l \in \Omega_Q(\mathcal{S}_j^{k-1})} f_{QT}(n_T(l, \mathcal{S}_j^{k-1})). \tag{2.8}$$

In the simple case when f_{QT} is of the form $f_{QT}(n) = \tau n$, this reduces to

$$\sum_{i=1}^{c_k} A_{i,j}^k = \tau N_{QT}(\mathcal{S}_j^{k-1}), \tag{2.9}$$

where $N_{QT}(\mathcal{S}_j^{k-1})$ denotes the number of (Q, T) edges in state $\mathcal{S}_j^{k-1}$.

Similarly, the entry in the ith row and jth column of matrix C^k is denoted by $C_{i,j}^k$ and gives the transition rate from $\mathcal{S}_j^{k+1}$ to $\mathcal{S}_i^k$. In the $\mathcal{C}^{k+1}$ class, there are c_{k+1} elements, and in the $\mathcal{C}^k$ class, there are c_k elements; hence, matrix C^k has c_k rows and c_{k+1} columns. The entry $C_{i,j}^k$ is non-zero only in the case when the states $\mathcal{S}_j^{k+1}$ and $\mathcal{S}_i^k$ differ at one position, say at position l. The transition rate depends on the number of status Q neighbours of node l, that is $C_{i,j}^k = h(\mathcal{S}_j^{k+1}, \mathcal{S}_i^k)$, yielding

$$C_{i,j}^k = \begin{cases} 0 & \mathcal{S}_j^{k+1} \text{ and } \mathcal{S}_i^k \text{ differ in more than one node} \\ f_{TQ}(n_Q(l, \mathcal{S}_j^{k+1})) & \mathcal{S}_j^{k+1} \text{ and } \mathcal{S}_i^k \text{ differ only for node } l, \end{cases} \tag{2.10}$$

where $n_Q(l, \mathcal{S})$ is the number of status Q neighbours of node l in state $\mathcal{S}$.

In the simple case of $f_{TQ}(n) = \gamma$, it follows that $C_{i,j}^k$ is either zero or γ. In state $\mathcal{S}_j^{k+1}$, $k+1$ nodes of the graph have status T; hence, in the jth column of matrix C^k there are $k+1$ entries that are equal to γ and the remaining entries are zero. Hence, for all $j \in \{1, 2, \ldots, c_{k+1}\}$ we have

$$\sum_{i=1}^{c_k} C_{i,j}^k = \gamma(k+1). \tag{2.11}$$

The matrix B^k is a diagonal matrix with c_k rows and columns. This is because B^k accounts only for the rate of $\mathcal{S}_i^k \to \mathcal{S}_j^k$ type transitions. The rate of a transition from $\mathcal{S}_i^k$ to $\mathcal{S}_j^k$ is zero if $\mathcal{S}_i \neq \mathcal{S}_j$. If $\mathcal{S}_i = \mathcal{S}_j$, then we get

$$B_{i,i}^k = -\sum_{j=1}^{c_{k+1}} A_{j,i}^{k+1} - \sum_{j=1}^{c_{k-1}} C_{j,i}^{k-1}, \tag{2.12}$$

because the sum of the entries in any column is 0.

These rules can be implemented computationally to automatically generate and numerically solve the full set of differential equations. The following exercises encourage the reader to apply these rules.

Exercise 2.4. Find the matrices A^k, B^k and C^k for the SIS dynamics on a line graph with $N = 3$ nodes (see Fig. 2.2a). Write down the full system of master equations and verify that there are 2^N equations.

| **Exercise 2.5.** Do the same for a line graph with $N = 4$ nodes (see Fig. 2.2b).

| **Exercise 2.6.** Do the same for a star graph with $N = 4$ nodes (see Fig. 2.2c).

Exercise 2.7. Write down the system of master equations for the QAQ dynamics given in Example 2.3 on a line graph with $N = 3$ nodes (see Fig. 2.2a).

Exercise 2.8. Write down the system of master equations for the voter-like model given in Example 2.4 on a line graph with $N = 3$ nodes (see Fig. 2.2a).

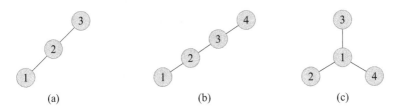

(a)   (b)   (c)

Fig. 2.2: Line with 3 (a) and 4 (b) nodes, respectively. (c) Star network with 4 nodes.

After formulating master equations for different binary dynamics and different graphs, we can write down the master equations for arbitrary binary dynamics on an arbitrary graph with $N = 3$ nodes. The state space consists of the states

$$\{QQQ, TQQ, QTQ, QQT, TTQ, TQT, QTT, TTT\}.$$

All potential transitions are depicted in Fig. 2.3. The rates depend on the process, which determines f_{QT} and f_{TQ}, and the graph, with adjacency matrix $G = (g_{ij})_{i,j=1,2,3}$, which determines how many neighbours of each status a given node has.

For example, for the transition from state QQT to QQQ node 3 moves from status T to Q. This rate depends on how many status Q neighbours it has. As both 1 and 2 have status Q, this is 0 if neither edge exists from node 3, 1 if only one edge exists and 2 if both edges exist. A convenient shorthand for the number of status Q neighbours 3 has, given that the state is QQT, is $g_{13} + g_{23}$. Thus, the transition rate from QQT to QQQ is $f_{TQ}(g_{13} + g_{23})$.

Similarly, the transition rate from state QQT to QTT happens at rate $f_{QT}(g_{32})$, because node 2 transitions from status Q to T, and this depends on whether node 2 is connected to the single node of status T, i.e. node 3. Determining the transition rates for all 24 possible transitions, indicated by arrows in Fig. 2.3, the master equations can be formulated as

$$\dot{X}_{QQQ} = f_{TQ}(g_{13} + g_{23})X_{QQT} + f_{TQ}(g_{12} + g_{32})X_{QTQ}$$
$$\qquad + f_{TQ}(g_{21} + g_{31})X_{TQQ} - 3f_{QT}(0)X_{QQQ},$$
$$\dot{X}_{QQT} = f_{QT}(0)X_{QQQ} + f_{TQ}(g_{12})X_{QTT} + f_{TQ}(g_{21})X_{TQT} - q_{QQT}X_{QQT},$$
$$\dot{X}_{QTQ} = f_{QT}(0)X_{QQQ} + f_{TQ}(g_{13})X_{QTT} + f_{TQ}(g_{31})X_{TTQ} - q_{QTQ}X_{QTQ},$$

$$\dot{X}_{TQQ} = f_{QT}(0)X_{QQQ} + f_{TQ}(g_{23})X_{TQT} + f_{TQ}(g_{32})X_{TTQ} - q_{TQQ}X_{TQQ},$$
$$\dot{X}_{QTT} = f_{TQ}(0)X_{TTT} + f_{QT}(g_{32})X_{QQT} + f_{QT}(g_{23})X_{QTQ} - q_{QTT}X_{QTT},$$
$$\dot{X}_{TQT} = f_{TQ}(0)X_{TTT} + f_{QT}(g_{31})X_{QQT} + f_{QT}(g_{13})X_{TQQ} - q_{TQT}X_{TQT},$$
$$\dot{X}_{TTQ} = f_{TQ}(0)X_{TTT} + f_{QT}(g_{21})X_{QTQ} + f_{QT}(g_{12})X_{TQQ} - q_{TTQ}X_{TTQ},$$
$$\dot{X}_{TTT} = f_{QT}(g_{21} + g_{31})X_{QTT} + f_{QT}(g_{12} + g_{32})X_{TQT}$$
$$\qquad\qquad + f_{QT}(g_{13} + g_{23})X_{TTQ} - 3f_{TQ}(0)X_{TTT},$$

where

$$q_{QQT} = f_{TQ}(g_{13} + g_{23}) + f_{QT}(g_{31}) + f_{QT}(g_{32}),$$
$$q_{QTQ} = f_{TQ}(g_{12} + g_{32}) + f_{QT}(g_{21}) + f_{QT}(g_{23}),$$
$$q_{TQQ} = f_{TQ}(g_{21} + g_{31}) + f_{QT}(g_{12}) + f_{QT}(g_{13}),$$
$$q_{QTT} = f_{QT}(g_{21} + g_{31}) + f_{TQ}(g_{13}) + f_{TQ}(g_{12}),$$
$$q_{TQT} = f_{QT}(g_{12} + g_{32}) + f_{TQ}(g_{23}) + f_{TQ}(g_{21}),$$
$$q_{TTQ} = f_{QT}(g_{13} + g_{23}) + f_{TQ}(g_{31}) + f_{TQ}(g_{32}).$$

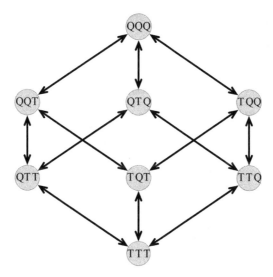

Fig. 2.3: The state space and all possible transitions for an arbitrary binary dynamics on an arbitrary graph with 3 nodes.

2.3.4 Master equations for SIR dynamics

The theory developed so far extends in a natural way to non-binary dynamics. To minimise technicalities, we consider just SIR disease propagation.

Example 2.6. As a first example, we write down the master equation for the simple network of two nodes connected by an edge. The state space is given by

$$\mathcal{C} = \{RR, RS, SR, SS, RI, IR, SI, IS, II\},$$

where the order indicates which node has each status. We will see later that it is useful to divide the states according to the number of I and S nodes. Let $\mathcal{C}^{i,j}$ denote the class of states in which there are i status I nodes and j status S nodes:

$$\mathcal{C}^{00} = \{RR\}, \quad \mathcal{C}^{01} = \{RS, SR\}, \quad \mathcal{C}^{02} = \{SS\},$$
$$\mathcal{C}^{10} = \{RI, IR\}, \quad \mathcal{C}^{11} = \{SI, IS\}, \quad \mathcal{C}^{20} = \{II\}.$$

The master equations follow by accounting for all possible transitions and their rates. They are

$$\dot{X}_{RR} = \gamma(X_{RI} + X_{IR}), \qquad \dot{X}_{SS} = 0, \qquad \dot{X}_{SI} = -(\gamma + \tau)X_{SI},$$
$$\dot{X}_{RS} = \gamma X_{IS}, \qquad \dot{X}_{RI} = \gamma X_{II} - \gamma X_{RI}, \qquad \dot{X}_{IS} = -(\gamma + \tau)X_{IS},$$
$$\dot{X}_{SR} = \gamma X_{SI}, \qquad \dot{X}_{IR} = \gamma X_{II} - \gamma X_{IR}, \qquad \dot{X}_{II} = \tau(X_{SI} + X_{IS}) - 2\gamma X_{II}.$$

Example 2.7. For a network with $N = 3$ nodes, the classes of the state space are

$$\mathcal{C}^{00} = \{RRR\}, \quad \mathcal{C}^{01} = \{RRS, RSR, SRR\}, \quad \mathcal{C}^{02} = \{RSS, SRS, SSR\},$$
$$\mathcal{C}^{03} = \{SSS\}, \quad \mathcal{C}^{10} = \{RRI, RIR, IRR\},$$
$$\mathcal{C}^{11} = \{RSI, RIS, SRI, SIR, IRS, ISR\}, \quad \mathcal{C}^{12} = \{SSI, SIS, ISS\},$$
$$\mathcal{C}^{20} = \{RII, IRI, IIR\}, \quad \mathcal{C}^{21} = \{SII, ISI, IIS\}, \quad \mathcal{C}^{30} = \{III\}.$$

In particular, for a line network with 3 nodes, the master equations are

$$\dot{X}_{RRR} = \gamma(X_{IRR} + X_{RRI} + X_{RIR}), \qquad \dot{X}_{RSS} = \gamma X_{ISS},$$
$$\dot{X}_{RRS} = \gamma(X_{IRS} + X_{RIS}), \qquad \dot{X}_{SRS} = \gamma X_{SIS},$$
$$\dot{X}_{RSR} = \gamma(X_{ISR} + X_{RSI}), \qquad \dot{X}_{SSR} = \gamma X_{SSI},$$
$$\dot{X}_{SRR} = \gamma(X_{SIR} + X_{SRI}), \qquad \dot{X}_{SSS} = 0,$$

for the $\mathcal{C}^{00}$, $\mathcal{C}^{01}$, $\mathcal{C}^{02}$ and $\mathcal{C}^{03}$ states;

$$\dot{X}_{RRI} = \gamma(X_{IRI} + X_{RII}) - \gamma X_{RRI}, \qquad \dot{X}_{SIR} = \gamma X_{SII} - (\gamma + \tau)X_{SIR},$$
$$\dot{X}_{RIR} = \gamma(X_{IIR} + X_{RII}) - \gamma X_{RIR}, \qquad \dot{X}_{IRS} = \gamma X_{IIS} - \gamma X_{IRS},$$
$$\dot{X}_{IRR} = \gamma(X_{IIR} + X_{IRI}) - \gamma X_{IRR} \qquad \dot{X}_{ISR} = \gamma X_{ISI} - (\gamma + \tau)X_{ISR},$$
$$\dot{X}_{RSI} = \gamma X_{ISI} - (\gamma + \tau)X_{RSI}, \qquad \dot{X}_{SSI} = -(\gamma + \tau)X_{SSI},$$
$$\dot{X}_{RIS} = \gamma X_{IIS} - (\gamma + \tau)X_{RIS}, \qquad \dot{X}_{SIS} = -(\gamma + 2\tau)X_{SIS},$$
$$\dot{X}_{SRI} = \gamma X_{SII} - \gamma X_{SRI}, \qquad \dot{X}_{ISS} = -(\gamma + \tau)X_{ISS},$$

for the $\mathcal{C}^{10}$, $\mathcal{C}^{11}$ and $\mathcal{C}^{12}$ states; and finally

$$\dot{X}_{RII} = \tau(X_{RSI} + X_{RIS}) + \gamma X_{III} - 2\gamma X_{RII}, \qquad \dot{X}_{SII} = \tau(X_{SSI} + X_{SIS}) - (2\gamma + \tau)X_{SII},$$
$$\dot{X}_{IRI} = \gamma X_{III} - 2\gamma X_{IRI}, \qquad \dot{X}_{ISI} = -(2\gamma + 2\tau)X_{ISI},$$
$$\dot{X}_{IIR} = \tau(X_{SIR} + X_{ISR}) + \gamma X_{III} - 2\gamma X_{IIR}, \qquad \dot{X}_{IIS} = \tau(X_{SIS} + X_{ISS}) - (2\gamma + \tau)X_{IIS},$$
$$\dot{X}_{III} = \tau X_{IIS} + 2\tau X_{ISI} + \tau X_{SII} - 3\gamma X_{III},$$

for the $\mathcal{C}^{20}$, $\mathcal{C}^{21}$ and $\mathcal{C}^{30}$ states. In general, the system can move from a $\mathcal{C}^{ij}$ state to a $\mathcal{C}^{i+1,j-1}$ state by infection or to a $\mathcal{C}^{i-1,j}$ state by recovery.

We emphasise that $\mathcal{C}^{ij}$ is not a state but a collection (or class) of states of the system. In Fig. 2.4, we give the possible paths that an arbitrary network with 4 nodes can take between the different classes. Although we group these states together into classes by how many individuals of each status there are, the transition rates into or out of different states in the same class may vary. Hence, the transition rates in the figure cannot be explicitly given without breaking the classes down into their individual states, as they depend on the choice of the state in a given class.

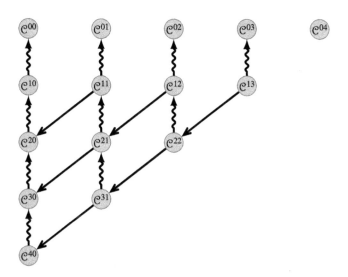

Fig. 2.4: Potential transitions for an SIR model for an arbitrary 4-node network. Straight lines correspond to transmissions and sinuous lines correspond to recoveries.

Exercise 2.9. Write down the full system of master equations for SIR dynamics on a triangle, i.e. on a complete graph with $N = 3$ nodes.

Finally, we investigate the structure of the matrix yielding the system of master equations. We introduce a coarser classification of the state space. Let $\mathcal{C}^i$ denote the class made up of states where there are i nodes of status I. Then, obviously

$$\mathcal{C}^i = \cup_{j=0}^{N-i} \mathcal{C}^{ij}$$

and the system can move from a state in class $\mathcal{C}^i$ to a state in class $\mathcal{C}^{i+1}$ by infection or to a state in class $\mathcal{C}^{i-1}$ by recovery. The number of states in class $\mathcal{C}^{ij}$ is $\binom{N}{i}\binom{N-i}{j}$; hence, a simple summation shows that the number of states in class $\mathcal{C}^i$ is $2^{N-i}\binom{N}{i}$. For example, in the case of a network with $N = 3$ nodes in the class $\mathcal{C}^0$, there are 8 states, in the class $\mathcal{C}^1$ there are 12 states, in the class $\mathcal{C}^2$ there are 6 states and in the class $\mathcal{C}^3$ there is a single state, as can be explicitly checked by using the list of states above.

To formulate the system of master equations, let $X^i(t)$ be a $2^{N-i}\binom{N}{i}$ dimensional vector, the coordinates of which yield the probability of the system being in the states of class $\mathcal{C}^i$, for $i = 0, 1, \ldots, N$. Since the possible transitions from class $\mathcal{C}^i$ are to classes $\mathcal{C}^{i+1}$ and $\mathcal{C}^{i-1}$, the system takes the form

$$\dot{X}^i = A^i X^{i-1} + B^i X^i + C^i X^{i+1}, \qquad i = 0, 1, \ldots, N, \qquad (2.13)$$

where A^0 and C^N are zero matrices. Thus, Eq. (2.13) is in the form (2.3), with a matrix P that can be written in block tridiagonal form as

$$P = \begin{pmatrix} B^0 & C^0 & 0 & 0 & \cdots & & 0 \\ A^1 & B^1 & C^1 & 0 & \vdots & & \vdots \\ 0 & A^2 & B^2 & C^2 & \vdots & & \vdots \\ 0 & 0 & A^3 & B^3 & \ddots & & 0 \\ \vdots & \cdots & \cdots & \ddots & \ddots & & C^{N-1} \\ 0 & \cdots & \cdots & 0 & A^N & & B^N \end{pmatrix}.$$

2.4 Lumping

We return to the case of general binary dynamics. The system of master equations given by (2.6) consists of 2^N linear differential equations, so the number of equations grows exponentially with N. This quickly becomes impractical or impossible to solve at large N. However, it is not always necessary to determine all probabilities. Particularly, when the number of nodes is large, the expected number or proportion of nodes of status Q and status T may be more useful. We denote these expected values at time t by $[Q](t)$ and $[T](t)$. They are expressed as

$$[Q](t) = \sum_{k=0}^{N} \left((N-k) \sum_{j=1}^{c_k} X_j^k(t) \right), \qquad [T](t) = \sum_{k=0}^{N} \left(k \sum_{j=1}^{c_k} X_j^k(t) \right).$$

There are different approaches to determine or approximate these expected values. Conceptually, the simplest way is through individual-based stochastic simulation, using methods such as the Gillespie algorithm [112, 113] (or other methods we present for SIS and SIR disease in Appendix A.1). However, this offers limited scope for a deeper understanding of the interactions between network topology and node dynamics, and results based on simulation are difficult to generalise. An alternative is the derivation of mean-field-like equations using pair or triple approximations [166, 256], heterogenous mean-field approximations [244], edge-based compartmental models [215, 222, 316] and effective degree-type models [115, 197, 300]. Chapters 4 to 7 give an extensive treatment of these modelling frameworks and show how different models are related. However, most of these models are approximate, and if we want an exact reduction, these mean-field-like approaches are insufficient. For networks with sufficient symmetry, the technique of lumping [100, 171, 268] can be used to reduce the exponential number of equations to a tractable number that are exact at a coarser scale. Ideally, the lumped system should have enough variables so that its solution provides relevant and significant information about the system. For example, the reduced system should be able to provide the expected number of nodes of each status.

Lumping, in general, is a method for coarse-graining Markov chains by partitioning, or lumping, the state space. The problem is to partition the state space in such a way that the Markov property is not lost in the aggregated process. Lumping of Markov chains is studied in [171, 268]; the particular case of random walks on networks is examined in [100]. Spectral properties of the transition matrix are related to the lumpability of the system [25, 155]. However, due to the exponentially large state space, in the case of network processes, these spectral methods have limited efficiency. This leads us to use another way to find partitions, namely exploiting the symmetries of the underlying network, as presented below.

The remainder of this section is structured as follows. First, we introduce a few definitions which help our analysis. Then, the idea of lumping is illustrated by motivating examples. Next, the lumping of linear systems of ODEs is dealt with in general. This is followed by the presentation of lumping system (2.6) and showing how this is related to the automorphisms (symmetries) of the graph. Finally, the lumping procedure is carried out for small networks and for arbitrarily large networks with special structure.

2.4.1 Partition of the state space

Consider a system with n states $\mathcal{S}_1$, $\mathcal{S}_2$, ..., $\mathcal{S}_n$. For the class of problems we study, we can represent the probability of state $\mathcal{S}_i$ by $X_i(t)$, with the vector $X = (X_1, X_2, \ldots, X_n)$ denoting all of the probabilities. We can think of transitions from $\mathcal{S}_r$ to $\mathcal{S}_j$ in terms of probability flowing between the states. The amount of this flux is given by $P_{jr}X_r$.

Sometimes, we do not need to know X_i for every state i, or calculating the full solution may simply be too difficult. In such cases, we would like to use a coarser scale. This means partitioning (or lumping) some states together, creating m classes

$\{\mathcal{C}_1, \mathcal{C}_2, \ldots, \mathcal{C}_m\}$ of states. To make our terminology clear, we introduce some definitions.

Definition 2.2 *Given a set of objects* $\Sigma = \{\mathcal{S}_1, \mathcal{S}_2, \ldots, \mathcal{S}_n\}$, *a* partition *of* Σ *is a set* $\mathcal{C}^* = \{\mathcal{C}_1, \mathcal{C}_2, \ldots, \mathcal{C}_m\}$ *of classes having two properties:*

- *Each class* $\mathcal{C}_j$ *is a nonempty subset of* Σ.
- *Every* $\mathcal{S}_i \in \Sigma$ *belongs to exactly one* $\mathcal{C}_j$.

Typically, we have some collection of indices for our objects, usually integers, although in Example 2.8 below the states are indexed by a list giving the status of every node. There is a clear relation between a partition of the states and a partition of the indices. It will often be useful to have a shorthand for the set of indices associated with a class in the partition.

Definition 2.3 *Given a set of objects indexed by some indices* $\Sigma = \{\mathcal{S}_{i_1}, \mathcal{S}_{i_2}, \ldots, \mathcal{S}_{i_n}\}$, *let* $\mathcal{C}^* = \{\mathcal{C}_1, \mathcal{C}_2, \ldots, \mathcal{C}_m\}$ *be a partition of* Σ. *The* induced partition *on the index set* $\{i_1, i_2, \ldots, i_n\}$ *is* $L = \{L_1, L_2, \ldots, L_m\}$, *where each* L_k *is the set of indices appearing in* $\mathcal{C}_k$. *That is,* $L_k = \{i_j : \mathcal{S}_{i_j} \in \mathcal{C}_k\}$. *We refer to* L_k *as the* induced class *of indices associated with* $\mathcal{C}_k$.

We write $Y_j = \sum_{i \in L_j} X_i$ to be the combined probability of all states in $\mathcal{C}_j$. In many cases, the correct choice of partition leads to a significant simplification of the equations. In such cases, we arrive at a linear system for Y_j having fewer equations.

2.4.2 A motivating example

Looking back at Example 2.1 of Section 2.1, we see that lumping several states together into larger classes can significantly reduce the number of equations. However, Exercise 2.2 shows that this does not always work. We will see that the success of a lumping depends strongly on the symmetries of the network. We now consider a more complex network structure to demonstrate more clearly how symmetries simplify the equations.

Example 2.8. We investigate an SIS disease spreading in a star network with four nodes. The states and flows between them are represented by the flow diagram in Fig. 2.5. For this specific case, we use the notation $\mathcal{S}_{ABCD}$ to represent the state in which the central node has status A, the leftmost node status B, the top node status C and the rightmost node status D, where these are each either S or I.

The probability of state $\mathcal{S}_{ABCD}$ is denoted X_{ABCD}. An arrow from state $\mathcal{S}_i$ to state $\mathcal{S}_j$ means that the system can move directly from $\mathcal{S}_i$ to $\mathcal{S}_j$, either through recovery of a node or a transmission. An arrow from X_i to X_j has an associated flux, and this flux appears in the equations as an additive term for $\dot{X}_j$ and a subtracted term for $\dot{X}_i$. Many arrows in Fig. 2.5 are bidirected, but some are not.

When we group those states that are symmetric together, we arrive at a partition made up of classes $\mathcal{C}^{ik}$, where the superscript i is 0 if the central node is susceptible and 1 if infected, and the superscript k gives the number of peripheral nodes that are infected.

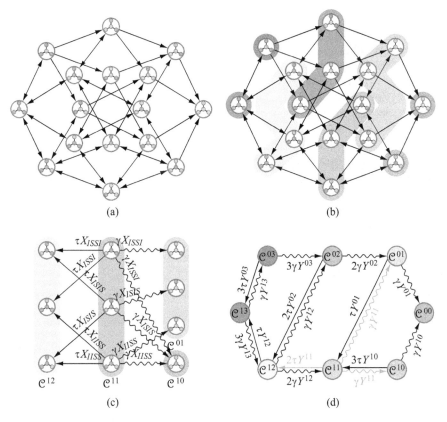

Fig. 2.5: Lumping for the spread of an SIS disease in a star with four nodes. Susceptible (○) and infected (◉) nodes are denoted by filled circles of different colours.
(a) The possible transitions in the original state space. (b) A partition formed by
grouping symmetric states together. (c) The fluxes from each state $\mathcal{S}$ in $\mathcal{C}^{11}$ to the
adjacent classes are proportional to the probability of $\mathcal{S}$, with the same coefficients
for all $\mathcal{S} \in \mathcal{C}^{11}$. (d) The final lumped system. In (c) and (d), recoveries are denoted
by sinusoidal paths. Note the relation between the highlighted fluxes out of $\mathcal{C}^{11}$ in
(d) and fluxes out of states within $\mathcal{C}^{11}$ in (c).

$$\mathcal{C}^{00} = \{\mathcal{S}_{SSSS}\}$$
$$\mathcal{C}^{01} = \{\mathcal{S}_{SISS}, \mathcal{S}_{SSIS}, \mathcal{S}_{SSSI}\}$$
$$\mathcal{C}^{10} = \{\mathcal{S}_{ISSS}\}$$
$$\mathcal{C}^{02} = \{\mathcal{S}_{SIIS}, \mathcal{S}_{SISI}, \mathcal{S}_{SSII}\}$$
$$\mathcal{C}^{11} = \{\mathcal{S}_{IISS}, \mathcal{S}_{ISIS}, \mathcal{S}_{ISSI}\}$$
$$\mathcal{C}^{03} = \{\mathcal{S}_{SIII}\}$$
$$\mathcal{C}^{12} = \{\mathcal{S}_{ISII}, \mathcal{S}_{IISI}, \mathcal{S}_{IIIS}\}$$
$$\mathcal{C}^{13} = \{\mathcal{S}_{IIII}\}$$

For notational convenience, we will use L_{ik} to denote the subscripts associated with
the states in $\mathcal{C}^{ik}$. Then, the probability Y^{ik} of the class $\mathcal{C}^{ik}$ is given by

$$Y^{ik} = \sum_{j \in L_{ik}} X_j$$

Consider the class $\mathcal{C}^{11}$ consisting of all states in which the central node and a single peripheral node are infected. The central node can transmit to either of the two remaining peripheral nodes, each with an associated rate τ. Looking in particular at $\mathcal{S}_{IISS}$, the transitions are to $\mathcal{S}_{IIIS}$ and $\mathcal{S}_{IISI}$. The flux from X_{IISS} to X_{IIIS} is τX_{IISS} and the flux to X_{IISI} is also τX_{IISS}. Thus, the total flux from X_{IISS} into $\mathcal{C}^{12}$ is $2\tau X_{IISS}$. Similarly, the fluxes from X_{ISIS} and X_{ISSI} into $\mathcal{C}^{12}$ are also proportional to X_{ISIS} and X_{ISSI}, respectively, with the same proportionality constant. Thus, the total flux from $\mathcal{C}^{11}$ to $\mathcal{C}^{12}$ is simply $2\tau Y^{11}$.

With the chosen partition, all classes $\mathcal{C}^{ij}$ have the property that, given another class $\mathcal{C}^{kl}$, the flow from any $\mathcal{S}_r \in \mathcal{C}^{ij}$ into the states in $\mathcal{C}^{kl}$ is proportional to X_r, and that same proportionality constant holds for every $\mathcal{S}_r \in \mathcal{C}^{ij}$. Thus, the total flux from $\mathcal{C}^{ij}$ to $\mathcal{C}^{kl}$ is proportional to Y^{ij} with the same constant of proportionality.

Because this holds for every pair of classes, we can reduce our system to just considering the classes, rather than the individual states. Our final equations are

$$\dot{Y}^{00} = \gamma(Y^{10} + Y^{01}),$$
$$\dot{Y}^{01} = 2\gamma Y^{02} + \gamma Y^{11} - (\gamma + \tau)Y^{01},$$
$$\dot{Y}^{10} = \gamma Y^{11} - (\gamma + 3\tau)Y^{10},$$
$$\dot{Y}^{02} = 3\gamma Y^{03} + \gamma Y^{12} - (2\gamma + 2\tau)Y^{02},$$

$$\dot{Y}^{11} = 3\tau Y^{10} + \tau Y^{01} + 2\gamma Y^{12} - (2\gamma + 2\tau)Y^{11},$$
$$\dot{Y}^{03} = \gamma Y^{13} - (3\gamma + 3\tau)Y^{03},$$
$$\dot{Y}^{12} = 3\gamma Y^{13} + 2\tau(Y^{02} + Y^{11}) - (3\gamma + \tau)Y^{12},$$
$$\dot{Y}^{13} = 3\tau Y^{03} + \tau Y^{12} - 4\gamma Y^{13}.$$

The system for the X variables would have 16 equations. Here, we have reduced this to 8, at the cost that we can no longer resolve individual states.

Exercise 2.10. Consider a different partition of the states in Fig. 2.5, where $\mathcal{C}^k$ is made up of all states with k infected nodes. Let Y_k be the probability that k nodes are infected. Show that we cannot write down a system of linear differential equations just in terms of Y_k. [Hint: Assume that initially $X_{ISSS} = 1$ and calculate the initial rate of change of Y_2. Repeat with $X_{SISS} = 1$.]

2.4.3 Lumping of linear systems

In Example 2.8, the lumping worked because it provided a partition such that for any two classes $\mathcal{C}_l$ and $\mathcal{C}_j$ we could express the total flow of probability from $\mathcal{C}_l$ to $\mathcal{C}_j$ as some constant times the amount of probability in $\mathcal{C}_l$. In fact, a stronger condition holds: each state $\mathcal{S}_r \in \mathcal{C}_l$ has the property that the combined flow from it to the states in $\mathcal{C}_j$ gave the same value, regardless of $\mathcal{S}_r$. This motivates the following definition.

Definition 2.4 *A Markov process is called* lumpable *if there is a partition of the states $\{\mathcal{S}_1, \mathcal{S}_2, \ldots, \mathcal{S}_n\}$ into a set of classes $\mathcal{C} = \{\mathcal{C}_1, \mathcal{C}_2, \ldots, \mathcal{C}_m\}$ such that for any two classes $\mathcal{C}_j$ and $\mathcal{C}_l$ the sum*

$$\overline{\mathcal{A}}_{jl} = \sum_{\mathcal{S}_i \in \mathcal{C}_j} h(\mathcal{S}_r, \mathcal{S}_i)$$

takes the same value for any $S_r \in \mathcal{C}_l$. Then, the partition $\{\mathcal{C}_1, \mathcal{C}_2, \ldots, \mathcal{C}_m\}$ is called a lumping of the state space.

The sum represents the combined transition rate from S_r to any state in $\mathcal{C}_j$.

We now explore the mathematical concept of lumping for a general system, which can be expressed in the form

$$\dot{X} = \mathcal{A}X,$$

where $\mathcal{A}$ is an arbitrary $n \times n$ matrix. If we think of the components of X as measuring the probability (or some other quantity) of the states of a system, then our goal is to partition the states in such a way that if we take Y to be a vector whose entries give the sum of X_i for each class, we can arrive at a new linear system for Y. The key detail that guaranteed this previously is that given two classes, $\mathcal{C}_l$ and $\mathcal{C}_j$, and any state $S_r \in \mathcal{C}_l$, the combined flux from S_r into all of the states in $\mathcal{C}_j$ is proportional to X_r, with the same proportionality constant holding for all $S_r \in \mathcal{C}_l$.

We now express this condition in terms of the matrix $\mathcal{A}$. The following definition goes back to the so-called Dynkin criterion [100].

Definition 2.5 *The linear system $\dot{X} = AX$ is called* lumpable *if there is a partition $L = \{L_1, L_2, \ldots, L_m\}$ of the set $\{1, 2, \ldots, n\}$ satisfying the following property: for any classes L_j and L_l, there exists a number $\overline{A}_{jl}$ such that*

$$\overline{A}_{jl} = \sum_{i \in L_j} A_{ir}, \text{ for } r \in L_l,$$

that is, the sum does not depend on r whenever $r \in L_l$. The $m \times m$ matrix $\overline{A}$ is called a lumping *of matrix $\mathcal{A}$, and the partition L is called a* lumping *of $\{1, 2, \ldots, n\}$.*

If the linear system corresponds to a Markov process, then either both the system and the process are lumpable or both are not. This follows by taking L to be the induced partition of the classes.

We will define $Y_j = \sum_{i \in L_j} X_i$ and set Y to be the vector whose entries are Y_j. Then, taking $\overline{A}$ we will show that

$$\dot{Y} = \overline{A}Y. \tag{2.14}$$

We derive equation (2.14) in several steps. First, consider a lumping $\{L_1, L_2, \ldots, L_m\}$ of $\{1, 2, \ldots, n\}$. We define the vectors U_j for $j = 1, \ldots, m$ such that the ith entry of U_j is 1 if $i \in L_j$ and 0 otherwise. Then, the dot product of U_j with X gives $Y_j = \sum_{i \in L_j} X_i$. Taking the matrix U whose jth row is U_j, we have

$$Y = UX.$$

where Y is the vector whose entries are Y_j.

The following simple result about $U\mathcal{A}$ holds.

Proposition 2.1 *If matrix $\overline{A}$ is a lumping of matrix $\mathcal{A}$ and U is as defined above, then $U\mathcal{A} = \overline{A}U$.*

Proof. The element in the jth row and rth column of $U\mathcal{A}$ is

$$(U\mathcal{A})_{jr} = \sum_{i=1}^{n} U_{ji}\mathcal{A}_{ir} = \sum_{i\in L_j} \mathcal{A}_{ir} = \overline{\mathcal{A}}_{jl},$$

where l is the index for which $r \in L_l$. The element in the jth row and rth column of $\overline{\mathcal{A}}U$ is

$$(\overline{\mathcal{A}}U)_{jr} = \sum_{k=1}^{m} \overline{\mathcal{A}}_{jk}U_{kr} = \overline{\mathcal{A}}_{jl},$$

where l is the index such that $r \in L_l$, since every column of U has a single unit entry, the rest being zero. Thus, the two expressions are equal.   □

We are now able to prove that $\dot{Y} = \overline{\mathcal{A}}Y$.

Proposition 2.2 *Let $\overline{\mathcal{A}}$ be a lumping of matrix $\mathcal{A}$ and let U be the matrix for which $U\mathcal{A} = \overline{\mathcal{A}}U$ holds. Based on this, we introduce the new, m-dimensional (lumped) variable $Y = UX$. This lumped variable satisfies the lumped linear ODE system $\dot{Y} = \overline{\mathcal{A}}Y$.*

Proof. We already have $Y = UX$ and U is a constant matrix. Thus, $\dot{Y} = U\dot{X}$, but $\dot{X} = \mathcal{A}X$. So we have $\dot{Y} = U\mathcal{A}X$. As $U\mathcal{A} = \overline{\mathcal{A}}U$, our system becomes $\dot{Y} = \overline{\mathcal{A}}UX$. Substituting for UX, we finally have

$$\dot{Y} = \overline{\mathcal{A}}Y.$$

□

The crucial step in lumping is finding the partition of the state space. This can either be derived intuitively through the symmetries of the network or by attempting to identify partitions that satisfy the properties stated in Definition 2.4 or 2.5. Once the partition is known, both $\overline{\mathcal{A}}$ and U follow. The lumpability condition on $\mathcal{A}$ is a non-trivial requirement, and not all systems will be lumpable.

It should be noted that the results we have proven are more general than what we need. For the systems we consider, we are always considering vectors X that represent probabilities. The proofs did not rely on this, and so the same procedure will work for more general systems, but we do not investigate this further.

2.4.4 The use of graph symmetries to lump a binary dynamic network model

We found the partition for the star network SIS system of Example 2.8 by identifying states that were symmetric and lumping them into a single class. In this section, we see that this approach works in general. It should be noted that this is not the only option for lumping a system to get simpler equations. As a trivial example for any disease, there is always the option to create a partition with just a single class $\mathcal{C}^*$ containing all states, then we would arrive at the equation $\dot{Y}^* = 0$. This is mathematically simple, but it has no informational value beyond stating that the quantity we are measuring (probability in this case) is preserved. There is a balance between simplicity and information content.

Our goal after lumping the states is to recover information about the number of nodes of each status. So all states that are lumped together must have the same number of nodes of each status. In this section, we show a procedure for finding lumpings that respect this property if there are symmetries. We focus our attention on binary processes to keep the indexing simple, but the approach is more general.

Consider now an arbitrary binary dynamic process spreading in a network, such as SIS disease. We will arbitrarily set the node statuses to be Q and T. For a network of N nodes, there are 2^N possible states. We recall the important partition

$$\{\mathcal{C}^0, \mathcal{C}^1, \dots, \mathcal{C}^N\},$$

where $\mathcal{C}^k$ denotes the class made up of all states with k nodes of status T. There are $c_k = \binom{N}{k}$ states in $\mathcal{C}^k$, denoted $\mathcal{S}_1^k$, $\mathcal{S}_2^k$, ..., $\mathcal{S}_{c_k}^k$. We set X_i^k to be the probability of state $\mathcal{S}_i^k$ and use X^k to be a vector whose entries are the probabilities of each state of $\mathcal{C}^k$. That is, $X^k = (X_1^k, X_2^k, \dots, X_{c_k}^k)$, and we use X to be the vector formed by taking the entries of X^0, then of X^1 and so forth until X^N.

We have already seen that we cannot necessarily write down a consistent system of equations if we choose $\mathcal{C}^0, \mathcal{C}^1, \dots, \mathcal{C}^N$ to be our partition (Exercises 2.2 and 2.10). We will have to refine this partition. However, equation (2.6) shows that for this partition, the equations for a given X^k can be expressed entirely in terms of X^{k-1}, X^k and X^{k+1}

$$\dot{X}^k = A^k X^{k-1} + B^k X^k + C^k X^{k+1}, \qquad k = 0, 1, \dots, N$$

for some matrices A^k, B^k and C^k, with B^k being a diagonal matrix. Then, $\dot{X} = PX$, where

$$P = \begin{pmatrix} B^0 & C^0 & 0 & \cdots & & 0 \\ A^1 & B^1 & C^1 & \cdots & & \vdots \\ 0 & A^2 & B^2 & \ddots & & \vdots \\ \vdots & \ddots & \ddots & \ddots & C^{N-1} \\ 0 & \cdots & \cdots & & A^N & B^N \end{pmatrix}.$$

The specific form of these submatrices depends on the network structure and the spreading binary process.

Definition 2.6 *Given a set and a partition of that set $J = \{J_1, J_2, \dots, J_{m_J}\}$, a second partition $L = \{L_1, L_2, \dots, L_{m_L}\}$ is a* refinement *of J if for every L_l there is a J_j such that $L_l \subseteq J_j$.*

If L is a refinement of J, and we denote all of the L_l that are a subset of J_j by L_1^j, L_2^j, $\ldots, L_{l_j}^j$, then these form a partition of J_j.

For simplicity, we will assume that the states are indexed from 1 to 2^N, so that if $i < j$, then the number of nodes in state S_i having status T is less than or equal to the number of nodes in S_j having status T. If the number of nodes with status T is the same, the ordering is arbitrary, but it is unchanging.

We now turn specifically to partitions of our state space. There are 2^N possible states $S_1, S_2, \ldots, S_{2^N}$.

Definition 2.7 *Given the partition of* $\{S_1, S_2, \ldots, S_{2^N}\}$ *into* $\{\mathcal{C}^0, \mathcal{C}^1, \ldots, \mathcal{C}^N\}$, *we say that a lumping* respects $\{\mathcal{C}^0, \mathcal{C}^1, \ldots, \mathcal{C}^N\}$ *if the partition for the lumping is a refinement of* $\{\mathcal{C}^0, \mathcal{C}^1, \ldots, \mathcal{C}^N\}$.

In Example 2.8, the lumping respected the partition $\{\mathcal{C}^0, \mathcal{C}^1, \mathcal{C}^2, \mathcal{C}^3, \mathcal{C}^4\}$, with $\mathcal{C}^1$, $\mathcal{C}^2$ and $\mathcal{C}^3$ each divided into two smaller classes. Our goal is to find a lumping $\overline{P}$ of P that respects the partition $\{\mathcal{C}^0, \mathcal{C}^1, \mathcal{C}^2, \ldots, \mathcal{C}^N\}$.

Consider a refinement of $\{\mathcal{C}^0, \mathcal{C}^1, \mathcal{C}^2, \ldots, \mathcal{C}^N\}$. For every k, let $\{\mathcal{C}_1^k, \mathcal{C}_2^k, \ldots, \mathcal{C}_{l_k}^k\}$ denote those classes in the refined partition that are a subset of $\mathcal{C}^k$. To test whether this partition provides a lumping, we check that Definition 2.4 is satisfied. Let $\mathcal{C}_j^h$ and $\mathcal{C}_l^k$ be any two classes in the partition of the state space. If the number of nodes of status T differ by two or more, then the flow between these states is zero. Alternately, if the number of nodes of each status is the same and the classes are not the same, then again the flow between these classes is zero. So the properties of the definition are immediately satisfied for these cases. Thus, we only need to consider $\mathcal{C}_j^h$ and $\mathcal{C}_l^k$ if the number of nodes in status T differ by exactly 1 or if $\mathcal{C}_j^h = \mathcal{C}_l^k$.

Lemma 2.8 *The binary Markov process described by equation (2.6) has a lumping that respects* $\{\mathcal{C}^0, \mathcal{C}^1, \ldots, \mathcal{C}^N\}$ *if each class* $\mathcal{C}^k$ *may be partitioned into* $\{\mathcal{C}_1^k, \mathcal{C}_2^k, \ldots, \mathcal{C}_{l_k}^k\}$ *and the following properties hold for any k:*

- *For any classes* $\mathcal{C}_l^{k-1}$ *and* $\mathcal{C}_j^k$, *there exists a number* $\overline{A}_{jl}^k$ *such that*

$$\overline{A}_{jl}^k = \sum_{S_i \in \mathcal{C}_j^k} h(S_r, S_i) \tag{2.15}$$

  for any $S_r \in \mathcal{C}_l^{k-1}$.
- *For any classes* $\mathcal{C}_l^{k+1}$ *and* $\mathcal{C}_j^k$, *there exists a number* $\overline{C}_{jl}^k$ *such that*

$$\overline{C}_{jl}^k = \sum_{S_i \in \mathcal{C}_j^k} h(S_r, S_i) \tag{2.16}$$

  for any $S_r \in \mathcal{C}_l^{k+1}$.

Proof. To prove this, we show that Definition 2.4 holds for any $\mathcal{C}_j^m$ and $\mathcal{C}_l^k$, that is, we just show that for any two classes $\mathcal{C}_j^m$ and $\mathcal{C}_l^k$, the sum $\sum_{\mathcal{S}_i \in \mathcal{C}_j^m} h(\mathcal{S}_r, \mathcal{S}_i)$ is the same for all $\mathcal{S}_r \in \mathcal{C}_l^k$. We break this into four parts:

- If $|m - k| \geq 2$, then at least two nodes are required to change status of the system to move from any $\mathcal{S}_r$ to $\mathcal{S}_i$. Thus, the sum is trivially 0.
- If $m = k - 1$, then satisfying the first condition is the same as satisfying Definition 2.4.
- If $m = k + 1$, then satisfying the second condition is the same as satisfying Definition 2.4.
- If $m = k$ and $\mathcal{C}_j^k \neq \mathcal{C}_l^k$, then the system cannot move between states in these classes because any change of state will change the number of nodes with status T. Thus, the only case remaining to check is if $\mathcal{C}_j^k = \mathcal{C}_l^k$. In this case, if $\mathcal{S}_i \neq \mathcal{S}_r$, the result is still zero, so the sum collapses to just a single term:

$$\sum_{\mathcal{S}_i \in \mathcal{C}_j^m} h(\mathcal{S}_i, \mathcal{S}_r) = h(\mathcal{S}_r, \mathcal{S}_r).$$

From equation (2.4) we have

$$h(\mathcal{S}_r, \mathcal{S}_r) = - \sum_{\mathcal{S}_i \in \mathcal{C}^{k+1}} h(\mathcal{S}_r, \mathcal{S}_i) - \sum_{\mathcal{S}_i \in \mathcal{C}_{k-1}} h(\mathcal{S}_r, \mathcal{S}_i).$$

Alternately, we can show this by referring to the diagonal element of B^k. If the conditions of the lemma hold, then this sum will be the same for all $\mathcal{S}_r \in \mathcal{C}_l^k$. $\square$

Equation 2.15 states that for any state $\mathcal{S}_r$ in the class $\mathcal{C}_l^{k-1}$, the total flux into $\mathcal{C}_j^k$ is proportional to the probability of state $\mathcal{S}_r$ with the same proportionality constant $\overline{A}_{jl}^k$. Equation (2.16) is equivalent, but for $\mathcal{C}_l^{k+1}$ to $\mathcal{C}_j^k$. Thus, for these cases the flux of probability from one class to another is simply a constant times the combined probability in the first class.

We are almost ready to prove the main result of this section, that if we place "symmetric" states into the same partition, then we arrive at a valid lumping. We first need to give a mathematical definition of what "symmetric" means in this context.

Definition 2.9 *Let $G = G(V, E)$ be a graph with vertices and edges given by sets $V(G)$ and $E(G)$, respectively. A bijection $\Phi : V(G) \rightarrow V(G)$ such that $(x, y) \in E(G)$ if and only if $(\Phi(x), \Phi(y)) \in E(G)$ is an* automorphism *of graph G. The set of all automorphisms of G, under the composition of maps, forms the* automorphism group *denoted by* $\mathrm{Aut}(G)$ *([82, 330]).*

In less mathematical language, we can think of the graph drawn on paper. Applying the function Φ to the node labels corresponds to relabelling the nodes, so that node x is now labelled with the label $\Phi(x)$ (see Fig. 2.6). If the result of replacing each node name x by $\Phi(x)$ is a graph that is identical to what would be seen after

	1	2	3	4	5
$A_1 = \text{id}$	1	2	3	4	5
$A_2 = \text{rot}$	5	1	2	3	4
$A_3 = \text{rot}$	4	5	1	2	3
$A_4 = \text{rot}$	3	4	5	1	2
$A_5 = \text{rot}$	2	3	4	5	1
$A_6 = \text{ref}$	1	5	4	3	2
$A_7 = \text{ref}$	3	2	1	5	4
$A_8 = \text{ref}$	5	4	3	2	1
$A_9 = \text{ref}$	2	1	5	4	3
$A_{10} = \text{ref}$	4	3	2	1	5

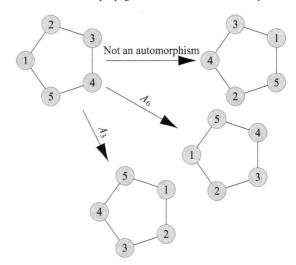

Fig. 2.6: (Left) The automorphism group of the cycle network with 5 nodes, composed from rotations and reflections. (Right) Examples of A_3, A_6 and a permutation that is not an automorphism.

moving each node $\Phi(x)$ to the current location of x (with any associated edges following), then Φ is an automorphism. It is important to note that if we perform one automorphism to G and then perform another to the result, the combined outcome is also an automorphism of the original graph.

If the nodes of a graph have some status associated with them, then we say that Φ takes the state S_i to state S_j if $S_i(l) = S_j(\Phi(l))$ for all l. In this case, we write $S_j = \Phi(S_i)$.

Definition 2.10 *Given the automorphisms of a graph, the* orbit *of a state S_j is the class of states of the form $\Phi(S_j)$ for all automorphisms Φ.*

If there is some Φ such that $S_i = \Phi(S_j)$, then the orbits of the two are identical.

We introduce an equivalence relation for $\mathcal{C}$, saying states are equivalent if they are in the same orbit. We call this the automorphism equivalence relation. The set of orbits form a partition of the states. If we list an orbit in some order $S_{j_1}, S_{j_2}, \ldots, S_{j_m}$ and then perform the same automorphism to each state, we get back the same class of states, but in a new order. So the automorphism only permutes the elements of an orbit.

Exercise 2.11.
    a. Show that all equivalent states have the same number of nodes of status T.
    b. Show that if we create a partition such that each class is made up of a set of equivalent states, then we have a refinement of $\{\mathcal{C}^0, \mathcal{C}^1, \ldots, \mathcal{C}^N\}$.

We can now formulate our main result connecting the automorphism group of the graph to the lumping of the Markov chain.

Theorem 2.11 *If we partition the states by creating classes made up of states that are equivalent under automorphisms, then the resulting partition yields a lumping that respects* $\{\mathcal{C}^0, \mathcal{C}^1, \ldots, \mathcal{C}^N\}$.

To prove this theorem, we will rely on the observation that if the flow from state $\mathcal{S}_{l_1}$ to $\mathcal{S}_{j_1}$ is cX_{l_1} for some constant rate c, and if Φ takes $\mathcal{S}_{l_1}$ to $\mathcal{S}_{l_2}$ and $\mathcal{S}_{j_1}$ to $\mathcal{S}_{j_2}$, then the flow from $\mathcal{S}_{l_2}$ to $\mathcal{S}_{j_2}$ is cX_{l_2}.

Proof. Let the partition be given as $\{\mathcal{C}^k_j \subset \mathcal{C}^k : j = 1, 2, \ldots, n_k, \ k = 0, 1, \ldots, N\}$. The assumption of the theorem can be formulated as follows: for any $k \in \{0, 1, \ldots, N\}$, $j \in \{1, 2, \ldots, n_k\}$ and $\mathcal{S}_r, \mathcal{S}_q \in \mathcal{C}^k_j$, there is an automorphism Φ, for which $\Phi(\mathcal{S}_r) = \mathcal{S}_q$. Lemma 2.8 will be applied to prove the statement. We will check that condition (2.15) holds. Equation (2.16) can be checked similarly. Let $\mathcal{C}^{k-1}_l$ and $\mathcal{C}^k_j$ be arbitrary classes in the partition and $\mathcal{S}_r \in \mathcal{C}^{k-1}_l$ be an arbitrary state. We show that the sum in (2.15) is independent of the choice of the state $\mathcal{S}_r$.

We first prove that the relation

$$h(\Phi(\mathcal{S}_r), \Phi(\mathcal{S}_i)) = h(\mathcal{S}_r, \mathcal{S}_i) \tag{2.17}$$

holds for any automorphism Φ. If $h(\mathcal{S}_r, \mathcal{S}_i) \neq 0$, then there is a single node x whose status is Q in $\mathcal{S}_r$ and T in $\mathcal{S}_i$, and all other nodes have the same status in both. The value of $h(\mathcal{S}_r, \mathcal{S}_i)$ is given by some function $f_{QT}(n_Q, n_T)$, where n_Q is the number of neighbours of x having status Q and n_T is the number of nodes with status T in state $\mathcal{S}_r$. Consider an automorphism Φ. Then, $y = \Phi(x)$ has status Q in $\Phi(\mathcal{S}_r)$ and T in $\Phi(\mathcal{S}_i)$. All other nodes have the same status in both. The number of neighbours of y with each status is again n_Q and n_T, so $h(\Phi(\mathcal{S}_r), \Phi(\mathcal{S}_i)) = f_{QT}(n_Q, n_T)$, which proves (2.17).

Now let $\mathcal{S}_q \in \mathcal{C}^{k-1}_l$ be an arbitrary but fixed state in its class. Let Φ be an automorphism taking $\mathcal{S}_r$ to $\mathcal{S}_q$, i.e. $\Phi(\mathcal{S}_r) = \mathcal{S}_q$. Using (2.17), we get

$$\sum_{\mathcal{S}_i \in \mathcal{C}^k_j} h(\mathcal{S}_r, \mathcal{S}_i) = \sum_{\mathcal{S}_i \in \mathcal{C}^k_j} h(\Phi(\mathcal{S}_r), \Phi(\mathcal{S}_i)) = \sum_{\mathcal{S}_i \in \mathcal{C}^k_j} h(\mathcal{S}_q, \Phi(\mathcal{S}_i)).$$

This shows that the sum is independent of which r is chosen, completing the proof. $\square$

2.5 Applications of lumping

We can now take the results of the previous section and use it to develop a recipe for lumping. We begin with the assumption that the network under consideration has symmetries, and that we are able to find them. In practice, there may not be any symmetries, or it may be difficult to identify them (see, e.g., Chapter 3 in [330]). Our approach is restricted to networks for which it is possible to find the symmetries.

Recipe for lumping: Given a network G, and a dynamic process spreading on the network, the steps to derive the reduced master equations are:

1. Identify a group of automorphisms (symmetries) of the network.
2. Partition the states so that two states are in the same class if there is an automorphism that maps one to the other (each is a subset of $\mathcal{C}^k$ for some k).
3. From each class $\mathcal{C}_l$, choose a single state $\mathcal{S}_r$. For each other class $\mathcal{C}_j \neq \mathcal{C}_l$, calculate

$$\overline{\mathcal{A}}_{jl} = \sum_{\mathcal{S}_i \in \mathcal{C}_j} h(\mathcal{S}_r, \mathcal{S}_i).$$

4. Define the diagonal element $\overline{\mathcal{A}}_{jj}$ so that the columns sum to zero: $\overline{\mathcal{A}}_{jj} = -\sum_{l \neq j} \overline{\mathcal{A}}_{jl}$.

Then if Y_j measures the total probability of being in $\mathcal{C}_j$, the lumped equations become

$$\dot{Y} = \overline{\mathcal{A}}Y,$$

where Y is the vector of Y_j.

Fig. 2.7: A method to generate the lumped equations.

We now apply these steps to several examples. Throughout, $\mathcal{C}^k$ denotes the set of states with k infected nodes for SIS disease (or with k "status T" nodes for other binary processes). For the final lumping partition, we use $\mathcal{C}_1^k, \mathcal{C}_2^k, \ldots, \mathcal{C}_{l_k}^k$ to denote the subclasses of $\mathcal{C}^k$. If there is no subclass, we use $\mathcal{C}_1^k = \mathcal{C}^k$.

2.5.1 Lumping for some small networks

We now show some applications of the recipe in Fig. 2.7 to small networks.

Example 2.9. Consider SIS dynamics on a line graph with $N = 3$ nodes, as in Fig. 2.2a. Then, the state space is $\{SSS, SSI, SIS, ISS, SII, ISI, IIS, III\}$, where, for example, SSI represents the state of the network in which the statuses of nodes 1 and 2 are S and the status of node 3 is I. The states SSI and ISS are equivalent via a reflection around the central node; hence, they are in the same lumping class. However, the state SIS is not equivalent to these as it cannot be mapped into either via a graph automorphism, since in this state the I node has two S neighbours, while in the two other states the I node has only a single S neighbour.

Thus, the class $\mathcal{C}^1 = \{SSI, SIS, ISS\}$ consists of two lumping classes, namely $\{SSI, ISS\}$ and $\{SIS\}$. Similar reasoning leads to the observation that the class $\mathcal{C}^2 = \{SII, ISI, IIS\}$ consists of two lumping classes, namely $\{SII, IIS\}$ and $\{ISI\}$. Thus, the lumping classes are $\mathcal{C}_1^0 = \{SSS\}$, $\mathcal{C}_1^1 = \{SSI, ISS\}$, $\mathcal{C}_2^1 = \{SIS\}$, $\mathcal{C}_1^2 = \{SII, IIS\}$, $\mathcal{C}_2^2 = \{ISI\}$ and $\mathcal{C}_1^3 = \{III\}$.

As an example, we consider $\mathcal{C}_1^2$ and arbitrarily choose *SII*. The states reachable from SII are III, SSI and SIS. We have $h(\text{SII,III}) = \tau$, $h(\text{SII,SSI}) = \gamma$ and

$h(\text{SII},\text{SIS}) = \gamma$. Thus, the flow from $\mathcal{C}_1^2$ to $\mathcal{C}_1^3$ is τY_1^2, the flow to $\mathcal{C}_1^1$ is γY_1^2 and the flow to $\mathcal{C}_2^1$ is γY_1^2. The terms corresponding to other starting states can be calculated similarly (see Exercise 2.13).

Once the process is complete, we have a single equation for each class. Thus, the full system of $2^3 = 8$ differential equations can be lumped to 6 equations. This is not a large gain, but for similar larger systems the reduction becomes significant.

Exercise 2.12. Using the full system of master equations in Exercise 2.4, for the 3-node line graph, write each Y as a linear combination of the probabilities of states in a given class. Then, differentiating this equation and using careful substitution, derive the lumped system for the above example.

Exercise 2.13. By using the recipe in Fig. 2.7, derive the lumped system for the 3-node line graph.

Choosing another process with binary dynamics, while keeping the same graph, gives the same lumping classes; only the coefficients in the lumped system change. This can be checked by solving the following exercise.

Exercise 2.14. Using the recipe in Fig. 2.7, write down the full system of master equations and the lumped system for QAQ dynamics on a line graph with $N = 3$ nodes.

In order to better understand how to find the lumping classes, it is useful to examine the lumping for some graphs with $N = 4$ nodes. The lumping recipe does not depend on the dynamics; hence, in the examples below the dynamic will not be specified: a general binary dynamics with two statuses Q and T will be used. For each graph, the reader is asked to formulate the lumped system for a given dynamics.

Example 2.10. Consider a general binary dynamics with two statuses Q and T on a complete graph with $N = 4$ nodes (see Fig. 2.8b). We again take $\mathcal{C}^k$ to denote the class of states with k nodes of status T. All four states in class $\mathcal{C}^1$ are equivalent via suitable graph automorphisms, since any permutation of the nodes is an automorphism. For example, state $QTQQ$ can be taken to state $QQQT$ by an automorphism Φ, for which $\Phi(2) = 4$, $\Phi(4) = 2$, $\Phi(1) = 1$ and $\Phi(3) = 3$. Thus, $\mathcal{C}^1$ is a single lumping class. Similarly, all six states in class $\mathcal{C}^2$ are equivalent via suitably chosen graph automorphisms. For example, state $QTQT$ is equivalent to state $QQTT$ via the automorphism Φ, for which $\Phi(2) = 3$, $\Phi(3) = 2$, $\Phi(1) = 1$ and $\Phi(4) = 4$; this Φ is not unique. Thus, $\mathcal{C}^2$ is a single lumping class. The same is true for $\mathcal{C}^3$; hence, the lumping classes are $\mathcal{C}_1^0 = \mathcal{C}^0$, $\mathcal{C}_1^1 = \mathcal{C}^1$, $\mathcal{C}_1^2 = \mathcal{C}^2$, $\mathcal{C}_1^3 = \mathcal{C}^3$ and $\mathcal{C}_1^4 = \mathcal{C}^4$. Thus, the full system of $2^4 = 16$ differential equations can be lumped to 5 equations on a complete graph.

Exercise 2.15. By using the recipe in Fig. 2.7, write down the lumped system for SIS dynamics on a complete graph with $N = 4$ nodes.

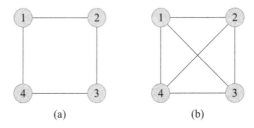

Fig. 2.8: Cycle (a) and fully connected (b) networks with 4 nodes.

Example 2.11. Consider a general binary dynamics with two statuses Q and T on a star graph with $N = 4$ nodes, as in Fig. 2.2c. The central node is numbered 1 and the leaves are numbered 2, 3 and 4. The states $QTQQ$, $QQTQ$ and $QQQT$ of the class $\mathcal{C}^1$ are all equivalent via suitable graph automorphisms, since for these states, any permutation of the nodes that keeps the central node fixed is an automorphism. For example, state $QTQQ$ is equivalent to state $QQTQ$ via an automorphism, Φ, for which $\Phi(2) = 3$, $\Phi(3) = 2$, $\Phi(1) = 1$ and $\Phi(4) = 4$. However, state $TQQQ$ is not equivalent to any of the above since no such automorphism exists. This is because, in the latter state, the node with status T has three neighbours of status Q, while in the other states it has a single neighbour of status Q. Thus, $\mathcal{C}^1$ consists of two lumping classes: $\{QTQQ, QQTQ, QQQT\}$ and $\{TQQQ\}$. Similarly, states in class $\mathcal{C}^2$ for which the central node is of status T are equivalent via appropriately chosen automorphisms. For example, state $TQTQ$ is equivalent to state $TQQT$ via the automorphism Φ, for which $\Phi(3) = 4$, $\Phi(4) = 3$, $\Phi(1) = 1$ and $\Phi(2) = 2$. Thus, $\mathcal{C}^2$ consists of two lumping classes: $\{TTQQ, TQTQ, TQQT\}$ and $\{QTTQ, QQTT, QTQT\}$. The class $\mathcal{C}^3$ can also be divided into two lumping classes; hence, the lumping classes are

$$\mathcal{C}_1^0 = \mathcal{C}^0, \quad \mathcal{C}_1^1 = \{QTQQ, QQTQ, QQQT\}, \quad \mathcal{C}_2^1 = \{TQQQ\},$$

$$\mathcal{C}_1^2 = \{QTTQ, QQTT, QTQT\}, \quad \mathcal{C}_2^2 = \{TTQQ, TQTQ, TQQT\},$$

$$\mathcal{C}_1^3 = \{QTTT\}, \quad \mathcal{C}_2^3 = \{TTTQ, TTQT, TQTT\}, \quad \mathcal{C}_1^4 = \mathcal{C}^4.$$

These are the partitions seen in Fig. 2.5. The full system of $2^4 = 16$ differential equations can be lumped to 8 equations on the star graph, as we saw in Example 2.8.

Exercise 2.16. By using the recipe in Fig. 2.7, write down the lumped system for the voter model (Example 2.4) on a star graph with $N = 4$ nodes.

Example 2.12. Consider a general binary dynamics with two statuses Q and T on a cycle graph with $N = 4$ nodes, as in Fig. 2.8a. All four states in class $\mathcal{C}^1$ are equivalent via a suitable rotation. For example, state $QTQQ$ can be taken to state $QQTQ$ via the automorphism Φ, for which $\Phi(2) = 3$, $\Phi(3) = 4$, $\Phi(4) = 1$ and $\Phi(1) = 2$. Thus, $\mathcal{C}^1$ forms a single lumping class. Similarly, there are four states in class $\mathcal{C}^2$ (i.e. $\{TTQQ, QTTQ, QQTT, TQQT\}$) that are equivalent via automorphisms, namely via rotations. For example, state $QTTQ$ is equivalent to state $QQTT$ by the rotation Φ, given by $\Phi(2) = 3$, $\Phi(3) = 4$, $\Phi(4) = 1$ and $\Phi(1) = 2$. Thus, $\mathcal{C}^2$ consists of two lumping classes: $\{TTQQ, QTTQ, QQTT, TQQT\}$ and $\{QTQT, TQTQ\}$. The class $\mathcal{C}^3$ forms a single lumping class, because its elements are all equivalent via rotations. Hence, the lumping classes are

$$\mathcal{C}_1^0 = \mathcal{C}^0, \quad \mathcal{C}_1^1 = \mathcal{C}^1, \quad \mathcal{C}_1^2 = \{TTQQ, QTTQ, QQTT, TQQT\},$$
$$\mathcal{C}_2^2 = \{QTQT, TQTQ\}, \quad \mathcal{C}_1^3 = \mathcal{C}^3, \quad \mathcal{C}_1^4 = \mathcal{C}^4.$$

Thus, the full system of $2^4 = 16$ differential equations can be lumped to 6 equations on a cycle graph with 4 nodes. We note that reflections are not needed in building up the lumping classes.

Exercise 2.17. By using the recipe in Fig. 2.7, write down the lumped system for SIS dynamics on a cycle graph with $N = 4$ nodes.

Example 2.13. Consider a general binary dynamics with two statuses Q and T on a line graph with $N = 4$ nodes, as in Fig. 2.2b. States $TQQQ$ and $QQQT$ of class $\mathcal{C}^1$ are equivalent via a reflection. States $QTQQ$ and $QQTQ$ of class $\mathcal{C}^1$ are also equivalent via a reflection. However, states $QQQT$ and $QTQQ$ are not equivalent. This is because in state $QQQT$, the node with status T has one neighbour of status Q, while in state $QTQQ$, it has two. There is no automorphism between the states. Thus, $\mathcal{C}^1$ consists of two lumping classes: $\{TQQQ, QQQT\}$ and $\{QTQQ, QQTQ\}$. Similarly, state $TQQT$ in class $\mathcal{C}^2$ is not equivalent to any other state of this class. In this state, both nodes of status T have a single neighbour, which is of status Q. However, state $TTQQ$ is equivalent to $QQTT$ via a reflection. Since the only nontrivial automorphism of the graph is the reflection, class $\mathcal{C}^2$ is divided into four lumping classes: $\{TQQT\}$, $\{QTTQ\}$, $\{TTQQ, QQTT\}$ and $\{TQTQ, QTQT\}$. Class $\mathcal{C}^3$ can be divided into two lumping classes; hence, the lumping classes are

$$\mathcal{C}_1^0 = \mathcal{C}^0, \quad \mathcal{C}_1^1 = \{TQQQ, QQQT\}, \quad \mathcal{C}_2^1 = \{QTQQ, QQTQ\}, \quad \mathcal{C}_1^2 = \{TQQT\}$$
$$\mathcal{C}_2^2 = \{QTTQ\}, \quad \mathcal{C}_3^2 = \{TTQQ, QQTT\}, \quad \mathcal{C}_4^2 = \{TQTQ, QTQT\},$$
$$\mathcal{C}_1^3 = \{TTTQ, QTTT\}, \quad \mathcal{C}_2^3 = \{TTQT, TQTT\}, \quad \mathcal{C}_1^4 = \mathcal{C}^4.$$

Thus, the full system of $2^4 = 16$ differential equations can be lumped to 10 equations on a line graph with 4 nodes.

Exercise 2.18. By using the recipe in Fig. 2.7, write down the lumped system for SIS dynamics on a line graph with $N = 4$ nodes.

Exercise 2.19. Determine the lumping classes for a general binary dynamics on a lollipop network (see Fig. 2.9a).

Exercise 2.20. Determine the lumping classes for a general binary dynamics on a toast network (see Fig. 2.9b).

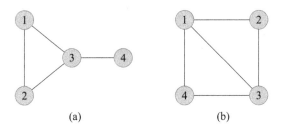

Fig. 2.9: (a) Lollipop and (b) toast networks with 4 nodes.

Exercise 2.21. Consider an SIR disease spreading on a single edge as in Example 2.6. Derive the lumped system.

Exercise 2.22. Consider SIR spread on a fully connected 3-node graph. The states are given in Example 2.7. Derive the lumped system.

Exercise 2.23. Consider SIR spread on the line network with three nodes. The states and master equations are given in Example 2.7. Derive the lumped system.

2.5.2 Lumping for some classes of networks of arbitrary size

In this section, we show applications of the general lumping theorem and demonstrate lumping for some arbitrarily large graphs with special symmetry structure.

Lumping for the complete network

First, we show that for a complete graph the 2^N-dimensional system given by (2.6) can be lumped to an $(N+1)$-dimensional system. The lumped system is well known in the literature, but so far as we are aware, it was first derived from the full 2^N-dimensional system in [288].

The automorphism group of the complete graph is the permutation group S_N, that is, any relabelling is an automorphism. Hence, the orbit of any element from $\mathcal{C}^k$ is equal to $\mathcal{C}^k$ itself. All states with k infected nodes can be lumped together. There are $N+1$ lumping classes: $\mathcal{C}^k$ for all $k \in \{0,1,\ldots,N\}$. We have

$$Y^k = \sum_{j=1}^{c_k} X_j^k, \quad k = 1,\ldots,N.$$

Proposition 2.3 *If G is a complete graph, then the Y^k functions satisfy the following differential equations:*

$$\dot{Y}^0 = f_{TQ}(N-1)Y^1 - Nf_{QT}(0)Y^0,$$

$$\dot{Y}^k = (k+1)f_{TQ}(N-k-1)Y^{k+1} + (N-k+1)f_{QT}(k-1)Y^{k-1}$$
$$- ((N-k)f_{QT}(k) + kf_{TQ}(N-k))Y^k, \quad for \quad k = 1,2,\ldots,N-1,$$
$$\dot{Y}^N = f_{QT}(N-1)Y^{N-1} - Nf_{TQ}(0)Y^N.$$

Proof. Using the steps of Fig. 2.7, we need only consider a single S_k chosen from each C^k. For simplicity, we choose the state $T \cdots TQ \cdots Q$, where nodes 1 through k have status T and the remaining $(N-k)$ nodes have status Q.

For state S_k, each of the $(N-k)$ nodes of status Q has k neighbours of status T. So the rate of moving from S_k to any state in C^{k+1} is $(N-k)f_{QT}(k)$. Similarly, each of the k neighbours of status T has $(N-k)$ neighbours of status k. Thus, the rate of moving from S_k to S_{k-1} is $kf_{TQ}(N-k)$. Thus

$$\overline{A}_{k+1,k} = (N-k)f_{QT}(k),$$
$$\overline{A}_{k-1,k} = kf_{TQ}(N-k)$$

and

$$\overline{A}_{k,k} = -(N-k)f_{QT}(k) - kf_{TQ}(N-k).$$

So the equations are

$$\dot{Y}^k = (k+1)f_{TQ}(N-k-1)Y^{k+1} + (N-k+1)f_{QT}(k-1)Y^{k-1}$$
$$- ((N-k)f_{QT}(k) + kf_{TQ}(N-k))Y^k.$$

For variables Y^N and Y^0, some terms on the right-hand side become zero, and we arrive at the equations claimed. □

Lumping for the star network

Consider a star-like network with $N > 2$ nodes. In this network, a single central node is connected to all other nodes with no further connections, as in Fig. 2.10a. Let the first node be the centre of the star. Thus, for example $TQQ \cdots Q$ denotes the state when the central node is of status T and the other nodes are of status Q. We will show that in the case of a star network, the 2^N-dimensional system as defined by equation (2.6) can be lumped to a $2N$-dimensional system for an arbitrary binary dynamics. The automorphism group of the star graph is the permutation group S_{N-1}: an automorphism must leave the central node unchanged but can permute the remaining $N-1$ nodes in an arbitrary way. Therefore, two states are equivalent via an automorphism if and only if the centre is of the same status and the number of non-central nodes of status T is the same. Hence, for $l = 1,2,\ldots,N-1$, class C^l of states of the graph, such that there are l nodes of status T, can be lumped into two classes: in the first class, the central node is T and there are $l-1$ non-central T nodes; in the second class, the central node is Q and there are l non-central T nodes.

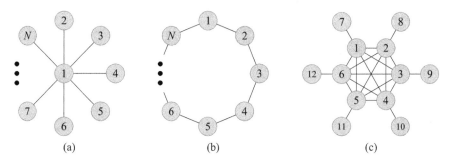

Fig. 2.10: (a) General star and (b) general cycle networks with N nodes. (c) House-hold network with 12 nodes (6 households).

In the case $l = 0$ and $l = N$, there is obviously only one class. This means that alto-gether there are $2 + 2(N-1) = 2N$ lumping classes: $\mathcal{C}^0 = \{\mathcal{S}^0\}$, $\mathcal{C}^N = \{\mathcal{S}^N\}$, and for $k \in \{1, 2, \ldots, N-1\}$, we have $\mathcal{C}_1^k = \{\mathcal{S}_j^k : \mathcal{S}_j^k(1) = T\}$, $\mathcal{C}_2^k = \{\mathcal{S}_j^k : \mathcal{S}_j^k(1) = Q\}$. So for $1 \le k \le N-1$, a subscript of 1 denotes that the central node has status T and a subscript of 2 denotes status Q. We use L_1^k and L_2^k to denote the induced parti-tions of $\{1, \ldots, c_k\}$ corresponding to $\mathcal{C}_1^k$ and $\mathcal{C}_2^k$. Thus, the lumped variables can be introduced as

$$Y^0 = X^0, \quad Y^N = X^N, \quad Y_1^k = \sum_{i \in L_1^k} X_i^k, \quad Y_2^k = \sum_{i \in L_2^k} X_i^k,$$

for $k = 1, \ldots, N-1$. Similarly to Proposition 2.3, we can prove:

Proposition 2.4 *Let G be a star graph of N nodes, with the central node labelled* 1. *Then*

$$\dot{Y}^0 = f_{TQ}(N-1)Y_1^1 + f_{TQ}(1)Y_2^1 - Nf_{QT}(0)Y^0,$$
$$\dot{Y}^N = f_{QT}(1)Y_1^{N-1} + f_{QT}(N-1)Y_2^{N-1} - Nf_{TQ}(0)Y^N$$

and, for $k = 1, 2, \ldots, N-1$,

$$\dot{Y}_1^k = (N-k+1)f_{QT}(1)Y_1^{k-1} + f_{QT}(k-1)Y_2^{k-1} + kf_{TQ}(0)Y_1^{k+1}$$
$$\quad - ((N-k)f_{QT}(1) + (k-1)f_{TQ}(0) + f_{TQ}(N-k))Y_1^k,$$
$$\dot{Y}_2^k = (N-k)f_{QT}(0)Y_2^{k-1} + f_{TQ}(N-k-1)Y_1^{k+1} + f_{TQ}(1)(k+1)Y_2^{k+1}$$
$$\quad - (f_{QT}(k) + (N-k-1)f_{QT}(0) + kf_{TQ}(1))Y_2^k.$$

Proof. We first address Y_1^k for $1 \le k \le N-1$. Taking $\mathcal{S}_l \in \mathcal{C}_1^k$, the central node has status T, and $(N-k)$ of the peripheral nodes have status Q. The remaining $k-1$

peripheral nodes have status T. The rate at which the central node changes status is $f_{TQ}(N-k)$, yielding a state in $\mathcal{C}_2^{k-1}$. The rate at which each of the $N-k$ status Q peripheral node changes status is $f_{QT}(1)$, yielding a state in $\mathcal{C}_1^{k+1}$. The rate at which the remaining $k-1$ status T peripheral nodes change status is $f_{TQ}(0)$, yielding a state in $\mathcal{C}_1^{k-1}$. So the total rate of flow of probability from $\mathcal{C}_1^k$ to $\mathcal{C}_2^{k-1}$ is $f_{TQ}(N-k)Y_1^k$, to $\mathcal{C}_1^{k+1}$ is $(N-k)f_{QT}(1)Y_1^k$ and to $\mathcal{C}_1^{k-1}$ is $(k-1)f_{TQ}(0)$.

We now consider $\mathcal{S}_2^k$, where the central node has status Q and there are k status T peripheral nodes and $(N-k-1)$ status Q peripheral nodes. Similar analysis gives that the total rate of flow of probability to $\mathcal{C}_1^{k+1}$ is $f_{QT}(1)$, to $\mathcal{C}_2^{k+1}$ is $(N-k-1)f_{QT}(0)$ and to $\mathcal{C}_2^{k-1}$ is $kf_{TQ}(1)$. Similar analysis applies to $\mathcal{C}^0$ and $\mathcal{C}^N$. Combining these results together yields the claimed equations. □

Lumping for the household network

Consider the simplest network with a so-called household structure of Fig. 2.10c. It consists of two types of nodes, inner and outer nodes. Outer nodes have only within- household connections, while inner nodes have both within-household connections, and connections to other households. We consider the simplest case where each household has two nodes, an inner and an outer node. The inner nodes of all households form a complete graph with $N/2$ nodes (N is an even number), and every outer node is connected to an inner node. Thus, the degree of all inner nodes is $\frac{N}{2}$ and the degree of all outer nodes is 1, as in Fig. 2.10c. It is possible to prove that for this household-type network, the 2^N-dimensional system given by (2.6) can be lumped to an $\binom{N/2+3}{3}$-dimensional system.

The automorphism group of this graph is the permutation group $\mathbf{S_{N/2}}$: an automorphism can permutate the inner nodes in an arbitrary way, and once the automorphism is given on the inner nodes, its effect on the outer nodes is determined uniquely. In order to determine the lumping classes, note first that there may be four different types of households in this graph: QT households, in which the inner node is Q and the outer node is T, TQ households, QQ households, and TT households. Therefore, two states of the whole graph are equivalent through an automorphism if and only if the number of QT-, TQ-, QQ- and TT-type households is the same in the two states. Hence, to obtain all different states we have to choose (with repetition) $N/2$ households out of the four different types. Thus, using the formula for the number of combinations with repetitions, the number of different states is $\binom{4+N/2-1}{N/2} = \binom{N/2+3}{3}$. Hence, states with the same number of QQ-, QT-, TQ- and TT-type households can be lumped into one newly defined lumped variable.

Lumping for the cycle network: non-trivial lumping

For completely connected and star networks, lumping can be carried out intuitively. However, intuition alone is prone to error, and thus, it is desirable to use the automorphism group to work out lumping classes rigorously. Even for a relatively simple network such as the cycle graph (C_N), where N nodes are connected in a close chain such that each node connects to the two nearest neighbours only, as in Fig. 2.10b, we can encounter surprises. For illustration, consider the $N = 5$ case.

The automorphisms of the cycle network is known as the *dihedral group* D_N, and can be given in terms of all possible *rotations* and *reflections* of the network (see, for example, the case $N = 5$ in the table in Fig. 2.6). There are N rotations and N reflections, so $|D_N| = 2N$. Here, D_5 is made up of five rotations (including the identity) and five reflections. We now look for the lumping classes $\mathcal{C}_i^k$ for $i \in \{1, 2, \ldots, c_k\}$, with c_k yet to be determined.

Each $\mathcal{C}_i^k$ is a subset of $\mathcal{C}^k$, so in particular for $k = 0$, the first lumping class is trivial $\mathcal{C}_1^0 = \{(QQQQQ)\}$. Now consider

$$\mathcal{C}^1 = \{(QQQQT), (QQQTQ), (QQTQQ), (QTQQQ), (TQQQQ)\}.$$

We can see that the orbit of the first element $(QQQQT)$ is

$$D_5((QQQQT)) = \{\Phi((QQQQT)) : \Phi \in D_5\} = \mathcal{C}^1.$$

Hence, $\mathcal{C}_1^1 = \mathcal{C}^1$. The situation changes when $\mathcal{C}^2$ is considered

$$\mathcal{C}^2 = \{(QQQTT), (QQTQT), (QTQQT), (TQQQT), (QQTTQ),$$
$$(QTQTQ), (TQQTQ), (QTTQQ), (TQTQQ), (TTQQQ)\}.$$

The orbit of the first element $(QQQTT) \in \mathcal{S}^2$ is

$$D_5((QQQTT)) = \{\Phi((QQQTT)) : \Phi \in D_5\} =$$
$$\{(QQQTT), (TQQQT), (QQTTQ), (QTTQQ), (TTQQQ)\} = \mathcal{C}_1^2.$$

The orbit of $(QQQTT)$ only captures 5 out of the 10 possible states in $\mathcal{C}^2$. The rotations and reflections map $(QQQTT)$ onto identical configurations. This increases the number of lumping classes, so the dimensionality reduction of the system is less significant. The remaining five elements form another lumping class

$$\mathcal{C}_2^2 = \{(QQTQT), (QTQQT), (QTQTQ), (TQQTQ), (TQTQQ)\}.$$

Continuing, four more lumping classes can be identified. This means that the original system with $2^5 = 32$ equations can be reduced to a system with only 8 equations. Using similar arguments, for $N = 6$ and $N = 7$, the exact systems can be lumped from 64 and 128 to 18 and 30 equations, respectively. Given that for the cycle graph $2^N > |\mathrm{Aut}(G)| = 2N$, the argument presented in Section 2.4.4 can be used to show that the number of lumping classes for the cycle graph is bounded from below by $\frac{2^N}{2N} = \frac{2^{N-1}}{N}$. This indicates that the number of equations in the lumped system is much larger than polynomial in N. It is interesting to note that in the case when the number of nodes is a prime number, $N = p$, then it can be shown that the number of lumping classes is $(2^{p-1} - 1)/p + 2^{(p-1)/2} + 1$. A similar formula is not known in the general case of an arbitrary value of N.

2.6 Conclusions and outlook

In this chapter, we have investigated network processes for which the status of a node can change in response to the status of its neighbours. These processes are relevant to understanding a number of important phenomena such as epidemic propagation, firing in neuronal networks, the voter model in social sciences or the Ising model in statistical physics. Such processes are controlled by the structure of the graph. We have provided a unified framework for arbitrary networks and arbitrary dynamics. The system of master equations serves as a theoretical basis for deriving and developing further exact and approximate models, e.g. the pair approximation model. Moreover, approximating models can be validated by comparing them to these exact models. The exact model also enables us to test simulation results. In Fig. 2.11, we show that for the fully connected and star networks and SIS dynamics, the lumped system is identical to results based on stochastic simulation. The lumped systems are obtained from Propositions 2.3 and 2.4 with node statuses $Q = S$ and $T = I$, and by substituting the actual transition rates as follows: $f_{SI} = \tau n$, where n denotes the number of infected neighbours and $f_{IS} = \gamma$. This not only confirms that lumping is correct, but provides strong evidence that the simulations are correctly implemented.

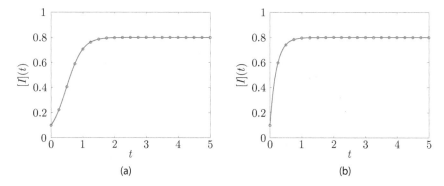

Fig. 2.11: Illustration of the perfect agreement between the lumped systems (solid lines) and stochastic simulations ($\circ$) for the (a) full and (b) star networks with $N = 1000$ and $\gamma = 1$, and with (a) $\tau = 0.005$ and (b) $\tau = 4.0$.

The master equations allow us to determine the time dependence of the probability of an arbitrary state, e.g. QTQ in a network with three nodes, meaning that nodes 1 and 3 are in status Q and node 2 is in status T. As a consequence, marginal probabilities (e.g. the probability of a node having a given status or the expected number of nodes with a given status) can be obtained, for example, the probability node 1 has status Q is $X_{QQQ} + X_{QQT} + X_{QTQ} + X_{QTT}$.

Spectral investigation of the transition matrix gives information about the long-term behaviour of the system [230, 312]. The transition matrix P, given in (2.3), has a zero eigenvalue, because the sum of the entries in each column is zero. The entries in the diagonal are negative and other entries are non-negative; hence, Gershgorin's theorem [109, 119] yields that all eigenvalues have real parts at most zero. The spectrum of the transition matrix has been studied in more detail in the case of SIS dynamics. For a complete graph, the system can be lumped to a tractable size. The spectrum of the resulting $(N + 1) \times (N + 1)$ matrix was studied in [230] in order to understand the quasi-steady state behaviour. Picard [252] proved that the matrix has a single zero eigenvalue (with eigenvector corresponding to the disease-free fully susceptible state) and all other eigenvalues are real and negative. We note that this is true for a more general class of tridiagonal matrices (see Theorem 8.2.6 in [99]).

Numerical investigation shows that one negative eigenvalue (denoted λ_1 and called the small eigenvalue) is very close to zero and the remaining negative eigenvalues (called large eigenvalues) are far from zero. As N or the infection rate increases, the small eigenvalue λ_1 converges rapidly to zero and the large eigenvalues decrease further; the spectral gap increases. This explains the appearance of a quasi-steady state, since the small eigenvalue dominates according to $\exp(\lambda_1 t)$. Thus, the time to extinction is of order $-1/\lambda_1$. Lumping enables us to determine λ_1 as a function of the infection rate τ even for relatively large graphs with a few hundred nodes. (In [312], graphs with at most $N = 13$ nodes are investigated.) In Fig. 2.12, $-1/\lambda_1$ is plotted for a range of τ values for a graph with $N = 500$ nodes and for recovery rate $\gamma = 1$.

It is well known that the threshold for the existence of the endemic equilibrium in the mean-field approximation for a complete graph is given by $N\tau = \gamma$; hence, for these parameter values the mean-field yields $\tau_c = 1/N\gamma = 0.002$ as a threshold value. In Fig. 2.12, one can see the well-known fact that the continuous-time Markov chain does not give a threshold value; on the other hand, $-1/\lambda_1$ depends on τ in a strongly nonlinear way when $\tau > \tau_c$. The inset with a logarithmic scale shows that in this range the time to extinction increases faster than exponentially, and becomes soon of order 10^6. Practically, this means that the quasi-steady state is almost equivalent to a steady state in the classical sense. In Fig. 2.12, one can see that around $\tau = 0.0027$ there is a threshold-like abrupt change in the value of the time to extinction. We emphasise again that these plots are based on using the exact master equations and exploit the possibility of lumping.

The lumped system also enables us to derive an explicit formula for the approximation of the quasi-steady state. Namely, the quasi-steady state can be obtained starting from the tridiagonal transition matrix by omitting its first row and column and changing the upper-left entry in such a way that the sum of the first column becomes zero. Then, the remaining $N \times N$ matrix does not have an absorbing state and its stationary state is the quasi-steady state of the original system. This stationary state is given by the eigenvector corresponding to the zero eigenvalue. For a tridiagonal matrix, this can be given explicitly. Carrying out this calculation, the quasi-steady state is

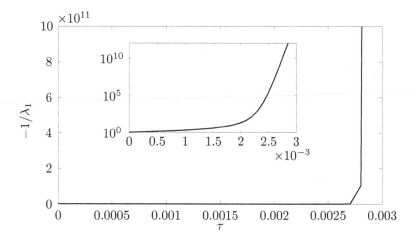

Fig. 2.12: Dependence of the time to extinction (that is of order $-1/\lambda_1$) on the per-edge infection rate τ for SIS dynamics on a complete graph with $N = 500$ and $\gamma = 1$ with a lin-log plot in the inset.

$$[I]^e = \sum_{k=0}^{N-1} (k+1)A_k \Bigg/ \sum_{k=0}^{N-1} A_k, \qquad (2.18)$$

where $A_0 = 1$ and

$$A_k = \frac{\tau^k (N-1)(N-2)\cdots(N-k)}{\gamma^k (k+1)}, \qquad k = 1,2,\ldots,N-1.$$

This formula will be used later in Chapter 4 to test the accuracy of mean-field approximations.

Unfortunately, the number of equations required for an exact description of the probabilities of each state grows exponentially with the size of the network. Automorphisms of the graph can be exploited to reduce the number of equations through lumping. We have identified the precise link between the lumpability of the equations resulting from network processes and the symmetries of the network as specified by the automorphism group of the network. In Table 2.1, we display a number of networks for which lumping can be carried out with success (i.e. the full system can be reduced to a more tractable system, which is still exact, and can be evaluated numerically.)

It is worth noting that lumping in general relies on the identification of the automorphism group of a network, which in itself is a formidable task. However, as seen from our example, it is not necessary to have the complete group in order to obtain a significant reduction. It is also feasible to consider the possibility of a procedure that would not be exact, but rather an approximate lumping.

In general, this top-down technique and lumping work best and are ideal for networks of small size or networks with many symmetries. In many instances, the long-term behaviour is of interest. Using this approach, dynamics with a single absorbing state (like the SIS) are difficult to analyse. For dynamics with more absorbing states (like the voter model or SIR), the final ratio of probabilities of the different absorbing states can be studied.

It is relatively straightforward to generalise this approach to transition rates which may depend on the density of different statuses in the immediate neighbourhood or beyond. This could include dependence on global properties, e.g. the total number of infected nodes, or modelling population-wide effects. This method applies similarly to hypergraphs [38, 111, 193] when classic edges are replaced by hyperedges. For this or for the former case, transition rates may depend non-linearly on the number of nodes of different statuses in the immediate neighbourhood or in the hyperedge.

	Network	Full system	Lumped system
	2 line	4	3
	3 line	8	6
SIS	4 line	16	10
	4 cycle	16	6
	Lollipop	16	12
	Toast	16	9
	Fully connected N	2^N	$N+1$
	Star N	2^N	$2N$
	2 line	9	6
SIR	3 line	27	18
	Fully connected 3	27	10
	Fully connected N	3^N	$(N+1)(N+2)/2$
	Star N	3^N	$3N(N+1)/2$

Table 2.1: The reduction in number of equations from the full to the lumped system for SIS and SIR epidemics and for a number of networks.

Chapter 3
Propagation models on networks: bottom-up

In this chapter, we present a different approach to deriving exact models. In Chapter 2, we began with equations for every possible state of the system and then aggregated them into a simpler form. Here, we begin by deriving separate equations for the status of each node. These typically depend on the states of pairs of nodes, so we introduce equations for the pairs, which in turn depend on triples. We build up equations at each level. For a typical network, the number of equations we obtain is too large to be tractable. Consequently, we introduce "closures", whereby terms corresponding to larger structures are represented in terms of smaller structures, in order to create a closed system of equations. In most cases, this representation involves an approximation, but in the case of SIR dynamics on trees or networks with *cut-vertices*, it is possible to reduce the number of equations considerably while keeping the model exact.

3.1 Illustrative examples

We begin by exploring Markovian SIR dynamics on a simple line network with three nodes (see Fig. 3.1a). The system starts at the level of nodes. We use $\langle S_i \rangle (t)$ to denote the probability that node i is susceptible at time t, $\langle I_i \rangle (t)$ to give the probability that i is infected at time t, and $\langle R_i \rangle (t)$ to denote the probability that it is recovered. We will use notation such as $(\langle I_i \rangle + \langle R_i \rangle)(t)$ to denote the probability that i is either infected or recovered at time t and $\langle S_i I_j \rangle (t)$ to denote the probability that i is susceptible and j is infected at time t.

Each node i can become infected if it is susceptible and at least one of its neighbours is infected (see, for example, Fig. 3.1b). For each infected neighbour, the rate at which i becomes infected is τ. So the probability that node 1 is in danger of being infected by node 2 is $\langle S_1 I_2 \rangle$. Thus, $\langle S_1 \rangle$ decreases at rate $\tau \langle S_1 I_2 \rangle$. Node 2 has two possible sources of infection, so its rate of infection is $\tau(\langle S_2 I_1 \rangle + \langle S_2 I_3 \rangle)$.

© Springer International Publishing AG 2017
I.Z. Kiss et al., *Mathematics of Epidemics on Networks*, Interdisciplinary Applied Mathematics 46, DOI 10.1007/978-3-319-50806-1_3

Node 3 becomes infected at rate $\tau\langle S_3 I_2\rangle$. Once infected, each node recovers at rate γ regardless of the state of any other node. Thus the flux out of $\langle I_i\rangle$ is $\gamma\langle I_i\rangle$. Collating this yields

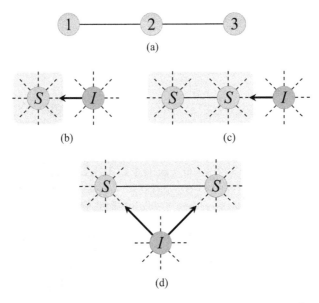

Fig. 3.1: (a) Simple line network with three nodes. The spread of infection through the network relies on the invasion of susceptible "clusters" (highlighted in grey), where "clusters" could mean a node, a pair, a triple (open or closed) and other constituent parts of the network. (b) The creation of new infected nodes is a direct result of the infection of susceptible nodes by their infected neighbours; this requires SI links and equations for these. (c, d) SI links, on the other hand, are created when SS pairs are invaded by infected nodes, which requires us to develop equations for triples such as SSI (open or closed). If one would be to write down evolution equations at the level of nodes, pairs, triples, etc., the procedure in (b)–(d) would need to continue until full network size is reached.

$$\langle\dot{S}_1\rangle = -\tau\langle S_1 I_2\rangle, \qquad \langle\dot{I}_1\rangle = \tau\langle S_1 I_2\rangle - \gamma\langle I_1\rangle, \qquad \langle\dot{R}_1\rangle = \gamma\langle I_1\rangle,$$
$$\langle\dot{S}_2\rangle = -\tau\langle I_1 S_2\rangle - \tau\langle S_2 I_3\rangle, \quad \langle\dot{I}_2\rangle = \tau\langle I_1 S_2\rangle + \tau\langle S_2 I_3\rangle - \gamma\langle I_2\rangle, \quad \langle\dot{R}_2\rangle = \gamma\langle I_2\rangle,$$
$$\langle\dot{S}_3\rangle = -\tau\langle I_2 S_3\rangle, \qquad \langle\dot{I}_3\rangle = \tau\langle I_2 S_3\rangle - \gamma\langle I_3\rangle, \qquad \langle\dot{R}_3\rangle = \gamma\langle I_3\rangle.$$

A closer inspection of the system reveals that equations for $\langle\dot{R}_i\rangle$s can be eliminated because $\langle R_i\rangle = 1 - \langle S_i\rangle - \langle I_i\rangle$ for each i and for all $t \geq 0$. This reduces the number of differential equations. However, we have new terms such as $\langle S_1 I_2\rangle$ and other pairs on the right-hand side. We need to know how they evolve, which requires additional equations. Using similar arguments, we find

$$\langle S_1 \dot{I}_2 \rangle = \tau \langle S_1 S_2 I_3 \rangle - (\tau + \gamma) \langle S_1 I_2 \rangle,$$

$$\langle \dot{I_1 S_2} \rangle = -\tau \langle I_1 S_2 I_3 \rangle - (\tau + \gamma) \langle I_1 S_2 \rangle,$$

$$\langle \dot{S_2 I_3} \rangle = -\tau \langle I_1 S_2 I_3 \rangle - (\tau + \gamma) \langle S_2 I_3 \rangle,$$

$$\langle I_2 \dot{S}_3 \rangle = \tau \langle I_1 S_2 S_3 \rangle - (\tau + \gamma) \langle I_2 S_3 \rangle.$$

These equations require further information about specific triples. Once we add the necessary triples to the equations, neglecting the $\langle R_i \rangle$ equations, our full system is

$$\langle \dot{S}_1 \rangle = -\tau \langle S_1 I_2 \rangle,$$

$$\langle \dot{I}_1 \rangle = \tau \langle S_1 I_2 \rangle - \gamma \langle I_1 \rangle,$$

$$\langle \dot{S}_2 \rangle = -\tau \langle I_1 S_2 \rangle - \tau \langle S_2 I_3 \rangle,$$

$$\langle \dot{I}_2 \rangle = \tau \langle I_1 S_2 \rangle + \tau \langle S_2 I_3 \rangle - \gamma \langle I_2 \rangle,$$

$$\langle \dot{S}_3 \rangle = -\tau \langle I_2 S_3 \rangle,$$

$$\langle \dot{I}_3 \rangle = \tau \langle I_2 S_3 \rangle - \gamma \langle I_3 \rangle,$$

$$\langle S_1 \dot{I}_2 \rangle = \tau \langle S_1 S_2 I_3 \rangle - (\tau + \gamma) \langle S_1 I_2 \rangle,$$

$$\langle \dot{I_1 S_2} \rangle = -\tau \langle I_1 S_2 I_3 \rangle - (\tau + \gamma) \langle I_1 S_2 \rangle,$$

$$\langle \dot{S_2 I_3} \rangle = -\tau \langle I_1 S_2 I_3 \rangle - (\tau + \gamma) \langle S_2 I_3 \rangle,$$

$$\langle I_2 \dot{S}_3 \rangle = \tau \langle I_1 S_2 S_3 \rangle - (\tau + \gamma) \langle I_2 S_3 \rangle,$$

$$\langle S_1 \dot{S}_2 I_3 \rangle = -(\tau + \gamma) \langle S_1 S_2 I_3 \rangle,$$

$$\langle I_1 \dot{S}_2 I_3 \rangle = -(2\tau + 2\gamma) \langle I_1 S_2 I_3 \rangle,$$

$$\langle I_1 \dot{S}_2 S_3 \rangle = -(\tau + \gamma) \langle I_1 S_2 S_3 \rangle.$$

This system has "only" 13 differential equations compared to the 27 of the top-down model. It is worth noting that by only including those terms that were needed to calculate $\langle S_i \rangle$ and $\langle I_i \rangle$, we are not able to directly calculate quantities such as $\langle I_1 I_2 \rangle$. We could add equations to the system if this were something we needed to calculate, but it is irrelevant to the calculation of $\langle S_i \rangle$ and $\langle I_i \rangle$.

Even though the number of equations is significantly reduced compared to the top-down model, the dependency of singles on pairs (see Fig. 3.1b), and of the pairs on triples (see Fig. 3.1c and Fig. 3.1d), and so on, leads to a rapid increase in the number of equations as the network size increases, and this approach becomes impractical. For a generic network, we would need to include terms with size equal to the number of nodes in the network. We will see ways in which to reduce the number of equations to a more tractable size, but first we investigate the relation between these bottom-up equations above and the top-down equations for the three node line network as given below,

$$\dot{X}_{SSS} = 0,$$

$$\dot{X}_{SSI} = -(\gamma + \tau) X_{SSI},$$

$$\dot{X}_{SIS} = -(\gamma + 2\tau) X_{SIS},$$

$$\dot{X}_{ISS} = -(\gamma + \tau) X_{ISS},$$

$$\dot{X}_{IIS} = \tau (X_{SIS} + X_{ISS}) - (2\gamma + \tau) X_{IIS},$$

$$\dot{X}_{ISI} = -(2\gamma + 2\tau) X_{ISI},$$

$$\dot{X}_{SII} = \tau (X_{SSI} + X_{SIS}) - (2\gamma + \tau) X_{SII},$$

$$\dot{X}_{III} = \tau (X_{IIS} + 2X_{ISI} + X_{SII}) - 3\gamma X_{III},$$

$$\dot{X}_{RSS} = \gamma X_{ISS},$$

$$\dot{X}_{SRS} = \gamma X_{SIS},$$

$$\dot{X}_{RRI} = \gamma (X_{IRI} + X_{RII}) - \gamma X_{RRI},$$

$$\dot{X}_{RIR} = \gamma (X_{IIR} + X_{RII}) - \gamma X_{RIR},$$

$$\dot{X}_{IRR} = \gamma (X_{IIR} + X_{IRI}) - \gamma X_{IRR},$$

$$\dot{X}_{RRS} = \gamma (X_{IRS} + X_{RIS}),$$

$$\dot{X}_{RSR} = \gamma (X_{ISR} + X_{RSI}),$$

$$\dot{X}_{SRR} = \gamma (X_{SIR} + X_{SRI}),$$

$$\dot{X}_{RRR} = \gamma (X_{IRR} + X_{RRI} + X_{RIR}),$$

$$\dot{X}_{IIR} = \tau (X_{SIR} + X_{ISR}) + \gamma X_{III} - 2\gamma X_{IIR},$$

$$\dot{X}_{IRI} = \gamma X_{III} - 2\gamma X_{IRI},$$

$$\dot{X}_{SSR} = \gamma X_{SSI},$$
$$\dot{X}_{RII} = \tau(X_{RSI} + X_{RIS}) + \gamma X_{III} - 2\gamma X_{RII},$$
$$\dot{X}_{RIS} = \gamma X_{ISI} - (\gamma + \tau)X_{RIS},$$
$$\dot{X}_{RSI} = \gamma X_{ISI} - (\gamma + \tau)X_{RSI},$$
$$\dot{X}_{IRS} = \gamma X_{IIS} - \gamma X_{IRS},$$
$$\dot{X}_{ISR} = \gamma X_{ISI} - (\gamma + \tau)X_{ISR},$$
$$\dot{X}_{SRI} = \gamma X_{SII} - \gamma X_{SRI},$$
$$\dot{X}_{SIR} = \gamma X_{SII} - (\gamma + \tau)X_{SIR},$$

where $X_{ABC}(t)$ denotes the probability of the network being in state ABC at time t. It is possible to derive the bottom-up equations from these. We begin by looking at $\langle S_2 \rangle$. Summing over all the states where the second node is susceptible gives

$$\langle S_2 \rangle = (X_{SSS} + X_{SSI} + X_{ISS} + X_{ISI} + X_{RSS} + X_{SSR} + X_{RSR} + X_{RSI} + X_{ISR})(t).$$

Thus, the derivative of $\langle S_2 \rangle$ can be found by summing the derivatives of the terms on the right-hand side. Many terms cancel (in particular all terms with γ). After some algebra, we arrive at

$$\langle \dot{S_2} \rangle(t) = -\tau(X_{ISS} + X_{ISI} + X_{ISR}) - \tau(X_{SSI} + X_{ISI} + X_{RSI}),$$

where we have separated terms so that we can make use of

$$\langle I_1 S_2 \rangle = X_{ISS} + X_{ISI} + X_{ISR} \quad \text{and} \quad \langle S_2 I_3 \rangle = X_{SSI} + X_{ISI} + X_{RSI}.$$

We arrive at $\langle \dot{S_2} \rangle = -\tau \langle I_1 S_2 \rangle - \tau \langle S_2 I_3 \rangle$, the bottom-up equation for $\langle S_2 \rangle$. A detailed analysis, and results of how one can use the top-down system to derive equations for subsystems, such as nodes, edges, triples, etc., is given in [279, 280, 283] and will not be discussed here. But we will now illustrate this by showing how seemingly complex bottom-up models can be manipulated to generate analytically and numerically tractable versions through the use of closures.

3.1.1 Closures: a succinct overview

The bottom-up models are unwieldy unless we can find simplifications which rely on expressing pairs in terms of singles, triples in terms of pairs and singles, and so on. Typically, we perform these simplifications by expressing higher order structures in terms of lower order ones. For example, if we are able to express all triples in terms of pairs and singles, e.g. if $\langle S_i S_j I_k \rangle = \langle S_i S_j \rangle \langle S_j I_k \rangle / \langle S_j \rangle$, then the total number of equations is $\mathcal{O}(N)$, provided that the number of edges scales as such. This yields systems simple enough to be studied numerically, and sometimes even analytically.

The main ideas of closures originate from statistical physics [175] and have been used widely in physical problems for many years [185]. More recently, they have become popular in models motivated by population dynamics problems in ecology [133, 165, 166, 207, 227, 256, 279, 280, 282, 283]. The common theme of all these approaches is to express higher order structures in terms of lower order ones, either exactly or approximately. This breaks the cascade whereby each structure depends on higher order ones. This approach relies on the assumption that at some scale, certain variables can be treated as independent [187]. To make this more formal, the approach of Keeling [165] and van Baalen [306] is followed, and correlation measures are introduced. These quantify the propensity of neighbouring sites to have identical or different statuses. To this end, let $\mathcal{C}_{A_iB_j}$ be the correlation between types A and B over the edge connecting nodes i and j. This can be written as

$$\mathcal{C}_{A_iB_j} = \frac{P(A_iB_j)}{P(A_i)P(B_j)} = \frac{\langle A_iB_j \rangle}{\langle A_i \rangle \langle B_j \rangle}.$$

Hence, $\mathcal{C}_{A_iB_j} = 1$ is equivalent to assuming independence. For both SIS and SIR, this is not the case, since infected nodes transmit to their susceptible neighbours, so infected nodes have an increased probability of being joined to other infected nodes ($\mathcal{C}_{I_iI_j} \geq 1$), while susceptibles and infecteds are negatively correlated ($\mathcal{C}_{S_iI_j}, \mathcal{C}_{I_iS_j} \leq 1$). In some sense, knowing that i is infected gives us information about the status of j, increasing our expectation that j is infected as well. Nevertheless, in our notation, this allows us to write the pair probability as

$$\langle A_iB_j \rangle = \langle A_i \rangle \langle B_j \rangle \mathcal{C}_{A_iB_j}. \tag{3.1}$$

If we assume independence at the level of pairs, we get

$$\langle A_iB_j \rangle \simeq \langle A_i \rangle \langle B_j \rangle, \tag{3.2}$$

Equation (3.1) can be generalised to higher order structures. For example, following van Baalen [306], the probability of an open triple, i.e. (i, j, k) with i and k not connected, can be written as

$$\langle A_iB_jC_k \rangle^< = \langle A_i \rangle \langle B_j \rangle \langle C_k \rangle \mathcal{C}_{A_iB_j} \mathcal{C}_{B_jC_k} \mathcal{T}^<_{A_iB_jC_k},$$

where $\mathcal{T}^<_{A_iB_jC_k}$ is the triple-level correlation (the superscript denotes that it is not a triangle). Note that if $\langle B_j \rangle = 0$, then this is 0. If $\langle B_j \rangle > 0$, then plugging in the corresponding expressions for the pair-level correlations and neglecting triple-level correlations (i.e. $\mathcal{T}^<_{A_iB_jC_k} = 1$) yields the approximation

$$\langle A_iB_jC_k \rangle^< \simeq \frac{\langle A_iB_j \rangle \langle B_jC_k \rangle}{\langle B_j \rangle}. \tag{3.3}$$

Here, and for the remainder of this book, if $\langle B_j \rangle = 0$, we take the right-hand side to be 0. A similar argument when i, j and k form a triangle (also called a closed triple) yields

$$\langle A_iB_jC_k \rangle^\triangle = \langle A_i \rangle \langle B_j \rangle \langle C_k \rangle \mathcal{C}_{A_iB_j} \mathcal{C}_{B_jC_k} \mathcal{C}_{A_iC_k} \mathcal{T}^\triangle_{A_iB_jC_k},$$

which, after some algebra, leads to

$$\langle A_iB_jC_k \rangle^\triangle \simeq \frac{\langle A_iB_j \rangle \langle B_jC_k \rangle \langle A_iC_k \rangle}{\langle A_i \rangle \langle B_j \rangle \langle C_k \rangle}. \tag{3.4}$$

This approach can be extended to higher orders and can be used for square lattices or structured networks, but here we will focus on the two most basic and more widely used closures, namely closures at pair, Eq. (3.2), and triple levels, Eqs. (3.3) & (3.4). Following on from [279, 280], the resulting closed systems will be referred to as individual-based and pair-based models, respectively.

We begin by looking at the pair-level closure on the line network with three nodes. The equations we derive are approximate. The pair closures are

$$\langle S_1 I_2 \rangle = \langle S_1 \rangle \langle I_2 \rangle, \quad \langle I_1 S_2 \rangle = \langle I_1 \rangle \langle S_2 \rangle,$$
$$\langle S_2 I_3 \rangle = \langle S_2 \rangle \langle I_3 \rangle \quad \text{and} \quad \langle I_2 S_3 \rangle = \langle I_2 \rangle \langle S_3 \rangle.$$

The identities above are not exact. To emphasise that we are solving a system obtained by using closures, we will use $\langle X_i \rangle$ to denote the approximation to $\langle S_i \rangle$, and $\langle Y_i \rangle$ to denote the approximation to $\langle I_i \rangle$. The closed equations become

$$\langle \dot{X}_1 \rangle = -\tau \langle X_1 \rangle \langle Y_2 \rangle, \qquad\qquad \langle \dot{Y}_1 \rangle = \tau \langle X_1 \rangle \langle Y_2 \rangle - \gamma \langle Y_1 \rangle,$$
$$\langle \dot{X}_2 \rangle = -\tau \langle Y_1 \rangle \langle X_2 \rangle - \tau \langle X_2 \rangle \langle Y_3 \rangle, \qquad \langle \dot{Y}_2 \rangle = \tau \langle Y_1 \rangle \langle X_2 \rangle + \tau \langle X_2 \rangle \langle Y_3 \rangle - \gamma \langle Y_2 \rangle,$$
$$\langle \dot{X}_3 \rangle = -\tau \langle Y_2 \rangle \langle X_3 \rangle, \qquad\qquad \langle \dot{Y}_3 \rangle = \tau \langle Y_2 \rangle \langle X_3 \rangle - \gamma \langle Y_3 \rangle.$$

Such approximations can be generalised for any network and have been considered extensively by [181]. These closures often give reasonable-to-excellent agreement with the true values of $\langle S_i \rangle$ and $\langle I_i \rangle$, but provide only an approximation to the exact system. Nevertheless, in this chapter we show that closures at the level of triples can be exact on tree networks, and we give a generalisation for a network with loops involving closures at an arbitrary level. In simple terms, for the line network, the following closures are exact for an SIR disease:

$$\langle S_1 S_2 I_3 \rangle = \frac{\langle S_1 S_2 \rangle \langle S_2 I_3 \rangle}{\langle S_2 \rangle}, \quad \langle I_1 S_2 I_3 \rangle = \frac{\langle I_1 S_2 \rangle \langle S_2 I_3 \rangle}{\langle S_2 \rangle} \text{ and } \langle I_1 S_2 S_3 \rangle = \frac{\langle I_1 S_2 \rangle \langle S_2 S_3 \rangle}{\langle S_2 \rangle}$$

(again taking these to be 0 if $\langle S_2 \rangle = 0$). The system can now be written as

$$\langle \dot{X}_1 \rangle = -\tau \langle X_1 Y_2 \rangle, \qquad\qquad \langle \dot{X_1 Y_2} \rangle = \tau \frac{\langle X_1 X_2 \rangle \langle X_2 Y_3 \rangle}{\langle X_2 \rangle} - (\tau + \gamma)\langle X_1 Y_2 \rangle,$$
$$\langle \dot{Y}_1 \rangle = \tau \langle X_1 Y_2 \rangle - \gamma \langle Y_1 \rangle,$$
$$\langle \dot{X}_2 \rangle = -\tau \langle Y_1 X_2 \rangle - \tau \langle X_2 Y_3 \rangle, \qquad \langle \dot{Y_1 X_2} \rangle = -\tau \frac{\langle Y_1 X_2 \rangle \langle X_2 Y_3 \rangle}{\langle X_2 \rangle} - (\tau + \gamma)\langle Y_1 X_2 \rangle,$$
$$\langle \dot{Y}_2 \rangle = \tau \langle Y_1 X_2 \rangle + \tau \langle X_2 Y_3 \rangle - \gamma \langle Y_2 \rangle, \qquad \langle \dot{X_2 Y_3} \rangle = -\tau \frac{\langle Y_1 X_2 \rangle \langle X_2 Y_3 \rangle}{\langle X_2 \rangle} - (\tau + \gamma)\langle X_2 Y_3 \rangle,$$
$$\langle \dot{X}_3 \rangle = -\tau \langle Y_2 X_3 \rangle,$$
$$\langle \dot{Y}_3 \rangle = \tau \langle Y_2 X_3 \rangle - \gamma \langle Y_3 \rangle, \qquad \langle \dot{Y_2 X_3} \rangle = \tau \frac{\langle Y_1 X_2 \rangle \langle X_2 X_3 \rangle}{\langle X_2 \rangle} - (\tau + \gamma)\langle Y_2 X_3 \rangle,$$

with two extra equations needed for the new variables introduced by the closures. These are

$$\langle X_1 \dot{X}_2 \rangle = -\tau \frac{\langle X_1 X_2 \rangle \langle X_2 Y_3 \rangle}{\langle X_2 \rangle}, \qquad \langle X_2 \dot{X}_3 \rangle = -\tau \frac{\langle Y_1 X_2 \rangle \langle X_2 X_3 \rangle}{\langle X_2 \rangle}.$$

We emphasise that for the three-node line network, if the identities $\langle S_i \rangle = \langle X_i \rangle$, $\langle I_i \rangle = \langle Y_i \rangle$ $(i = 1, 2, 3)$, $\langle S_1 I_2 \rangle = \langle X_1 Y_2 \rangle$, $\langle I_1 S_2 \rangle = \langle Y_1 X_2 \rangle$, $\langle S_2 I_3 \rangle = \langle X_2 Y_3 \rangle$ and $\langle I_2 S_3 \rangle = \langle Y_2 X_3 \rangle$ hold at $t = 0$, they will hold for all $t \geq 0$.

3.2 General bottom-up model

We now consider the general equations for a bottom-up model of SIS and SIR dynamics in an arbitrary network with N nodes. We assume that no node has an edge to itself, but we allow for node i to have an edge to node j having some weight g_{ij}. Typically, $g_{ij} = 1$ if there is an edge from j to i, and 0 otherwise, but the model formulation works for any directed and weighted networks, and thus we can consider $g_{ij} \in [0, \infty)$ for $i, j = 1, 2, \ldots, N$. We can use the *adjacency matrix* $G = (g_{ij})_{i,j=1,2,\ldots,N}$ to represent the network. We assume that the transmission rate from i to j is τg_{ij}. Each node may have its own recovery rates γ_i $(i = 1, 2, \ldots, N)$. We note that some of the results in the chapter may require various further constraints on the network, such as binary weights (i.e. 0 or 1) or all weights rescaled so that the maximum weight is less than or equal to 1, or that the network is strongly connected. These will be specified on a case-by-case basis.

We now derive the SIS and SIR versions of the bottom-up model, truncating the equations after pairs.

3.2.1 SIS dynamics

In the SIS model, if we know $\langle I_i \rangle$, then we know $\langle S_i \rangle = 1 - \langle I_i \rangle$. So we will start by writing down an equation just for the rate of change of $\langle I_i \rangle$. This equation will depend on $\langle S_i I_j \rangle$, so we write down an equation for that as well. In turn this depends on other pairs, which we also track. Some equations depend on triples, which we do not track. We get

$$\langle \dot{I}_i \rangle = \tau \sum_{j=1}^{N} g_{ij} \langle S_i I_j \rangle - \gamma_i \langle I_i \rangle, \tag{3.5a}$$

$$\langle \dot{S_i I_j} \rangle = \tau \sum_{k=1, k\neq i}^{N} g_{jk} \langle S_i S_j I_k \rangle - \tau \sum_{k=1, k\neq j}^{N} g_{ik} \langle I_k S_i I_j \rangle$$
$$\qquad - \tau g_{ij} \langle S_i I_j \rangle - \gamma_j \langle S_i I_j \rangle + \gamma_i \langle I_i I_j \rangle, \tag{3.5b}$$

$$\langle \dot{I_i I_j} \rangle = \tau \sum_{k=1, k\neq i}^{N} g_{jk} \langle I_i S_j I_k \rangle + \tau \sum_{k=1, k\neq j}^{N} g_{ik} \langle I_k S_i I_j \rangle - (\gamma_i + \gamma_j) \langle I_i I_j \rangle$$

$$+ \tau g_{ij}\langle S_i I_j \rangle + \tau g_{ji}\langle I_i S_j \rangle, \tag{3.5c}$$

$$\langle \dot{S_i S_j} \rangle = -\tau \sum_{k=1,k\neq j}^{N} g_{ik}\langle I_k S_i S_j \rangle - \tau \sum_{k=1,k\neq i}^{N} g_{jk}\langle S_i S_j I_k \rangle, \tag{3.5d}$$

where $i, j = 1, 2, \ldots, N$ and $i \neq j$. The derivation of the equations follows naturally by taking into account the network structure and transitions allowed by the dynamics. For example, in the $\langle S_i I_j \rangle$ equation, terms such as $\tau g_{jk}\langle S_i S_j I_k \rangle$ (with $k \neq i$) contribute positively and represent infections coming from outside the pair. Similarly, SI pairs can change type either by infection from within the pair, $-\tau g_{ij}\langle S_i I_j \rangle$, or from outside, given by terms such as $-\tau g_{ik}\langle I_k S_i I_j \rangle$ (with $k \neq j$), or by recovery of the infected node. We do not need an equation for $\langle I_i S_j \rangle$ since it equals $\langle S_j I_i \rangle$. The process of writing down equations in such a systematic way is directly related to the Bogoliubov–Born–Green–Kirkwood–Yvon (BBGKY) hierarchy developed in statistical physics.

3.2.2 SIR dynamics

Deriving the SIR model is very similar. We need equations for both $\langle S_i \rangle$ and $\langle I_i \rangle$, but we can deduce $\langle R_i \rangle = 1 - \langle S_i \rangle - \langle I_i \rangle$ from these. The SIR dynamics are given by

$$\langle \dot{S_i} \rangle = -\tau \sum_{j=1}^{N} g_{ij}\langle S_i I_j \rangle, \tag{3.6a}$$

$$\langle \dot{I_i} \rangle = \tau \sum_{j=1}^{N} g_{ij}\langle S_i I_j \rangle - \gamma_i \langle I_i \rangle, \tag{3.6b}$$

$$\langle \dot{S_i I_j} \rangle = \tau \sum_{k=1,k\neq i}^{N} g_{jk}\langle S_i S_j I_k \rangle - \tau \sum_{k=1,k\neq j}^{N} g_{ik}\langle I_k S_i I_j \rangle$$
$$- \tau g_{ij}\langle S_i I_j \rangle - \gamma_j \langle S_i I_j \rangle, \tag{3.6c}$$

$$\langle \dot{S_i S_j} \rangle = -\tau \sum_{k=1,k\neq j}^{N} g_{ik}\langle I_k S_i S_j \rangle - \tau \sum_{k=1,k\neq i}^{N} g_{jk}\langle S_i S_j I_k \rangle, \tag{3.6d}$$

where $i, j = 1, 2, \ldots, N$ and $i \neq j$. Regarding the exact systems (3.5) and (3.6), or the systems built up by including larger terms, some remarks can be made:

1. The equations emerge naturally, starting from the variables we are interested in (typically $\langle S_i \rangle$ and $\langle I_i \rangle$ for each node i) and building up to higher order structures. The equations need not account for every possible configuration of states. For example, for the SIR dynamics, the exact equations do not require knowledge of pairs such as $\langle II \rangle$.
2. All triples appearing can be written as $\langle ASB \rangle$, with $A, B \in \{S, I\}$. In a tree network with SIR dynamics, this means that nodes to the "left" and "right" of the

central node have not yet communicated, since the only way is through the central susceptible node. Thus, conditional on the central node being susceptible, their statuses are independent.
3. The closures may require extra variables. For example, if we close at the level of triples, we expect $\langle S_i S_j I_k \rangle = \langle S_i S_j \rangle \langle S_j I_k \rangle / \langle S_j \rangle$. So we need evolution equations for SS pairs.

We emphasise that systems (3.5) for an SIS disease and (3.6) for an SIR disease are exact [129, 181, 280, 281, 283]. However, they are not closed. Both systems rely on knowing triples, but they do not predict how those triples change in time. If we include equations for triples in the same manner, they will depend on quadruples. This cascade continues up to the network size, at which point the system is exact and complete, but the number of equations is typically far too large to be analytically or numerically tractable. A solution to this is to introduce closures, as explained before. This leads us to some questions:

1. At what level, e.g. pairs, triples, etc., should the closure be applied?
2. How well will the solutions of the closed system compare to results based on the original exact system?
3. How do all the above depend on the structure/topology of the network and properties of the dynamics?

These will be addressed in a methodical way in the following sections.

3.3 Differential inequalities and a comparison theorem for ODEs

Here, we will provide a mathematical intermezzo into the fundamentals of differential inequalities and will present a comparison theorem for ODE systems. A detailed and comprehensive study of differential inequalities and comparison theory can be found in the book by Szarski [296]. Based on these results, ordering relations between the exact, bottom-up and closed systems can be established. This allows us to prove rigorously that a given approximation is an upper or lower bound to the exact model. Generally, the comparison problem can be stated as follows.

Problem: Consider the ODE $\dot{x}(t) = f(x(t))$ and the differential inequality $\dot{y}(t) \leq f(y(t))$ with a given differentiable function $f : \mathbb{R}^n \to \mathbb{R}^n$ subject to initial conditions satisfying $y(0) \leq x(0)$. The task is to find conditions on f ensuring that $y(t) \leq x(t)$ for $t \geq 0$. Statements of this type are called comparison theorems. In what follows, the ordering relation for vectors is used in the following sense:

$$u \leq v, \text{ if } u_i \leq v_i \text{ for } i = 1, 2, \ldots, n, \qquad u < v, \text{ if } u_i < v_i \text{ for } i = 1, 2, \ldots, n.$$

Brief history of the problem: The theory of ordinary differential inequalities was originated by Kamke and Müller's work [160, 228]. A sufficient condition on f for the desired inequality to hold is the *Kamke–Müller condition*, which is equivalent to requiring that if $u \leq v$ and $u_i = v_i$ then $f_i(u) \leq f_i(v)$ for $i = 1, 2, \ldots, n$.

This condition essentially means that the function in the ith coordinate, f_i is increasing in all coordinates x_j for $j \neq i$. However, this leads to an alternative sufficient condition, which if satisfied allows us to call the f function *cooperative*. More precisely, f is called *cooperative* if

$$\partial_j f_i \geq 0 \text{ for } i, j = 1, 2, \ldots, n \text{ and } i \neq j.$$

It can be shown that if f is cooperative in a convex domain, then it satisfies the Kamke–Müller condition. Cooperative systems generate monotone dynamical systems that are dealt with in the book chapter by Hirsch and Smith [142] and the book by Smith [290]. For these dynamical systems, the ordering relation is preserved by the dynamics, i.e. $p \leq q$ implies $\varphi(t, p) \leq \varphi(t, q)$ for $t \geq 0$. This enables one to characterise the long-term behaviour of the system and to develop the Poincaré–Bendixson theory for higher dimensional phase spaces. In order to make it clearer how the general result emerges, and how the added degree of complexity of going from one-dimensional to the multidimensional case can be tackled, the following observations are made.

Observation 1. For a one-dimensional linear system, the comparison theorem reduces to Gronwall's lemma [250]. Namely, if $\dot{u}(t) \leq au(t)$, then $u(t) \leq u(0) \exp(at)$. This can be proved by multiplying the inequality $\dot{u}(t) - au(t) \leq 0$ by $\exp(-at)$ and integrating from 0 to t. Slightly more rigorously, we can state that if $\dot{u} \leq Lu$ in the interval $[t_1, t_2]$, then $u \leq u(t_1) e^{L(t - t_1)}$ throughout this interval. This idea can be generalised to a one-dimensional non-linear system.

Observation 2. Let $f : \mathbb{R} \to \mathbb{R}$ be a differentiable function (in fact, Lipschitz continuity is enough), let $\dot{x}(t) = f(x(t))$ and $\dot{y}(t) \leq f(y(t))$. Then, $y(0) \leq x(0)$ implies that $y(t) \leq x(t)$ for $t \geq 0$. This can be proved by using the Lipschitz continuity of f and applying Gronwall's lemma to the function $u(t) = x(t) - y(t)$. This can be formulated more rigorously as follows. Recall that a function h satisfies the Lipschitz condition if $|h(x) - h(y)| \leq L|x - y|$.

Lemma 3.1 *Let $f, g : \mathbb{R} \to \mathbb{R}$ be two continuous functions which satisfy the Lipschitz condition, with $g \leq f$. If $x(t)$ and $y(t)$ are the solutions of the differential equations $\dot{x}(t) = f(x(t))$ and $\dot{y}(t) = g(y(t))$, respectively, and if $y(t_0) \leq x(t_0)$, then $y(t) \leq x(t)$ for all $t \geq t_0$.*

Proof. Let us assume that there exists $t_2 \geq t_0$ such that $x(t_2) > y(t_2)$. Let t_1 be the last time point before t_2 where $x(t_1) = y(t_1)$ holds, and hence $x(t) < y(t)$ for all $t \in (t_1, t_2)$. Consider the function $u(t) = y(t) - x(t)$ on the $t \in [t_1, t_2]$ interval. Its derivative yields

$$\dot{u}(t) = g(y(t)) - f(x(t)) \leq f(y(t)) - f(x(t)) \leq L(y(t) - x(t)),$$

where L is the Lipschitz constant. Hence, Gronwall's lemma yields

$$u(t) = y(t) - x(t) \le u(t_1)e^{L(t-t_1)} = 0,$$

since $u(t_1) = y(t_1) - x(t_1) = 0$. This implies that $y(t) \le x(t)$ for all $t \in [t_1, t_2]$. However, this is a contradiction. $\square$

We note that all differentiable functions in a closed interval $[t_1, t_2]$ satisfy the Lipschitz condition.

For the transition from the one- to the multidimensional case, the following observations highlight potential pitfalls in assuming a simple generalisation and offer a viable solution to extending the comparison results to the multidimensional case.

Observation 3.a We can show that the comparison theorem does not hold for an arbitrary multidimensional system by considering two simple two-dimensional systems,

$$(\text{System 1}) \begin{cases} \dot{x}_1 = x_2, \\ \dot{x}_2 = 1 - x_1, \end{cases} \quad \text{and} \quad (\text{System 2}) \begin{cases} \dot{y}_1 = y_2, \\ \dot{y}_2 = -y_1. \end{cases}$$

It is clear that if $x_1 = y_1$ and $x_2 = y_2$, then $\dot{x}_1 = \dot{y}_1$ and $\dot{y}_2 \le \dot{x}_2$. However, the conclusion that $y_2 \le x_2$ for all later times does not hold true. The initial conditions $x_1(\pi/2) = y_1(\pi/2) = 0$ and $x_2(\pi/2) = y_2(\pi/2) = 1$ yield $x_1 = 1$, $x_2 = 0$ and $y_1 = \sin(t)$, $y_2 = \cos t$. It can be seen that $y_2(t) \not\le x_2(t)$ for some t, with $t > \pi/2$.

Observation 3.b The Kamke–Müller condition, however, is sufficient to imply that if $\dot{x}(t) = f(x(t))$ and the strict differential inequality $\dot{y}(t) < f(y(t))$ holds and $y(0) < x(0)$, then $y(t) < x(t)$ for $t \ge 0$. The main idea of proof is the following. Assuming that the statement is not true, let T be the first point where the inequality $y(t) < x(t)$ is violated, that is $x_i(T) = y_i(T)$ holds at least for one coordinate i, and $y(t) < x(t)$ for $t < T$. Then, by using the Kamke–Müller condition for f_i, we have

$$\dot{y}_i(T) < f_i(y(T)) \le f_i(x(T)) = \dot{x}_i(T).$$

This contradicts that $x_i(T) = y_i(T)$ and $x_i(t) < y_i(t)$ for $t < T$. This observation gives rise to the following general comparison result [160, 228].

Theorem 3.2 *Assume that f satisfies the Kamke–Müller condition. If $\dot{x}(t) = f(x(t))$ and the differential inequality $\dot{y}(t) \le f(y(t))$ holds and $y(0) \le x(0)$, then $y(t) \le x(t)$ for $t \ge 0$.*

This statement above can be reduced to the framework of Observation 3.b by introducing the functions y^n satisfying the strict inequality $\dot{y}^n(t) < f(y^n(t)) + a_n e$, where (a_n) is a sequence tending to zero, and $e = (1, 1, \ldots, 1) \in \mathbb{R}^n$. Then, the Observation shows that $y^n(t) < x(t)$, and it can be proved that $y^n(t) \to y(t)$ as $n \to \infty$; hence, $y(t) \le x(t)$. The detailed proof can be found in Chapter 2 of the book by Szarski

[296]: see Lemma 6.1 and Theorems 8.1 and 9.3. This idea of the proof is generalised to an arbitrary ordering defined by a cone, as in Theorem 3.2 in the book
chapter by Hirsch and Smith [142]. This result will be used to establish rigorous
ordering results between exact and closed models later in this chapter.

3.4 Closures at the level of pairs and triples for SIS dynamics

We return to the analysis of systems closed at the level of pairs and triples. We take
the simplest model, closed at the level of pairs, and investigate the steady states
and their stability. For a fully connected network, we compare the model to the
exact model from which it is derived and the classic compartmental SIS model. The
performance of the closed systems is explored for a number of different networks
using simulation.

3.4.1 The individual-based model

We start by noting that "individual-based model" as used in this book does not refer
to agent-based or explicit network-based stochastic simulation. It simply expresses
that equations are formulated at node level and only involve nodes or individuals.
The simplest closure is at the level of pairs. This yields the reduced system below.

SIS individual-based model

$$\langle \dot{Y_i} \rangle = \left(\tau \sum_{j=1}^{N} g_{ij}(1 - \langle Y_i \rangle)\langle Y_j \rangle \right) - \gamma_i \langle Y_i \rangle, \tag{3.7}$$

where we again emphasise that $\langle X_i \rangle = 1 - \langle Y_i \rangle$ and $\langle Y_i \rangle$ denote approximations to $\langle S_i \rangle$ and $\langle I_i \rangle$, respectively. Given a network which may be theoreti-
cal/synthetic or reconstructed from real data, we can solve system (3.7) nu-
merically to obtain a first approximation to the true stochastic SIS epidemic
model. An implemented version of this can be found as supplementary material in [129]. This closed model requires N differential equations.

This and similar models have been used extensively [60, 255, 279, 280, 310,
312]. In [255], for example, the leading eigenvalue of the adjacency matrix of the
network and the parameters of the disease transmission model combine to give a
threshold condition which separates the stable disease-free steady state from the
epidemic or endemic stage. However, the crucial assumption of the independence
of neighbours, which is only an approximation of the true transmission dynamics
on the network, was not fully explained. While the relative simplicity allows us to

derive some analytical or semi-analytical results, such models are only approximate and tend to perform well when the ratio τ/γ is large or small, and when networks are relatively well connected.

Exercise 3.1. Using SIS dynamics and the network consisting of one single edge:
   a. Write down the bottom-up exact and the corresponding reduced system with closures at the level of pairs;
   b. Solve these equations numerically, using a programming language of your choice, and establish whether the probability of nodes being infected in the reduced/closed model is overestimated or underestimated.

Exercise 3.2. Using SIS dynamics and the line network with three nodes (Fig. 3.1a):
   a. Write down the bottom-up exact and the corresponding reduced system with closures at the level of pairs;
   b. Solve these equations numerically, using a programming language of your choice, and establish whether the probability of nodes being infected in the reduced/closed model is overestimated or underestimated.

We continue by analysing some properties of the individual-based model.

3.4.2 The individual-based model overestimates the true probability of nodes being infected

Given the insight from the previous section on differential inequalities, we can give the following rigorous result.

Theorem 3.3 *Consider a (possibly weighted and directed) network G, the exact bottom-up SIS model given by system* (3.5) *and the equivalent individual-based closed system* (3.7). *Assuming that both models start with identical initial conditions,* $\langle S_i \rangle(0) = 1 - \langle Y_i \rangle(0)$, $\langle I_i \rangle(0) = \langle Y_i \rangle(0)$ $(i = 1, 2, \ldots, N)$, *it follows that* $\langle I_i \rangle(t) \le Y_i(t)$ *for* $i = 1, 2, \ldots, N$ *and* $t \ge 0$.

Proof. We start from the exact system,

$$\langle \dot{I}_i \rangle = \tau \sum_{j=1}^{N} g_{ij} \langle S_i I_j \rangle - \gamma_i \langle I_i \rangle = \tau \sum_{j=1}^{N} g_{ij} \langle S_i \rangle \langle I_j \rangle - \gamma_i \langle I_i \rangle + \tau \sum_{j=1}^{N} g_{ij} \left(\langle S_i I_j \rangle - \langle S_i \rangle \langle I_j \rangle \right)$$

$$= \tau \sum_{j=1}^{N} g_{ij} (1 - \langle I_i \rangle) \langle I_j \rangle - \gamma_i \langle I_i \rangle + \tau \sum_{j=1}^{N} g_{ij} \left(\langle S_i I_j \rangle - \langle S_i \rangle \langle I_j \rangle \right)$$

$$\le \tau \sum_{j=1}^{N} g_{ij} (1 - \langle I_i \rangle) \langle I_j \rangle - \gamma_i \langle I_i \rangle, \tag{3.8}$$

where we have used the fact that Markovian SIS epidemics are non-negatively correlated, i.e. $\langle S_i I_j \rangle \le \langle S_i \rangle \langle I_j \rangle$ for $t \ge 0$ (see [60]) (and also assumed that this result holds for directed and weighed networks). While this is sufficient for the proof, we conjecture that $\langle S_i I_j \rangle < \langle S_i \rangle \langle I_j \rangle$ as long as the system is away from a "pure state" (i.e. $\langle I_i \rangle (0)$ is either 0 or 1) and has not been absorbed by the "all nodes susceptible" state. The closed individual-based system takes the form

$$\langle \dot{Y}_i \rangle = \tau \sum_{j=1}^{N} g_{ij}(1 - \langle Y_i \rangle)\langle Y_j \rangle - \gamma_i \langle Y_i \rangle.$$

Introducing the function $f : \mathbb{R}^N \to \mathbb{R}^N$ with coordinate functions

$$f_i(x) = \tau \left(\sum_{j=1}^{N} g_{ij}(1 - x_i)x_j \right) - \gamma_i x_i,$$

it can immediately be seen that the solution of the closed system satisfies the differential equation $\dot{x}(t) = f(x(t))$, and the solution of the exact system satisfies the differential inequality $\dot{y}(t) \leq f(y(t))$, with both systems starting from the same initial condition. Differentiating the coordinate functions f_i with respect to x_j, we get $\partial_j f_i(x) = \tau g_{ij}(1 - x_i) \geq 0$, since $x_i \in [0,1]$. Hence, the system is cooperative in the unit cube $[0,1]^N$, which is convex. As a result, f satisfies the Kamke–Müller condition. Therefore, the general comparison Theorem 3.2 implies that $\langle I_i \rangle (t) \leq \langle Y_i(t) \rangle$ for $i = 1, 2, \ldots, N$ and $t \geq 0$. □

Exercise 3.3. Using SIS dynamics and the network consisting of one single edge:
  a. Use the exact system to show that if $\langle S_1 I_2 \rangle (0) = 1$, then $\langle I_1 \rangle (0) = 0$, $\langle \dot{I}_1 \rangle (0) = \tau$, $\langle \ddot{I}_1 \rangle (0) = -\tau^2 - 2\tau\gamma$, $\langle \dddot{I}_1 \rangle (0) = \tau^3 + 4\tau^2\gamma + 3\tau\gamma^2$.
  b. Use the exact system to show that if $\langle I_1 S_2 \rangle (0) = 1$, then $\langle I_1 \rangle (0) = 1$, $\langle \dot{I}_1 \rangle (0) = -\gamma$, $\langle \ddot{I}_1 \rangle (0) = \gamma^2$, $\langle \dddot{I}_1 \rangle (0) = \tau^2\gamma - \gamma^3$.
  c. Use the approximate system to show that if $y_1(0) = 0$ and $y_2(0) = 1$, then $y_1(0) = 0$, $\dot{y}_1(0) = \tau$, $\ddot{y}_1(0) = -\tau^2 - 2\tau\gamma$, $\dddot{y}_1(0) = \tau^3 + 5\tau^2\gamma + 3\tau\gamma^2$.
  d. Use the approximate system to show that if $y_1(0) = 1$ and $y_2(0) = 0$, then $y_1(0) = 1$, $\dot{y}_1(0) = -\gamma$, $\ddot{y}_1(0) = \gamma^2$, $\dddot{y}_1(0) = 2\tau^2\gamma - \gamma^3$.
  e. Use a Taylor series expansion around zero for both $\langle I_1 \rangle (t)$ and $y_1(t)$ to show that immediately after zero, but close to it, $\langle I_1 \rangle (t) < y_1(t)$.

Exercise 3.4. Using SIS dynamics and the line network with three nodes (see Fig. 3.2a):
  a. Use the exact system to show that if $\langle S_1 I_2 S_3 \rangle (0) = 1$, then $\langle I_1 \rangle (0) = 0$, $\langle \dot{I}_1 \rangle (0) = \tau$, $\langle \ddot{I}_1 \rangle (0) = -\tau^2 - 2\tau\gamma$, $\langle \dddot{I}_1 \rangle (0) = \tau^3 + 4\tau^2\gamma + 3\tau\gamma^2$.
  b. Use the approximate system to show that if $y_1(0) = 0$, $y_2(0) = 1$ and $y_3(0) = 0$, then $y_1(0) = 0$, $\dot{y}_1(0) = \tau$, $\ddot{y}_1(0) = -\tau^2 - 2\tau\gamma$, $\dddot{y}_1(0) = \tau^3 + 5\tau^2\gamma + 3\tau\gamma^2$.
  c. Use a Taylor series expansion around zero for both $\langle I_1 \rangle (t)$ and $y_1(t)$ to show that immediately after zero, but close to it, $\langle I_1 \rangle (t) < y_1(t)$.

Exercise 3.5. (This exercise concerns the correlations between pairs and singles and addresses this by an example for a small, specific network followed by the case of a general network.)

    a. Using SIS dynamics and the network consisting of one single edge, show that if $\langle S_1 I_2 \rangle(0) = 1$, then $f(t) = \langle S_1 I_2 \rangle - \langle S_1 \rangle \langle I_2 \rangle$ satisfies the following identities: $f(0) = 0$, $f'(0) = 0$ and $f''(0) = -\tau \gamma$. What does this imply about the correlations immediately after $t = 0$?

    b. For a general network and for the exact bottom-up model, show that if $f(t) = (\langle S_i I_j \rangle - \langle S_i \rangle \langle I_j \rangle)(t)$, then its derivative at $t = 0$ satisfies

$$
\begin{aligned}
f'(t) = &-2\gamma f(t) - (\langle S_i I_j \rangle \langle S_j \rangle + \langle I_i S_j \rangle \langle S_i \rangle) \\
&+ \tau \sum_{k, k \neq i} g_{jk}(\langle S_i S_j I_k \rangle - \langle S_i \rangle \langle S_j I_k \rangle) \\
&- \tau \sum_{k, k \neq j} g_{ik}(\langle I_k S_i I_j \rangle - \langle I_j \rangle \langle S_i I_k \rangle).
\end{aligned}
$$

Assuming that immediately after $t = 0$, $(\langle S_i S_j I_k \rangle - \langle S_i \rangle \langle S_j I_k \rangle)(t) < 0$ and $(\langle I_k S_i I_j \rangle - \langle I_j \rangle \langle S_i I_k \rangle)(t) > 0$, what does this imply about the sign of $f(t)$ for $t > 0$, given that the system starts from a pure state at $t = 0$, and hence with $f(0) = 0$?

3.4.3 Steady states of the individual-based model

We now consider possible steady states of system (3.7), assuming that $\gamma_i = \gamma$ for all i. A steady state is denoted as $(y_1, y_2, \ldots, y_N)$ with $\langle Y_i \rangle \to y_i$. Using $\langle \dot{Y}_i \rangle(t \to \infty) = 0$ for $i = 1, 2, \ldots, N$, the system becomes:

$$
\tau \sum_{j=1}^{N} g_{ij}(1 - y_i)y_j - \gamma y_i = 0. \tag{3.9}
$$

This can be rewritten as

$$
y_i = \frac{\tau \sum g_{ij} y_j}{\gamma + \tau \sum g_{ij} y_j}.
$$

Its trivial solution, $y_i = 0$ for all i, is called the disease-free steady state. If there is a solution with each y_i satisfying $0 \leq y_i \leq 1$ and $y_j > 0$ for at least one j, then we say we have an endemic steady state. The model predicts that if infection is at this level, it will remain at this level. The implicit equations above can be perceived as a recurrence relation and one can prove the existence of an endemic steady state under some conditions.

Theorem 3.4 *Given a directed and weighted network, G, let $k_i = \sum_j g_{ij}$ be the degree of node i (or the sum of incoming weights to node i). If $\gamma < \tau k_i$ for all*

$i = 1, 2, \ldots, N$, then an endemic state exists. Moreover, there is a $c \leq 1 - \frac{\gamma}{\tau k_i}$ for all i such that $c \leq y_i \leq 1$ holds for all i.

Proof. Let $T : \mathbb{R}_+^N \to \mathbb{R}_+^N$ be a continuous mapping given by

$$T_i(x) = \frac{\tau \sum_j g_{ij} x_j}{\gamma + \tau \sum_j g_{ij} x_j}. \tag{3.10}$$

It is straightforward to see that T maps the N-dimensional unit cube, $[0,1]^N$, to itself. Moreover, T maps $\mathbb{R}_+^N$ to $[0,1]^N$. In order to prove the existence of the endemic steady state, it will be sufficient to show that there is a $c > 0$ such that if $x_j \geq c$ for all j, then $T_i(x) \geq c$ for all i. Hence, we set out to prove the existence of such a constant c. Now $T_i(x) \geq c$ is equivalent to $(1-c)\tau \sum_j g_{ij} x_j \geq \gamma c$. However, assuming that $x_j \geq c$ for all j, an intermediate term can be found and the inequality can be extended to give

$$(1-c)\tau \sum_j g_{ij} x_j \geq (1-c)\tau c k_i \geq \gamma c.$$

Therefore, a sufficient condition for the desired inequality to hold is $c \leq 1 - \frac{\gamma}{\tau k_i}$, and provided that $1 - \frac{\gamma}{\tau k_i} > 0$, c can be chosen to be $\min_i(1 - \frac{\gamma}{\tau k_i}) \geq 1 - \frac{\gamma}{\tau k_{min}}$.

We now use Brouwer's fixed-point theorem [53]. The continuous function T maps a compact convex set, $[c,1]^N$, to itself, so it has a fixed point in $[c,1]^N$. Since $c > 0$, this fixed point has $y_i \geq c$ for all i. Thus, an endemic steady state exists. $\square$

This result only proves the existence of an endemic steady state with non-zero prevalence at each i. It does not prove that the endemic state is stable, nor does it prove that if $\gamma/\tau k_i \geq 1$ for some i, then there is no endemic steady state. For example, if a node has degree $k_i = 0$, it will eventually recover and never be reinfected. Thus, in equilibrium, $y_i = 0$. However, the remainder of the network may still have an endemic equilibrium.

It is important to note that results on bounds on the spectral radius of non-negative matrices [54] allow us to write

$$\min_i(k_i) \leq \Lambda_{max}(G) \leq \max_i(k_i),$$

where $\Lambda_{max}(G)$ is the largest eigenvalue of the adjacency matrix of G, provided that the network is strongly connected so as to guarantee that the largest eigenvalue refers to the entire network. Using this result, the condition above on the existence of a non-zero endemic state, $\frac{\gamma}{\tau k_i} < 1$, can be extended to take into account the spectral radius of the adjacency matrix G, $\Lambda_{max}(G)$. Namely, the following inequality holds:

$$\frac{\tau}{\gamma} > \frac{1}{k_{min}} \geq \frac{1}{\Lambda_{max}(G)},$$

where $k_{min} = \min_i(k_i)$. As we shall see, the condition for the stability of the non-zero endemic steady state is $\frac{\tau}{\gamma} \geq \frac{1}{\Lambda_{max}(G)}$, which means that this result just misses out on linking the change in stability of the endemic state with its actual existence. However, for homogenous or regular networks with all degrees equal to d, the inequality above becomes $\frac{\tau}{\gamma} > \frac{1}{d} = \frac{1}{\Lambda_{max}(G)}$.

In fact, using more involved calculations, it can be shown that the endemic steady state exists for any network when $\frac{\tau}{\gamma} > \frac{1}{\Lambda_{max}(G)}$. The proof can be found in [192]. Here, we only recall the result.

Theorem 3.5 *Given a directed, weighted, and strongly connected network, G, let $\Lambda_{max}(G)$ be the largest eigenvalue of the adjacency matrix of G. If $\gamma < \tau\Lambda_{max}(G)$, then an endemic (nonzero) steady state exists. Moreover, all of its coordinates are positive. If $\gamma > \tau\Lambda_{max}(G)$, then there is no endemic steady state.*

The uniqueness of the endemic steady state can be verified by using elementary calculations. We prove this below following the proof of Theorem 3.1 in [192].

Theorem 3.6 *Let G be a directed, weighted and strongly connected network. If system (3.9), giving the steady state of system (3.7), has a non-zero solution, then this is unique in the cube $(0, 1]^N$.*

Proof. It can be shown by elementary arguments that all coordinates of an endemic steady state are positive, that is $y_i > 0$ and $z_i > 0$ for all i, the details of which are given in the proof of Lemma 3.2 in [192]. Assume that there are two non-zero steady states $y, z \in (0, 1]^N$. Then, they differ at least in one of their coordinates. We can assume without loss of generality that this is the first coordinate and that $z_1 > y_1$. If there are differences at several coordinates of y and z, then choose the coordinate i for which z_i/y_i is the largest. Again, we can assume $i = 1$, that is $z_1/y_1 \geq z_j/y_j$ for all $j \neq 1$. Using the steady-state equation (3.9) with $i = 1$, we have

$$\tau(1 - y_1) \sum_{j=1}^{N} g_{ij}y_j - \gamma y_1 = 0.$$

Multiplying this equation by z_1/y_1 and using $z_1/y_1 \geq z_j/y_j$ for all $j \neq 1$ leads to

$$\tau(1 - y_1) \sum_{j=1}^{N} g_{ij}z_j - \gamma z_1 \leq 0.$$

Finally, using $z_1 > y_1$ yields

$$\tau(1 - z_1) \sum_{j=1}^{N} g_{ij}z_j - \gamma z_1 < 0,$$

which is a contradiction since the left-hand side is zero, because it is exactly the steady state equation (3.9) for z with $i = 1$. □

Equipped with these results and in line with [312], we proceed to make some important qualitative observations about the steady states of the system. The implicit steady state equations, Eq. (3.9), can be conveniently rearranged to give

$$y_i = \frac{\tau \sum_{j=1}^{N} g_{ij} y_j}{\tau \sum_{j=1}^{N} g_{ij} y_j + \gamma} = 1 - \frac{1}{1 + \frac{\tau}{\gamma} \sum_{j=1}^{N} g_{ij} y_j}. \tag{3.11}$$

A close inspection of the equation above reveals the following important observations about the qualitative behaviour of the steady state:

1. $y_i = 0$ is a trivial solution,
2. for large values of $\frac{\tau}{\gamma}$, if i has non-zero degree, then $y_i = 1 - \mathcal{O}(\frac{\gamma}{\tau})$,
3. for $\gamma = 0$, all nodes which are reachable from the initially infected nodes will eventually become infected, with $y_i = 1$ being the steady state,
4. as shown in [192, 312], in a strongly connected network, either $y_i = 0$ for all i or none of the y_is are zero, and
5. the non-zero steady state infection probability of any node i can be expressed as a continued fraction, as follows:

$$y_i = 1 - \cfrac{1}{1 + \alpha k_i - \alpha \sum_{j=1}^{N} \cfrac{g_{ij}}{1 + \alpha k_j - \alpha \sum_{k=1}^{N} \cfrac{g_{jk}}{1 + \alpha k_l - \alpha \sum_{q=1}^{N} \cfrac{g_{lq}}{1 + \alpha k_q - \cdots}}}},$$

where $\alpha = \tau/\gamma$.

The continued fraction expansion and some bounds on y_i can be derived as follows. Based on (3.11), the steady state infection probability is

$$y_i = 1 - \frac{1}{1 + \alpha \sum_j g_{ij} y_j} = 1 - \frac{1}{1 + \alpha k_i - \alpha \sum_j g_{ij} (1 - y_j)}$$

because $k_i = \sum_j g_{ij}$. Using Eq. (3.10), this equation can also be written as $y_i = T_i(y)$. Hence, the solution can be obtained by iteration. Let the zeroth iterate be $y_i^{(0)} = 1$ and then define $y_i^{(n+1)} = T(y^{(n)})$. Therefore, the first iterate is

$$y_i^{(1)} = 1 - \frac{1}{1 + \alpha k_i},$$

which is the first term in the continued fraction expansion. Calculating $y_i^{(2)} = T(y^{(1)})$ leads to the second term in the continued fraction expansion, and in general, $y_i^{(n)}$ is the nth term in the expansion. It can be shown by induction that $y_i^{(n+1)} < y_i^{(n)}$, and hence the terms in the continued fraction expansion form a decreasing sequence which is bounded from below by zero. Thus, each $y_i^{(n)}$ sequence converges. The limit is a fixed point of T, which is a steady state. Depending on the values of the parameters τ and γ, this may be the disease-free steady state.

The terms of the continued fraction expansion serve as upper bounds for the steady state as can be verified as follows. Simple algebra shows that the mapping T is monotone in the sense that $x \leq y$ implies $T(x) \leq T(y)$, where the ordering relation between vectors is understood coordinate-wise, i.e. $x \leq y$ means $x_i \leq y_i$ for all i and $T(x) \leq T(y)$ means $T_i(x) \leq T_i(y)$ for all i. The steady state is in the unit cube, i.e. $y \leq y^{(0)}$ (the latter is defined as $y_i^{(0)} = 1$ for all i). Hence, by the monotonicity of the mapping T, we have $y = T(y) \leq T(y^{(0)}) = y^{(1)}$, i.e. the steady state is bounded by the first term of the continued fraction expansion

$$y_i \leq 1 - \frac{1}{1 + \alpha k_i}.$$

Simple induction shows that $y \leq y^{(n)}$, i.e. the steady state is bounded by the nth term of the continued fraction expansion. Since the iterates $y^{(n)}$ form a decreasing sequence, they give a sequence of improving upper bounds. For example, the second term gives the upper bound

$$y_i \leq 1 - \frac{1}{1 + \alpha k_i - \alpha \sum_j \frac{g_{ij}}{1 + \alpha k_j}} \leq 1 - \frac{1}{1 + \alpha k_i}.$$

3.4.4 Stability of the steady states of the individual-based model

Disease-free steady state

The trivial disease-free steady state is given by $y = (y_1, y_2, \ldots, y_N) = (0, 0, \ldots, 0)$. Based on Eq. (3.7), it is straightforward to see that the linearisation of the closed model around this steady state leads to solving the eigenvalue problem

$$\det(\tau G - \gamma I - \lambda I) = 0,$$

where G is the adjacency matrix of the network and I is the N-dimensional identity matrix. From basic linear algebra, it follows that if the largest eigenvalue of G is $\Lambda_{\max}(G)$, then the largest eigenvalue of $\tau G - \gamma I$ is simply $\tau \Lambda_{\max}(G) - \gamma$. Hence, the critical condition or threshold for stability is given by

$$\alpha = \frac{\tau}{\gamma} \leq \frac{1}{\Lambda_{\max}(G)} = \alpha_c, \tag{3.12}$$

where it is implicitly assumed that the network is strongly connected with the largest eigenvalue referring to the entire network.

Transcritical bifurcation and the emergence of the endemic steady state

First, we recall Theorem 4.1 in [59]. Consider a general system of ODEs with a parameter ϕ:

$$\frac{dx}{dt} = f(x,\phi), \; f : \mathbb{R}^N \times \mathbb{R} \to \mathbb{R}^N \; \text{and} \; f \in \mathbb{C}^2(\mathbb{R}^N \times \mathbb{R}). \tag{3.13}$$

Without loss of generality, it is assumed that 0 is an equilibrium for this system for all values of the parameter ϕ, that is $f(0,\phi) = 0$ for all ϕ.

Theorem 3.7 *Let $A = D_x f(0,0) = \frac{\partial f_i}{\partial x_j}(0,0)$ be the linearisation of system (3.13) around the equilibrium 0, with ϕ evaluated at 0. Assume that zero is a simple eigenvalue of A, with a corresponding non-negative right eigenvector w and left eigenvector v, and that all other eigenvalues of A have negative real parts. Let f_k be the kth component of f, and define a and b as*

$$a = \sum_{k,i,j=1}^N v_k w_i w_j \frac{\partial^2 f_k}{\partial x_i \partial x_j}(0,0), \qquad b = \sum_{k,i=1}^N v_k w_i \frac{\partial^2 f_k}{\partial x_i \partial \phi}(0,0). \tag{3.14}$$

Then, a and b determine the local dynamics of (3.13) around 0. In particular, if $a < 0$ and $b > 0$, then when ϕ changes from negative to positive, 0 changes from stable to unstable. Correspondingly, a negative unstable equilibrium becomes positive and locally asymptotically stable.

The other cases $a > 0, b > 0$, or $a < 0, \; b < 0$ or $a > 0, \; b < 0$ can also be analysed, but here we focus on the $a < 0, \; b > 0$ case. This result can be applied to characterise the behaviour of the individual-based model around the disease-free steady state.

Theorem 3.8 *Given an undirected, weighted (with $g_{ij} = g_{ji}$) and connected network G, a transcritical bifurcation occurs in Eq. (3.7) at the critical value $\frac{\tau}{\gamma} = \frac{1}{\Lambda_{max}}$, where Λ_{max} is the largest eigenvalue of G, and at the disease-free steady state $(\langle Y_1 \rangle, \langle Y_2 \rangle, \ldots, \langle Y_N \rangle) = (0, 0, \ldots, 0)$. More precisely, for $\frac{\tau}{\gamma}$ close enough to the critical value, we have*

1. *for $\frac{\tau}{\gamma} < \frac{1}{\Lambda_{max}(G)}$ only the trivial disease-free steady state is stable and the endemic steady state does not exist,*
2. *for $\frac{\tau}{\gamma} > \frac{1}{\Lambda_{max}(G)}$ both the trivial disease-free and a unique endemic state exist with the first being unstable and the second stable.*

Proof. We rescale time to use $t' = \gamma t$ in Eq. (3.7). This yields

$$\dot{Y}_k = \frac{\tau}{\gamma}\left(\sum_{i=1}^N g_{ki}(1 - Y_k)Y_i\right) - Y_k = f_k\left(Y, \frac{\tau}{\gamma}\right), \tag{3.15}$$

with $\phi = \frac{\tau}{\gamma} - \frac{1}{\Lambda_{max}}$. Due to how $A = \frac{1}{\Lambda_{max}}G - I$ depends on G, it follows that $\frac{1}{\Lambda_{max}}\Lambda_{max} - 1 = 0$ is an eigenvalue of A with the corresponding right eigenvector w

satisfying $Aw = 0$, which is equivalent to $Gw = \Lambda_{\max}w$, i.e. w is the right eigenvector of G corresponding to the largest eigenvalue $\Lambda_{\max}$. Because G is a non-negative matrix, the Perron–Frobenius theorem [108] states that G has a positive largest eigenvalue, $\Lambda_{\max}(G)$ with a corresponding eigenvector whose elements are all positive. Since G is symmetric, $Gw = \Lambda_{\max}w$ is equivalent to $w^T G^T = \Lambda_{\max}w^T$, and it follows that the left eigenvector of G is in fact $v = w^T$.

Let $\left(\frac{\partial f_k}{\partial Y_1}, \frac{\partial f_k}{\partial Y_2}, \dots, \frac{\partial f_k}{\partial Y_N} \right)$ be the kth row of the Jacobian of system (3.15); it is given by

$$\left(\frac{\tau}{\gamma}g_{k1}(1-Y_k), \frac{\tau}{\gamma}g_{k2}(1-Y_k), \dots, -\frac{\tau}{\gamma}\sum_i g_{ki}Y_i - 1, \dots, \frac{\tau}{\gamma}g_{k1}(1-Y_N) \right), \quad (3.16)$$

which when evaluated at $Y = (0,0,\dots,0)$ gives rise to $J = \frac{\tau}{\gamma}G - I$, where I is the N-dimensional identity matrix. If $\frac{\tau}{\gamma} = \frac{1}{\Lambda_{\max}}$, then $J = A$. Let $M_k = \left(\frac{\partial^2 f_k}{\partial Y_i \partial Y_j} \right)_{i,j=1,2,\dots,N}$ be the matrix of mixed derivatives so that

$$M_k = \begin{pmatrix} 0 & \cdots & 0 & -\frac{\tau}{\gamma}g_{k1} & 0 & \cdots & 0 \\ 0 & \cdots & 0 & -\frac{\tau}{\gamma}g_{k2} & 0 & \cdots & 0 \\ \vdots & \ddots & \vdots & \vdots & \vdots & \ddots & 0 \\ -\frac{\tau}{\gamma}g_{k1} & \cdots & -\frac{\tau}{\gamma}g_{k,k-1} & -\frac{\tau}{\gamma}g_{k,k} & -\frac{\tau}{\gamma}g_{k,k+1} & \cdots & -\frac{\tau}{\gamma}g_{kN} \\ \vdots & \ddots & \vdots & \vdots & \vdots & \ddots & \vdots \\ 0 & \cdots & 0 & -\frac{\tau}{\gamma}g_{kN} & 0 & \cdots & 0 \end{pmatrix}. \quad (3.17)$$

Multiplying M_k with the transpose of w from the left, i.e. $w^T = (w_1, w_2, \dots, w_N)$, and with w from the right yields

$$w^T M_k w = \sum_{i,j=1}^{N} w_i w_j \frac{\partial^2 f_k}{\partial Y_i \partial Y_j} = -2\frac{\tau}{\gamma} w_k \sum_{i=1}^{N} g_{ki} w_i.$$

Hence, plugging this into the expression for a in (3.14) gives

$$a = -2\frac{\tau}{\gamma} v_k w_k \sum_{i=1}^{N} g_{ki} w_i,$$

which due to w's positivity leads to $a < 0$.

Differentiating $\left(\frac{\partial f_k}{\partial Y_1}, \frac{\partial f_k}{\partial Y_2}, \dots, \frac{\partial f_k}{\partial Y_N} \right)$ with respect to $\frac{\tau}{\gamma}$ gives $(g_{k1}, g_{k2}, \dots, 0, \dots, g_{kN})$ with zero in the kth position. Since this is the kth row of G, it follows that

$$\left(\frac{\partial^2 f_k}{\partial Y_i \partial \left(\frac{\tau}{\gamma} \right)} \right)_{k,i=1,2,\dots,N} = G.$$

Hence, b in (3.14) can be written as

$$b = \langle v, Gw \rangle = \langle v, \Lambda_{\max} w \rangle = \Lambda_{\max} |v|^2 > 0,$$

since $w = v^T$, and where $\langle \cdot, \cdot \rangle$ is the dot product and $|v| = \sqrt{v_1^2 + v_2^2 + \cdots + v_N^2}$. Our result follows from Theorem 3.7. $\square$

Endemic steady state

The result above gives limited information about the levels of infection that nodes achieve at the endemic steady state. To address this, a different approach is needed.

We define $\text{diag}(y)$ to be the diagonal matrix whose diagonal is y and set $\alpha = \tau/\gamma$. Then, Eq. (3.9) becomes a vector–matrix equation

$$\frac{1}{\alpha} y = Gy - \text{diag}(y)Gy,$$

We assume $\alpha = \alpha_c + \varepsilon$ and write $y = \varepsilon a x + \varepsilon^2 b z + \mathcal{O}(\varepsilon^3)$, where the unknowns x and z are unit vectors and a and b are scalars. Note that $1/(\alpha_c + \varepsilon) = 1/\alpha_c - \varepsilon/\alpha_c^2$ to order ε. At order ε and ε^2, we have

$$\mathcal{O}(\varepsilon): \qquad \frac{a}{\alpha_c} x = aGx,$$

$$\mathcal{O}(\varepsilon^2): \qquad \frac{b}{\alpha_c} z - \frac{a}{\alpha_c^2} x = bGz - a^2 \text{diag}(x)Gx.$$

At order ε, we see that x is an eigenvalue of G and $1/\alpha_c$ is the eigenvalue. From the Perron–Frobenius theorem [108], we can conclude that the only eigenvector of G whose entries are all non-negative is the eigenvector of the largest (positive) eigenvalue. Thus, we conclude $1/\alpha_c = \Lambda_{\max}(G)$. Unfortunately, this gives us no information about a, and so while we know the relative magnitudes of the leading order entries for y, we do not yet know their actual magnitudes.

To find a, we first define w to be the unit eigenvector of G^T corresponding to $1/\alpha_c$. We multiply the order ε^2 equation from the left by w^T. Because $w^T/\alpha_c = w^T G$, the terms with z cancel. We are left with

$$-\frac{a}{\alpha_c^2}(w \cdot x) = -a^2 \left(w^T \text{diag}(x)Gx \right).$$

We replace Gx by x/α_c and solve for a in terms of eigenvectors of G:

$$a = \frac{w \cdot x}{\alpha_c \left(w^T \operatorname{diag}(x)x \right)}.$$

As w, x and α_c are all known, this determines a. Note that if G is symmetric (the graph is undirected), then $w = x$ and $w \cdot x = 1$. Then,

$$a = \frac{1}{\alpha_c \sum x_i^3}.$$

Special case of the individual-based SIS model

It is worth noting that for homogenous or regular networks and for particular initial conditions, system (3.7) can be considerably reduced, from N equations to one. More precisely, the following result holds.

Proposition 3.1 *Consider the initial value problem (IVP) given by system (3.7) defined on a regular network where each node has n links. Furthermore, assume that the initial conditions are $\langle Y_i \rangle (0) = i_0$ with $0 \le i_0 \le 1$ for $i = 1, 2, \ldots, N$. The solution of this IVP is given by $\langle Y_i \rangle = \langle Y \rangle$ for all i, with $\langle Y \rangle$ being the solution of*

$$\langle \dot{Y} \rangle = \tau n (1 - \langle Y \rangle)\langle Y \rangle - \gamma \langle Y \rangle, \tag{3.18}$$

with the initial condition $\langle Y \rangle (0) = i_0$.

Proof. It is enough to check that the solution of system (3.18) is a solution of system (3.7), that is setting $\langle Y_i \rangle = \langle Y \rangle$ for all i. If this is the case, then based on the existence and uniqueness of the solutions of IVPs the result follows. Hence, let us plug $\langle Y_i \rangle = \langle Y \rangle$ into system (3.7). This yields

$$\langle \dot{Y_i} \rangle = \langle \dot{Y} \rangle = \tau n (1 - \langle Y \rangle)\langle Y \rangle - \gamma \langle Y \rangle = \tau \sum_{j=1}^{N} g_{ij}(1 - \langle Y_i \rangle)\langle Y_j \rangle - \gamma_i \langle Y_i \rangle, \tag{3.19}$$

where we used the facts that $\langle Y \rangle$ is a solution of system (3.18) and that each node has exactly n neighbours. We note that we had to require that all nodes have the same recovery rate. $\square$

Exercise 3.6. Consider the individual-based model in the case of a single edge:

$$\langle \dot{Y_1} \rangle = \tau g_{12}(1 - Y_1)Y_2 - \gamma Y_1, \tag{3.20}$$

$$\langle \dot{Y_2} \rangle = \tau g_{21}(1 - Y_2)Y_1 - \gamma Y_2, \tag{3.21}$$

where one can assume that $g_{11} = g_{22} = 0$ and that $g_{12} = g_{21} = 1$.

a. Rescale time by using $s = \gamma t$ and show that the equations can be rewritten to give

$$\langle \dot{Y_1} \rangle = \frac{\tau}{\gamma} g_{12}(1 - Y_1)Y_2 - Y_1, \tag{3.22}$$

$$\langle \dot{Y_2} \rangle = \frac{\tau}{\gamma} g_{21}(1 - Y_2)Y_1 - Y_2. \tag{3.23}$$

b. Find the Jacobian around $(Y_1, Y_2) = (0,0)$ and show that the eigenvalues are $\lambda_1 = -1 + \frac{\tau}{\gamma}$ and $\lambda_2 = -1 - \frac{\tau}{\gamma}$.

c. Re-evaluate the Jacobian at $\lambda_1 = 0 \Leftrightarrow \frac{\tau}{\gamma} = 1$ and show that it has a zero eigenvalue with a right eigenvector equal to $w = (1,1)^T$ and a left eigenvector $v = (1,1)$.

d. Based on Theorem 3.8, show that $a = -4$ and $b = 2$.

3.4.5 Relation between the exact, individual-based and classic compartmental SIS model

For a fully connected network, where the initial condition is identical over all nodes, Proposition 3.1 implies that Eq. (3.7) reduces to

$$\langle \dot{Y} \rangle = \tau(N-1)\langle Y \rangle(1 - \langle Y \rangle) - \gamma \langle Y \rangle,$$

where $\langle Y \rangle(t) = \langle Y_i \rangle(t)$ for all $i = 1, 2, \ldots, N$ and the $(N-1)$ factor accounts for the existence of all-to-all connections. Hence, the prevalence in the whole population is given by $N\langle Y \rangle$, and this obeys the ODE

$$N\langle \dot{Y} \rangle = \tau N\langle Y \rangle(N - N\langle Y \rangle)\frac{N-1}{N} - \gamma N\langle Y \rangle$$

$$\leq \tau N\langle Y \rangle(N - N\langle Y \rangle) - N\gamma \langle Y \rangle, \tag{3.24}$$

where the last term is equivalent to the equation of the prevalence in the classic compartmental setting given by the equation

$$\dot{\tilde{I}}(t) = \tau \tilde{I}(N - \tilde{I}) - \gamma \tilde{I}.$$

However, Lemma 3.1 allows us to conclude that

$$N\langle Y\rangle(t) \leq \tilde{I}(t) \text{ for } t \geq 0,$$

which means that the classic mean-field model bounds the individual-based model from above.

Now, starting from the forward Kolmogorov equations on a fully connected graph, as defined in Proposition 2.3, the following inequality for the expected prevalence holds,

$$\begin{aligned}
\frac{dE[I(t)]}{dt} &= (\tau N - \gamma)E[I(t)] - \tau E[I^2(t)] \\
&= \tau E[I(t)(N - I(t))] - \gamma E[I(t)] \\
&= \tau E[I(t)]E[N - I(t)] - \gamma E[I(t)] + \tau \text{Cov}(I(t), N - I(t)) \\
&\leq \tau E[I(t)]E[N - I(t)] - \gamma E[I(t)],
\end{aligned}$$

where we used that in this case $\text{Cov}(I(t), N - I(t)) \leq 0$ [60]. Now due to the $(N-1)/N$ factor in Eq. (3.24), the relation between $E[I(t)]$ and $N\langle Y\rangle$ is not evident. However, for large values of N indeed $E[I(t)] \leq N\langle Y\rangle(t)$ for $t \geq 0$. Thus, consolidating these results yields

$$E[I(t)] \leq N\langle Y\rangle \leq \tilde{I}(t) \text{ for } t \geq 0. \tag{3.25}$$

It is worth noting that a numerical test of the above, via solving the corresponding ODE systems, is non-trivial due to initial conditions. For the classic compartmental and individual-based models, the fraction of the population and the probability of a node being infected can easily be made equivalent. However, setting initial conditions in the exact Kolmogorov equations requires more care. For example, starting with a proportion p of the population being infected, i.e. pN individuals, translates to each node being infected with probability p, but this requires that the exact system is started in a configuration where the probability of having k infectious nodes at $t = 0$ in a network of size N should be given by $\binom{N}{k}p^k(1 - p)^{N-k}$. Hence, the equivalence in initial conditions is made via the mean of the binomial distribution with N trials and probability p, chosen so that pN is an integer. Finally, we note that since the exact model converges to the mean-field model [29] as N increases, Eq. (3.25) shows that all three models become equivalent as $N \to \infty$.

Exercise 3.7. Based on Proposition 2.3, the Kolmogorov equations for a fully connected network are

$$\dot{x}^k(t) = \tau(N - k + 1)(k - 1)x^{k-1}(t) - (\tau(N - k)(k) + \gamma k)x^{k-1}(t) + \gamma(k + 1)x^{k+1}(t),$$

where $k = 0, 1, \ldots, N$, and with obvious adjustments for $k = 0$ and $k = N$. Show that $E[I(t)] = \sum k x^k$ satisfies

$$\frac{dE[I(t)]}{dt} = (\tau N - \gamma)E[I(t)] - \tau E[I^2(t)].$$

3.4.6 Closures at the level of triples for SIS dynamics and numerical examples

As shown in the introduction, closures can be applied at triple level, and for this purpose one can use closures given by Eqs. (3.3) & (3.4). Compared to the individual-based model, this increases the number of equations significantly and will make mathematical analysis extremely challenging, if not impossible. However, as will be shown in Figs. 3.3 & 3.7, the resulting pair-based model leads to better agreement with simulation results for all networks considered. Given a network, either theoretical/synthetic or reconstructed based on real data, the exact system (3.5) can be closed at the level of triples, using Eqs. (3.3) & (3.4) for open or closed triples. These closures require adding an equation for $\langle X_i X_j \rangle$. The resulting system can be solved numerically to obtain an approximation to the true stochastic SIS epidemic model. An implemented version of this can be found as supplementary material in [129] and at the webpage given in the Preface. The equations to be numerically evaluated are given below.

SIS pair-based model

$$\langle \dot{Y}_i \rangle = \tau \left(\sum_{j=1}^{N} g_{ij} \langle X_i Y_j \rangle \right) - \gamma_i \langle Y_i \rangle, \tag{3.26a}$$

$$\langle \dot{X_i Y_j} \rangle = \tau \sum_{k=1,k \neq i}^{N} g_{jk} \frac{\langle X_i X_j \rangle \langle X_j Y_k \rangle}{\langle X_j \rangle} - \tau \sum_{k=1,k \neq j}^{N} g_{ik} \frac{\langle Y_k X_i \rangle \langle X_i Y_j \rangle}{\langle X_i \rangle}$$
$$- \tau g_{ij} \langle X_i Y_j \rangle - \gamma_j \langle X_i Y_j \rangle + \gamma_i \langle Y_i Y_j \rangle, \tag{3.26b}$$

$$\langle \dot{X_i X_j} \rangle = -\tau \sum_{k=1,k \neq i}^{N} g_{jk} \frac{\langle X_i X_j \rangle \langle X_j Y_k \rangle}{\langle X_j \rangle} - \tau \sum_{k=1,k \neq j}^{N} g_{ik} \frac{\langle Y_k X_i \rangle \langle X_i X_j \rangle}{\langle X_i \rangle}$$
$$+ \gamma_i \langle Y_i X_j \rangle + \gamma_j \langle X_i Y_j \rangle, \tag{3.26c}$$

with the additional relations

$$\langle X_i \rangle = 1 - \langle Y_i \rangle, \quad \langle Y_i Y_j \rangle = 1 - \langle X_i X_j \rangle - \langle X_i Y_j \rangle - \langle X_j Y_i \rangle, \quad \langle Y_i X_j \rangle = \langle X_j Y_i \rangle. \tag{3.26d}$$

If $\langle X_j \rangle = 0$, we replace closures of the form $\langle A_i X_j \rangle \langle X_j Y_k \rangle / \langle X_j \rangle$ with 0.

This closed model requires $N + 3|E|$ differential equations, where $|E|$ is the number of edges in the network. N equations are needed at node level, since $\langle X_i \rangle = 1 - \langle Y_i \rangle$. $\langle X_i X_j \rangle$ pairs only require $|E|$ equations. However, $\langle X_i Y_j \rangle$ pairs require $2|E|$ equations since each pair needs to be considered as directed. We can solve for $\langle Y_i Y_j \rangle$ in terms of the other variables. Finally, pairs such $\langle Y_k X_i \rangle$ in closures are identical to $\langle X_i Y_k \rangle$ and no further equations are required. If i, j and k form a triangle, closures such as $\frac{\langle X_i X_j \rangle \langle X_j Y_k \rangle}{\langle X_j \rangle}$ are replaced by $\frac{\langle X_i X_j \rangle \langle X_j Y_k \rangle \langle X_i Y_k \rangle}{\langle X_i \rangle \langle X_j \rangle \langle Y_k \rangle}$.

In this section, we provide extensive numerical examples of the performance of the individual- and pair-based models given by systems (3.7) and (3.26) for the SIS and SIR dynamics, respectively. This is done by comparing output from the closed systems to results from exact lumped systems or agent-based stochastic simulation. For networks with a high degree of structural symmetry, such as fully connected and star, we compare directly to the exact systems. For all other cases, such as 1d and 2d lattices, regular random, Erdős–Rényi random, bimodal and scale-free-like networks, we use simulation. We wish to highlight some basic principles that need to be taken into account when making such comparisons, and we draw some conclusions about the performance of the closed models.

It is well known and widely accepted that comparison of stochastic and deterministic models is challenging, due to the very different underlying mathematical principles. In particular, an ODE model will always show an epidemic outbreak above the critical epidemic threshold, while a stochastic model can still become extinct or, at best, it will be able to maintain a quasi-stationary state for a considerable time before reaching the absorbing state [230, 232]. Below the threshold, a similar principle exists in that stochastic models can still display small-scale growth even if the ODE will always yield a decreasing prevalence level. The closer to the threshold, the more significant these differences are. Hence, for our comparisons, we will focus mainly on above-threshold regimes, with a few exceptions. Of course, individual realisations can still die out due to stochasticity or can be much faster/slower compared to the general trend of other realisations. Therefore, the choice of parameters is such that the need to time-shift epidemics [151, 222] or remove non-epidemic realisations is eliminated or reduced to a minimum. Hence, guided by theoretical results and normal constraints when comparing stochastic to deterministic models, we consider scenarios where good agreement between models is likely.

Furthermore, we consider the impact of initial conditions when dealing with individual-based models and specifying the probability of nodes being susceptible or otherwise at time $t = 0$. To observe this requirement, we typically use a single realisation of the network and each simulation is started with the same initially infected nodes. This is in line with the ingredients needed for closed models, namely a concrete, single network and a given number of infectious nodes labelled precisely. Otherwise, simulating epidemics on different network realisations of the same network-generating algorithm or the same number of initially infected nodes but in different places on the network would possibly require to average over the solution of different ODEs, which correspond to different networks or initial conditions.

ODE models tend to perform better if the transmission rate is either very large or very small compared to the recovery rate. For small levels of transmission, disease spread across an edge rarely happens. Thus, an edge that transmits is unlikely to transmit again for a long time, increasing the independence of common neighbours of a node. So the network structure will play a lesser role and the agreement will be good since approximating many independent Poisson process is straightforward. At the other extreme, where transmission is high compared to recovery, most nodes become infected. When one node recovers, most of its neighbours are infected, and so its rate of reinfection is approximately τ times its degree, irrespective of the local topology, decreasing the dependence on the specific local structure. Thus, most of our examples will concentrate on above-threshold epidemics but still resulting in a mild-to-medium overall epidemic and thus being away from the very high transmissibility regime.

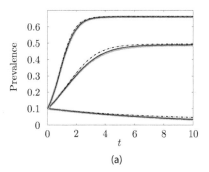

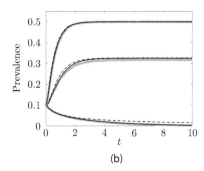

(a)

(b)

Fig. 3.2: Expected prevalence from the SIS model is plotted to compare the exact lumped systems (thick grey lines) for (a) fully connected (see Proposition 2.3) and (b) star network (see Proposition 2.4) with the corresponding pair-based (solid black lines) and individual-based (dash-dotted black lines) models. Parameter values are: $N = 100$, $\gamma = 1$ and from top to bottom in both figures (a) $\tau = 0.03$, $\tau = 0.02$ and $\tau = 0.01$, and (b) $\tau = 1$, $\tau = 0.5$ and $\tau = 0.1$. In both cases, epidemics started with ten randomly chosen nodes, excluding the central node in case of the star.

While the system closed at the level of pairs is simple to write down and implement for both SIS and SIR dynamics, the system closed at the level of triples is more complex, as apart from equations for nodes, equations for edges are also needed. Even though conceptually and mathematically it is straightforward to write down general equations, computationally it is more challenging to correctly track all edges in the necessary states. Sharkey [280] and Hadjichrysanthou & Sharkey [129] provide *Matlab* code for systems with closures at the level of triples for SIR and SIS, respectively and at the webpage given in the Preface.

SIS on fully connected and star networks

In Fig. 3.2, we begin by comparing the exact model, given by ODE systems, to the individual- and pair-based models. As shown earlier in this chapter, the numerical result confirms that the individual-based model provides an overestimate of the overall level of infection in the network. The error is mild for both fully connected (see Fig. 3.2a) and star (see Fig. 3.2b) networks and, as explained earlier, the overestimation is milder for high transmissibility. The numerical result shows clearly that the pair-based model is superior to the individual-based model as it leads to better agreement with the exact system.

SIS on other networks

In Fig. 3.3, we consider a range of different network topologies. While the line and lattice network structures are predetermined, random regular and scale-free-like networks are created according to the Configuration Model [225] (see also Chapter 1).

Our aim here is to give a flavour, and establish general principles as to how the approximate or closed systems perform, rather than providing an exhaustive numerical investigation. It is again clear that closures at the level of triples outperform closures at the level of pairs, i.e. the pair-based model results in a more accurate approximation of the true or simulated stochastic model. Both models, however, overestimate the prevalence compared to simulation results. The pair-based model seems to always lead to a model that approximates the quasi-stationary state well, even if it overestimates the average prevalence in the growing stage (see Figs. 3.3a & 3.3b). However, the individual-based model can grossly over predict the quasi-stationary or endemic equilibrium (see Figs. 3.3a & 3.3d).

The triple closure for SIS, while not exact, yields excellent agreement in many situations even when transmissibility is not high. Apart from the 1d and 2d lattices, the performance of the pair-based model is excellent and it approximates the time evolution of the simulated prevalence well.

Special case of the SIS model closed at the triple level

It is worth noting that for homogenous or regular networks and for particular initial conditions, system (3.26) can be considerably reduced. More precisely, the following result holds.

Proposition 3.2 *Consider the initial value problem (IVP) given by system (3.26) defined on a regular network where each node has n links. Furthermore, assume that the initial conditions for singles are $\langle Y_i \rangle(0) = i_0$, with $0 \leq i_0 \leq 1$ and $\langle X_i \rangle(0) = 1 - \langle Y_i \rangle(0) = 1 - i_0 = s_0$ for $i = 1, 2, \ldots, N$. For pairs where $g_{ij} \neq 0$, we set $\langle X_i Y_j \rangle(0) = s_0 i_0$, $\langle Y_i Y_j \rangle(0) = i_0^2$ and $\langle X_i X_j \rangle(0) = s_0^2$ for $i, j = 1, 2, \ldots, N$. The solution of this IVP is given by $\langle Y_i \rangle = \langle Y \rangle$ for all i and $(\langle X_i Y_j \rangle, \langle Y_i Y_j \rangle, \langle X_i X_j \rangle) = (\langle XY \rangle, \langle YY \rangle, \langle XX \rangle)$ for $i, j = 1, 2, \ldots, N$ if $g_{ij} \neq 0$, with $(\langle Y \rangle, \langle XY \rangle, \langle YY \rangle, \langle XX \rangle)$ being the solution of*

$$\langle \dot{Y} \rangle = +\tau n \langle XY \rangle - \gamma \langle Y \rangle, \quad (with \ \langle X \rangle = 1 - \langle Y \rangle), \tag{3.27a}$$

$$\langle \dot{XY} \rangle = \tau(n-1) \frac{\langle XX \rangle \langle XY \rangle}{\langle X \rangle} - \tau(n-1) \frac{\langle YX \rangle \langle XY \rangle}{\langle X \rangle}$$
$$- (\tau + \gamma) \langle XY \rangle + \gamma \langle YY \rangle, \tag{3.27b}$$

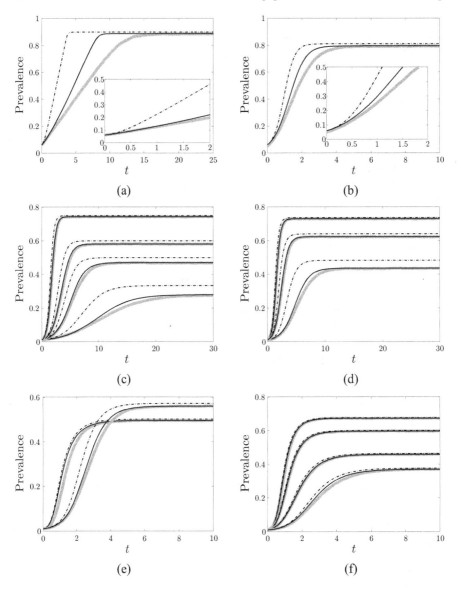

Fig. 3.3: Expected prevalence from the SIS model with simulation (thick grey lines), pair-based (solid black lines) and individual-based (dash-dotted black lines) plotted for the following networks: (a) line with $\tau = 5$; (b) 2d square lattice with $\tau = 1.5$; (c) regular random with $\langle k \rangle = 10$, $\tau = 0.15, 0.20, 0.25, 0.4$ from bottom to top; (d) Erdős–Rényi with $\tau = 0.4$, $\langle k \rangle = 5, 7.5, 10$ from bottom to top (e) bimodal with $\tau = 0.25$ and (i) $N_1 = 800$, $k_1 = 4$, $N_2 = 200$, $k_2 = 34$ (rising faster) and (ii) $N_1 = N_2 = 500$, $k_1 = 5$, $k_2 = 15$, with both having $\langle k \rangle = 10$; and (f) scale-free-like ($P(k) = c/k^2$, with $k = 7, 8, \ldots, 109$) with $\tau = 0.08, 0.1, 0.15, 0.2$ from bottom to top. For networks (a) and (b), $N = 100$, while for (c) to (f), $N = 1000$. For all cases, $\gamma = 1$, and 1000 simulations were run on a single network realisation, starting with the same 6 centrally infected and 10 randomly infected nodes for networks (a, b) and (c–f), respectively.

$$\langle \dot{YY} \rangle = \tau(n-1)\frac{\langle YX \rangle \langle XY \rangle}{\langle X \rangle} + \tau(n-1)\frac{\langle YX \rangle \langle XY \rangle}{\langle X \rangle}$$
$$- 2\gamma \langle YY \rangle + \tau(\langle XY \rangle + \langle YX \rangle), \tag{3.27c}$$

$$\langle \dot{XX} \rangle = -\tau(n-1)\frac{\langle YX \rangle \langle XX \rangle}{\langle X \rangle} - \tau(n-1)\frac{\langle XX \rangle \langle XY \rangle}{\langle X \rangle}$$
$$+ \gamma(\langle YX \rangle + \langle XY \rangle), \tag{3.27d}$$

with the initial condition

$$(\langle Y \rangle, \langle XY \rangle, \langle YY \rangle, \langle XX \rangle)(0) = (i_0, s_0 i_0, i_0^2, s_0^2).$$

Proof. We begin by emphasising that the main implication of the initial conditions is that all edges in the network are equivalent at time $t = 0$. This, coupled with the regularity of the network, makes the evolution equations for the nodes identical. This provides an intuitive explanation of why the result holds. The proof is similar to that of Proposition 3.1. It is again sufficient to check that the solution of system (3.27) is a solution of system (3.26). If this is the case, then based on the existence and uniqueness of the solutions of IVPs, the result follows. Hence, we plug

$$(\langle Y_i \rangle, \langle X_i X_j \rangle, \langle X_i Y_j \rangle, \langle Y_i X_j \rangle, \langle Y_i Y_j \rangle) = (\langle Y \rangle, \langle XX \rangle, \langle XY \rangle, \langle YX \rangle, \langle YY \rangle)$$

into system (3.26). At node level, this leads to

$$\langle \dot{Y_i} \rangle = \langle \dot{Y} \rangle = \tau n \langle XY \rangle - \gamma \langle Y \rangle = \tau \sum_{j=1}^{N} g_{ij} \langle X_i Y_j \rangle - \gamma \langle Y_i \rangle, \tag{3.28}$$

since we set all $\langle X_i Y_i \rangle$s to $\langle XY \rangle$ and used that $(\langle XY \rangle, \langle Y \rangle)$ is a solution of system (3.27). Due to the network being regular, the sum could be written as $n \times \langle XY \rangle$. We note that we had to require that the recovery rate is the same for all nodes. We now focus on the equations for pairs and consider $\langle X_i Y_j \rangle$. This leads to

$$\langle \dot{X_i Y_j} \rangle = \langle \dot{XY} \rangle = \tau(n-1)\frac{\langle XX \rangle \langle XY \rangle}{\langle X \rangle} - \tau(n-1)\frac{\langle YX \rangle \langle XY \rangle}{\langle X \rangle} - (\tau+\gamma)\langle XY \rangle + \gamma \langle YY \rangle$$
$$= \tau \sum_{k=1,k\neq i}^{N} g_{jk}\frac{\langle X_i X_j \rangle \langle X_j Y_k \rangle}{\langle X_j \rangle} - \tau \sum_{k=1,k\neq j}^{N} g_{ik}\frac{\langle Y_k X_i \rangle \langle X_i Y_j \rangle}{\langle X_i \rangle}$$
$$- \tau \langle X_i Y_j \rangle - \gamma(\langle X_i Y_j \rangle - \langle Y_i Y_j \rangle),$$

where we have used that both nodes i and j, at the end of a pair, have exactly $(n-1)$ leftover links where infection can either generate or destroy certain pair combinations. The same arguments can be used to derive all remaining equations. □

An important consequence of this result is that the reduced system (3.27) can be rescaled to obtain the pairwise model discussed in Chapter 4. This can be done by introducing some new population-level quantities as follows: $[A] = N\langle A \rangle$ and $[AB] = nN\langle AB \rangle$, with $A, B \in \{S, I\}$, where N is the total number of nodes and nN is the total number of links (counted twice). Under this change of variables, system (3.27) is equivalent to the pairwise model

$$[\dot{S}] = \gamma[I] - \tau[SI], \tag{3.29a}$$

$$[\dot{SI}] = \gamma([II] - [SI]) + \tau\left(\zeta\frac{[SS][SI]}{[S]} - \zeta\frac{[IS][SI]}{[S]} - [SI]\right), \tag{3.29b}$$

$$[\dot{SS}] = 2\gamma[SI] - 2\tau\zeta\frac{[IS][SS]}{[S]}, \tag{3.29c}$$

$$[\dot{II}] = 2\tau[SI] - 2\gamma[II] + 2\tau\zeta\frac{[IS][SI]}{[S]}, \tag{3.29d}$$

where $\zeta = \frac{n-1}{n}$, and we used that $[SI] = [IS]$.

3.5 Closures at the level of pairs for SIR dynamics

We now consider SIR disease spreading in networks by writing down equations at the level of nodes. This requires closures at the level of pairs and thus leads to a fairly simple system where the final epidemic size can be given analytically as a system of implicit equations.

3.5.1 The individual-based SIR model

We begin by introducing closure at the level of pairs. This yields the reduced system given below.

SIR individual-based model

$$\langle \dot{X}_i \rangle = -\tau \sum_{j=1}^{N} g_{ij} \langle X_i \rangle \langle Y_j \rangle, \tag{3.30a}$$

$$\langle \dot{Y}_i \rangle = \tau \left(\sum_{j=1}^{N} g_{ij} \langle X_i \rangle \langle Y_j \rangle \right) - \gamma_i \langle Y_i \rangle, \tag{3.30b}$$

where the ODEs can be implemented and numerically evaluated with code for this purpose found in the supplementary material of [280]. This closed model requires $2N$ equations.

This and similar models have been used extensively [279, 280], and their relative simplicity makes the derivation of analytical or semi-analytical results possible. However, as with SIS disease, such models are only approximate and tend to perform well only when networks are relatively well connected or when the infection rate is high with respect to the recovery rate, or *vice-versa*. It is worth noting that the linearisation of this system around the trivial disease-free steady state leads to a $2N \times 2N$ matrix, which is composed of four $N \times N$ block matrices. Assuming an ordering of $\langle X \rangle$s followed by $\langle Y \rangle$s shows that the top and bottom-left blocks evaluate to zero, while the bottom-right block is identical to that obtained when using the individual-based SIS model. Hence, the threshold in the two individual-based models is identical [255]. This demonstrates the error in the approximations leading to this model. As we will see in Chapters 4 and 5, the thresholds from pairwise SIS and SIR models are different and in line with intuition, whereby, all being equal, SIS epidemics spread more readily (see also Section 3.5.3).

> **Exercise 3.8.** Using SIR dynamics and the network consisting of one single edge:
> **a**. Write down the bottom-up exact and the corresponding reduced system with closures at the level of pairs.
> **b**. Solve these equations numerically, using a programming language of your choice, and establish whether the reduced/closed model over or underestimates the probability of nodes being infected.

> **Exercise 3.9.** Using SIR dynamics and the line network with three nodes, Fig. 2.2a:
> **a**. Write down the bottom-up exact and the corresponding reduced systems with closures at the level of pairs and triples.
> **b**. Solve these equations numerically, using a programming language of your choice, and establish whether the reduced/closed model over or underestimates the probability of nodes being infected.

> **Exercise 3.10.** Formulate and prove the equivalent of Proposition 3.1 but for the individual-based SIR model.

3.5.2 The final epidemic size based on the individual-based SIR model

It is possible to derive an implicit equation for the final epidemic size directly from the individual-based model (3.30). To simplify the notation, let $\langle A_i \rangle = a_i$, where $A \in \{X, Y, Z\}$, with $\langle Z_i \rangle$ denoting the probability that node i is recovered. Obviously, $(\langle X_i \rangle + \langle Y_i \rangle + \langle Z_i \rangle)(t) = 1$ for $t \geq 0$. The initial conditions are such that $z_i(0) = 0$ and $x_i(0) + y_i(0) = 1$. The equation for x_i can be rearranged to give

$$\dot{x}_i + (\tau \sum_j g_{ij} y_j) x_i = 0,$$

which can be rewritten to yield

$$\frac{d}{dt} \left(x_i e^{\int_0^t \tau \sum_j g_{ij} y_j(\hat{t}) d\hat{t}} \right) = 0.$$

This can be solved to obtain

$$x_i = x_i(0)e^{-\int_0^t \tau \sum_j g_{ij} y_j(\hat{t})d\hat{t}}.$$

Noting that $\dot{z}_i = \gamma y_i$ leads to $z_i(t) = \gamma \int_0^t y_i(\hat{t})d\hat{t}$ or $\int_0^t y_i(\hat{t})d\hat{t} = z_i/\gamma$. Now using this, the equation for x_i is equivalent to

$$x_i = x_i(0)e^{-\tau \sum_j g_{ij} \int_0^t y_j(\hat{t})d\hat{t}} = x_i(0)e^{-\tau \sum_j g_{ij} z_j(t)/\gamma}.$$

Letting $t \to \infty$, and using that $y_i(\infty) = 0$, implies that $x_i \to 1 - z_i$, and this leads to

$$1 - z_i(\infty) = x_i(0)e^{-(\tau/\gamma)\sum_j g_{ij} z_j(\infty)}. \tag{3.31}$$

Rearranging the above gives the node-level implicit final epidemic size formula

$$z_i(\infty) = 1 - x_i(0)e^{-(\tau/\gamma)\sum_j g_{ij} z_j(\infty)}, \tag{3.32}$$

where $i = 1, 2, \ldots, N$. This implicit formula can be regarded as a map from $\mathbb{R}^N$ to $\mathbb{R}^N$ whose fixed point will yield the final epidemic size, namely $\sum_{i=1}^N z_i(\infty)$. All the above is subject to appropriate parameter values that exclude the trivial zero solution when $x_i(0) = 1$.

3.5.3 SIS: an upper bound on SIR

It is worthwhile to compare SIS and SIR dynamics in the case where both processes run on the same network and have the same initial conditions. We start from the exact equations of the node-level infection. For SIS, and following [313], we rewrite the equations for $\langle \dot{I}_i \rangle$ as follows:

$$\langle \dot{I}_i \rangle = \tau \sum_j g_{ij} \langle S_i I_j \rangle - \gamma \langle I_i \rangle$$

$$= \tau \sum_j g_{ij} \langle I_j \rangle - \gamma \langle I_i \rangle - \tau \sum_j g_{ij} \langle I_i I_j \rangle, \tag{3.33}$$

where we used the fact that $\langle I_j \rangle = \langle S_i I_j \rangle + \langle I_i I_j \rangle$. Similarly for SIR dynamics, one can use $\langle I_j \rangle = \langle S_i I_j \rangle + \langle I_i I_j \rangle + \langle R_i I_j \rangle$ in order to write

$$\langle \dot{I}_i \rangle = \tau \sum_j g_{ij} \langle S_i I_j \rangle - \gamma \langle I_i \rangle$$

$$= \tau \sum_j g_{ij} \langle I_j \rangle - \gamma \langle I_i \rangle - \tau \sum_j g_{ij} (\langle I_i I_j \rangle + \langle R_i I_j \rangle), \tag{3.34}$$

noting that $\langle \cdot \rangle$ should in fact be replaced by $\langle \cdot \rangle_{SIS}$ and $\langle \cdot \rangle_{SIR}$ for SIS and SIR, respectively. Due to the positivity of $\tau \sum_j g_{ij} \langle R_i I_j \rangle$, one is tempted to conclude that $\langle \dot{I}_i \rangle_{SIR} \le \langle \dot{I}_i \rangle_{SIS}$. Furthermore, this relation between the first derivatives suggests that the probability of any node being infected at a given time is higher for SIS than for SIR dynamics. However, this argument alone is not sufficient to justify the claim (see the counterexample in Observation 3a of Section 3.3). While numerically confirmed and widely accepted, this result needs a more rigorous mathematical justification and, to our knowledge, this remains an open question.

3.6 General closures at higher levels for SIR dynamics for networks with and without loops

Intuition and numerical tests suggest that closures at the level of triples are exact for SIR dynamics on tree networks. Hence, the bottom-up model leads to a numerically computable exact representation with a large but tractable number of non-linear differential equations. In fact, if the epidemic starts with a single infectious node, the proof of this statement is rather simple. Infection initiated at a single node on a tree graph will proceed in a linear fashion. Consequently, triples such as $\langle I_i S_j I_k \rangle$ and $\langle R_i S_j I_k \rangle$ are not seen, as infection must have passed through node j. Furthermore, $\langle S_j I_k \rangle = \langle S_i S_j I_k \rangle + \langle I_i S_j I_k \rangle + \langle R_i S_j I_k \rangle$, and using that $\langle I_i S_j I_k \rangle = \langle R_i S_j I_k \rangle = 0$ yields $\langle S_j I_k \rangle = \langle S_i S_j I_k \rangle$, and hence the closure is

$$\langle S_i S_j I_k \rangle = \langle S_j I_k \rangle.$$

Therefore, technically speaking, the triples can be given in terms of pairs alone or, alternatively, we note that if k has been infected, then $\langle S_j \rangle = \langle S_i S_j \rangle + \langle I_i S_j \rangle + \langle R_i S_j \rangle$, where $\langle I_i S_j \rangle = \langle R_i S_j \rangle = 0$, since infection to node i must pass through node j. This then allows us to write that $\langle S_i S_j I_k \rangle = \langle S_j I_k \rangle = \frac{\langle S_i S_j \rangle \langle S_j I_k \rangle}{\langle S_j \rangle}$. Applying the above, the reduced/closed system for a tree with a single index case is given by

$$\langle \dot{S}_i \rangle = -\tau \sum_{j=1}^{N} g_{ij} \langle S_i I_j \rangle,$$

$$\langle \dot{I}_i \rangle = \tau \sum_{j=1}^{N} g_{ij} \langle S_i I_j \rangle - \gamma_i \langle I_i \rangle,$$

$$\langle S_i I_j \rangle = \tau \sum_{k=1, k \neq i}^{N} g_{jk} \langle S_j I_k \rangle - \tau g_{ij} \langle S_i I_j \rangle - \gamma_j \langle S_i I_j \rangle.$$

If we want an exact model, numerical experiments from small networks suggest that loops, such as closed triples or other types, should not be approximated using the constituent parts. The challenge is that if a node is in a loop, the statuses of its neighbours may have correlations that must be considered. It turns out that closing in a way in which loops remain intact leads to exact systems [181]. Identifying nodes at which these closures are possible allows us to link network properties to the existence of exact closures. In particular, it is important to recall the definition of a cut-vertex in a graph.

Definition 3.9 *Let $G = \{V, E\}$ be a connected network. A node v is called a* cut-vertex *iff $V \setminus \{v\}$ is disconnected.*

For our purposes, we are interested in cut-vertices, i.e. single nodes whose removal leads to disconnected components or subnetworks. The second edge property of interest is the concept of a bridge.

Definition 3.10 *Let $G = \{V, E\}$ be a connected graph. An edge $e \in E(G)$ is called a* bridge *iff its removal increases the number of connected components.*

It follows that an edge is a bridge iff it is not contained in any cycle, and any end node of a bridge with degree greater than 1 is a cut-vertex.

For example, in Fig. 3.4, nodes $\{1\}, \{2\}, \{3\}, \{4\}$, and $\{5\}$ are cut-vertices, and edge $(4, 5)$ is a bridge. The basic idea we will use here is that if a cut vertex u is initially susceptible, then as long as it remains susceptible, information cannot cross from one side of the node to the other. It turns out that this special property of cut-vertices is in fact strongly related to the concept of the "test node", as used in Chapter 6 in order to define the edge-based compartmental model [221, 222]. During the process of generating the full bottom-up model, this observation manifests itself by cut-vertices only appearing in the susceptible state when part of a triple, or subnetworks involving more nodes. For example, in the bottom-up model for SIR dynamics on a 3-node line network, Fig. 3.1a, with the central node or node 2 being a cut-vertex, the only triples needing equations are: $\langle S_1 S_2 I_3 \rangle$, $\langle I_1 S_2 S_3 \rangle$ and $\langle I_1 S_2 I_3 \rangle$. In all of these, node 2 is in status S. Similarly, for the lollipop network in Fig. 2.9a, see also [181], at the level of quadruplets the bottom-up model only requires equations for $\langle I_1 I_2 S_3 I_4 \rangle$, $\langle I_1 S_2 S_3 I_4 \rangle$, $\langle S_1 I_2 S_3 I_4 \rangle$, $\langle I_1 I_2 S_3 S_4 \rangle$, $\langle I_1 S_2 S_3 S_4 \rangle$, $\langle S_1 I_2 S_3 S_4 \rangle$ and $\langle S_1 S_2 S_3 I_4 \rangle$, where node 3, the cut-vertex, is always in status S.

Intuition tells us that there is no need to generate further equations and that even the equations for the combinations of nodes and statuses as shown above are redundant since these can be written in terms of the constituent parts. For example, for the line and lollipop networks, we could write the following exact relations:

$$\langle I_1 S_2 I_3 \rangle = \frac{\langle I_1 S_2 \rangle \langle S_2 I_3 \rangle}{\langle S_2 \rangle} \text{ and } \langle S_1 I_2 S_3 I_4 \rangle = \frac{\langle S_1 I_2 S_3 \rangle \langle S_3 I_4 \rangle}{\langle S_3 \rangle}.$$

It is also helpful to think of the consequences of removing one or more cut-vertices. In the case of the line network with three nodes, upon removal of the central node, or effectively duplicating it, the network decomposes into two edges. Similarly, the lollipop decomposes into a triangle and an edge with no further links between these components. The decomposition process is illustrated in Fig. 3.4b for a toy network. We will make this rigorous and show how to use it in order to obtain simplifications of bottom-up models while keeping them exact.

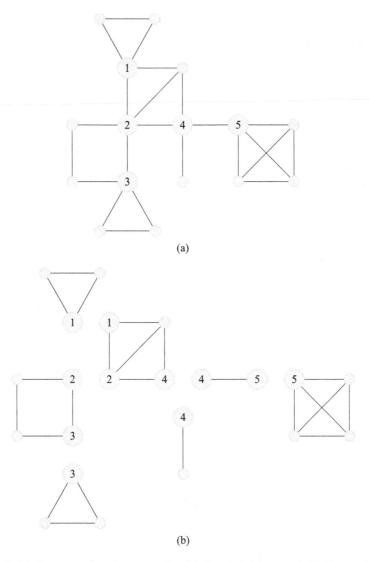

(a)

(b)

Fig. 3.4: (a) An example of a network with 5 cut-vertices and (b) the equivalent network upon decomposition into subnetworks by the removal of cut-vertices. The decomposed network has 7 subgraphs/subnetworks of 5 different types: edge, triangle, cycle of size four, toast (square with one diagonal) and a fully connected square. The five cut-vertices, 1, 2, 3, 4 and 5, belong to $Ind(1) = 2, Ind(2) = 2, Ind(3) = 2, Ind(4) = 3$ and $Ind(5) = 2$ subnetworks, respectively.

3.6.1 The relationship between closures and structural network properties

Based on the intuition gained from the closures on simple networks and their link to the structural properties of the network, via cut-vertices, the result on closures in [283] can be generalised. Moreover, this result formalises the link between closures and the structural properties of the network.

Theorem 3.11 *Let $G = \{V, E\}$ be a network with a set of vertices V and a set of edges E. Consider a connected subset of vertices $F = \{v_1, v_2, \ldots, v_k\} \subseteq V$, and assume that there is a $v_i \in F$, a cut-vertex in G. Provided that every path from a node in $F_1 = \{v_1, v_2, \ldots, v_{i-1}\}$ to a node in $F_2 = \{v_{i+1}, v_{i+2}, \ldots, v_k\}$ goes through v_i, then*

$$\langle Z_{v_1} \cdots Z_{v_{i-1}} S_{v_i} Z_{v_{i+1}} \cdots Z_{v_k} \rangle \langle S_{v_i} \rangle = \langle Z_{v_1} \cdots Z_{v_{i-1}} S_{v_i} \rangle \langle S_{v_i} Z_{v_{i+1}} \cdots Z_{v_k} \rangle, \qquad (3.35)$$

where $Z \in \{S, I, R\}$ for all $v_j \neq v_i$, and $\langle \cdot \rangle$ denotes the probability of a given set of nodes having the given statuses at time t.

We note that when $\langle S_{v_i} \rangle = 0$, both sides of Eq. (3.35) equal zero and the closure holds trivially. Further:

1. F could be the entire vertex set or a strict subset of it, and
2. The sets F_1 and F_2 do not have to be connected components of $G \setminus \{v_i\}$. For example, if removing v_i breaks G into three connected components, F_1 might consist of two of them.

Proof. We assume $\langle S_{v_i} \rangle \neq 0$. By definition of conditional probabilities,

$$\langle Z_{v_1} \cdots Z_{v_{i-1}} S_{v_i} Z_{v_{i+1}} \cdots Z_{v_k} \rangle = \langle Z_{v_1} \cdots Z_{v_{i-1}} S_{v_i} Z_{v_{i+1}} \cdots Z_{v_k} | S_{v_i} \rangle \langle S_{v_i} \rangle. \qquad (3.36)$$

The conditional probability term on the right-hand side can be written as

$$\langle Z_{v_1} \cdots Z_{v_{i-1}} S_{v_i} Z_{v_{i+1}} \cdots Z_{v_k} | S_{v_i} \rangle = \langle Z_{v_1} \cdots Z_{v_{i-1}} S_{v_i} | S_{v_i} \rangle \langle S_{v_i} Z_{v_{i+1}} \cdots Z_{v_k} | S_{v_i} \rangle, \qquad (3.37)$$

because any path between F_1 and F_2 must travel through the cut-vertex v_i. Given that v_i is susceptible, transmission via this route has not occurred, and thus the projection of the system state on the subgraphs spanned by F_1 and F_2 must be independent. Combining Eqs. (3.36–3.37) and using that

$$\langle Z_{v_1} \cdots Z_{v_{i-1}} S_{v_i} | S_{v_i} \rangle = \langle Z_{v_1} \cdots Z_{v_{i-1}} S_{v_i} \rangle / \langle S_{v_i} \rangle,$$
$$\langle S_{v_i} Z_{v_{i+1}} \cdots Z_{v_k} | S_{v_i} \rangle = \langle S_{v_i} Z_{v_{i+1}} \cdots Z_{v_k} \rangle / \langle S_{v_i} \rangle,$$

gives

$$\frac{\langle Z_{v_1} \cdots Z_{v_{i-1}} S_{v_i} Z_{v_{i+1}} \cdots Z_{v_k} \rangle}{\langle S_{v_i} \rangle} = \frac{\langle Z_{v_1} \cdots Z_{v_{i-1}} S_{v_i} \rangle \langle S_{v_i} Z_{v_{i+1}} \cdots Z_{v_k} \rangle}{\langle S_{v_i} \rangle^2}, \qquad (3.38)$$

which is equivalent to the general closure specified in Eq. (3.35). □

Note that if one of the subgraphs has its own cut-vertex, we can repeat this process with that subnetwork. It is straightforward to see that this result automatically implies that we can derive exact closed models for tree graphs (see Subsection 3.6.3), line networks of arbitrary length, lollipop (Fig. 2.9a), general star (Fig. 2.10a) and star triangle (Fig. 3.6b) networks (see Table 3.1 for the number of equations required in the closed exact system). Further generalisations of this result can be found in [281].

3.6.2 Feasibility of generalised closure and examples

In Fig. 3.5, we give a recipe for using Theorem 3.11 to derive a reduced exact system of equations yielding $\langle S_i \rangle$, $\langle I_i \rangle$ and $\langle R_i \rangle$ for every node i in a network.

Example 3.1. Consider the lollipop network in Fig. 2.9a. We follow the steps of the recipe in Fig. 3.5. There is only one cut-vertex, node 3. The network decomposes into a triangle of nodes 1, 2 and 3 and a single edge of nodes 3 and 4. We begin with the equations for individual nodes:

$$\langle \dot{S}_1 \rangle = -\tau \left(\langle S_1 I_2 \rangle + \langle S_1 I_3 \rangle \right), \qquad \langle \dot{S}_3 \rangle = -\tau \left(\langle I_1 S_3 \rangle + \langle I_2 S_3 \rangle + \langle S_3 I_4 \rangle \right),$$

$$\langle \dot{I}_1 \rangle = -\gamma \langle I_1 \rangle + \tau \left(\langle S_1 I_2 \rangle + \langle S_1 I_3 \rangle \right), \qquad \langle \dot{I}_3 \rangle = -\gamma \langle I_3 \rangle + \tau \left(\langle I_1 S_3 \rangle + \langle I_2 S_3 \rangle + \langle S_3 I_4 \rangle \right),$$

$$\langle R_1 \rangle = 1 - \langle S_1 \rangle - \langle I_1 \rangle, \qquad \langle R_3 \rangle = 1 - \langle S_3 \rangle - \langle I_3 \rangle,$$

$$\langle \dot{S}_2 \rangle = -\tau \left(\langle I_1 S_2 \rangle + \langle S_2 I_3 \rangle \right), \qquad \langle \dot{S}_4 \rangle = -\tau \langle I_3 S_4 \rangle,$$

$$\langle \dot{I}_2 \rangle = -\gamma \langle I_2 \rangle + \tau \left(\langle I_1 S_2 \rangle + \langle S_2 I_3 \rangle \right), \qquad \langle \dot{I}_4 \rangle = -\gamma \langle I_4 \rangle + \tau \langle I_3 S_4 \rangle,$$

$$\langle R_2 \rangle = 1 - \langle S_2 \rangle - \langle I_2 \rangle, \qquad \langle R_4 \rangle = 1 - \langle S_4 \rangle - \langle I_4 \rangle.$$

These equations depend on some pairs, namely all arrangements of edges with one susceptible and one infected node. We need additional equations for these. We first look specifically at the equation for $\langle S_1 I_3 \rangle$. An $S_1 I_3$ edge can be created from an $S_1 S_3$ edge if infection reaches node 3 from node 2 or 4, and it is removed either by transmission to node 1 from node 3 or 2 or recovery of node 3. For this edge, we get

$$\langle \dot{S}_1 I_3 \rangle = \tau \left(\langle S_1 I_2 S_3 \rangle + \langle S_1 S_3 I_4 \rangle \right) - (\tau + \gamma) \langle S_1 I_3 \rangle - \tau \langle S_1 I_2 I_3 \rangle,$$

$$= \tau \left(\langle S_1 I_2 S_3 \rangle + \frac{\langle S_1 S_3 \rangle \langle S_3 I_4 \rangle}{\langle S_3 \rangle} \right) - (\tau + \gamma) \langle S_1 I_3 \rangle - \tau \langle S_1 I_2 I_3 \rangle,$$

where we have used Theorem 3.11 and the fact that node 3 is a cut-vertex to replace $\langle S_1 S_3 I_4 \rangle$. We can now write down equations for the remaining pairs involved in the singles equations, replacing any triple that crosses between the two subnetworks. These are listed below:

$$\langle \dot{S}_1 I_2 \rangle = \tau \langle S_1 S_2 I_3 \rangle - (\tau + \gamma) \langle S_1 I_2 \rangle - \tau \langle S_1 I_2 I_3 \rangle,$$

$$\langle \dot{I}_1 S_2 \rangle = \tau \langle S_1 S_2 I_3 \rangle - (\tau + \gamma) \langle I_1 S_2 \rangle - \tau \langle I_1 S_2 I_3 \rangle,$$

Recipe for the generalised closure: Given a network G with vertices V and edges E, our goal is to derive a system of equations that exactly gives $\langle S_i \rangle(t)$, $\langle I_i \rangle(t)$, and $\langle R_i \rangle(t)$ for every node i in the network. The following recipe allows us to find closures that do not require any approximations.

1. Identify the cut-vertices $C = \{v_{i_1}, v_{i_2}, \dots, v_{i_L}\}$. One algorithm to do this is the *depth-first search* algorithm of [278] which runs in polynomial time in $(|E| + |V|)$.
2. Decompose the network into connected subnetworks where the subnetworks overlap only in the cut-vertices. The subnetworks are chosen so that none of them can be disconnected by removal of a vertex, as in Fig. 3.4. Denote the subnetworks $G_1, G_2, \dots, G_P$, where $G_p = \{V_p, E_p\}$ with $p = 1, 2, \dots, P$.
3. For each node i in the network, we can find

$$\langle \dot{S}_i \rangle = -\tau \sum_j g_{ij} \langle S_i I_j \rangle,$$

$$\langle \dot{I}_i \rangle = -\gamma \langle I_i \rangle + \tau \sum_j g_{ij} \langle S_i I_j \rangle,$$

$$\langle R_i \rangle = 1 - \langle S_i \rangle - \langle I_i \rangle.$$

We can find similar equations for the derivatives of all the pairs that arise in these equations. These pairs depend on triples. In turn the triples depend on quadruples. A hierarchy of equations forms.

4. As this hierarchy is built, only write down new variables if they are required by some variables already found. If a term appears corresponding to nodes that do not fit within a single subnetwork, then it will involve a susceptible cut vertex. Use Theorem 3.11 to express this in terms of other variables.

Fig. 3.5: A method to generate an exact system of equations using generalised closures determined by the cut vertices.

$$\langle S_2 \dot{I}_3 \rangle = \tau \left(\langle I_1 S_2 S_3 \rangle + \frac{\langle S_2 S_3 \rangle \langle S_3 I_4 \rangle}{\langle S_3 \rangle} \right) - (\tau + \gamma) \langle S_2 I_3 \rangle - \tau \langle I_1 S_2 I_3 \rangle,$$

$$\langle I_1 \dot{S}_3 \rangle = \tau \langle S_1 I_2 S_3 \rangle - (\tau + \gamma) \langle I_1 S_3 \rangle - \tau \left(\langle I_1 I_2 S_3 \rangle + \frac{\langle I_1 S_3 \rangle \langle S_3 I_4 \rangle}{\langle S_3 \rangle} \right),$$

$$\langle I_2 \dot{S}_3 \rangle = \tau \langle I_1 S_2 S_3 \rangle - (\tau + \gamma) \langle I_2 S_3 \rangle - \tau \left(\langle I_1 I_2 S_3 \rangle + \frac{\langle I_2 S_3 \rangle \langle S_3 I_4 \rangle}{\langle S_3 \rangle} \right),$$

$$\langle S_3 \dot{I}_4 \rangle = -(\tau + \gamma) \langle S_3 I_4 \rangle - \tau \left(\frac{\langle I_1 S_3 \rangle \langle S_3 I_4 \rangle}{\langle S_3 \rangle} + \frac{\langle I_2 S_3 \rangle \langle S_3 I_4 \rangle}{\langle S_3 \rangle} \right),$$

$$\langle I_3 \dot{S}_4 \rangle = \tau \left(\frac{\langle I_1 S_3 \rangle \langle S_3 S_4 \rangle}{\langle S_3 \rangle} + \frac{\langle I_2 S_3 \rangle \langle S_3 S_4 \rangle}{\langle S_3 \rangle} \right) - (\tau + \gamma) \langle I_3 S_4 \rangle.$$

A few additional pairs appear in these equations and so we need more equations. These are given below:

$$\langle S_1 \dot{S}_3 \rangle = -2\tau \langle S_1 I_2 S_3 \rangle - \tau \frac{\langle S_1 S_3 \rangle \langle S_3 I_4 \rangle}{\langle S_3 \rangle},$$

$$\langle S_2\dot{S}_3\rangle = -2\tau\langle I_1 S_2 S_3\rangle - \tau\frac{\langle S_2 S_3\rangle\langle S_3 I_4\rangle}{\langle S_3\rangle},$$

$$\langle S_3\dot{S}_4\rangle = -\tau\big(\langle I_1 S_3\rangle + \langle I_2 S_3\rangle\big)\frac{\langle S_3 S_4\rangle}{\langle S_3\rangle}.$$

Note that $\langle S_1 S_2\rangle$ is not needed, nor is any edge involving a recovered node. These equations for pairs introduce some triples. We now write down the equations for these triples. These involve quadruples that cross between the two components. Again, we can reduce these quadruples using singles, pairs and triples, and we find

$$\langle S_1\dot{S}_2 I_3\rangle = \tau\frac{\langle S_1 S_2 S_3\rangle\langle S_3 I_4\rangle}{\langle S_3\rangle} - (2\tau+\gamma)\langle S_1 S_2 I_3\rangle,$$

$$\langle S_1\dot{I}_2 I_3\rangle = \tau\left(\langle S_1 S_2 I_3\rangle + \frac{\langle S_1 I_2 S_3\rangle\langle S_3 I_4\rangle}{\langle S_3\rangle}\right) - 2(\tau+\gamma)\langle S_1 I_2 I_3\rangle,$$

$$\langle S_1\dot{I}_2 S_3\rangle = -(2\tau+\gamma)\langle S_1 I_2 S_3\rangle - \tau\frac{\langle S_1 I_2 S_3\rangle\langle S_3 I_4\rangle}{\langle S_3\rangle},$$

$$\langle I_1\dot{S}_2 I_3\rangle = \tau\left(\langle S_1 S_2 I_3\rangle + \frac{\langle I_1 S_2 S_3\rangle\langle S_3 I_4\rangle}{\langle S_3\rangle}\right) - 2(\tau+\gamma)\langle I_1 S_2 I_3\rangle,$$

$$\langle I_1\dot{S}_2 S_3\rangle = -(2\tau+\gamma)\langle I_1 S_2 S_3\rangle,$$

$$\langle I_1\dot{I}_2 S_3\rangle = \tau\big(\langle S_1 I_2 S_3\rangle + \langle I_1 S_2 S_3\rangle\big) - 2(\tau+\gamma)\langle I_1 I_2 S_3\rangle - \tau\frac{\langle I_1 I_2 S_3\rangle\langle S_3 I_4\rangle}{\langle S_3\rangle}.$$

In writing down these equations, one additional triple appeared $\langle S_1 S_2 S_3\rangle$, so we complete our system with

$$\langle S_1\dot{S}_2 S_3\rangle = -\tau\frac{\langle S_1 S_2 S_3\rangle\langle S_3 I_4\rangle}{\langle S_3\rangle}.$$

Had we not used the closures, we would have needed to write equations for the following variables:

1. **Singles:** $\langle S_1\rangle$, $\langle I_1\rangle$, $\langle S_2\rangle$, $\langle I_2\rangle$, $\langle S_3\rangle$, $\langle I_3\rangle$, $\langle S_4\rangle$, $\langle I_4\rangle$;
2. **Pairs:** $\langle S_1 I_2\rangle$, $\langle S_1 I_3\rangle$, $\langle I_1 S_2\rangle$, $\langle S_2 I_3\rangle$, $\langle I_1 S_3\rangle$, $\langle I_2 S_3\rangle$, $\langle S_3 I_4\rangle$, $\langle I_3 S_4\rangle$;
3. **Triples:** $\langle S_1 S_2 I_3\rangle$, $\langle S_1 I_2 I_3\rangle$, $\langle S_1 I_2 S_3\rangle$, $\langle I_1 S_2 I_3\rangle$, $\langle I_1 S_2 S_3\rangle$, $\langle I_1 I_2 S_3\rangle$, $\langle I_1 S_3 I_4\rangle$, $\langle I_2 S_3 I_4\rangle$, $\langle I_2 S_3 S_4\rangle$, $\langle I_1 S_3 S_4\rangle$, $\langle S_2 S_3 I_4\rangle$, $\langle S_1 S_3 I_4\rangle$;
4. **Quadruples:** $\langle I_1 I_2 S_3 I_4\rangle$, $\langle S_1 I_2 S_3 I_4\rangle$, $\langle I_1 S_2 S_3 I_4\rangle$, $\langle I_1 I_2 S_3 S_4\rangle$, $\langle S_1 I_2 S_3 S_4\rangle$, $\langle I_1 S_2 S_3 S_4\rangle$, $\langle S_1 S_2 S_3 I_4\rangle$.

The resulting system over the 35 variables above (i.e. 8 singles, 8 pairs, 12 triples and 7 quadruples) emerges naturally, uses no closures and only contains variables that arise until full system size is reached. We call this the natural full system (NFS), and such systems have far fewer equations compared to the system over all node and node status combinations, i.e. the master equations. In contrast, the reduced system (RS), where we used Theorem 3.11, leads to far fewer equations (see Table 3.1).

Exercise 3.11. Consider the equations from Example 3.1.

   a. How many differential equations would be needed to write down the full master equations for the SIR model on the lollipop network? (We do not need a differential equation for the state of all four nodes recovered because the sum of probabilities is conserved.)

   b. How many differential equations are needed using the closures in Example 3.1?

Exercise 3.12. Now consider the lollipop network but add two new nodes 5 and 6, which join with each other, and node 4 to create a triangle. Consider the closures described in Fig. 3.5.

   a. Explain why the equations involving nodes 1, 2 and 3 remain the same.

   b. What are the new equations for the edge joining nodes 3 and 4?

   c. What new variables are needed to account for nodes 4, 5 and 6 (do not derive the equations)?

   d. How many total differential equations are needed for this closed system, and how does this compare to the full master equations?

We now investigate the number of equations that the recipe produces for a given network G having L cut-vertices, with the network decomposing into P subnetworks. Let $\mathrm{Ind}(v)$ denote the number of subnetworks a cut-vertex v appears in (see Fig. 3.4).

Let $m_1, m_2, \ldots, m_P$ denote the number of equations required for the corresponding subnetwork. This number depends on the structure of the network. For example, a single edge requires 4 equations at node level and 3 equations at edge level (i.e. (SS), (SI), (IS)), and this gives a total of 7 equations for an edge. Similar reasoning for a triangle subnetwork yields 6 equations at node level, $3 \times 3 = 9$ at edge level and $2^3 - 1 = 7$ at triple level (the III state and the states with R nodes are dynamically unimportant), yielding 22 equations. Furthermore, a cycle graph with 4 nodes, Fig. 2.8a, needs 45 equations, and the toast network, Fig. 2.9b, needs 57 equations (see [181]).

Given this, an upper bound for the number of equations needed to describe the epidemic dynamics exactly is given by

$$N_{\mathrm{eq}}(G) = \sum_{p=1}^{P} m_p - 2 \sum_{l=1}^{L} (\mathrm{Ind}(v_{i_l}) - 1).$$

The formula simply sums the number of equations needed for all subnetworks and adjusts this to account for repeated equations caused by cut-vertices being part of multiple subnetworks. The reason $N_{\mathrm{eq}}(G)$ may only be a bound can be seen from Example 3.1, which did not require an equation for $\langle S_1 S_2 \rangle$ (but did require one for $\langle S_2 S_3 \rangle$ and $\langle S_1 S_3 \rangle$). Similarly, other fully susceptible arrangements at a higher level may not be needed. While the formula could be further improved, the exact overestimate depends in a non-trivial way on the structure of the network and the epidemic dynamics. Our investigations show that removing the unnecessary equations induced by fully susceptible states will not considerably reduce the number of equations.

In what follows, we apply this bound to networks ranging from small toy networks to networks of arbitrary size but with some structural symmetry. This is done to bound the number of equations in the closed system and see how this changes or scales with the number of nodes and edges.

1. *Lollipop network with 4 nodes (Table 3.1):* Upon the removal of the single cut-vertex, v, the network decomposes into an edge and a triangle with $Ind(v) = 2$. Thus, $P = 2, L = 1, m_{edge} = 7$ and $m_{triangle} = 22$, giving $N_{eq}(G) = 7 + 22 - 2 \times (2 - 1) = 27$. The true number in the RS is 26 since $S_1 S_2$ does not require an equation.

2. *Line network of size N (Table 3.1):* In a line network there are $(N - 2)$ cut-vertices and the network decomposes into $(N - 1)$ subnetworks, which are all edges. Each cut-vertex is member of 2 subnetworks. Hence, $P = N - 1$ and $L = (N - 2)$, and for each cut-vertex we have $Ind(v_i) = 2$, with $i = 1, 2, \ldots, N - 2$. Therefore, the upper bound for the number of equations is $N_{eq}(G) = 7 \times (N - 1) - 2 \times (N - 2) \times (2 - 1) = 5N - 3$.

3. *Star network of size N (Fig. 2.10a):* A star network has a single cut-vertex and its removal leads to $(N - 1)$ subnetworks, all edges, with $P = (N - 1)$ and $L = 1$. The cut-vertex, v, belongs to $(N - 1)$ subnetworks and thus $Ind(v) = (N - 1)$. Hence, $N_{eq}(G) = 7 \times (N - 1) - 2 \times ((N - 1) - 1) = 5N - 3$.

4. *Tree networks of size N:* Both the line network and the star network are special cases of tree networks and in general for these $N_{eq}(G) = 5N - 3$.

5. *Star triangle N (Fig. 3.6b):* A star triangle network has an odd number of vertices and a single cut-vertex whose removal leads to $M = (N - 1)/2$ triangles. Thus, all subnetworks are triangles and the single cut-vertex is part of M subnetworks. Thus, $N_{eq}(G) = 22 \times M - 2 \times (M - 1) = 20M + 2 = 10N - 8$.

6. *Networks with loops of maximum size three (Fig. 3.6a):* In this case, there could be many cut-vertices which depending on network structure could be part of many or few subnetworks. Hence, we neglect the negative terms completely and rely on the observation that upon decomposing the network, only edges and triangles remain. Assuming that $|E|$ denotes the number of edges and T the number of uniquely counted triangles, the upper bound becomes $N_{eq}(G) = 7 \times (|E| - 3 \times T) + 22 \times T$, which for most networks is still $\mathcal{O}(N)$.

Exercise 3.13. For a line network with N nodes, show that the natural full system has $(3N^2 - N + 2)/2$ equations (see Table 3.1).

In some cases, a more empirical approach may give a tighter bound or the bound itself can be derived more readily. For example, for an arbitrary tree network, $N_{eq}(G)$ can be calculated by taking into account that only equations up to edges need to be written down, as all triples can be closed. Hence, equations for all nodes and edges amount to $2 \times V + 3 \times |E|$. In a tree network, the number of edges is $|E| = (N - 1)$ and this leads to $5N - 3$. Furthermore, for networks consisting of only non-overlapping triangles and classic edges, see Fig. 3.6a, a stronger result can be given.

Theorem 3.12 *Consider a network with N nodes, $|E|$ edges, T triangles and no larger loops. The number of equations needed to fully describe the system dynamics is at most $2N + 3|E| + 7T$, which in turn is bounded above by $10N$.*

	NFS	RS
(triangle diagram)	27	17
(triangle-edge diagram)	35	26
(line diagram)	$(3N^2 - N + 2)/2$	$5N - 3$
(star diagram)	not computed	$5N - 3$
(star triangle diagram)	not computed	$19M + 2$

Table 3.1: Reduction in the number of equations due to closures in a number of networks. N denotes the number of nodes in the line and star networks and M denotes the number of triangles in a star triangle network. NFS stands for the natural full system and RS for the reduced system.

Proof. The $2N$ contribution follows from the equations at node level, namely each node may be susceptible, infected or recovered, but because these probabilities sum to 1 only two equations are needed.

If the network decomposes as in Fig. 3.4, then each edge must be considered in states SI, IS and SS. We do not need to consider the II state or any state involving a recovered node because these will not be needed for any nodes. This adds an extra $3 \times |E|$ to the equation count. Similarly, for a stand-alone triangle, we consider $2^3 - 1 = 7$ states which do not involve any recovered nodes or the fully infected state. This brings the total number of equations to the value claimed. However, we note that this is an overestimate since there may be SS edges which could be neglected as occurred in Example 3.1.

Since two edges from each triangle contribute to creating the spanning tree of the network, while the extra edges from each triangle can be captured by the triangle count, it follows that the number of edges in such a network can be expressed as $|E| = N - 1 + T$. Furthermore, the number of triangles is bounded above by $N/2$ as the addition of a new triangle requires 2 extra nodes at all points. Hence, summarising, we can write

$$2 \times N + 3 \times |E| + 7 \times T = 5 \times N + 10 \times T - 3 \leq 10N.$$

□

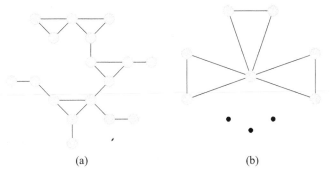

(a)   (b)

Fig. 3.6: (a) Example of a network composed of non-overlapping triangles and classic edges and (b) star triangle network.

The results above are summarised in Table 3.1. As alluded to earlier, further small reductions in the number of equations can be achieved by removing equations corresponding to various fully susceptible states. In this spirit, the $20 \times M + 2$ equations for the star triangle network, see Fig. 3.6b, can be reduced by M if the equations for the probabilities of all the leaf edges being in the SS state are removed. Similar opportunistic observations can be made but the gain is not significant.

We note that for the decomposition of a network, there are two extreme scenarios: (a) the network has many cut-vertices and a large number of subnetworks of a few different types and (b) the network decomposes to relatively few distinct subnetwork, but of large sizes. The tree networks are a good example for scenario (a), where the only subnetwork is the edge, where each edge requires only 3 equations. In this case, there are few equations per subnetwork but many subnetworks. More structured networks will typically have distinct subnetworks of larger sizes, which, for an exact description, will require a larger number of equations. Here, we have fewer subnetworks, but many more equations per subnetwork. More importantly, it is non-trivial to find a simple relation between subnetworks and the number of equations needed for an exact description, and this may require further attention and work. It is straightforward to see that the desirable scenario for an exact representation is scenario (a), and it is likely that in this case an exact description is possible. Complexity quickly increases from 22 equations needed for a closed triangle to 57 equations for a subnetwork equivalent to a toast network (see [181] for equation counts for these and other small networks). Thus, both scenarios require a large number of equations. Generating and implementing the equations needed for an exact description is prone to error and thus we highly recommend the development of an algorithmic approach, where equations can be generated automatically rather than manually. The description above illustrates clearly that the family of networks with many cut-vertices is more amenable to this approach, and it is likely that for networks with few cut-vertices, the task of writing down an exact system is out of reach.

3.6.3 Exact, numerically computable model on tree graphs

Given a tree graph, the exact system (3.6) can be closed at the level of triples, which are all open, using Eq. (3.3). The resulting system can be solved numerically to obtain an exact representation of the true stochastic SIR epidemic model. An implemented version of this can be found as supplementary material in [280] and on the webpage given in the Preface. The equations are as follows:

SIR pair-based model

$$\langle \dot{X_i} \rangle = -\tau \sum_{j=1}^{N} g_{ij} \langle X_i Y_j \rangle, \tag{3.39a}$$

$$\langle \dot{Y_i} \rangle = \tau \sum_{j=1}^{N} g_{ij} \langle X_i Y_j \rangle - \gamma_i \langle Y_i \rangle, \tag{3.39b}$$

$$\langle \dot{X_i Y_j} \rangle = \tau \sum_{k=1,k\neq i}^{N} g_{jk} \frac{\langle X_i X_j \rangle \langle X_j Y_k \rangle}{\langle X_j \rangle} - \tau \sum_{k=1,k\neq j}^{N} g_{ik} \frac{\langle Y_k X_i \rangle \langle X_i Y_j \rangle}{\langle X_i \rangle}$$
$$- \tau g_{ij} \langle X_i Y_j \rangle - \gamma_j \langle X_i Y_j \rangle, \tag{3.39c}$$

$$\langle \dot{X_i X_j} \rangle = -\tau \sum_{k=1,k\neq j}^{N} g_{ik} \frac{\langle Y_k X_i \rangle \langle X_i X_j \rangle}{\langle X_i \rangle} - \tau \sum_{k=1,k\neq i}^{N} g_{jk} \frac{\langle X_i X_j \rangle \langle X_j Y_k \rangle}{\langle X_j \rangle}, \tag{3.39d}$$

where due to Theorem 3.35 closures such as $\frac{\langle X_i X_j \rangle \langle X_j Y_k \rangle}{\langle X_j \rangle}$ are exact. This closed model requires $2N + 3|E|$ equations. $2N$ equations are needed at node level, since the probability of node i being recovered is given by $1 - \langle X_i \rangle - \langle Y_i \rangle$. $\langle X_i X_j \rangle$ pairs only require $|E|$ equations; however, $\langle X_i Y_j \rangle$ pairs require $2|E|$ equations since each pair needs to be considered as directed. Finally, pairs such $\langle Y_k X_i \rangle$ in closures are in fact identical to $\langle X_i Y_k \rangle$ and no further equations are required.

    In a tree network, all nodes (except those with degree one) are cut-vertices, ensuring that system (3.39) is exact. However, if closed triples (triangles) or other cycles are present, the closures become approximations. We can improve the approximation by accounting for closed triples using closure (3.4). To do this, terms such as $\frac{\langle X_i X_j \rangle \langle X_j Y_k \rangle}{\langle X_j \rangle}$ and $\frac{\langle Y_k X_i \rangle \langle X_i Y_j \rangle}{\langle X_i \rangle}$ are replaced by $\frac{\langle X_i X_j \rangle \langle X_j Y_k \rangle \langle X_i Y_k \rangle}{\langle X_i \rangle \langle X_j \rangle \langle Y_k \rangle}$ and $\frac{\langle Y_k X_i \rangle \langle X_i Y_j \rangle \langle Y_k Y_j \rangle}{\langle Y_k \rangle \langle X_i \rangle \langle Y_j \rangle}$, respectively, whenever nodes i, j and k form a closed triple. These new closures require extra variables, such as $\langle Y_k Y_j \rangle$, and the number of equations in the new system will increase.

Exercise 3.14. Formulate and prove the equivalent of Proposition 3.2 but for the SIR model closed at the level of triples.

3.6.4 Numerical examples of the performance of closed systems for SIR disease

In Fig. 3.7, we repeat an almost identical experiment to that shown in Fig. 3.3 for SIS dynamics. As before, closures at a higher level lead to better approximations with exact results for tree networks, such as the line network shown in Fig. 3.7a. As our theoretical results show, for such networks the closure provides an exact model. Except for the square lattice, see Fig. 3.7b, the agreement between the pair-based model and simulations is excellent. This is as expected since all networks are generated according to the Configuration Model and thus with a limited amount of clustering or longer loops. The individual-based model continues to perform in a much inferior way, especially for structured networks such as the 1d and 2d lattice, see Figs. 3.7a & 3.7b, and for more sparsely connected networks, or when the transmission rate leads to small or moderate epidemics.

Exercise 3.15. Using SIS and SIR dynamics and the networks consisting of one single edge and a simple line with three nodes, complement your findings in Exercises 3.1, 3.2, 3.8 and 3.9 and complete the following tasks:
  a. write down the pair-based system for both dynamics and networks;
  b. solve these equations numerically and check the relation between the exact, individual-based and pair-based models.

3.7 Conclusions and outlook

Formulating bottom-up models and their analytical and numerical analysis is a well-developed area of research for both SIS [60, 129, 310, 312, 313] and SIR [181, 279–281, 283] dynamics. Perhaps the most striking result is that for finite networks and for the Markovian SIR epidemics it is often possible to develop a numerically tractable reduced but exact model (see also [181, 281, 283]). In contrast to top-down models, bottom-up models show more flexibility and can be adapted to numerous different scenarios and networks. As shown in the chapter, for highly structured or symmetric networks such as the line and star triangle networks, it may be possible to exploit the network structure and write down a closed but exact system, or at least find an upper bound for the number of equations needed for such a system [181].

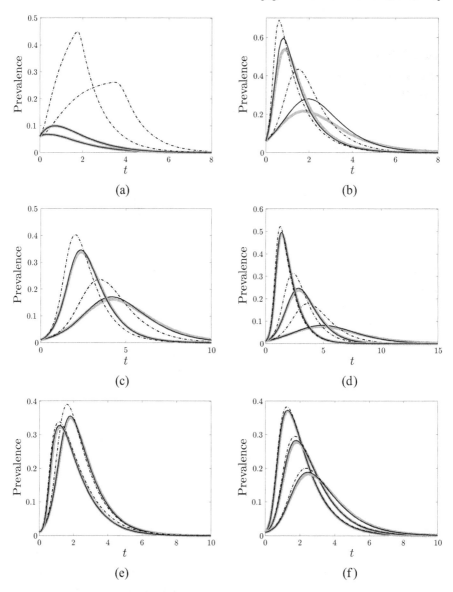

Fig. 3.7: Expected prevalence from simulations (thick grey lines), pair-based (solid black lines) and individual-based (dash-dotted black lines) models plotted for the following networks: (a) line with $\tau = 5$ (bottom) and $\tau = 10$ (top); (b) 2d square lattice with $\tau = 1.5$ (bottom) and $\tau = 4$ (top); (c) regular random with $\langle k \rangle = k = 10$, $\tau = 0.25$ (bottom) and $\tau = 0.4$ (top); (d) Erdős–Rényi with $\tau = 0.4$ and $\langle k \rangle = 5, 7.5, 15$, from bottom to top (e) bimodal with $\tau = 0.4$ and (i) $N_1 = 800$, $k_1 = 4$, $N_2 = 200$, $k_2 = 34$ (rising faster) and (ii) $N_1 = N_2 = 500$, $k_1 = 5$, $k_2 = 15$ with both having $\langle k \rangle = 10$; (f) scale-free-like ($P(k) = c/k^2$, with $k = 7, 8, \ldots, 109$) with $\tau = 0.1, 0.15, 0.2$ from bottom to top. For networks (a) and (b), $N = 100$, while for (c) to (f), $N = 1000$. For all cases, $\gamma = 1$, and for all networks 1000 simulations were run on a single network realisation, starting with the same 6 centrally infected and 10 randomly infected nodes for networks (a,b) and (c–f), respectively.

While for SIS dynamics both the individual- and pair-based models perform well for many different networks and parameter combinations, it is likely that a similar exactness result will not hold. Obviously, for a fully connected or complete network, the classic compartmental model becomes exact in the limit of large network size. While individual-based models are not exact, they allow us to link characteristics of the network to epidemic severity indicators such as the endemic equilibrium or even the threshold. It is worth noting that, for example, the exactness results for tree networks for SIR dynamics can be exploited further to generate exact models even on non-tree networks by an opportunistic placement of the initial conditions, which renders the network tree-like [181, 283].

When closed models are known not to be exact but perform well, there are little or no results on rigorous error estimates, and this line of research could potentially still hold many novel and relevant results. Starting with simple toy networks could provide important insight into what can realistically be achieved on this front. As the analysis and findings suggest, the closed models presented here work well for Configuration Model-like networks, or networks without cliques/cycles or structure. For clustered networks, or networks with longer cycles, some adaptation of the closures is possible, as we showed for connected triples/triangles, but in general the agreement between closed and exact or simulated models suffers, especially if clustering is pushed to high values.

As a general rule, if the individual-based and pair-based models are in close agreement, then there is relatively little relevant information contained in knowing the pairs. As such, it is unlikely that having better information about larger structures would significantly improve the results. So close agreement between individual-based and pair-based models suggests a good approximation of the true dynamics.

Chapter 4
Mean-field approximations for homogeneous networks

As seen in Chapters 2 and 3, because of the high-dimensionality of exact mathematical models describing spreading processes on networks, the models are often neither tractable nor numerically solvable for networks of realistic size. We can avoid this fundamental difficulty by refocusing our attention on expected population-scale quantities, such as the expected prevalence or the expected number of edges where one node is susceptible while the other is infectious. This opens up a range of possibilities to formulate so-called mean-field models (i.e. typically low-dimensional ODEs or PDEs) that are widespread in the physics and mathematical biology literature. These are used to approximate stochastic processes, with the potential to be exact in the large system or "thermodynamic" limit, see (for example, [151, 166, 167, 243] and the literature overview in Section 4.8).

Figure 4.1 shows the outcomes of many stochastic realisations of SIS and SIR epidemics on an Erdős–Rényi network with average degree $\langle K \rangle = 30$. The realisations form the cloud plot, which highlights the stochasticity of the spreading processes. The same figure also shows the average prevalence based on all realisations together with predictions of the simplest mean-field approximations, system (4.8) for the SIS case and (4.9) for the SIR case. The derivation and analysis of mean-field models capable of describing such average behaviours is the focus of a significant part of this book. Many such models have been and can be derived. The simplest models typically require stronger assumptions. We set out to present these models in a unified framework, highlighting the motivation and applicability of these models. Some of these models may be network- or process-specific and only perform well for certain networks or processes, e.g. giving good results for SIR dynamics but not for SIS.

The typical recipe for deriving a mean-field model is to first identify some quantities of practical interest and derive equations for how the average value of these quantities change in time. Often, these equations rely on new variables. We iterate the process, deriving equations for the new variables. Occasionally, this process terminates quickly, yielding a small, self-consistent system of equations. Other times,

© Springer International Publishing AG 2017   117
I.Z. Kiss et al., *Mathematics of Epidemics on Networks*, Interdisciplinary Applied
Mathematics 46, DOI 10.1007/978-3-319-50806-1_4

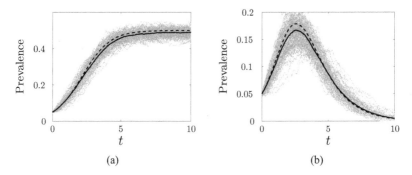

Fig. 4.1: The time dependence of the expected number of infected nodes for an Erdős–Rényi random graph from 100 agent-based stochastic simulations (grey curves) and from the mean-field system closed at the level of pairs (black dashed curves) for (a) an SIS epidemic and (b) an SIR epidemic. The averages of the simulations are shown with black solid curves. The parameter values are $N = 1000$, $\langle K \rangle = 30$, recovery rate $\gamma = 1$ and infection rate $\tau = 1/15$. We note that as the number of nodes increases, the spread of simulation around the mean decreases.

the iteration would continue until the full system size is reached; then, we typically "close" the model by approximating the newly introduced variables in terms of already existing ones. There is some "art" in the process: sometimes, there are options for what variables or approximations to use, and the choice determines the accuracy and complexity of the model.

This chapter and the following chapter systematically develop mean-field models, starting with the most basic ones that usually work for simpler networks, e.g. regular and Erdős–Rényi, and moving towards the more sophisticated models that can handle networks with high degree heterogeneity (which are the focus of Chapter 5), preferential mixing or clustering. We present results concerning steady states, stability and model consistency for many mean-field models to aid the readers' understanding of this important area. We note that these models have been extended to account for more complex structural properties of networks, such as households and networks with arbitrary subgraph distribution as well as directed or weighted networks. These will not be dealt with in the book, but relevant references are given in the concluding section of this chapter.

4.1 Exact, unclosed models

For models of infectious disease, the variables of interest are typically the average number of nodes having a given status. We denote these by $[S](t)$, $[I](t)$ and $[R](t)$ for susceptible, infected and recovered, respectively. The expected number of edges connecting nodes of status A to nodes of status B is denoted by $[AB](t)$, with $A, B \in \{S, I, R\}$. The expected number of triples, where a middle B node is connected to a node of status A and to a node of status C, is denoted by $[ABC](t)$. A formal definition of these quantities is given in the next subsection. Note that there is an implicit direction, so each edge is counted twice. An edge from a susceptible node to an infected node is counted towards $[SI]$, but it will also be counted towards $[IS]$.

4.1.1 The variables of mean-field models: population-level counts

To define the population-level counts, we start from the random variable $X_i(t)$ that determines the type of node i at time t, e.g. $X_i(t) = I$ if node i is infected at time t. The above expected values can be defined formally as

$$[A](t) = \sum_{i=1}^{N} P(X_i(t) = A),$$

where $A \in \{S, I, R\}$. The expected number of edges in a given status can be defined similarly as follows:

$$[AB](t) = \sum_{i=1}^{N} \sum_{j=1}^{N} g_{ij} P(X_i(t) = A, X_j(t) = B),$$

with $A, B \in \{S, I, R\}$. It is important to note here that edges connecting two suscep-tible nodes contribute twice to the $[SS]$ count, since the pairs (i, j) and (j, i) are both counted in the $[SS]$ class when nodes i and j are susceptible. As an example, con-sider the graph with two infected and two susceptible nodes, shown in Fig. 4.2. The nodes are labelled clockwise from the top left, with nodes 1 and 2 being infected. Let us for a moment assume that this is a snapshot of a given realisation of the epi-demic. We determine the different counts denoting them also by $[\cdot]$, although these are not expectations. In the situation given in Fig. 4.2, we have $[I] = 2$ and $[S] = 2$. Counting the pairs, we find that $(3, 2)$, $(4, 2)$ and $(4, 1)$ are of SI type and, hence $[SI] = 3$. The II pairs are $(1, 2)$ and $(2, 1)$; therefore, $[II] = 2$. In a similar way, we get $[IS] = 3$ and $[SS] = 2$; the total number of pairs is 10. Let us finally turn to the counting of triples. The number of ABC triples is defined in general as

$$[ABC](t) = \sum_{i=1}^{N} \sum_{j=1}^{N} \sum_{k=1}^{N} g_{ij} g_{jk} P(X_i(t) = A, X_j(t) = B, X_k(t) = C),$$

with $A, B, C \in \{S, I, R\}$. Using this definition, the number of SSI triples in the graph shown in Fig. 4.2 is $[SSI] = 3$, namely the SSI triples are $(3, 4, 1)$, $(3, 4, 2)$ and $(4, 3, 2)$. The number of other triples can be similarly obtained as $[ISS] = 3$, $[IIS] = 3$, $[SII] = 3$, $[ISI] = 2$ and $[SIS] = 2$; there are 16 triples altogether.

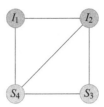

Fig. 4.2: Counting the number of pairs and triples.

The expected values introduced above obey some conservation relations. The simplest one, for the SIS dynamics, is based on the fact that $P(X_i(t) = S) + P(X_i(t) = I) = 1$ for any $i = 1, 2, \ldots, N$, since any node is either susceptible or infected. This immediately implies that

$$[S](t) + [I](t) = N$$

for any time instant t. For the SIR epidemic, the corresponding equation is $[S](t) + [I](t) + [R](t) = N$. The conservation relation for the pairs is based on the simple fact that for the SIS epidemic, a pair can be in one of the following four statuses: SS, SI, IS or II. Hence, $P(X_i(t) = S, X_j(t) = S) + P(X_i(t) = S, X_j(t) = I) + P(X_i(t) = I, X_j(t) = S) + P(X_i(t) = I, X_j(t) = I) = 1$, implying

$$[SS](t) + [SI](t) + [IS](t) + [II](t) = \sum_{i=1}^{N} \sum_{j=1}^{N} g_{ij} = Nn,$$

where $n = \langle k \rangle$ is the average degree defined in Section 1.2.2. We note that n is used to denote the average degree in this chapter for the sake of brevity and to follow the widely used notation of pairwise models. The definition of pairs immediately implies that $[SI] = [IS]$; hence, the above relation can be formulated as

$$[SS](t) + 2[SI](t) + [II](t) = Nn.$$

Similar arguments lead to further pair conservation relations in the SIS case, namely

$$[SS](t) + [SI](t) = n_S(t)[S](t), \quad [SI](t) + [II](t) = n_I(t)[I](t),$$

where $n_S(t)$ and $n_I(t)$ denote the average degree of susceptible and infected nodes, respectively. The pair conservation relations in the SIR case can be formulated similarly. Returning to the SIS case, triple conservation relations can also be formulated. The relation

$$[SSI](t) + [ISI](t) = (n_S(t) - 1)[SI](t)$$

will play an important role. Besides deriving it formally from the triple definitions, one can argue as follows to prove the relation. Taking an arbitrary SI edge, the S node has $(n_S - 1)$ further neighbours; hence, the total number of triples containing an SI pair is $(n_S - 1)[SI]$. On the other hand, the same quantity can be obtained as the sum of those triples that contain an SI pair, namely $[SSI] + [ISI]$. In a similar way, one can derive further triple conservation relations, for example, $[SSS](t) + [ISS](t) = (n_S(t) - 1)[SS](t)$ and $[SIS](t) + [IIS](t) = (n_I(t) - 1)[IS](t)$.

4.1.2 Exact differential equations for the singles and pairs

We use heuristic arguments here to derive exact differential equations for the expected number of nodes and edges in given statuses. Later in this chapter, we derive these from the master equations, system (2.6), consisting of 2^N equations governing the full SIS dynamics. We sacrifice having information about the precise statuses of each node to get a much smaller system of equations for the expected numbers of nodes in each status. The main parameters of the epidemic processes are the infection and recovery rates denoted by τ and γ, respectively.

Theorem 4.1 *For the SIS epidemic on an arbitrary network (undirected and not weighted), the expected values $[S]$ and $[I]$ satisfy the following system*

$$[\dot{S}] = \gamma[I] - \tau[SI], \tag{4.1a}$$
$$[\dot{I}] = \tau[SI] - \gamma[I]. \tag{4.1b}$$

The proof of the theorem is presented in Section 4.6. The differential equations can be obtained heuristically following the top two flow diagrams in Fig. 4.3. The rate of transmission to S nodes is τ times the number of SI edges. The rate of recovery of infectious nodes back into a susceptible status is γ times the number of infectious nodes. Thus, the rate of change of $[S]$ is $\gamma[I] - \tau[SI]$. We similarly find the $[I]$ equation. For an SIR epidemic, the I nodes do not become susceptible again; instead their status changes to R. The equations take a different form.

Theorem 4.2 *For the SIR epidemic on an arbitrary network (undirected and not weighted), the expected values $[S]$, $[I]$ and $[R]$ satisfy the following system*

$$[\dot{S}] = -\tau[SI], \tag{4.2a}$$
$$[\dot{I}] = \tau[SI] - \gamma[I], \tag{4.2b}$$
$$[\dot{R}] = \gamma[I]. \tag{4.2c}$$

We note that the variables in the SIS and SIR systems thus far are not independent because of conservation relations, taking the form $[S] + [I] = N$ in the SIS case and $[S] + [I] + [R] = N$ in the SIR case. Hence, for the exact systems one of the equations could be omitted in both cases. However, for systems with approximate closures, the conservation laws do not hold automatically; their validity may depend on the choice of the closure. Therefore, differential equations for all singles and pairs that are needed to get a self-contained system will be considered. This will always be followed by investigating whether conservation laws hold as they enable us to reduce the system.

As illustrated in Fig. 4.3, the dynamics of the expected number of S nodes ($[S]$) and I nodes ($[I]$), i.e. singles, depends on the number of SI pairs ($[SI]$); hence, the system depends on pairs, for which we need additional equations. Similarly, the number of pairs depends on the number of triples. For example, the number of SS pairs decreases due to infection from outside the pair, i.e. it changes proportionally

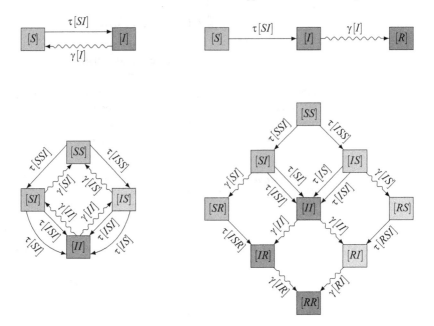

Fig. 4.3: Flow diagrams showing the flux between compartments of singles (top) and compartments of pairs (bottom). The SIS case is on the left and SIR on the right. In the compartments of pairs, solid lines denote infections coming from within the pair (with a rate depending on a pair) or from outside the pair (with a rate depending on a triple), and the wiggle lines denote a recovery. The colour indicates the status of the "first" node in the edge. Symmetry allows us to conclude that some of the variables (lighter shade, on the right of each diagram) must equal the symmetric version (e.g. $[RS] = [SR]$), so we do not need to directly calculate both.

to the number of SSI triples ($[SSI]$) with rate $2\tau[SSI]$ (recall the SS pair is counted twice). In an SI or IS pair, the infected node can recover with rate γ; hence, in the case of an SIS epidemic, the number of SS pairs increases at rate $\gamma([SI] + [IS])$. Since the numbers of SI and IS pairs are equal, we can use $2\gamma[SI]$. Extending this simple heuristic reasoning to SI and II pairs and by accounting for all within- and outside-pair transitions, we arrive at the following theorems.

Theorem 4.3 *For the SIS epidemic on an arbitrary network (undirected and not weighted) the expected values of* $[S]$, $[I]$, $[SI]$, $[II]$, *and* $[SS]$ *satisfy the following system of differential equations*

$$[\dot{S}] = \gamma[I] - \tau[SI], \tag{4.3a}$$

$$[\dot{I}] = \tau[SI] - \gamma[I], \tag{4.3b}$$

$$[\dot{SI}] = \gamma([II] - [SI]) + \tau([SSI] - [ISI] - [SI]), \tag{4.3c}$$

$$[\dot{SS}] = 2\gamma[SI] - 2\tau[SSI], \tag{4.3d}$$

$$[\dot{II}] = -2\gamma[II] + 2\tau([ISI] + [SI]). \tag{4.3e}$$

This result can also be derived directly from the master equations, system (2.6) (see [299]).

As before, for SIR epidemics the I nodes do not become susceptible again; hence, the terms due to recovery show up in different equations. We find the following.

Theorem 4.4 *For the SIR epidemic on an arbitrary network (undirected and not weighted), the expected values of* [S], [I], [SI] *and* [SS] *satisfy the following system of differential equations:*

$$[\dot{S}] = -\tau[SI], \tag{4.4a}$$

$$[\dot{I}] = \tau[SI] - \gamma[I], \tag{4.4b}$$

$$[\dot{R}] = \gamma[I], \tag{4.4c}$$

$$[\dot{SI}] = -\gamma[SI] + \tau([SSI] - [ISI] - [SI]), \tag{4.4d}$$

$$[\dot{SS}] = -2\tau[SSI], \tag{4.4e}$$

The differential equations for [II], [SR], [IR] and [RR] can be formulated similarly, but we leave them out because the other variables do not depend on them and they are not generally of epidemiological interest.

4.2 Closures at the pair and triple level and the resulting models

We now present the ideas leading to closures of the systems above.

4.2.1 Closures

We now assume that the network is homogeneous: each node has the same degree n. There are $[I]$ infected nodes, making up a proportion $[I]/N$ of the population. Assuming that infected nodes are distributed randomly, an average susceptible node has $n[I]/N$ infected neighbours. This assumption makes the closed system inexact since infected nodes are more likely to be in contact with other infected nodes because of how infection propagates. Using this assumption, however, the total number of SI edges is approximated by

$$[SI] \approx \frac{n}{N}[S][I]. \tag{4.5}$$

This relation is referred to as a pair closure since it replaces a term for a pair to yield a "closed" system of equations. Using this, the dependence on higher order moments in system (4.1) is avoided. This closure is used also for the SIR epidemic in system (4.2). We note that sometimes $\frac{n}{N-1}$ is used instead of $\frac{n}{N}$, because after choosing a susceptible node, the remaining population contains only $N-1$ nodes.

To improve our accuracy, we need a better accounting of the fact that infected nodes are not uniformly distributed. We take system (4.3) and close it by assuming an algebraic expression for $[SSI]$ and for $[ISI]$ in terms of singles and pairs. To derive this, we start again with a susceptible node and determine what proportion of the edges starting from this node lead to infected or susceptible nodes. The total number of edges starting from susceptible nodes is $n[S]$. The total number of SI edges is $[SI]$; hence, a proportion $[SI]/n[S]$ of the edges starting from susceptible nodes lead to infected nodes. Similarly, the ratio of edges leading to susceptible nodes is $[SS]/n[S]$. Thus, if we choose a susceptible node u and two neighbours v and w (arbitrarily calling v "first"), the probability that v is susceptible and w infected is $[SS][SI]/n^2[S]^2$. There are $n(n-1)$ ways to choose v and w. Thus, the expected number of SSI triples is $[S]n(n-1)[SS][SI]/n^2[S]^2 = (n-1)[SS][SI]/n[S]$. Assuming that these are uniformly distributed, we conclude that

$$[SSI] \approx \frac{n-1}{n}\frac{[SS][SI]}{[S]}. \tag{4.6}$$

Similar analysis leads to the closure

$$[ISI] \approx \frac{n-1}{n}\frac{[SI]^2}{[S]}. \tag{4.7}$$

As before, this is only an approximation: for an SIS epidemic, a previously infected, but recently recovered node will tend to have more infected neighbours than other susceptible nodes. Thus, the distribution of infected neighbours to susceptible nodes is not truly uniform. Further, for an SIR epidemic, if there are many short cycles, then one neighbour of a susceptible node u being infected is correlated with other neighbours of u being infected, so again the distribution is not uniform. If there are few short cycles, we will see that this closure is accurate for SIR epidemics.

4.2.2 Closed systems

We now write out the closed systems. We first apply closure (4.5) to system (4.1). Since the closure is only approximate, the variables in the closed system are, strictly speaking, different from the original ones, so we use a different notation: $[S]_f$ and $[I]_f$ are the approximations given by the closed system.

SIS homogeneous mean-field at single level

$$[\dot{S}]_f = \gamma[I]_f - \tau\frac{n}{N}[S]_f[I]_f, \tag{4.8a}$$

$$[\dot{I}]_f = \tau\frac{n}{N}[S]_f[I]_f - \gamma[I]_f. \tag{4.8b}$$

The subscript "f" refers to "first" since this can be considered as the first approximation when the expected number of pairs is expressed in terms of the expected number of singles. The system and variables resulting by applying the triple closure will be referred to as the second approximation. In the case of the SIR epidemic, the simplest closed system takes the form below.

SIR homogeneous mean-field at single level

$$[\dot{S}]_f = -\tau \frac{n}{N}[S]_f[I]_f, \tag{4.9a}$$

$$[\dot{I}]_f = \tau \frac{n}{N}[S]_f[I]_f - \gamma[I]_f, \tag{4.9b}$$

$$[\dot{R}]_f = \gamma[I]_f. \tag{4.9c}$$

It is worth noting that these closures lead to the usual SIS and SIR disease models. These equations are an approximation for epidemics on static networks. However, if we consider an alternate problem, in which nodes select a new set of random neighbours at each moment, our assumption that $[SI] \approx \frac{n}{N}[S][I]$ is correct. So these equations are correct if we consider a system in which nodes are quickly changing their neighbours. Thus, the difference between these closed equations and the full equations is due to the duration of partnerships creating correlations between the status of neighbouring nodes.

We now apply the triple closures (4.6) and (4.7) to system (4.3). This is the second approximation, so we use the notation $[S]_s$, $[I]_s$, $[SI]_s$, $[SS]_s$ and $[II]_s$, yielding the system below.

SIS homogeneous pairwise

$$[\dot{S}]_s = \gamma[I]_s - \tau[SI]_s, \tag{4.10a}$$

$$[\dot{I}]_s = \tau[SI]_s - \gamma[I]_s, \tag{4.10b}$$

$$[\dot{SI}]_s = \gamma([II]_s - [SI]_s) + \tau \frac{n-1}{n} \frac{[SI]_s([SS]_s - [SI]_s)}{[S]_s} - \tau[SI]_s, \tag{4.10c}$$

$$[\dot{SS}]_s = 2\gamma[SI]_s - 2\tau \frac{n-1}{n} \frac{[SI]_s[SS]_s}{[S]_s}, \tag{4.10d}$$

$$[\dot{II}]_s = -2\gamma[II]_s + 2\tau \frac{n-1}{n} \frac{[SI]_s^2}{[S]_s} + 2\tau[SI]_s. \tag{4.10e}$$

We will show later that this system conserves certain quantities; hence, we do not need all of the differential equations to determine the number of susceptible and infected nodes. In the next section, it will be shown that two differential equations are enough to determine all the unknown functions, because there is one conservation relation for singles and two for pairs.

For SIR epidemics, the same closures are applied to system (4.4), leading to the following.

SIR homogeneous pairwise

$$[\dot{S}]_s = -\tau[SI]_s,$$ (4.11a)

$$[\dot{I}]_s = \tau[SI]_s - \gamma[I]_s,$$ (4.11b)

$$[\dot{R}]_s = \gamma[I]_s,$$ (4.11c)

$$[\dot{SI}]_s = -\gamma[SI]_s + \tau\frac{n-1}{n}\frac{[SI]_s([SS]_s - [SI]_s)}{[S]_s} - \tau[SI]_s,$$ (4.11d)

$$[\dot{SS}]_s = -2\tau\frac{n-1}{n}\frac{[SI]_s[SS]_s}{[S]_s},$$ (4.11e)

The system could be augmented by further equations if one had a reason to be interested in the values of pairs such as $[II]$, $[SR]$, $[IR]$ or $[RR]$.

When the closed systems are solved, initial conditions are needed for all model variables. Typically, the initial number of susceptible, infected and recovered nodes are given; these can be used as initial conditions in the mean-field models at single level. In the case of pairwise models, further initial conditions are needed for the initial number of pairs. Assuming that the different types of nodes are distributed randomly initially, the initial condition for AB pairs can be given as

$$[AB]_0 = \frac{n}{N}[A]_0[B]_0,$$

where $[A]_0$ and $[B]_0$ denote the initial number of nodes of status A and B, respectively. Obviously, this is not the only choice for the initial conditions of the pairs. For example, correlations in the initial position of infected nodes can be built in, but these will dampen in time.

4.2.3 Clustered pairwise model

The closures above fail to describe clustered graphs in a satisfactory way. The clustering coefficient of a network is simply the ratio of triangles to triples (both open and closed). More formally, it can be defined by using the adjacency matrix G of the network. Let the network be undirected without weights; that is, G is symmetric and each entry is 0 or 1. For a matrix A, we will use the notations $\|A\| = \sum_{i=1}^{N}\sum_{j=1}^{N} a_{ij}$ and $\text{Tr}(A) = \sum_{i=1}^{N} a_{ii}$ for the trace of the matrix. The number of pairs, i.e. edges in our terminology, is $\|G\|$. It can be shown that the number of triples is $\|G^2\| - \text{Tr}(G^2)$ and the number of triangles is $\text{Tr}(G^3)$. The clustering coefficient of the network is defined as the ratio of the number of triangles and triples:

$$\phi = \frac{\text{Tr}(G^3)}{\|G^2\| - \text{Tr}(G^2)}.$$

In Chapter 3, we have shown that the bottom-up approach can handle certain classes of clustered networks, but the number of resulting equations can be forbidding. Hence, it seems more natural to consider mean-field and percolation models to capture this extra degree of complexity in the structure of the network. Below, we give a succinct overview of some extensions. In the spirit of this chapter, we start with extensions of the pairwise model. Here, we present two closures that attempt to capture clustering within the framework of pairwise models. The classical closure for clustered graphs was first introduced in [166, 256], the idea of which will be presented now. Take an SI edge and consider the $(n-1)$ other neighbours of the susceptible node. The average number of those neighbours that are not connected to the infected node is $(1-\phi)(n-1)$, while the average number of the neighbours connected to it is $\phi(n-1)$. For those that are not connected to the I node, we can apply the original idea of closure, i.e. we can say that the proportion of S neighbours is $[SS]/n[S]$, and hence the number of SSI triples given by these neighbours is

$$[SI](1-\phi)(n-1)\frac{[SS]}{n[S]} = (1-\phi)\frac{n-1}{n}\frac{[SS][SI]}{[S]}.$$

Consider now those neighbours of the S node that are connected to the I node as well. The proportion of susceptible nodes among these neighbours is scaled with the correlation $C_{SI} = \frac{N}{n}\frac{[SI]}{[S][I]}$, leading to

$$[SI]\phi(n-1)\frac{[SS]}{n[S]}C_{SI} = \phi\frac{n-1}{n}\frac{[SS][SI]}{[S]}\frac{N}{n}\frac{[SI]}{[S][I]}.$$

Thus, adding the above expressions we approximate the total number of SSI triples in a network with clustering coefficient ϕ as

$$[SSI] \approx \frac{n-1}{n}\frac{[SS][SI]}{[S]}\left((1-\phi)+\phi\frac{N}{n}\frac{[SI]}{[S][I]}\right).$$

By introducing C_{II} in the same manner, the number of ISI triples can be similarly approximated as

$$[ISI] \approx \frac{n-1}{n}\frac{[SI]^2}{[S]}\left((1-\phi)+\phi\frac{N}{n}\frac{[II]}{[I]^2}\right).$$

Intuitively, the scaling factors or correlations measure the propensity of nodes with certain statuses to be more or less likely to be neighbours than if they were randomly distributed in the network. The performance of these closure relations can be investigated by comparison to simulations. However, there is a theoretical issue concerning them, namely, the closed systems given by these relations do not preserve the conservation relations at the level of pairs. That is, even if $[SS] + [SI] = n[S]$ and $[SI] + [II] = n[I]$ initially, these equations fail later if the closures are used. These discrepancies can be fixed by using a modified scaling factor instead of the correlation C_{SI}. In [149], the scaling factor

$$\frac{C_{AI}}{p_{S|S}C_{SI} + p_{I|S}C_{II}}$$

is introduced, where $p_{A|S} = \frac{[AS]}{n[S]}$ for $A \in \{S, I\}$. This yields the closure approximation

$$[ASI] \approx \frac{n-1}{n} \frac{[AS][SI]}{[S]} \left(1 - \phi + \phi \frac{n[S][I][AI]}{([SS][I] + [S][II])[A][SI]}\right).$$

Applying these closure relations in system (4.3), we obtain a closed clustered pairwise model for SIS epidemic dynamics. Similarly, the closures can be used in system (4.4) to get a clustered pairwise model for SIR epidemic dynamics. Note that system (4.4) must be augmented by an equation for $[II]$ in order to apply this closure.

Other developments in this area include the continuation of the pairwise equations beyond pairs by writing down equations for triples, which now will include quadruples [152]. The complexity grows since triples themselves can be open or closed, and quadruples can be of the following types:

The network structure is accounted for by writing down closures separately for each quadruple type, hoping to more accurately capture the local structure. The performance of this model seems to be mildly better than that of the simple pairwise; however the exploration of this model is far from trivial as it relies on a good parametrisation of the network model, where quadruples can be controlled well. The dimensionality of such a model increases and therefore the insight gained may be less compared to that gained from simpler models.

Clustering can have a significant impact on the epidemic threshold, final epidemic size or endemic state. For example, it is widely accepted that the value of the transmission rate needed to generate an epidemic is larger for networks which are clustered when compared to an equivalent network with the same degree distribution but no clustering. Where neighbours of a given node are likely to be connected leads to local depletion of susceptibles where transmission events are effectively wasted [122, 166, 213, 214].

Perhaps the most significant progress in this direction has been made by using a percolation theory approach [117, 213, 237]. However, the analytical tractability of these models requires the consideration of specific classes of networks. In this case, the networks usually are built based on allocating a number of triangle corners or hyperstubs to each node according to some degree distribution, which later on, with some probability, will either decompose into two classical edges or will go on to form triangles. This construction allowed the authors to use the probability generating function machinery coupled with percolation theory (see Chapter 6). This then led to finding analytical expressions for the size of the giant component and the location of the percolation threshold.

Several further developments of such models have been proposed. First of all, Karrer and Newman [164] have extended the network models to take into account arbitrary distributions of subgraphs and investigate the size of the giant component, the location of the phase transition at which the giant component appears, and percolation properties for both site and bond percolation on networks generated by the model. On the other hand, two other significant developments can be noted. In [318], the authors have successfully extended the edge-based compartmental model (EBCM) framework of Chapter 6 to provide a compact mean-field model that gives excellent agreement with simulations and accurately describes the temporal evolution of the disease. Finally, combining the results in [164] and [318], Ritchie et al. [265] extended the EBCM framework further to model SIR epidemics on graphs with arbitrary subgraph distributions. This is also a mean-field model that describes the temporal evolution of the epidemic and provides a strong framework for modelling SIR disease spread on clustered networks.

4.3 Analysis of the closed systems

4.3.1 SIS homogeneous mean-field equations at the single level

Consider first the SIS system closed at the level of pairs, i.e. system (4.8). Note that adding the two equations, we get that $[S]_f(t) + [I]_f(t)$ is constant in time. If the initial condition satisfies $[S]_f(0) + [I]_f(0) = N$, then by using $[S]_f(t) = N - [I]_f(t)$, the system can be reduced to the single equation

$$[\dot{I}]_f = \tau \frac{n}{N}(N - [I]_f)[I]_f - \gamma [I]_f.$$

We can analyse the behaviour of this dynamical system. It has a disease-free steady state $[I]_f^{df} = 0$ and another fixed point at $[I]_f^e = N(1 - \frac{\gamma}{n\tau})$. This second point is only biologically meaningful if it is positive, i.e. $\gamma < n\tau$, in which case it gives an endemic equilibrium. Differentiating the right-hand side of the differential equation with respect to $[I]_f$, i.e. linearising the right-hand side, one obtains that for $\gamma > n\tau$ the disease-free state is stable, while for $\gamma < n\tau$ the endemic state is stable. Thus at $\gamma = n\tau$ a transcritical bifurcation occurs, the disease-free steady state loses its stability, and the stable endemic steady state appears. Moreover, these stabilities are global for $N > [I]_f > 0$. For $\gamma > n\tau$, the right-hand side is negative, i.e. $[I]_f$ is decreasing in time and converges to zero, starting from any meaningful initial condition. In the case $\gamma < n\tau$, the right-hand side changes sign from positive to negative at the endemic steady state, which means that all solutions starting in the interval $(0, N]$ converge to the endemic steady state. Summarising, we have the following analytical results about the closed system (4.8).

Analytical results for SIS homogeneous mean-field equations at the single level

- Conservation of singles: $[S]_f(t) + [I]_f(t) = N$.
- Reducibility to a single equation: $[\dot{I}]_f = \tau \frac{n}{N}(N - [I]_f)[I]_f - \gamma[I]_f$.
- Transcritical bifurcation at $\gamma = n\tau$: trivial steady state loses its stability and a stable endemic steady state appears.
- Global behaviour for $\gamma > n\tau$: all solutions converge to the disease-free steady state $[I]_f^{df} = 0$.
- Global behaviour for $\gamma < n\tau$: all solutions converge to the endemic steady state $[I]_f^e = N(1 - \frac{\gamma}{n\tau})$.

4.3.2 SIR homogeneous mean-field equations at the single level

Consider now the system closed at the level of pairs for SIR propagation, i.e. system (4.9). Note that adding the three differential equations, we get that $[S]_f(t) + [I]_f(t) + [R]_f(t)$ is constant in time. If the initial condition satisfies $[S]_f(0) + [I]_f(0) + [R]_f(0) = N$, then by using $[R]_f(t) = N - [S]_f(t) - [I]_f(t)$, we can reduce the system to the first two equations. Hence, the analysis reduces to studying the two-dimensional phase portrait of the system.

Obviously, we have $[S]_f(t) + [I]_f(t) \leq N$; hence, the phase portrait will be determined in the triangle given by $0 \leq [S]_f(t) + [I]_f(t) \leq N$. This system has infinitely many steady states, namely the points $(S^*, 0)$, with any $S^* \in [0, N]$. Linearisation shows that one eigenvalue of the Jacobian is $\lambda_1 = 0$ at every steady state; this is a consequence of having a line of steady states. The other eigenvalue is $\lambda_2 = \frac{n\tau}{N}S^* - \gamma$. The steady state $(S^*, 0)$ is unstable when $\lambda_2 > 0$, that is when $S^* > N\frac{\gamma}{n\tau}$. Since $S^* \in [0, N]$, unstable steady states occur in the case $\frac{\gamma}{n\tau} < 1$, i.e. the bifurcation is at $n\tau = \gamma$.

The differential equation of $[S]_f$ shows that, except at the equilibria, $[S]_f$ is decreasing, while $[I]_f$ may increase first and later decrease. The switch occurs where the derivative of $[I]_f$ is zero, i.e. $[S]_f = N\frac{\gamma}{n\tau}$. Because $[S]_f < N$, this is only possible if $n\tau > \gamma$ (consistent with the observations above).

The phase portrait can be determined simply by inspecting the signs of the derivatives and drawing the direction field. In the case $n\tau > \gamma$, the trajectories go down and to the right above the line $S = N\frac{\gamma}{n\tau}$ and down and to the left below it as in the left panel of Fig. 4.4. In the case $n\tau < \gamma$, the trajectories go down and to the left at any point of the triangle as in the right panel of Fig. 4.4. All trajectories converge to one of the steady states situated along the left boundary of the triangle. In (a) there are orbits that emerge from one fixed point and approach another; these do not exist in (b). The trajectory's limit depends on its initial condition. Thus in contrast to the SIS case, the long-time behaviour of the system depends on the initial condition.

Perhaps surprisingly, the equation of the trajectories can be given analytically. For simplicity, we assume $[R]_f(0) = 0$, so $[S]_f(0) + [I]_f(0) = N$. We can rewrite Eq. (4.9a) as $[\dot{S}]_f + \tau\frac{n}{N}[S]_f[I]_f = 0$. Using an integrating factor, we get

$$\frac{d}{dt}\left(e^{\tau\frac{n}{N}\int_0^t [I]_f(\hat{t})d\hat{t}}[S]_f(t)\right) = 0.$$

Note that $\int_0^t [I]_f(\hat{t})d\hat{t} = [R]_f(t)/\gamma$, so $\frac{d}{dt}\left(e^{\tau\frac{n}{N\gamma}[R]_f}[S]_f\right) = 0$. Integrating this yields $[S]_f = S_0 e^{-\tau\frac{n}{N\gamma}[R]_f}$. Because $[S]_f + [I]_f + [R]_f = N$, we can express $[S]_f$ and $[I]_f$ in terms of $[R]_f$ as

$$[S]_f = S_0 e^{-\tau\frac{n}{N\gamma}[R]_f}$$
$$[I]_f = N - S_0 e^{-\tau\frac{n}{N\gamma}[R]_f} - [R]_f.$$

The trajectories in Fig. 4.4 were plotted using this formula. We can use this to derive an equation for the final epidemic size $[R]_f^\infty = \lim_{t\to\infty}[R]_f(t)$. Using that $[I]_f \to 0$ as $t \to \infty$, we conclude

$$[R]_f^\infty = N - S_0 e^{-\tau\frac{n}{N\gamma}[R]_f^\infty}. \tag{4.12}$$

This implicit relation cannot be solved analytically, but it can be solved numerically, for example by iteration. Its solution yields the final epidemic size $[R]_f^\infty$. Note that this depends on the initial number of susceptible nodes. Summarising, the following analytical results for the closed system (4.9) can be given.

Analytical results for SIR homogeneous mean-field equations at the single level

- Conservation of singles: $[S]_f(t) + [I]_f(t) + [R]_f(t) = N$.
- Reducibility to two differential equations: (4.9a) and (4.9b).
- There are infinitely many steady states: $(S^*, 0)$ with any $S^* \in [0, N]$. Bifurcation occurs at $\gamma = n\tau$. If $\gamma > n\tau$, then all steady states are stable, while for $\gamma < n\tau$ the steady states with $S^* > N\frac{\gamma}{n\tau}$ are unstable.
- Global behaviour: all solutions converge to a steady state $([S]_f^\infty, [I]_f^\infty, [R]_f^\infty)$ with $[I]_f^\infty = 0$, $[R]_f^\infty + [S]_f^\infty = N$, and the final epidemic size $[R]_f^\infty$ is determined by Exercise 4.1. This steady state depends on the initial condition.

Exercise 4.1. Repeat the derivation of (4.12) to find $[S]_f(t)$ and $[I]_f(t)$ in terms of the initial conditions and $[R]_f(t)$ if $[R]_f(0) \neq 0$.

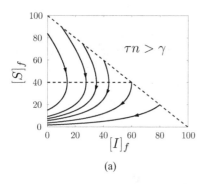
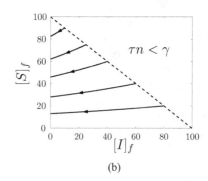

(a)   (b)

Fig. 4.4: Phase portrait of the mean-field equation of an SIR epidemic in the case of (a) $\tau n > \gamma$ and (b) $\tau n < \gamma$. The parameter values are $N = 100$, $n = 5$, $\gamma = 1$ and (a) $\tau = 0.5$, (b) $\tau = 0.1$. The line $[S]_f = N\frac{\gamma}{n\tau}$ is also shown in the left panel.

4.3.3 SIS homogeneous pairwise equations

Consider the system closed at the level of triples in the case of SIS dynamics, i.e. system (4.10). Note that adding the first two equations, we get that $[S]_s(t) + [I]_s(t)$ is constant in time. If the initial condition satisfies $[S]_s(0) + [I]_s(0) = N$, then

$$[S]_s(t) + [I]_s(t) = N.$$

Hence, the equation for $[\dot{S}]_s$ can be replaced with $[S]_s(t) = N - [I]_s(t)$, reducing the system to four differential equations. Alternatively, one may choose to keep the $[S]_s$ equation and dispose of the second equation. Moreover, pair conservation holds in the closed system, meaning that $2[SI]_s + [SS]_s + [II]_s$ is constant in time, where we keep in mind that $[SI]_s = [IS]_s$. The conservation relation can be verified by showing that the derivative of the sum is 0. This enables us to reduce the system further to three differential equations, by omitting one of the equations for $[SI]_s$, $[SS]_s$ or $[II]_s$. The total number of pairs can be expressed in terms of the average degree n:

$$2[SI]_s + [SS]_s + [II]_s = nN.$$

If this relation holds initially, it remains true for all time.

We have seen that in the exact system, the pair conservation relation

$$[SS] + [SI] = n[S]$$

also holds. We will prove in Section 4.7 that this conservation relation is preserved if the simplest closures (4.6) and (4.7) are used. Similarly,

$$[SI]_s + [II]_s = n[I]_s.$$

This enables further reduction to two differential equations: only one of the three variables $[SI]_s$, $[SS]_s$ and $[II]_s$ is needed. It is useful to choose $[S]_s$ and $[SS]_s$ as the main variables, as this results in a system which is easier to analyse. We have the reduced system

$$[\dot{S}]_s = \gamma[I]_s - \tau[SI]_s,$$
$$[\dot{SS}]_s = 2\gamma[SI]_s - 2\tau\frac{n-1}{n}\frac{[SI]_s[SS]_s}{[S]_s},$$

with the algebraic relations below yielding the remaining variables as

$$[I]_s = N - [S]_s, \quad [SI]_s = n[S]_s - [SS]_s, \quad [II]_s = nN - 2[SI]_s - [SS]_s.$$

Substituting these into the differential equations yields

$$[\dot{S}]_s = \gamma N - (\gamma + n\tau)[S]_s + \tau[SS]_s, \tag{4.13a}$$
$$[\dot{SS}]_s = 2(n[S]_s - [SS]_s)\left(\gamma - \tau(n-1)\frac{[SS]_s}{n[S]_s}\right). \tag{4.13b}$$

This two-dimensional system can be studied by elementary phase plane analysis, carried out in Section 4.7. This analysis shows that there is a disease-free steady state, given by $[S]_s^{df} = N$, $[SS]_s^{df} = nN$, which is stable if $\tau(n-1) < \gamma$, while for $\tau(n-1) > \gamma$ there is a stable endemic steady state, given by $[S]_s^e = N\frac{\gamma(n-1)}{\tau n(n-1)-\gamma}$, $[SS]_s^e = \frac{\gamma n}{\tau(n-1)}[S]_s^e$. Moreover, the stability is global in both cases, i.e. all solutions with realisable initial conditions converge to the stable steady state. The fact that the solutions of the reduced two-dimensional systems converge to the steady states implies that the solutions of the original five-dimensional system (4.10) converge to a steady state. In the case of $\tau(n-1) < \gamma$, these converge to the disease-free steady state, and for $\tau(n-1) > \gamma$ they tend to the endemic steady state.

The correlation $C_{AB} = \frac{N}{n}\frac{[AB]}{[A][B]}$ plays an important role in understanding the dynamic of the propagation process. The pairwise system (4.10) enables us to compute correlations of different types of nodes. The correlations given by this approximating system will be denoted by C_{AB}^s, where the superscript refers to the fact that these correlations are computed from the solutions of the second approximation. Assuming uniformly distributed initial infections, Section 4.7 will show that for any positive finite time,

$$\frac{n}{N}[S]_s[I]_s > [SI]_s, \quad [SS]_s > \frac{n}{N}[S]_s^2, \quad [II]_s > \frac{n}{N}[I]_s^2 \tag{4.14}$$

implying the following inequalities for the correlations

$$C_{SI}^s < 1, \quad C_{SS}^s > 1, \quad C_{II}^s > 1.$$

This means that susceptible nodes are more likely to be connected to other susceptible nodes and infected nodes are more likely to be connected to other infected nodes. This is due to the process itself: if newly infected nodes were selected randomly from all nodes, we would see no correlation of neighbours, but because they are selected from those nodes with infected neighbours, a correlation develops. Moreover, based on the inequalities in (4.14) it will be proved that the prevalence obtained from the second (pairwise) approximation is always bounded from above by the first (single level) approximation, that is $[I]_s < [I]_f$. To conclude, the analytical results derived from the closed system (4.10) are summarised below.

Analytical results for SIS homogeneous pairwise equations

- Conservation of singles and pairs:

$$[S]_s + [I]_s = N, \quad 2[SI]_s + [SS]_s + [II]_s = nN,$$
$$[SS]_s + [SI]_s = n[S]_s, \quad [SI]_s + [II]_s = n[I]_s.$$

- Reducibility to the two-dimensional system (4.13).
- Transcritical bifurcation at $\gamma = \tau(n-1)$: the disease-free steady state loses its stability and a stable endemic steady state appears.
- Global behaviour for $\gamma > \tau(n-1)$: all solutions converge to the disease-free steady state

$$[I]_s^{df} = 0, \ [S]_s^{df} = N, \ [SI]_s^{df} = 0, \ [SS]_s^{df} = nN, \ [II]_s^{df} = 0.$$

- Global behaviour for $\gamma < \tau(n-1)$: all solutions converge to the endemic steady state

$$[I]_s^e = Nn \frac{\gamma - \tau(n-1)}{\gamma - \tau n(n-1)}, \ [S]_s^e = N \frac{\gamma(n-1)}{\tau n(n-1) - \gamma}, \ [SI]_s^e = \frac{\gamma}{\tau}[I]_s^e, \quad (4.15)$$

$$[SS]_s^e = \frac{\gamma n}{\tau(n-1)}[S]_s^e, \ [II]_s^e = [I]_s^e \left(1 + \frac{\gamma(n-1)}{\tau n} \frac{[I]_s^e}{[S]_s^e}\right).$$

- The correlations $C_{AB}^s = \frac{N}{n} \frac{[AB]_s}{[A]_s[B]_s}$ satisfy the inequalities

$$C_{SI}^s < 1, \quad C_{SS}^s > 1, \quad C_{II}^s > 1.$$

- The prevalence obtained from the second approximation is bounded from above by the first approximation, that is $[I]_s < [I]_f$.

4.3.4 SIR homogeneous pairwise equations

Consider the system for the SIR dynamics closed at the level of triples, i.e. system (4.11). Note that adding the first three equations, we get that $[S]_s(t) + [I]_s(t) + [R]_s(t)$ is constant in time. If the initial condition satisfies $[S]_s(0) + [I]_s(0) + [R]_s(0) = N$, then for all time

$$[S]_s(t) + [I]_s(t) + [R]_s(t) = N.$$

Hence, the $[\dot{R}]_s$ equation can be replaced with $[R]_s(t) = N - [S]_s(t) - [I]_s(t)$, reducing the system to four differential equations to analyse. Similarly to the mean-field equation at single level, there are infinitely many steady states, for which $[I]_s = 0$ and $[SI]_s = 0$. The differential equations for $[S]_s$ and $[SS]_s$ show that they are decreasing, but they cannot go negative. Hence, they have a long time limit, corresponding to one of these steady states. The precise final state depends on the initial condition. Here, we assume that initially there are no recovered nodes and that susceptible and infected nodes are distributed randomly at the initial time. Using a subscript 0 to denote the initial condition and assuming $[S]_0 \in (0, N)$, we get $[I]_0 = N - [S]_0$, $[SI]_0 = \frac{n}{N}[S]_0[I]_0$ and $[SS]_0 = \frac{n}{N}[S]_0^2$.

Equations (4.11a), (4.11d) and (4.11e) do not depend on $[I]_s$, so they can be solved without solving (4.11b). This enables us to derive relations between these three variables by a manipulation of their differential equations. In Section 4.7, we show that the variables $[SI]_s$ and $[SS]_s$ can be expressed in terms of $[S]_s$ as

$$[SS]_s = \frac{n}{N}[S]_s^2 Q^2, \tag{4.16a}$$

$$[SI]_s = n[S]_s \left(1 + \frac{\gamma}{\tau}(1 - Q) - \frac{[S]_s}{N} Q^2 \right), \tag{4.16b}$$

where $Q = [S]_0^{1/n}[S]_s^{-1/n}$. These equations can be used to determine the steady state value of $[S]_s$ and the final epidemic size, i.e. the steady state value of $[R]_s$. These will be denoted by $[S]_s^\infty = \lim_{t \to \infty} [S]_s(t)$ and $[R]_s^\infty = \lim_{t \to \infty} [R]_s(t)$. Taking the limit $t \to \infty$ in Eq. (4.16b) and using that the steady state value of $[SI]_s$ is $[SI]_s^\infty = 0$, leads to an implicit equation for $[S]_s^\infty$,

$$n[S]_s^\infty \left(1 + \frac{\gamma}{\tau}\left(1 - \left(\frac{[S]_0}{[S]_s^\infty}\right)^{\frac{1}{n}}\right) - \frac{[S]_s^\infty}{N}\left(\frac{[S]_0}{[S]_s^\infty}\right)^{\frac{2}{n}} \right) = 0$$

leading to

$$N(\tau + \gamma)([S]_s^\infty)^{\frac{2}{n}} = N\gamma[S]_0^{\frac{1}{n}}([S]_s^\infty)^{\frac{1}{n}} + \tau[S]_s^\infty[S]_0^{\frac{2}{n}}. \tag{4.17}$$

This equation cannot be solved analytically; however, its numerical solution, obtained, for example, by iteration, yields the steady state value of $[S]_s$, and then the final epidemic size via $[R]_s^\infty = N - [S]_s^\infty$.

The stability of the disease-free steady state will be studied in Example 4.4, where it will be shown that it is stable when $\tau(n-2) < \gamma$ and unstable when $\tau(n-2) > \gamma$. To conclude, the analytical results based on the closed system (4.11) are summarised below.

Analytical results for SIR homogeneous pairwise equations

- Conservation of singles: $[S]_f(t) + [I]_f(t) + [R]_f(t) = N$.
- Reducibility to four differential equations: (4.11a), (4.11b), (4.11d), and (4.11e).
- The disease-free steady state is stable when $\tau(n-2) < \gamma$ and unstable when $\tau(n-2) > \gamma$.
- Global behaviour: there are infinitely many steady states. Solutions converge to steady states $([S]_s^\infty, [I]_s^\infty, [R]_s^\infty)$ with $[I]_s^\infty = 0$ and $[R]_s^\infty + [S]_s^\infty = N$.
- The steady state depends on the initial condition, $[S]_s^\infty$ is determined by Eq. (4.17), and the final epidemic size is $[R]_s^\infty = N - [S]_s^\infty$.

4.4 Basic reproductive ratio

One of the most commonly studied (and over-interpreted) quantities of mathematical epidemiology is the basic reproductive ratio, R_0. It is heuristically defined to be the average number of new infections caused by individuals that are infected shortly after disease introduction in a completely susceptible population [4, 78]. The individuals we average over are chosen according to the way the disease chooses individuals. So, for example, higher degree nodes tend to occur with higher weight than lower degree nodes, and degree 0 nodes will not be involved. If $R_0 > 1$, then an epidemic is possible (but not guaranteed), while if $R_0 < 1$, the disease will die out.

For the simple compartmental SIS and SIR models which are equivalent to systems (4.8) and (4.9), the models assume that when an infected individual interacts with a neighbour, that neighbour is randomly chosen from the population at that moment. Thus, these equations model a scenario in which an infected individual transmits at a constant rate throughout its infectious period, and at least early in the spread effectively all transmissions are successful as almost all individuals are susceptible. Because the average infection duration is $1/\gamma$, this means that the average number of new infections caused per infectious individual is $R_0 = \beta/\gamma$, where $\beta = \tau n$ is the total transmission rate from one node. However, once we account for the fact that nodes remain connected to their neighbours, we observe that the rate at which infected nodes cause infection reduces with time: for a highly infectious disease, a node may run out of susceptible neighbours before recovering.

In defining R_0, we referred to the underlying stochastic process rather than the deterministic approximation. So when we discuss R_0 for a deterministic model, the model must exactly map to a stochastic process for the definition to make sense.

Sometimes, the mapping involves simplifications. Other times, the deterministic approximation may be easy to analyse, but the mapping may be complicated, or it may be difficult to find R_0 for the stochastic model. For some stochastic models (such as SIS in a network with a degree distribution where $P(k) \sim k^{-\alpha}$ for any $\alpha > 1$ [63]), the concept of R_0 may break down while a deterministic approximation may still yield a useful prediction. Thus, it is useful to have a concept of R_0 which can be applied to deterministic models.

In this section, we discuss a "growth-based" reproductive ratio that is appropriate for a deterministic model. Rather than calculating how many infections a node causes during its infectious period, we will think of a short interval and calculate how many infections happen for every recovery. The calculation is based on the observation that the rate of change of I is the rate at which new individuals are infected minus the rate at which they recover. For the simple compartmental models, the total instantaneous rate of recovery is γI and the total instantaneous rate of infection when all individuals are susceptible is βI. The ratio of the infection rate to the recovery rate is

$$\overline{R}_0 = \frac{\beta}{\gamma}$$

which is the same as R_0 calculated from the stochastic model.

To help our analysis, we introduce a definition:

Definition 4.5 *Given a deterministic model of disease spread in a population, if the linearisation at the disease-free equilibrium predicts $I(t) \approx ce^{\lambda t}$, then λ is the Malthusian parameter.*

Generally λ is the largest eigenvalue of the Jacobian matrix. However, we need to note that in many cases we might have a conserved quantity, for example we might have $[S] + [I] + [R] = N$. In such a case, we can eliminate an equation for one of $[S]$, $[I]$ or $[R]$. If we do not do this change of variables, the Jacobian matrix will have a zero eigenvalue for every such relation and the eigenvector corresponds to a change in N. These eigenvalues do not show up in the growth of I, and so they are ignored when finding the Malthusian parameter.

Exercise 4.2. Consider system (4.9).
   a. We consider the equilibrium $[S]_f = N$, $[I]_f = [R]_f = 0$.
      i. Find the Jacobian matrix (without using $[S]_f + [I]_f + [R]_f = N$).
      ii. Find the eigenvalues λ_1, λ_2 and λ_3 and eigenvectors v_1, v_2, and v_3.
      iii. The solution near the equilibrium is $([S]_f, [I]_f, [R]_f)(t) = (N, 0, 0) + c_1 e^{\lambda_1 t} v_1 + c_2 e^{\lambda_2 t} v_2 + c_3 e^{\lambda_3 t} v_3$. Find $[I]_f(t)$ as t grows.
      iv. Find the Malthusian parameter.
   b. We now reduce the dimension of the system and repeat.
      i. Show that $[S]_f(t) + [I]_f(t) + [R]_f(t)$ is constant.
      ii. Replace $[S]_f$ by $N - [I]_f - [R]_f$ and eliminate the equation for $[\dot{S}]_f$. Find the new Jacobian matrix at the equilibrium $[I]_f = [R]_f = 0$.
      iii. Compare the eigenvalues with those found in the first part.
      iv. Find the Malthusian parameter.

We generalise our calculation of $\overline{R}_0$ with the following definition.

Definition 4.6 *Given a deterministic model of disease spread in a population, the growth-based basic reproductive ratio $\overline{R}_0$ is the ratio of the total rate of infection to the total rate of recovery during the early exponential phase of the epidemic.*

We will calculate this under the assumption that the recovery rate is uniformly γ for every infected individual. We make two observations:

- We can write the total rate of new infections as $\dot{I} + \gamma I$.
- Early in the growth, after initial transients have died out, but before population-wide susceptible depletion is significant, we can approximate I as $ce^{\lambda t}$, where λ is the largest eigenvalue of the Jacobian at the fully susceptible equilibrium.

From this, it follows that at early time, the total rate of new infections is $(\lambda + \gamma)I$. Thus, the ratio of the total rate of infections to the total rate of recovery is $(\lambda + \gamma)I/\gamma I = (\lambda/\gamma) + 1$. Thus, the following theorem holds.

Theorem 4.7 *Let X represent a deterministic model of disease spread in a population with uniform recovery rate γ, if λ is the Malthusian parameter, then*

$$\overline{R}_0^X = \frac{\lambda}{\gamma} + 1. \tag{4.18}$$

If we know the distribution of infection intervals (delay between a node's infection and when it infects others), then a calculation of the type given in [319] will allow us to translate between $\overline{R}_0$ and R_0. It is important to note that when $R_0 = 1$, so does $\overline{R}_0$ for the corresponding deterministic model, but in general they are not equal. The $R_0 = \overline{R}_0 = 1$ case is important as it represents the transition from epidemics being impossible to epidemics being possible.

The distinction is subtle: R_0 represents the result of choosing an early time and calculating the expected number of new infections caused over the infectious period of individuals that become infected at that time. In contrast, $\overline{R}_0$ represents the result of choosing an early time and calculating the ratio of the average instantaneous per infected individual rate of causing infection to the average instantaneous per infected individual recovery rate. That is, R_0 is an average over an infectious period, while $\overline{R}_0$ is an average over all individuals infected at a given time.

The fact that $\overline{R}_0 \neq R_0$ results from the fact that the rate at which an individual causes new infections may differ early and late in the individual's infection period. In a growing epidemic, the infectious individuals are disproportionately recent infections (and thus $\overline{R}_0$ is more heavily weighted towards "younger" infections), while in a decaying outbreak there are relatively fewer recent infections. For our network models, once an individual transmits to a neighbour, that neighbour becomes (at least temporarily) immune to new infections. Thus, an individual is more infectious early in his/her infectious period than late and so $\overline{R}_0 > R_0 > 1$ above the epidemic threshold, with $\overline{R}_0 < R_0 < 1$ below the epidemic threshold. In other models, for example if there is an early less-infectious incubation period, an epidemic could have $R_0 > \overline{R}_0 > 1$.

We now look at some examples.

Example 4.1. The homogeneous mean-field approximation for the SIS model is given by a single equation, Eq. (4.8b), after the substitution $[S]_f = N - [I]_f$. The disease-free steady state is $[I]_f = 0$ and the Jacobian is the 1×1 matrix $J = \tau n - \gamma$. Hence, the Malthusian parameter, i.e. its only eigenvalue, is $\lambda = \tau n - \gamma$. Therefore, (4.18) leads to

$$\overline{R}_0^{MF} = \frac{\tau n}{\gamma},$$

which equals R_0. Equality occurs because an individual is equally infectious at all stages of infection. Epidemics are predicted to occur when $\overline{R}_0^{MF} > 1$, which is equivalent to $\tau n > \gamma$, the condition for instability of the disease-free steady state found in Section 4.3.1.

Example 4.2. The homogeneous mean-field approximation for SIR epidemic is given by system (4.9). The disease-free steady state is $[S]_f = N$, $[I]_f = 0$. The Jacobian is the 2×2 matrix

$$J = \begin{pmatrix} 0 & \tau n \\ 0 & \tau n - \gamma \end{pmatrix}.$$

Its eigenvalues are 0 and $\lambda = \tau n - \gamma$. Thus, the Malthusian parameter is $\lambda = \tau n - \gamma$; therefore, (4.18) leads to the same formula as in the SIS case:

$$\overline{R}_0^{MF} = \frac{\tau n}{\gamma}.$$

Example 4.3. The homogeneous SIS pairwise approximation is given by system (4.10). We can eliminate the $[S]_s$ and $[II]_s$ equations using $[S]_s = N - [I]_s$ and $[II]_s = Nn - 2[SI]_s - [SS]_s$. The disease-free steady state is $[I]_s = 0$, $[SI]_s = 0$, $[SS]_s = Nn$ and the Jacobian is the 3×3 matrix

$$J = \begin{pmatrix} -\gamma & \tau & 0 \\ 0 & \tau(n-2) - 3\gamma & -\gamma \\ 0 & 2\gamma - 2\tau(n-1) & 0 \end{pmatrix}.$$

Its eigenvalues are $-\gamma$ and the roots of the quadratic polynomial $\lambda^2 - \lambda(\tau(n-2) - 3\gamma) + 2\gamma^2 - 2\tau\gamma(n-1)$. Calculating the roots of this quadratic, Eq. (4.18) yields the growth-based basic reproductive ratio as

$$\overline{R}_0^{PW} = \frac{\tau(n-2) - \gamma + \sqrt{(\tau(n-2) - 3\gamma)^2 + 8\gamma(\tau(n-1) - \gamma)}}{2\gamma}. \tag{4.19}$$

Elementary algebra shows that the epidemic threshold $\overline{R}_0^{PW} = 1$ is equivalent to $\tau(n-1) = \gamma$, which is the condition of a transcritical bifurcation, i.e. the disease-free steady state loses its stability and the endemic steady state appears when $\overline{R}_0^{PW} = 1$.

Example 4.4. The homogeneous pairwise approximation in the case of an SIR dynamics is given by the system (4.11). Ignoring $[II]_s$, as it is not needed, the disease-free steady state is $[S]_s = N$, $[I]_s = 0$, $[SI]_s = 0$, $[SS]_s = Nn$. The Jacobian is the 4×4 matrix

$$J = \begin{pmatrix} 0 & 0 & -\tau & 0 \\ 0 & -\gamma & \tau & 0 \\ 0 & 0 & \tau(n-2)-\gamma & 0 \\ 0 & 0 & -2\tau(n-1) & 0 \end{pmatrix}.$$

Its eigenvalues are 0, 0, $-\gamma$, $\tau(n-2) - \gamma$. Hence, (4.18) leads to the growth-based basic reproductive ratio

$$\overline{R}_0^{\text{PW}} = \frac{\tau(n-2)}{\gamma}.$$

This agrees with the formula of [166] Section 5. For the underlying stochastic process, the probability of transmitting before recovering is $\tau/(\tau + \gamma)$ and a newly infected individual early in the epidemic has $n - 1$ susceptible neighbours (it cannot reinfect its source of infection). So the value of R_0 based on the average number of transmissions an individual causes is $\tau(n-1)/(\tau+\gamma)$.

Exercise 4.3. Show that $\overline{R}_0^{\text{PW}} > 1$, $\overline{R}_0^{\text{PW}} = 1$, or $\overline{R}_0^{\text{PW}} < 1$ iff $R_0 = \tau(n-1)/(\tau+\gamma)$ satisfies the same inequality.

An alternative derivation of R_0 from continuous-time models is given in [79, 308] through deriving a "next-generation matrix". In general, this R_0 will not agree with $\overline{R}_0$, but will give an identical threshold.

We end this section with a comment on the applicability of calculations relating R_0 to the final size of an epidemic. Although researchers often place considerable focus on how the final size of an epidemic is related to R_0, and we will derive several such relations, we must be careful. There is consistency across epidemiological models that having $R_0 < 1$ means epidemics are not possible from a small introduction and having $R_0 > 1$ means they are possible. However, there is no consistency in how the final size is related to R_0 for different models, in particular if the population is heterogeneous. In this case, the average individual being infected early in an epidemic is different from the average individual being exposed late in the epidemic. Thus, we should not anticipate that an averaged quantity early in the epidemic has much bearing on the average outcome for the entire epidemic. For practical application, these relations must be regarded as guidelines rather than as quantitative predictions.

4.5 Comparison of mean-field models to simulation and exact master equations

The mean-field systems closed at the level of pairs and triples involve approximations. Hence, their accuracy has to be verified. In this section, we numerically compare the time dependence of the prevalence derived from the closed or approximate models to exact master equations and to results based on simulation. The difference between approximate and exact models can be analytically estimated by stochastic and functional analytical techniques. These will be discussed in Chapter 10. The comparisons that can be carried out are not exhaustive or complete because the size of the exact system increases exponentially with the size of the network. This means that even a numerical solution of the exact system cannot be found or computed for large networks.

Therefore, the following options for comparison are available: (a) networks of small size, (b) networks with special structural symmetries when the exact system is lumpable and (c) stochastic simulation which replaces the solution of the exact master equations. The first option is not ideal since generally the closures rely on statistical independence of some statuses and, by its nature, is not expected to apply to small networks. The closed systems are expected to perform better in larger networks. The comparisons in this section will be carried out for the two latter cases.

The performance of mean-field models may depend on system parameters, namely the size of the network N, the values of the parameters describing the epidemic dynamics, i.e. per contact infection (τ) and recovery (γ) rates, and the network structure. As mentioned above, comparison to exact master equations is possible only for special networks. Therefore, the comparison for different values of N, τ and γ will be done for fully connected and star networks, while the dependence on the structure of the graph will be investigated using explicit stochastic simulation.

4.5.1 Comparison to exact master equations, dependence on system size and infection parameters

The closed systems we have developed so far assume a homogeneous degree distribution. We will test how they perform in comparison with exact systems. In particular, we will use the exact $\mathcal{O}(N)$ systems for fully connected and star networks from Chapter 2. Because of our homogeneity assumption, we expect good agreement with the fully connected networks, but not with the star networks. First, we investigate how the quality of the agreement depends on N, τ and γ for completely connected networks. This is followed by showing that the agreement is significantly worse for a star network, which in some sense is furthest from homogeneous.

Let us consider the SIS epidemic on a complete graph with N nodes. The system of master equations can be lumped to a system with $(N+1)$ variables $x_0, x_1, \ldots, x_N$, where $x_k(t)$ is the probability that there are k infected nodes at time t. This linear system is formulated in Proposition 2.3 and can be solved numerically. The expected number of infected nodes is $[I](t) = \sum_{k=0}^{N} k x_k(t)$. As a numerical comparison, the time dependence of the expected number of infected nodes is determined from the exact lumped system, the mean-field system at the level of singles (4.8) and the pairwise model (4.10). Thus, the following three functions are compared in this section: $[I](t)$, $[I]_f(t)$ and $[I]_s(t)$.

We first study the dependence on the disease dynamics parameters. The system behaviour depends only on the ratio τ/γ. So we arbitrarily fix $\gamma = 1$ and vary

τ. The analytical study of the system closed at the level of pairs revealed that the system behaviour changes at the critical value $\tau_c = \gamma/n$, where a transcritical bifurcation occurs. This is the point at which the disease-free steady state loses its stability and a stable endemic steady state appears. Hence, we investigate how the performance of the mean-field approximation changes in a neighbourhood of τ_c the predicted critical value of τ. Four values of τ are chosen: $\tau = 0.9\tau_c$, $\tau = \tau_c$, $\tau = 1.2\tau_c$ and $\tau = 1.5\tau_c$. The expected prevalences, $[I](t)$, $[I]_f(t)$, and $[I]_s(t)$, are plotted for a fixed system size with $N = 200$. The results are shown in Fig. 4.5. Close to the predicted critical value, the agreement is poor, but farther away it improves significantly. Further numerical comparisons show that for $\tau > 2\tau_c$ and $\tau < 0.8\tau_c$, the three curves almost coincide. Moreover, the pairwise system yields a slightly better approximation, and this effect will become more marked for other network types.

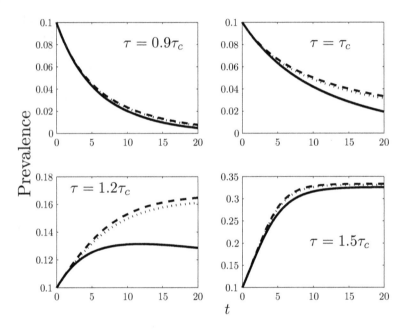

Fig. 4.5: The time dependence of the expected number of infected nodes based on the exact system for a complete graph, $[I](t)$ (solid lines), and mean-field systems closed at the level of pairs, $[I]_f(t)$ (dashed lines), and triples, $[I]_s(t)$ (dotted lines). The parameter values are $N = 200$ and $\gamma = 1$, with the τ values shown in each subfigure.

The expected prevalence obtained from the exact master equation tends to zero as time goes to infinity for any value of τ. This is because the fully susceptible state is the only absorbing state of the Markov chain. However, if the graph is large enough and τ is significantly larger than its predicted critical value, the time to reach the absorbing state is exponentially long. In the stochastic model, there is a so-called "quasi-steady" state that corresponds to the endemic steady state of the mean-field-like approximations. This was investigated in detail in [6, 7, 230]. In fact, the behaviour of the mean-field-like approximations and the exact master equations is qualitatively different. The mean-field approximations predict a transcritical bifurcation, while the exact master equations exhibit no bifurcation. The appearance of the quasi-steady state is not a sudden change in the behaviour of the exact system, and it can be perceived as a transition rather than bifurcation. Figure 4.5 shows that the behaviour change happens gradually from $\tau = 0.9\tau_c$ to $\tau = 1.5\tau_c$. If these computations are carried out with higher values of N, then a similar phenomenon can be observed close to the critical τ value. However, as N tends to infinity, the region of τ where the transition happens shrinks to a zero width.

Let us now investigate the effect of network size N. In order to make a quantitative comparison, rather than a visual comparison of the time-dependent prevalence, we compare the prevalence level in the endemic steady state. Of course, we choose a value for τ for which the endemic steady state is stable, e.g. $\tau = 2\tau_c$. The quasi-steady state, $[I]^e$, of the exact master equation can be determined analytically following (2.18). The steady state of the system closed at the level of pairs is

$$[I]^e_f = N\left(1 - \frac{\gamma}{n\tau}\right)$$

as shown in the beginning of Section 4.3. The steady state of the system closed at the level of triples is given in (4.15) as

$$[I]^e_s = Nn\frac{\gamma - \tau(n-1)}{\gamma - \tau n(n-1)}.$$

Figure 4.6 compares the endemic steady state values $[I]^e_f$ and $[I]^e_s$ to the exact value $[I]^e$ for different values of N. The value of τ is fixed at $\tau = 2\tau_c$; however, we have to keep in mind that, varying N, the critical value $\tau_c = \gamma/n$ changes due to the dependence of degree on N, $n = N - 1$. Hence, τ depends on N as $\tau = 2\gamma/(N-1)$, with γ fixed at $\gamma = 1$ as above. Varying the value of N, the approximation errors of the first and second approximations at the steady state are defined as

$$\frac{1}{N}\left|[I]^e_f - [I]^e\right|, \quad \frac{1}{N}\left|[I]^e_s - [I]^e\right|.$$

The mean-field approximations improve as N increases. The accuracy of the approximation is close to C/N as seen in Fig. 4.6, with the constant C depending on the epidemic parameters. The order of the approximation error is the same for the pair and triple closures; however, the constant C is smaller for the triple closure.

In Chapter 10, we rigorously show that the steady state for a complete graph obtained from the pair closure satisfies the inequality $\left|[I]^e_f/N - [I]^e/N\right| \leq C/N$, with an appropriate constant C. Since the triple closure also yields an error of $\mathcal{O}(1/N)$, it raises the natural question of whether more accurate closures are possible. In [176], new closures are introduced based on the assumption that the distribution of x_k (the probability of k infections) is binomial or close to normal. Numerical tests suggest

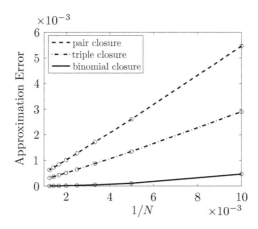

Fig. 4.6: The approximation errors for the SIS endemic equilibrium in a complete network: $\frac{1}{N}\left|[I]_f^e - [I]^e\right|$ (pair closure, dashed line), $\frac{1}{N}\left|[I]_s^e - [I]^e\right|$ (triple closure, dashed-dotted line) and $\frac{1}{N}\left|[I]_b^e - [I]^e\right|$ (binomial closure, solid line) plotted for different values of system sizes N, with $\gamma = 1$ and $\tau = 2\gamma/(N-1)$.

that the approximation error is $\mathcal{O}\left(1/N^2\right)$; however, a formal proof is missing. The approximation error obtained by using the binomial closure is also shown in Fig. 4.6. The steady state for this closure is

$$[I]_b^e = N\frac{Nq^2 - 1}{Nq - 1},$$

where $q = 1 - \frac{\gamma}{N\tau}$. Unfortunately, such closures can be applied only when the master equation has a form similar to that given in Proposition 2.3, i.e. the variables are functions $x_k(t)$, $k = 0, 1, \ldots, N$. For a complete graph, the lumped system takes this form. Several other random networks can be approximated well in this way, see [229], but in general this is not possible.

Exercise 4.4. For a star network, the master equations of an SIS epidemic can be lumped to a system of size $2N$ as given in Proposition 2.4. Solving this system with a numerical ODE solver, determine $[I](t)$, the expected value of the number of infected nodes, and compare it to $[I]_f(t)$ and $[I]_s(t)$ for $\tau = 0.9\tau_c$, $\tau = \tau_c$, $\tau = 1.1\tau_c$ and $\tau = 1.5\tau_c$, as done or illustrated in Fig. 4.5. Note that the predicted critical value of τ in this case is $\tau_c = \gamma/n$, where $n = (2N-2)/N$ is the average degree of the star graph. Observe that for a strongly heterogeneous network, such as the star graph, the performance of the mean-field models is significantly weaker.

4.5.2 Comparison to simulation, dependence on network structure

We carry out a systematic study of the performance of mean-field models as follows: SIS and SIR dynamics are considered on Configuration Model random networks with N nodes. The expected number of infected nodes is determined from individual-based stochastic simulation as described in Section 1.3.1. The average of several simulations, approximating the expected value of the prevalence, is denoted now by $[I](t)$. This is compared numerically to the prevalence given by the mean-field system closed at the level of pairs, Eqs. (4.8) and (4.9), and triples, systems (4.10) and (4.11). Hence, the comparison amounts to assessing how well $[I]_f(t)$ and $[I]_s(t)$ approximate $[I](t)$. This analysis focuses on mapping out the effect of the following parameters:

- transmission and recovery rates τ and γ, respectively;
- system size N;
- average degree $\langle K \rangle$;
- variance of the degree distribution $\langle K^2 \rangle - \langle K \rangle^2$.

This comparisons are carried out for the following four network types:

- regular random networks,
- bimodal random networks,
- Erdős–Rényi random networks,
- configuration random networks with power law degree distribution.

The network construction is described in Section 1.2.3. We now study how the epidemic dynamics depends on parameters in random regular (homogeneous) networks. Again, the system behaviour depends only on the ratio τ/γ; hence, we arbitrarily set $\gamma = 1$ and vary τ. The performance of the mean-field approximation is studied in a neighbourhood of the predicted critical value $\tau_c = \gamma/\langle K \rangle$. Four values of τ were chosen at $\tau = 0.9\tau_c$, $\tau = \tau_c$, $\tau = 1.1\tau_c$ and $\tau = 1.5\tau_c$, and the functions $[I](t)$, $[I]_f(t)$ and $[I]_s(t)$ are plotted in Fig. 4.7 for a system of size $N = 1000$. Close to the predicted critical value, the agreement is poorer when compared to infection rates farther from τ_c. Further numerical comparison shows that for $\tau > 2\tau_c$ and for $\tau < 0.8\tau_c$, the pairwise model provides excellent agreement with results from simulations and significantly outperforms the mean-field model closed at the level of pairs.

The agreement between simulation and mean-field models also depends on the average degree. In general, agreement is better for denser networks. This is illustrated in Fig. 4.8, where again $[I](t)$, $[I]_f(t)$ and $[I]_s(t)$ are shown for $\langle K \rangle = 50$ in the main figure and for $\langle K \rangle = 5$ in the inset. The value of τ in both cases is twice the predicted critical value $\tau_c = \gamma/\langle K \rangle$.

The mean-field approximations used here are based on the homogeneity assumption: each node has degree $\langle K \rangle$. Thus, we expect poorer agreement if the graph is not regular. As a first step, we consider an Erdős–Rényi random graph and investigate the effect of $\langle K \rangle$ on the accuracy of the ODE approximations. Figure 4.9 shows $[I](t)$, $[I]_f(t)$ and $[I]_s(t)$ for $\langle K \rangle = 50$ in the main figure and for $\langle K \rangle = 5$ in the inset.

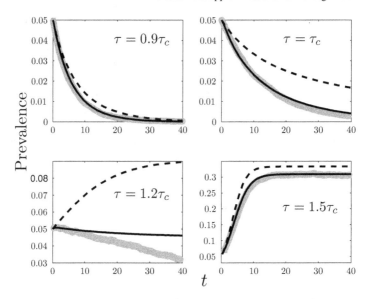

Fig. 4.7: The time dependence of the expected number of infected nodes in an SIS epidemic for a regular random network based on the average of 200 individual-based stochastic simulations $[I](t)$ (gray thick lines), and mean-field systems closed at the level of pairs, $[I]_f(t)$ (dashed lines), and triples, $[I]_s(t)$ (solid lines). The parameter values are $N = 1000$, $\langle K \rangle = 20$ and $\gamma = 1$; the τ values are shown in each subfigure.

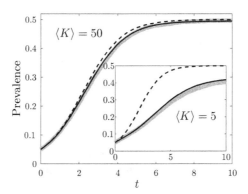

Fig. 4.8: The time dependence of the expected number of infected nodes in an SIS epidemic for a regular random network based on the average of 200 individual-based stochastic simulations $[I](t)$ (grey thick lines), and mean-field systems closed at the level of pairs, $[I]_f(t)$ (dashed lines), and triples, $[I]_s(t)$ (solid lines), for a denser graph with $\langle K \rangle = 50$ and a sparser graph with $\langle K \rangle = 5$ (inset). The parameter values are $N = 1000$, $\gamma = 1$ and $\tau = 2\tau_c = 2\gamma/\langle K \rangle$.

As described in Section 1.2.3, the parameter p is chosen so that $\langle K \rangle = (N-1)p$ yields the desired average degree $\langle K \rangle$. In both cases, τ is twice the predicted critical value $\tau_c = \gamma/\langle K \rangle$. Two observations can be made: (a) agreement is poorer than for a regular random graph and (b) agreement is better for the denser network.

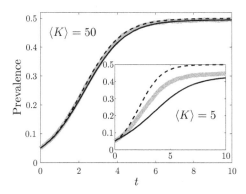

Fig. 4.9: The time dependence of the expected number of infected nodes in an SIS epidemic for an Erdős–Rényi random network based on the average of 200 individual-based stochastic simulations $[I](t)$ (grey thick lines), and mean-field systems closed at the level of pairs, $[I]_f(t)$ (dashed lines), and triples, $[I]_s(t)$ (solid lines), for a denser graph with $\langle K \rangle = 50$ and a sparser graph with $\langle K \rangle = 5$ (inset). The parameter values are $N = 1000$, $\gamma = 1$ and $\tau = 2\tau_c = 2\gamma/\langle K \rangle$.

We illustrate the effect of degree heterogeneity with a bimodal random network having only two different degrees. Let us consider the case when $N_1 = N/2$ nodes have high degree, d_1, while $N_2 = N/2$ nodes have low degree, d_2. Figure 4.10 compares $[I](t)$, $[I]_f(t)$ and $[I]_s(t)$ using networks with the same average degree, with different gaps between the values of the low and high degrees. In the main figure, the degree distribution is closer to homogeneous with $d_1 = 18$ and $d_2 = 22$, while in the inset, the degree distribution is strongly heterogeneous with $d_1 = 5$ and $d_2 = 35$. As expected, agreement is poor in the strongly heterogeneous case. This highlights the need for extending the homogenous mean-field models to handle heterogeneous networks.

The heterogeneity of the graph can be characterised by the variance of the degree distribution, $\langle K^2 \rangle - \langle K \rangle^2$. For a regular random network, with each node having degree n, the variance is $n^2 - n^2 = 0$. For a bimodal random network having half of its nodes of degree d_1 and half of degree d_2, the average degree is $\langle K \rangle = (d_1 + d_2)/2$, and the second moment is $\langle K^2 \rangle = (d_1^2 + d_2^2)/2$. Hence, the variance is $\langle K^2 \rangle - \langle K \rangle^2 = (d_1 - d_2)^2/4$. For an Erdős–Rényi random network with parameter p, the average degree is $\langle K \rangle = (N-1)p$ and the variance is $\langle K^2 \rangle - \langle K \rangle^2 = (N-1)p(1-p)$ since its degree distribution is binomial. For a random graph with power law degree distribution $p_k = Ck^{-\alpha}$, the average degree is $\langle K \rangle = \frac{\alpha-1}{\alpha-2}$ if $\alpha > 2$; otherwise it is infinite. The second moment is $\langle K^2 \rangle = \frac{\alpha-1}{\alpha-3}$ if $\alpha > 3$; otherwise it is infinite.

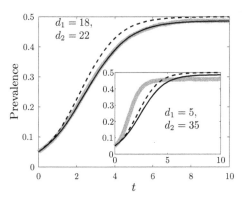

Fig. 4.10: The time dependence of the expected number of infected nodes in an SIS epidemic for bimodal random networks based on the average of 200 individual-based stochastic simulations $[I](t)$ (grey thick lines), and mean-field systems closed at the level of pairs, $[I]_f(t)$ (dashed lines), and triples, $[I]_s(t)$ (solid lines). The graph is nearly homogeneous in the main panel $d_1 = 18$ and $d_2 = 22$, and it is strongly heterogeneous in the inset, with $d_1 = 5$ and $d_2 = 35$. The parameter values are $N = 1000$, $\langle K \rangle = 20$, $\gamma = 1$ and $\tau = 2\tau_c = 2\gamma/\langle K \rangle$.

Therefore for $\alpha > 3$ the variance is given by $\frac{(\alpha-1)(2\alpha+1)}{(\alpha-3)(\alpha-2)^2}$ and for $2 \leq \alpha \leq 3$ it is infinite. However, here we use a power law with a cutoff such that the degrees vary between k_{min} and k_{max}. For this case, the variance of the degree distribution is computed directly from the formula $\langle K^2 \rangle - \langle K \rangle^2$.

We investigate the effect of network structure as follows. We construct regular, bimodal, Erdős–Rényi and power law random networks with $N = 1000$ nodes and with the same average degree $\langle K \rangle = 20$. The power law is used with $\alpha = 2$ and with the degree varying from $k_{min} = 7$ to $k_{max} = 110$. This choice of the minimal degree ensures that the average degree is approximately 20. The maximum degree is chosen in such a way that without the cutoff the probability of having a node with higher degree would be less than $1/N$, i.e. the expected number of such nodes is less than 1 in a network with N nodes. An SIS epidemic with infection rate $\tau = 2\tau_c = 2\gamma/\langle K \rangle$ is simulated on all networks. The variances of the degree distributions of the networks are as follows: zero for the regular random network, 10 for the bimodal network, 4.4 for the Erdős–Rényi network and 19.1 for the power law network. The prevalence obtained from the average of 200 simulations is plotted together with those obtained from the two mean-field models in Fig. 4.11. Higher variance in the degree distribution leads to faster early epidemic growth. This is because at early times the higher degree nodes are more likely to become infected, and in turn, they infect more neighbours. The pairwise ODE agrees well with the simulation for regular random networks and yields a reasonable approximation for Erdős–Rényi networks, since the degree heterogeneity for this network is not significant. However, the simulations cannot be captured by homogeneous mean-field models for the strongly heterogeneous bimodal and power law networks.

We now investigate the performance of the mean-field and pairwise models for SIR epidemics. These are given in systems (4.9) and in (4.11). Starting from the

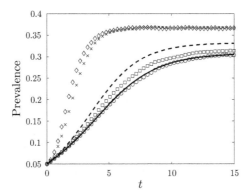

Fig. 4.11: The time dependence of the expected number of infected nodes in an SIS epidemic for regular (circle), bimodal (asterix), Erdős–Rényi (square) and power law random networks (diamond) based on 200 individual-based stochastic simulation compared with the mean-field systems closed at the level of pairs, $[I]_f(t)$ (dashed lines), and triples, $[I]_s(t)$ (solid lines). The parameter values are $N = 1000$, $\langle K \rangle = 20$, $\gamma = 1$ and $\tau = 2\tau_c = 2\gamma/\langle K \rangle$. For the bimodal network, $N_1 = N_2 = N/2$, $d_1 = 5$ and $d_2 = 35$. For the power law network, $\alpha = 2$, $k_{\min} = 7$ and $k_{\max} = 110$.

homogeneous case, a regular random network is considered and the functions $[I](t)$, $[I]_f(t)$ and $[I]_s(t)$ are shown in Fig. 4.12 for $\langle K \rangle = 50$ in the main figure and for $\langle K \rangle = 5$ in the inset. The value of τ in both cases is the double of the predicted critical value $\tau_c = \gamma/\langle K \rangle$. The pairwise model yields very good agreement in both cases, while the homogeneous mean-field model performs reasonably well only for the larger value of the average degree.

Consider now SIR epidemics on an Erdős–Rényi random network. Figure 4.13 shows $[I](t)$, $[I]_f(t)$ and $[I]_s(t)$ for $\langle K \rangle = 50$ in the main figure and for $\langle K \rangle = 5$ in the inset. In both cases, τ is twice the predicted critical value $\tau_c = \gamma/\langle K \rangle$. Neither the pairwise nor the mean-field model yields good agreement in the case of low average degree due to the heterogeneity in the degree distribution of the network.

To complete the systematic comparison of simulation results to output from the two mean-field models, the reader is invited to carry out the following exercises. For algorithm descriptions and ready-to-run source codes accompanying the book, see Appendix A.1 and the following webpage:

https://springer-math.github.io/Mathematics-of-Epidemics-on-Networks/

Exercise 4.5. Create an Erdős–Rényi random network with $N = 1000$ nodes and average degree $\langle K \rangle = 20$. Use the SIS dynamics and determine $[I](t)$ by averaging over 100 simulations. Compare $[I](t)$ to $[I]_f(t)$ and $[I]_s(t)$ for $\tau = 0.9\tau_c$, $\tau = \tau_c$, $\tau = 1.2\tau_c$ and $\tau = 1.5\tau_c$, as shown in Fig. 4.7 for a regular random network. The predicted critical value of τ is $\tau_c = \gamma/\langle K \rangle$.

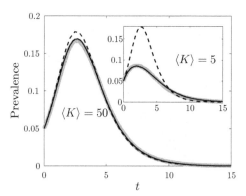

Fig. 4.12: The time dependence of the expected number of infected nodes in an SIR epidemic for a regular random network based on the average of 200 individual-based stochastic simulations $[I](t)$ (grey thick lines), and mean-field systems closed at the level of pairs, $[I]_f(t)$ (dashed lines), and triples, $[I]_s(t)$ (solid lines), for a denser network with $\langle K \rangle = 50$ and a sparser network with $\langle K \rangle = 5$ (inset). The parameter values are $N = 1000$, $\gamma = 1$ and $\tau = 2\tau_c = 2\gamma/\langle K \rangle$.

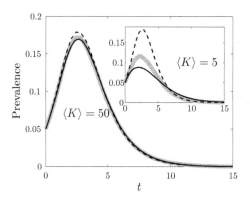

Fig. 4.13: The time dependence of the expected number of infected nodes in an SIR epidemic for an Erdős–Rényi random network based on the average of 200 individual-based stochastic simulations $[I](t)$ (grey thick lines), and mean-field systems closed at the level of pairs, $[I]_f(t)$ (dashed lines), and triples, $[I]_s(t)$ (solid lines), for a denser network with $\langle K \rangle = 50$ and a sparser network with $\langle K \rangle = 5$ (inset). The parameter values are $N = 1000$, $\gamma = 1$ and $\tau = 2\tau_c = 2\gamma/\langle K \rangle$.

Exercise 4.6. Create a bimodal random network with $N = 1000$ nodes, where $N_1 = N_2 = N/2$ and $d_1 = 5$, $d_2 = 35$. Use the SIS dynamics and determine $[I](t)$ by averaging over 100 simulations. Compare $[I](t)$ to $[I]_f(t)$ and $[I]_s(t)$ for $\tau = 0.9\tau_c$, $\tau = \tau_c$, $\tau = 1.2\tau_c$ and $\tau = 1.5\tau_c$, as shown in Fig. 4.7 for a regular random network. The predicted critical value of τ is $\tau_c = \gamma/\langle K \rangle$.

Exercise 4.7. Investigate the effect of system size N on the performance of mean-field models for regular random networks. Create regular random networks of size $N = 100$, $N = 200$, $N = 400$ and $N = 800$, with each node of degree $n = 10$. Use the SIS and the SIR dynamics and determine $[I](t)$ by averaging over 100 simulations. Compare $[I](t)$ to $[I]_f(t)$ and $[I]_s(t)$ for each system size separately.

Exercise 4.8. Investigate the effect of system size N on the performance of mean-field models for bimodal random networks. Create bimodal random networks of size $N = 100$, $N = 200$, $N = 400$ and $N = 800$. Take $N_1 = N_2 = N/2$ and use the degrees $d_1 = 5$ and $d_2 = 15$. Use the SIS and the SIR dynamics and determine $[I](t)$ by averaging over 100 simulations. Compare $[I](t)$ to $[I]_f(t)$ and $[I]_s(t)$ for each system size separately.

Exercise 4.9. Create two bimodal 1000-node random networks. Take $N_1 = N_2 = N/2$ and choose the degrees such that one has average degree $\langle K \rangle = 5$ and the other has average degree $\langle K \rangle = 25$ but both have the same variance $\langle K^2 \rangle - \langle K \rangle^2$. The predicted critical value of τ is $\tau_c = \gamma/\langle K \rangle$. Take $\tau = 2\tau_c$. Use the SIS and the SIR dynamics and determine $[I](t)$ by averaging over 100 simulations. Compare $[I](t)$ to $[I]_f(t)$ and $[I]_s(t)$ for both networks.

Exercise 4.10. Create a bimodal 1000-node random network with $N_1 = N_2 = N/2$ and a 1000-node power law random graph (with some k_{min} and k_{max}). Choose the parameters so that the networks have the same average degree and same variance in the degree distribution. The predicted critical value of τ is $\tau_c = \gamma/\langle K \rangle$. Take $\tau = 2\tau_c$. Use the SIS and the SIR dynamics and determine $[I](t)$ by averaging over 100 simulations. Compare $[I](t)$ to $[I]_f(t)$ and $[I]_s(t)$ for both networks.

Exercise 4.11. Create two 1000-node power law random graphs with different α values. The predicted critical value of τ is $\tau_c = \gamma/\langle K \rangle$. Take $\tau = 2\tau_c$. Use the SIS and the SIR dynamics and determine $[I](t)$ by averaging over 100 simulations. Compare $[I](t)$ to $[I]_f(t)$ and $[I]_s(t)$ for both networks.

Exercise 4.12. Create two 1000-node bimodal random networks, one with $N_1 = N_2 = N/2$, the other one with $N_1 = N/4$ and $N_2 = 3N/4$, keeping the average degree and variance at the same value. The predicted critical value of τ is $\tau_c = \gamma/\langle K \rangle$. Take $\tau = 2\tau_c$. Use the SIS and the SIR dynamics and determine $[I](t)$ by averaging over 100 simulations. Compare $[I](t)$ to $[I]_f(t)$ and $[I]_s(t)$ for both networks.

4.6 Derivation of mean-field models from master equations

The exact forms of the homogeneous mean-field equations at the level of singles and pairs for SIS and SIR epidemic were given in Section 4.1 in Theorems 4.1, 4.2, 4.3, and 4.4. The equations were formulated using heuristic arguments. In this section, we derive them directly from the master equations. We focus on the SIS dynamics and on the differential equations of the singles, following the proof in [288].

The SIR case can be dealt with similarly; the equations for the pairs are derived in [299]; however, this derivation is more technical and it is not presented here. In the second part of this section, the exact form of the mean-field equations at the level of singles for more general dynamics is derived.

4.6.1 Derivation of the homogeneous mean-field model at the single level for the SIS epidemic

The equations of Theorem 4.1 for the SIS epidemic take the form

$$[\dot{S}] = \gamma[I] - \tau[SI],$$
$$[\dot{I}] = \tau[SI] - \gamma[I].$$

To derive these, recall that $X_j^k(t)$ was defined to be the probability the system is in state $\mathcal{S}_j^k$ for $j = 1, \ldots, c_k$ at time t, and state $\mathcal{S}_j^k$ is one of the $c_k = \binom{N}{k}$ states having k infected nodes. Recall that the superscript k is an index, not an exponent. Then,

$$[I](t) = \sum_{k=0}^{N} \sum_{j=1}^{c_k} k X_j^k(t), \quad [S](t) = \sum_{k=0}^{N} \sum_{j=1}^{c_k} (N-k) X_j^k(t) \tag{4.20}$$

are the expected number of infected and susceptible nodes given by the master equation (2.6), and

$$[SI](t) = \sum_{k=0}^{N} \sum_{j=1}^{c_k} N_{SI}(\mathcal{S}_j^k) X_j^k(t) \tag{4.21}$$

is the expected value of the number of links connecting infected and susceptible nodes, where $N_{SI}(\mathcal{S}_j^k)$ is the number of links connecting infected and susceptible nodes in state $\mathcal{S}_j^k$.

In order to make the proof more accessible, we denote the vector $(X_1^k, X_2^k, \ldots, X_{c_k}^k)^T$ by X^k and recall from Section 2.3.3 that the master equation takes the form

$$\dot{X}^k = A^k X^{k-1} + B^k X^k + C^k X^{k+1}, \quad k = 0, 1, \ldots, N, \tag{4.22}$$

where A^0 and C^N are zero matrices. Moreover, according to (2.9) the A^k matrices satisfy the identities

$$\sum_{i=1}^{c_k} A_{i,j}^k = \tau N_{SI}(\mathcal{S}_j^{k-1}), \quad k = 0, 1, \ldots, N. \tag{4.23}$$

According to (2.11), the C^k matrices satisfy the identities

$$\sum_{i=1}^{c_k} C_{i,j}^k = \gamma(k+1), \quad k = 0, 1, \ldots, N \tag{4.24}$$

and according to (2.12), the elements in the diagonal matrices B^k satisfy

$$B_{i,i}^k = -\sum_{j=1}^{c_{k+1}} A_{j,i}^{k+1} - \sum_{j=1}^{c_{k-1}} C_{j,i}^{k-1}. \tag{4.25}$$

We can now prove Theorem 4.1.

Proof. Consider the row matrix $e_k = (1,1,\ldots,1)$ with c_k columns. Then, $\sum_{j=1}^{c_k} X_j^k = e_k X^k$ (here X^k is a column vector). Hence, from (4.20) we obtain

$$[I](t) = \sum_{k=0}^{N} k e_k X^k, \qquad [S](t) = \sum_{k=0}^{N} (N-k) e_k X^k. \tag{4.26}$$

We will expand $[\dot{I}]$ term by term, rearrange the result and show that $-\gamma[I] + \tau[SI]$ emerges. Using our notation, (4.25) becomes

$$B_{i,i}^k = -(e_{k+1}A^{k+1})_i - (e_{k-1}C^{k-1})_i,$$

and using that B^k is a diagonal matrix $B_{i,i}^k = (e_k B^k)_i$ holds, and hence,

$$(e_k B^k)_i = -(e_{k+1}A^{k+1})_i - (e_{k-1}C^{k-1})_i$$

which is true for all $i = 1,\ldots,c_k$. Thus,

$$e_{k+1}A^{k+1} + e_k B^k + e_{k-1}C^{k-1} = 0 \tag{4.27}$$

holds for all $k = 0,1,\ldots,N$. We note that when indices are out of the relevant range (i.e. A^{N+1} and C^{-1}), matrices should be set to zero. Differentiating $[I](t)$ and using (4.22), we obtain

$$
\begin{aligned}
[\dot{I}] &= \sum_{k=0}^{N} k e_k \dot{X}^k = \sum_{k=0}^{N} k e_k (A^k X^{k-1} + B^k X^k + C^k X^{k+1}) \\
&= \sum_{k=1}^{N} k e_k A^k X^{k-1} + \sum_{k=0}^{N} k e_k B^k X^k + \sum_{k=0}^{N-1} k e_k C^k X^{k+1} \\
&= \sum_{k=0}^{N-1} (k+1) e_{k+1} A^{k+1} X^k + \sum_{k=0}^{N} k e_k B^k X^k + \sum_{k=1}^{N} (k-1) e_{k-1} C^{k-1} X^k \\
&= \sum_{k=0}^{N} \left((k+1) e_{k+1} A^{k+1} + k e_k B^k + (k-1) e_{k-1} C^{k-1} \right) X^k.
\end{aligned}
$$

Now upon using (4.27), we obtain

$$[\dot{I}] = \sum_{k=0}^{N} \left(e_{k+1}A^{k+1} - e_{k-1}C^{k-1} \right) X^k.$$

The statement for $[I](t)$ follows from the proposition below. The proof for $[S](t)$ is similar.

Proposition 4.1 *For the matrices in the master equation* (4.22), *the following identities hold:*

1. $e_{k-1}C^{k-1} = \gamma k e_k,$
2. $\sum_{k=0}^{N} e_{k-1}C^{k-1}X^k = \gamma[I],$
3. $\sum_{k=0}^{N} e_{k+1}A^{k+1}X^k = \tau[SI].$

Proof. 1. According to (4.24), for all $j \in \{1,2,\ldots,c_k\}$, the following equality holds

$$(e_{k-1}C^{k-1})_j = \sum_{i=1}^{c_{k-1}} C_{i,j}^{k-1} = \gamma k,$$

implying $e_{k-1}C^{k-1} = \gamma k e_k$, since the jth coordinate of the left- and right-hand side
are equal.

2. The second statement follows directly from the first part of the Proposition and
from (4.26).

3. According to (4.23), for all $j \in \{1,2,\ldots,c_k\}$, the following equality holds

$$(e_{k+1}A^{k+1})_j = \sum_{i=1}^{c_{k+1}} A_{i,j}^{k+1} = \tau N_{SI}(\mathcal{S}_j^k),$$

which, by using (4.21), leads to

$$\sum_{k=0}^{N} e_{k+1}A^{k+1}X^k = \sum_{k=0}^{N} \sum_{j=1}^{c_k} (e_{k+1}A^{k+1})_j X_j^k = \tau \sum_{k=0}^{N} \sum_{j=1}^{c_k} N_{SI}(\mathcal{S}_j^k)X_j^k(t) = \tau[SI].$$

□

Thus, the exact mean-field equations at single level are derived from master equations for the SIS dynamic. This is extended to more general dynamics in the next
subsection.

4.6.2 Mean-field models for arbitrary dynamics

Recall first the formulation for a general dynamics from Chapter 2. Let $\{Q_1, Q_2,
\ldots, Q_m\}$ be the possible statuses of a node. Then, a state of the network can be
specified by an N-tuple $(q_1, q_2, \ldots, q_N)$, where $q_k \in \{Q_1, Q_2, \ldots, Q_m\}$ for all k. The
rate of transition

$$(q_1, q_2, \ldots, q_k, \ldots, q_N) \rightarrow (q_1, q_2, \ldots, q_k', \ldots, q_N)$$

is given by $f_{Q_i Q_j}(n_1, n_2, \ldots, n_m)$, where $q_k = Q_i$ and $q_k' = Q_j$, that is node k goes
from status Q_i to status Q_j and has n_l neighbours in status Q_l for $l = 1,2,\ldots,m$.
(We note that the status of at most one node can change during any transition.)

Thus, the node dynamics are specified by the functions $f_{Q_i Q_j}$, which yield the transition rate for a node to change from status Q_i to status Q_j with a neighbourhood configuration given by $(n_1, n_2, \ldots, n_m)$.

Let us briefly summarise how the functions $f_{Q_i Q_j}$ are given for some well-known node dynamics. Those (i, j) pairs for which the $Q_i \rightarrow Q_j$ transition cannot occur will not be listed. In fact, $f_{Q_i Q_j}$ is considered to be zero in those cases:

- SIS epidemic with infection rate τ and recovery rate γ:

$$f_{SI}(n_S, n_I) = \tau n_I, \quad f_{IS}(n_S, n_I) = \gamma.$$

- SIR epidemic with infection rate τ and recovery rate γ:

$$f_{SI}(n_S, n_I, n_R) = \tau n_I, \quad f_{IR}(n_S, n_I, n_R) = \gamma.$$

- Network of neurones with two statuses, quiescent (Q) and active (A):

$$f_{QA}(n_Q, n_A) = \omega \tanh(n_A), \quad f_{AQ}(n_Q, n_A) = \alpha.$$

- Voter model with two statuses A and B:

$$f_{AB}(n_A, n_B) = a \frac{n_B}{n_A + n_B}, \quad f_{BA}(n_A, n_B) = b \frac{n_A}{n_A + n_B}.$$

- Simplified voter model (each node has the same number of links) with two statuses A and B:

$$f_{AB}(n_A, n_B) = a n_B, \quad f_{BA}(n_A, n_B) = b n_A.$$

- Rumour spread with three statuses, ignorants (X), spreaders (Y), and stiflers (Z):

$$f_{XY}(n_X, n_Y, n_Z) = \tau n_Y, \quad f_{YZ}(n_X, n_Y, n_Z) = \gamma + \alpha(n_Y + n_Z).$$

Before continuing, we note an important group of transition functions which will lead to linear differential equations.

Definition 4.8 *A transition function* $f_{Q_i Q_j}(n_1, n_2, \ldots, n_m) = r + \sum c_m n_m$ *having only zeroth or first-order terms is called a* linear transition function.

Our aim now is to derive the exact form of the mean-field system for a general process. Carefully investigating the proof for the SIS case in the previous subsection reveals that Eqs. (4.23) and (4.24) play an important role in proving Proposition 4.1. The key point is that f_{SI} and f_{IS} are linear transition functions. Considering the examples above, we find that the transition functions are linear for the SIS and SIR epidemic and for the simplified voter and rumour spread models, but non-linear for the neuronal activation and voter models.

First, let us derive the exact mean-field equations for the case of a binary dynamics with linear transition functions. Let Q and T denote the two statuses and assume that the transition rates for the $Q \rightarrow T$ and $T \rightarrow Q$ transitions are

$$f_{QT}(n_Q, n_T) = r^0_{QT} + r^Q_{QT} n_Q + r^T_{QT} n_T, \qquad (4.28a)$$

$$f_{TQ}(n_Q, n_T) = r^0_{TQ} + r^Q_{TQ} n_Q + r^T_{TQ} n_T. \qquad (4.28b)$$

Denoting by $[Q]$ and $[T]$ the expected value of the number of nodes in status Q and T, respectively, and denoting by $[QQ]$, $[QT]$ and $[TT]$ the expected number of links connecting nodes of the given types, we can derive the following exact mean-field equations for the singles.

Theorem 4.9 *For an arbitrary binary dynamics with linear transition functions given in system (4.28) and for an arbitrary network (undirected and not weighted), the expected numbers $[Q]$ and $[T]$ satisfy the following system:*

$$[\dot{Q}] = r^0_{TQ}[T] + r^Q_{TQ}[QT] + r^T_{TQ}[TT] - r^0_{QT}[Q] - r^Q_{QT}[QQ] - r^T_{QT}[TQ], \qquad (4.29a)$$

$$[\dot{T}] = r^0_{QT}[Q] + r^Q_{QT}[QQ] + r^T_{QT}[TQ] - r^0_{TQ}[T] - r^Q_{TQ}[QT] - r^T_{TQ}[TT]. \qquad (4.29b)$$

The terms in the differential equations can be explained by heuristic arguments as follows. A node in status Q comes from a T node (independently from its neighbours) with rate r^0_{TQ}, or via the influence of its neighbours with status Q with rate r^Q_{TQ}, and finally, from T nodes via the impact of its neighbours with status T with rate r^T_{TQ}. These transitions explain the positive terms in the differential equation of $[Q]$. The negative terms in the equation are obtained similarly, representing transition back to T. The right-hand side of the second differential equation is simply the right-hand side of the first one but with negative sign. Besides this heuristic argument, one can follow the lines of the proof in the previous subsection for a rigorous proof. The reader is encouraged to work out the details.

Based on the binary dynamics, we derive the exact mean-field system for a process with three statuses first, and then for arbitrarily many statuses. The three statuses are denoted by Q, T and R. The dynamics are specified by the functions $f_{Q_i Q_j}(n_Q, n_T, n_R)$, where $Q_i, Q_j \in \{Q, T, R\}$. There are 6 transition functions altogether. We assume these transition functions take the simple form

$$f_{Q_i Q_j}(n_Q, n_T, n_R) = r^0_{ij} + r^Q_{ij} n_Q + r^T_{ij} n_T + r^R_{ij} n_R, \quad Q_i, Q_j \in \{Q, T, R\}. \qquad (4.30)$$

Similarly to the binary dynamics, nodes in status Q are created from T nodes (independently from its neighbours) with rate r^0_{TQ}; from R nodes (independently from its neighbours) with rate r^0_{RQ}; from T nodes via the impact of its Q, T or R neighbours with rates r^Q_{TQ}, r^T_{TQ} and r^R_{TQ}, respectively; and from R nodes via the effect of its Q, T or R neighbours with rates r^Q_{RQ}, r^T_{RQ} and r^R_{RQ}, respectively. Careful bookkeeping yields

$$\begin{aligned} [\dot{Q}] = &\ r^0_{TQ}[T] + r^0_{RQ}[R] - r^0_{QT}[Q] - r^0_{QR}[Q] \\ &+ r^Q_{TQ}[QT] + r^T_{TQ}[TT] + r^R_{TQ}[RT] + r^Q_{RQ}[QR] + r^T_{RQ}[TR] + r^R_{RQ}[RR] \\ &- r^Q_{QT}[QQ] - r^T_{QT}[TQ] - r^R_{QT}[RQ] - r^Q_{QR}[QQ] - r^T_{QR}[TQ] - r^R_{QR}[RQ]. \end{aligned}$$

Instead of writing down the differential equations for $[T]$ and $[R]$ as well, we switch to the general notation, where the statuses are denoted by Q_1, Q_2 and Q_3 and the coefficients in the transition functions are denoted by r_{ij}^1, r_{ij}^2 and r_{ij}^3. This means that (4.30) takes the form

$$f_{Q_iQ_j}(n_1,n_2,n_3) = r_{ij}^0 + \sum_{l=1}^{3} r_{ij}^l n_l$$

and the above differential equation for Q_1 can be given as

$$[\dot{Q}_1] = \sum_{j=1}^{3} r_{j1}^0 [Q_j] + \sum_{j=1}^{3}\sum_{l=1}^{3} r_{j1}^l [Q_l Q_j] - [Q_1] \sum_{j=1}^{3} r_{1j}^0 - \sum_{j=1}^{3}\sum_{l=1}^{3} r_{1j}^l [Q_1 Q_l],$$

where $r_{11}^l = 0$ by definition. Based on the above differential equation, corresponding to the 3-status dynamics, one can get the differential equation for the case of an arbitrary process.

Theorem 4.10 *Consider an arbitrary node process with m node statuses $Q_1, Q_2, \ldots, Q_m$ and with linear transition functions given as*

$$f_{Q_iQ_j}(n_1,n_2,\ldots,n_m) = r_{ij}^0 + \sum_{l=1}^{m} r_{ij}^l n_l$$

evolving on an arbitrary network (undirected and not weighted). The expected values $[Q_i]$ for $i = 1,2,\ldots,m$ satisfy the following system

$$[\dot{Q}_i] = \sum_{j=1}^{m} r_{ji}^0 [Q_j] + \sum_{j=1}^{m}\sum_{l=1}^{m} r_{ji}^l [Q_l Q_j] - [Q_i] \sum_{j=1}^{m} r_{ij}^0 - \sum_{j=1}^{m}\sum_{l=1}^{m} r_{ij}^l [Q_i Q_l],$$

where $r_{ii}^l = 0$ by definition.

The proof of the theorem follows the same arguments as those shown in the previous subsection for the SIS epidemic. However, due to being more technical it is not presented here. Based on the ideas above, one can derive differential equations for the pairs in the case of a general process with linear transition functions. That would be the generalisation of Theorems 4.3 and 4.4 for a general process. Theorem 4.10 can be applied to determine the mean-field differential equations for different processes, as shown by the exercises below.

Exercise 4.13. Derive the exact mean-field system at the level of singles for the SIR epidemic using Theorem 4.10.

Exercise 4.14. Derive the exact mean-field system at the level of singles for the simplified voter model using Theorem 4.10.

Exercise 4.15. Derive the exact mean-field system at the level of singles for the rumour spread model using Theorem 4.10.

4.7 Detailed analytical study of pairwise models

We now do a detailed analysis of the equations we derived for the systems closed at the level of triples. Throughout, we focus just on the biologically relevant values. Our initial conditions must have non-negative numbers of nodes and edges, and the number of edges of the various types, SS, SI, etc., must be realizable. If we say, for example, that an equilibrium exists, we mean that it exists with biologically relevant values. Solutions to our equations that begin with biologically relevant values maintain biologically relevant values.

4.7.1 Homogeneous SIS pairwise model

Our first goal is to prove the following pair conservation relation.

Proposition 4.2 *In system* (4.10), *the relations*

$$[SS]_s + [SI]_s = n[S]_s, \quad [SI]_s + [II]_s = n[I]_s$$

hold.

Proof. In order to simplify our calculation, we introduce

$$A(t) = [SS]_s(t) + [SI]_s(t) - n[S]_s(t), \quad B(t) = [SI]_s(t) + [II]_s(t) - n[I]_s(t). \quad (4.31)$$

Differentiating these equations and using the differential equations in system (4.10) leads to

$$\dot{A} = \gamma B - \tau([SSI]_s + [ISI]_s - (n-1)[SI]_s),$$
$$\dot{B} = -\gamma B + \tau([SSI]_s + [ISI]_s - (n-1)[SI]_s),$$

where the short notations $[SSI]_s = \frac{n-1}{n} \frac{[SS]_s [SI]_s}{[S]_s}$ and $[ISI]_s = \frac{n-1}{n} \frac{[SI]_s^2}{[S]_s}$ have been employed. Multiplying the equation for A by $[SI](n-1)/n[S]$, we get $[SSI]_s + [ISI]_s - (n-1)[SI]_s = -\frac{n-1}{n} \frac{[SI]_s}{[S]_s} A$. Introducing $U = \frac{n-1}{n} \frac{[SI]_s}{[S]_s}$, the system above becomes

$$\dot{A} = \gamma B + \tau U A,$$
$$\dot{B} = -\gamma B - \tau U A.$$

Thus, the pair (A, B) satisfies a system of homogeneous linear differential equations. The function U is continuous (assuming that $[S]_s \neq 0$); hence, the solution of the system is unique. Constant zero functions are solutions of the homogeneous system; therefore, if $A(0) = 0$ and $B(0) = 0$, then $A(t) = 0$ and $B(t) = 0$ for all t. This means that $[SS]_s(t) + [SI]_s(t) = n[S]_s(t)$ and $[SI]_s(t) + [II]_s(t) = n[I]_s(t)$ hold for all t if these are true at the initial time instant. □

Consider now the reduced system that was derived from (4.10) by exploiting the conservation of singles and pairs, that is

$$[\dot{S}]_s = \gamma N - (\gamma + n\tau)[S]_s + \tau[SS]_s, \qquad (4.32a)$$

$$[\dot{SS}]_s = 2(n[S]_s - [SS]_s)\left(\gamma - \tau(n-1)\frac{[SS]_s}{n[S]_s}\right). \qquad (4.32b)$$

We now carry out a detailed phase plane analysis of this two-dimensional system. First we prove a proposition about the local behaviour at the steady states.

Proposition 4.3 *In system* (4.10), *a transcritical bifurcation occurs at* $\gamma = \tau(n-1)$.

- *If* $\tau(n-1) < \gamma$, *then there is no endemic steady state and the disease-free steady state is asymptotically stable.*
- *If* $\tau(n-1) > \gamma$, *then the endemic steady state is asymptotically stable and the disease-free steady state is unstable.*

Proof. First, we find the steady states. At steady state, one of the factors in the right-hand side of the second equation is zero. If the first factor is zero, i.e. $n[S]_s = [SS]_s$, then we get the disease-free steady state, namely the first equation yields $[S]_s^{df} = N$, and then $[SS]_s^{df} = nN$. If the second factor is zero, that is $\gamma n[S]_s = \tau(n-1)[SS]_s$, we get the endemic steady state, namely from the first equation $[S]_s^e = N\frac{\gamma(n-1)}{\tau n(n-1)-\gamma}$, and then $[SS]_s^e = \frac{\gamma n}{\tau(n-1)}[S]_s^e$. The number of infected nodes in the endemic steady state is $[I]_s^e = N - [S]_s^e = Nn\frac{\tau(n-1)-\gamma}{\tau n(n-1)-\gamma}$. Hence, for $\tau(n-1) < \gamma$, the endemic steady state does not exist for relevant values, since either $[I]_s^e$ or $[S]_s^e$ would be negative.

To investigate the stability, we linearise the system at the steady states. The Jacobian matrix at the disease-free steady state is

$$J^{df} = \begin{pmatrix} -\gamma - \tau n & \tau \\ 2n(\gamma - (n-1)\tau) & -2(\gamma - (n-1)\tau) \end{pmatrix}.$$

The trace of the Jacobian is $\mathrm{Tr}(J^{df}) = -3\gamma + (n-2)\tau$, and its determinant is $\det(J^{df}) = 2\gamma(\gamma - (n-1)\tau)$. If $\tau(n-1) < \gamma$, the trace is negative, and the determinant is positive; It follows that the eigenvalues have negative real part; hence the disease-free steady state is asymptotically stable. For $\tau(n-1) > \gamma$, the determinant is negative; hence, an eigenvalue has positive real part and the disease-free steady state is unstable.

The Jacobian matrix at the endemic steady state is

$$J^e = \begin{pmatrix} -\gamma - \tau n & \tau \\ -\frac{2n\gamma}{(n-1)\tau}(\gamma - (n-1)\tau) & 2(\gamma - (n-1)\tau) \end{pmatrix}.$$

The trace of the Jacobian is $\mathrm{Tr}(J^e) = \gamma - (3n-2)\tau$, and its determinant is $\det(J^e) = \frac{2}{n-1}(\gamma - n(n-1)\tau)(\gamma - (n-1)\tau)$. If $\tau(n-1) > \gamma$, the trace is negative and the determinant is positive; hence, the endemic steady state is asymptotically stable. For $\tau(n-1) < \gamma$, the endemic steady state does not exist. $\square$

Our analysis is now completed by showing that the stability results in the above proposition are global.

Proposition 4.4 *The global behaviour of system* (4.10) *changes at* $\tau(n-1) = \gamma$.

- If $\tau(n-1) < \gamma$, then all solutions converge to the disease-free steady state.
- If $\tau(n-1) > \gamma$, then all solutions converge to the endemic steady state.

Proof. We prove this by investigating the direction field of the two-dimensional system (4.32). Observe first that the triangle with vertices $(0,0)$, $(N,0)$ and (N,nN) in the phase plane $([S]_s, [SS]_s)$ is positively invariant, i.e. trajectories cannot leave it. This can be seen by checking that on the boundary trajectories point inwards. For example, at the lower boundary where $[SS]_s = 0$ we have $[\dot{SS}]_s = 2n\gamma[S]_s > 0$, i.e. at this boundary, trajectories move upwards. Thus, the global behaviour of the system has to be studied in this triangle, corresponding to biologically relevant values.

The direction field of the system is determined by the null clines $[\dot{S}]_s = 0$ and $[\dot{SS}]_s = 0$, shown with dashed lines in Fig. 4.14. The first is a straight line that passes through the top corner (N, nN) of the triangle and divides the triangle into two parts. In the left part trajectories move to the right and in the right part they move to the left. The null cline $[\dot{SS}]_s = 0$ consists of two straight lines passing through the origin. One of them coincides with the left boundary line of the triangle, namely the line given by $[SS]_s = n[S]_s$. The other line is given by $(n-1)\tau[SS]_s = \gamma n[S]_s$, which lies outside the triangle if $\tau(n-1) < \gamma$, i.e. when there is no endemic steady state.

Figure 4.14 shows the direction fields in the two cases. In the left panel, $\tau(n-1) < \gamma$; hence, $[\dot{SS}]_s > 0$ in the whole triangle and thus trajectories tend to the disease-free steady state (N, nN), the top vertex. The right panel shows the case $\tau(n-1) > \gamma$. One of the straight lines giving $[\dot{SS}]_s = 0$ divides the triangle into two parts. In the top part, $[\dot{SS}]_s < 0$; therefore, trajectories move down. In the bottom region, $[\dot{SS}]_s > 0$; hence, trajectories move up. It is clear from the figure that there is no periodic orbit and there is a single stable steady state. Thus, all trajectories tend to the endemic steady state, which is inside the triangle. This elementary phase plane analysis completes the proof. □

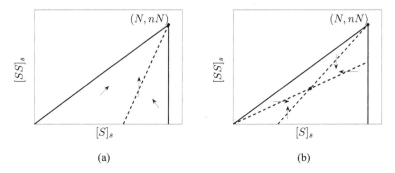

Fig. 4.14: Direction field in the $([S]_s, [SS]_s)$ plane of the pairwise equation of an SIS epidemic for (a) $\tau(n-1) < \gamma$ and (b) $\tau(n-1) > \gamma$.

System (4.10) enables us to estimate the correlations $C_{AB}^s = \frac{N}{n}\frac{[AB]_s}{[A]_s[B]_s}$. This is based on the following proposition.

Proposition 4.5 *The solution of system (4.10), subject to initial conditions based on random mixing*

$$[SI]_s(0) = \frac{n}{N}[S]_s(0)[I]_s(0), \quad [SS]_s(0) = \frac{n}{N}[S]_s^2(0), \quad [II]_s(0) = \frac{n}{N}[I]_s^2(0),$$

satisfies the following relations for any positive time

$$\frac{n}{N}[S]_s[I]_s - [SI]_s = [SS]_s - \frac{n}{N}[S]_s^2 = [II]_s - \frac{n}{N}[I]_s^2 > 0.$$

Proof. To make calculations simpler, we introduce the time-dependent functions

$$A = \frac{n}{N}[S]_s[I]_s - [SI]_s, \quad B = [SS]_s - \frac{n}{N}[S]_s^2, \quad C = [II]_s - \frac{n}{N}[I]_s^2.$$

Differentiating these functions and using the differential equations in system (4.10) leads to

$$\dot{A} = -\gamma(A+C) + \tau([ISI]_s - [SSI]_s + [SI]_s + \frac{n}{N}[S]_s[SI]_s - \frac{n}{N}[I]_s[SI]_s),$$

$$\dot{B} = -2\gamma A + \tau(2\frac{n}{N}[S]_s[SI]_s - 2[SSI]_s),$$

$$\dot{C} = -2\gamma C + \tau(2[ISI]_s + 2[SI]_s - 2\frac{n}{N}[I]_s[SI]_s),$$

where we used $[SSI]_s = \frac{n-1}{n}\frac{[SS]_s[SI]_s}{[S]_s}$ and $[ISI]_s = \frac{n-1}{n}\frac{[SI]_s^2}{[S]_s}$. Using Proposition 4.2 yields $[SSI]_s + [ISI]_s = (n-1)[SI]_s$. Therefore, the coefficients of τ in all of the three equations above can be simplified to $U = 2\frac{n}{N}[S]_s[SI]_s - 2[SSI]_s$. Thus, the above system takes the form

$$\dot{A} = -\gamma(A+C) + \tau U,$$
$$\dot{B} = -2\gamma A + \tau U,$$
$$\dot{C} = -2\gamma C + \tau U.$$

The equations for A, B and C are similar. We have $A(0) = B(0) = C(0) = 0$. Because they are initially equal, their derivatives are also initially equal. Extending this argument, it is straightforward to show that the unique solution is $A = B = C$, where A solves $\dot{A} = -2\gamma A + \tau U$. Thus, it remains to prove that $A(t) > 0$ for all positive t. Observe that using the closure, U can be written as

$$U = 2B\frac{[SI]_s}{[S]_s} + 2\frac{[SI]_s[SS]_s}{n[S]_s}.$$

The differential equation $\dot{A} = -2\gamma A + \tau U$ for A takes the form

$$\dot{A}(t) = p(t)A(t) + q(t),$$

with $p(t) = 2\tau \frac{[SI]_s(t)}{[S]_s(t)} - 2\gamma$ and $q(t) = 2\tau \frac{[SI]_s(t)[SS]_s(t)}{n[S]_s(t)}$. An integrating factor gives

$$A(t) = \int_0^t e^{P(s)-P(t)} q(s)ds,$$

where P is an integral of p, i.e. $P'(t) = p(t)$ and the initial condition $A(0) = 0$ was used. Since q is positive, A is also positive, completing the proof. $\square$

This proposition can be used to prove that the prevalence $[I]_s$ obtained from the second approximation is less than that given by the first approximation.

Proposition 4.6 *The prevalence $[I]_s$ obtained from system (4.10) and $[I]_f$ given by (4.8) satisfy the inequality $[I]_s \leq [I]_f$.*

Proof. Subtracting (4.10a) from (4.8a) and adding $0 = \tau n([S]_s[I]_s - [S]_s[I]_s)/N$ yields

$$[\dot{I}]_f - [\dot{I}]_s = -\gamma([I]_f - [I]_s) + \tau \frac{n}{N}([S]_f[I]_f - [S]_s[I]_s) + \tau\left(\frac{n}{N}[S]_s[I]_s - [SI]_s\right).$$

Applying the identities $[S]_f + [I]_f = N$ and $[S]_s + [I]_s = N$ in the middle term of the right-hand side, this equation can be written as

$$[\dot{I}]_f - [\dot{I}]_s = d([I]_f - [I]_s) + A,$$

where $d = -\gamma + n\tau - \frac{n}{N}([I]_f + [I]_s)$ and $A = \frac{n}{N}[S]_s[I]_s - [SI]_s$. Integrating this linear differential equation and using the initial condition $[I]_f(0) = [I]_s(0)$ yields

$$[I]_f(t) - [I]_s(t) = \int_0^t e^{D(s)-D(t)} A(s)ds,$$

where D is the integral of d, i.e. $D'(t) = d(t)$. According to Proposition 4.5, $A(s) > 0$; hence, the right-hand side is positive, which completes the proof. $\square$

4.7.2 Homogeneous SIR pairwise model

We will derive analytical formulas for $[SI]_s$ and $[SS]_s$ in terms of $[S]_s$ as it was given by system (4.16), starting from system (4.11). The initial conditions are assumed to be $[S]_0 \in (0,N)$, $[I]_0 = N - [S]_0$, $[SI]_0 = \frac{n}{N}[S]_0[I]_0$, $[SS]_0 = \frac{n}{N}[S]_0^2$, $[II]_0 = \frac{n}{N}[I]_0^2$. Dividing Eq. (4.11e) by Eq. (4.11a), we get

$$\frac{d[SS]_s}{d[S]_s} = 2\frac{n-1}{n}\frac{[SS]_s}{[S]_s}.$$

Integrating this equation and using the initial conditions $[S]_s(0) = [S]_0$, $[SS]_s(0) = [SS]_0$ implies

$$[SS]_s = \frac{n}{N}[S]_s^2 \left(\frac{[S]_0}{[S]_s}\right)^{\frac{2}{n}},$$

which proves (4.16a). Now, in order to prove (4.16b) divide Eq. (4.11d) by Eq. (4.11a) to yield

$$\frac{d[SI]_s}{d[S]_s} = \frac{n-1}{n}\frac{[SI]_s}{[S]_s} - \frac{n-1}{n}\frac{[SS]_s}{[S]_s} + \frac{\tau+\gamma}{\tau}.$$

Substituting $[SS]_s$ into this equation leads to

$$\frac{d[SI]_s}{d[S]_s} = \frac{n-1}{n}\frac{[SI]_s}{[S]_s} - \frac{n-1}{N}[S]_s\left(\frac{[S]_0}{[S]_s}\right)^{\frac{2}{n}} + \frac{\tau+\gamma}{\tau},$$

which is a linear differential equation for $[SI]_s$, with $[S]_s$ as the independent variable. Solving this linear equation subject to the initial condition $[SI]_0 = \frac{n}{N}[S]_0[I]_0$ yields the desired Eq. (4.16b).

4.8 Conclusions and outlook

In this chapter, we presented and analysed the most frequently used low-dimensional ODEs for expected population-level quantities such as the expected number of nodes and edges with different statuses. The systems of differential equations written at the level of pairs were introduced in the nineties by, among others, Matsuda et al. [207] Keeling et al. [170], and Rand [256]. The closures for triples, both for unclustered and clustered networks, were derived by investigating correlations in [256] and developed further for SIR epidemics in [166]. Differential equations for triples closed at the level of quadruples were introduced and compared to simulations in [152].

The closed homogeneous pairwise models were investigated by using dynamical system methods in [166] for the SIR epidemic, in [90] for the SIS epidemic and in [170, 256] for the SEIR epidemic with birth and death where oscillatory and chaotic behaviour may also occur. These studies were extended to the case when disease spread is possible not only through the contact network but also between any two nodes in the population [89]. The basic reproductive ratio, R_0, has been widely studied for mean-field models, using various definitions. In [90, 148], R_0 is defined based on the early growth rate of the prevalence. For the SIR homogeneous pairwise model, R_0 is determined in [166] and in [214, 301] for clustered populations. For the

SIS dynamics, and also for clustered networks, R_0 is studied in [149]. Different definitions of R_0 for the pairwise model are introduced and studied in [302]. Bauch introduced the so-called invasory pair approximation to derive an expression for the basic reproductive ratio when the contact network is a lattice [31]. A precise definition of R_0 for a general compartmental disease transmission model based on a system of ODEs is presented in [308].

Modelling epidemic propagation on networks by homogeneous mean-field models is reviewed in [167], while extensions of such models have also been developed for contact tracing [91, 149], households [10, 11, 17, 118, 148], directed or weighted networks [257, 258, 282] and control strategies such as vaccination [10, 11, 118, 148, 150].

The exact form of mean-field models can be derived from the full system of master equations. Master equations for different network processes are given in a general and abstract way in [256, 306]. The master equations are given in a detailed and rigorous form in [288], where the derivation of the homogeneous mean-field model is given. The homogeneous pairwise model is derived from master equations in [299]. The master equations of the SIS and SIR epidemics for a complete graph are investigated in [169], where the structure of the spectrum of the matrix yielding the linear system of master equations is also studied and the final epidemic size for the SIR model is determined.

Chapter 5
Mean-field approximations for heterogeneous networks

Section 4.5 showed that the homogeneous mean-field approximations cannot capture the system behaviour for networks with heterogeneous degree distributions. The heterogeneity in degree can significantly affect disease dynamics [254]. This requires more sophisticated models. In this chapter, we formulate mean-field models in terms of population-level counts of the number of nodes with a given degree and status. This chapter addresses:

- Exact unclosed systems: mean-field models at the level of singles and pairs for both SIS and SIR epidemics;
- Closures and closed versions of exact systems;
- Compact pairwise models for both SIS and SIR dynamics;
- Super-compact pairwise models for both SIS and SIR dynamics;
- Model analysis: epidemic threshold, endemic state, and final epidemic size;
- Comparison of the above models with individual-based stochastic simulation;
- Proofs to back up analytical results; and
- Effective degree and compact effective degree models.

5.1 Exact, unclosed models

In the spirit of mean-field models for homogenous networks, we now focus on the expected number of nodes with a given status and degree. We denote the number of susceptible, infected and recovered nodes of degree k by $[S_k](t)$, $[I_k](t)$ and $[R_k](t)$, respectively. In order to capture degree heterogeneity formally, let us introduce N_k as the number of nodes of degree k in the network for $k = 1, 2, \ldots, M$, where M is the maximum degree and $N_1 + N_2 + \cdots + N_M = N$. The degree distribution of the

© Springer International Publishing AG 2017
I.Z. Kiss et al., *Mathematics of Epidemics on Networks*, Interdisciplinary Applied
Mathematics 46, DOI 10.1007/978-3-319-50806-1_5

graph is then given as $p_k = \frac{N_k}{N}$, so that the average degree and the second moment of the degree distribution are

$$\langle K \rangle = \frac{1}{N} \sum_{k=1}^{M} k N_k, \quad \langle K^2 \rangle = \frac{1}{N} \sum_{k=1}^{M} k^2 N_k. \tag{5.1}$$

Denoting by $X_i(t)$ the random variable that gives the status of node i at time t, for example $X_i(t) = I$ if node i is infected at time t, the expected value of $[A_k]$ can be defined formally as the sum over all nodes i having degree k of the probability that node i has status A:

$$[A_k](t) = \sum_{i:k_i=k} P(X_i(t) = A),$$

where k_i is the degree of node i, $A \in \{S, I, R\}$. It is obvious, and can be shown formally, that $\sum_{k=1}^{M} [A_k] = [A]$ (every node with status A shows up in exactly one $[A_k]$ summation) and $[S_k](t) + [I_k](t) + [R_k](t) = N_k$ for all $k = 1, 2, \ldots, M$ (with probability 1 any given degree k node has one of the three statuses). The expected number of edges having endpoints of given statuses and degrees can be defined similarly as follows:

$$[A_k B_l](t) = \sum_{i:k_i=k} \sum_{j:k_j=l} g_{ij} P(X_i(t) = A, X_j(t) = B)$$

with $A, B \in \{S, I, R\}$. As before, each edge is counted twice. Some coarse graining, such as $[A_k B] = \sum_{l=1}^{M} [A_k B_l]$, will be particularly useful later in the chapter. For example, $[S_k I](t)$ denotes the expected value of the number of edges connecting susceptible nodes of degree k to infected nodes (with arbitrary degree) at time t. Using this notation for the SIS epidemic, we can find a result similar to Theorem 4.1.

Theorem 5.1 *For the SIS epidemic on an arbitrary network (undirected and not weighted), for $k = 1, 2, \ldots, M$, the expected values $[S_k]$ and $[I_k]$ satisfy*

$$[\dot{S}_k] = \gamma[I_k] - \tau[S_k I], \tag{5.2a}$$

$$[\dot{I}_k] = \tau[S_k I] - \gamma[I_k]. \tag{5.2b}$$

Note that summing these equations for $k = 1, 2, \ldots, M$ leads to the homogeneous mean-field equations (4.1). The differential equations can be explained intuitively as follows. The number of susceptible nodes of degree k decreases proportionally to the number of $S_k I$ edges with rate τ (due to infection) and increases proportionally to the number of infected nodes of degree k with rate γ (due to recovery). Similar reasoning yields the differential equation for $[I_k]$. In the case of the SIR epidemic, the I nodes do not become susceptible again, so the recovery terms do not appear in the S equations. Thus, the SIR equations take the form given in the theorem below.

Theorem 5.2 *For the SIR epidemic on an arbitrary network (undirected and not weighted), for $k = 1, 2, \ldots, M$, the expected values $[S_k]$, $[I_k]$ and $[R_k]$ satisfy*

$$[\dot{S}_k] = -\tau[S_k I], \tag{5.3a}$$

$$[\dot{I}_k] = \tau[S_k I] - \gamma[I_k], \tag{5.3b}$$

$$[\dot{R}_k] = \gamma[I_k]. \tag{5.3c}$$

Similarly to the homogeneous case, differential equations can be derived for the expected number of pairs in terms of triples. We will write

$$[S_k I] = \sum_{l=1}^{M} [S_k I_l], \tag{5.4}$$

and derive differential equations for the expected values of $[S_k I_l]$. There is flux into $[S_k I_l]$ out of $[I_k I_l]$ because of recovery (rate $\gamma[I_k I_l]$) and from triples $[S_k S_l I]$ by transmission (rate $\tau[S_k S_l I]$). The number of these pairs decreases proportionally to triples of the form $[I S_k I_l]$ because of transmission from the infected node outside the pair (rate $\tau[I S_k I_l]$) and, finally, it decreases also due to within-pair infection and recovery (rate $(\tau + \gamma)[S_k I_l]$). These considerations lead to the differential equation for $[S_k I_l]$, as given below. Extending this simple heuristic reasoning to SS and II pairs, the following theorem can be obtained.

Theorem 5.3 *For the SIS epidemic on an arbitrary network (undirected and not weighted), the expected values of $[S_k]$, $[I_k]$, $[S_k I_l]$, $[S_k S_l]$ and $[I_k I_l]$ satisfy the following system of differential equations:*

$$[\dot{S}_k] = \gamma[I_k] - \tau[S_k I], \tag{5.5a}$$

$$[\dot{I}_k] = \tau[S_k I] - \gamma[I_k], \tag{5.5b}$$

$$[\dot{S_k I_l}] = \gamma([I_k I_l] - [S_k I_l]) + \tau([S_k S_l I] - [I S_k I_l] - [S_k I_l]), \tag{5.5c}$$

$$[\dot{S_k S_l}] = \gamma([S_k I_l] + [I_k S_l]) - \tau([S_k S_l I] + [I S_k S_l]), \tag{5.5d}$$

$$[\dot{I_k I_l}] = \tau([S_k I_l] + [I_k S_l]) - 2\gamma[I_k I_l] + \tau([I S_k I_l] + [I_k S_l I]), \tag{5.5e}$$

where $[S_k I]$ is given in (5.4) and $k, l = 1, 2, \ldots, M$.

This system can be rigorously derived directly from the master equations (2.6). Figure 5.1 gives a pictorial interpretation of the system. Summing the equations over $k, l = 1, 2, \ldots, M$ leads to the homogeneous pairwise system (4.3), but loses information about triples, which will be needed when we close these equations.

As noted before, for the SIR epidemic the I nodes do not become susceptible again; hence, the terms corresponding to recovery appear in different equations. The differential equations take the form given in the theorem below.

Theorem 5.4 *For the SIR epidemic on an arbitrary network (undirected and not weighted), the expected values of $[S_k]$, $[I_k]$, $[R_k]$, $[S_k I_l]$ and $[S_k S_l]$ satisfy the following system of differential equations:*

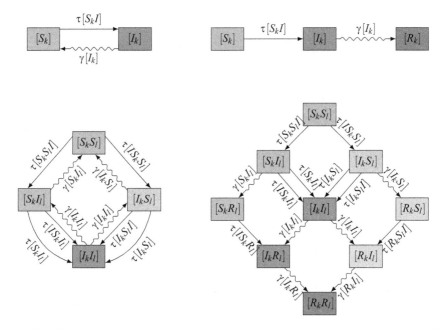

Fig. 5.1: Flow diagrams showing the flux between compartments of singles (top) and compartments of pairs (bottom), assuming heterogeneous degrees in contrast with the homogeneous case in Fig. 4.3. The SIS case is on the left and SIR on the right. The colour indicates the status of the "first" node in the edge. Symmetry allows us to avoid calculating some quantities (lighter shade).

$$[\dot{S}_k] = -\tau[S_kI], \tag{5.6a}$$

$$[\dot{I}_k] = \tau[S_kI] - \gamma[I_k], \tag{5.6b}$$

$$[\dot{R}_k] = \gamma[I_k], \tag{5.6c}$$

$$[\dot{S_kI_l}] = -\gamma[S_kI_l] + \tau([S_kS_lI] - [IS_kI_l] - [S_kI_l]), \tag{5.6d}$$

$$[\dot{S_kS_l}] = -\tau([S_kS_lI] + [IS_kS_l]), \tag{5.6e}$$

where $[S_kI]$ is given in (5.4) and $k,l = 1,2,\ldots,M$.

These equations are illustrated in Fig. 5.1. The $[I_kI_l]$ equations are left out as they are not needed to find any of the single terms.

5.2 Closures at the pair and triple level and the resulting models

In this section, the heuristic arguments leading to closures for networks with homogeneous degree distribution are extended in order to close the above four heterogeneous systems. These are mainly based on the work of Pastor-Satorras and

Vespignani [243] (as well as earlier work by Nold, Hethcote & Yorke, and May & Anderson [140, 141, 208, 241]) for mean-field models at single level and Eames and Keeling for pairwise models [90].

5.2.1 Closures

When the network was assumed to be homogeneous, the starting idea was to assume that infected nodes are randomly distributed and use $[I]/N$ as the probability that a neighbour of a susceptible node is infected. Thus, an average susceptible node has $n[I]/N$ infected neighbours, leading to the closure $[SI] \approx \frac{n}{N}[S][I]$, where $n = \langle K \rangle$ is a brief notation for the average degree. A naive generalisation of this argument might use the closure $[S_kI] \approx \frac{k}{N}[S_k][I]$, simply applying the previous reasoning to suscep- tible nodes of degree k. However, the information about the degree of the infected nodes cannot be neglected: the probability a random edge of a node connects to a susceptible or infected neighbour is proportional to the number of edges of infected or susceptible nodes. Consider a susceptible node of degree k whose stubs connect randomly to the $\sum_{l=1}^{M} lN_l$ stubs in the network. The number of stubs starting from infected nodes is $\sum_{l=1}^{M} l[I_l]$; hence, the probability that a randomly chosen stub is connected to a stub of an infected node is

$$\pi_I = \sum_{l=1}^{M} l[I_l] \Bigg/ \sum_{l=1}^{M} lN_l .$$

Therefore, the average number of edges connecting susceptible degree-k nodes to infected nodes can be approximated as

$$[S_kI] \approx k[S_k]\pi_I. \tag{5.7}$$

This relation can be used to close the exact systems (5.2) and (5.3) to produce solv- able self-contained systems. We refer to this as "pair closure" because it allows us to express pairs in terms of singles, thereby creating a closed system.

The closure is introduced in an alternative way in [90], using $[S_kI] = \sum_{l=1}^{M}[S_kI_l]$. We start with the more general closure

$$[S_kI_l] \approx n_{kl} \frac{[S_k]}{N_k} \frac{[I_l]}{N_l}, \tag{5.8}$$

where n_{kl} is the total number of edges connecting nodes of degree k with nodes of degree l. The ratios $[S_k]/N_k$ and $[I_l]/N_l$ give the probability that the degree k node is susceptible, while the degree l node is infected, respectively. This relies on the assumption that susceptible and infected nodes within each degree class are randomly distributed. If we assume n_{kl} follows from random neighbour selection (as in Configuration Model networks), then $n_{kl} = klN_kN_l/\sum_{j=1}^{M} jN_j$. Hence,

$$n_{kl} \frac{[S_k]}{N_k} \frac{[I_l]}{N_l} = k[S_k]\, l[I_l] \Big/ \sum_{j=1}^{M} jN_j \, ,$$

yielding

$$[S_k I] = \sum_{l=1}^{M} [S_k I_l] \approx k[S_k] \sum_{l=1}^{M} l[I_l] \Big/ \sum_{j=1}^{M} jN_j = k[S_k]\pi_I.$$

So closures (5.7) and (5.8) are equivalent under random neighbour selection.

The advantage of using n_{kl} is that it allows us to depart from the random mixing assumptions and to consider networks with preferential mixing, e.g. assortative (nodes of similar degrees connect more readily) or disassortative (nodes of differing degree connect more readily) [27, 39, 90, 179].

The closures at the level of triples for the pairwise systems (5.5) in the SIS case, and (5.6) in the SIR case, are based on the same idea as for homogeneous networks. However, in this case the degree of the middle node in the triple is used instead of the average degree of the network. As shown in [90], this leads to the closure relations

$$[A_l S_k I_m] \approx \frac{k-1}{k} \frac{[A_l S_k][S_k I_m]}{[S_k]}, \quad [A_l S_k I] \approx \frac{k-1}{k} \frac{[A_l S_k][S_k I]}{[S_k]}. \qquad (5.9)$$

This assumes that the neighbours of susceptible degree k nodes are interchangeable. We will see that this can be an accurate closure in the large network limit for the SIR case, but for the SIS case the neighbours of a susceptible degree k node that was previously infected are not interchangeable with the neighbours of susceptible degree k nodes that have not yet been infected. We refer to this as the "triple closure" because it allows us to take triples and express them in terms of smaller quantities, thereby creating a closed system.

5.2.2 Closed systems

Using the above closures, the exact systems can be closed. We first apply the pair closure (5.7) to the SIS system (5.2). Let $[S_k]_f$ and $[I_k]_f$ denote the functions given by the system of differential equations below. It should be highlighted that the systems derived using this closure assume that all neighbours of all nodes are interchangeable. Thus, they are equivalent to making the implicit assumption that all nodes change their neighbours rapidly, so that for each transmission event they have a newly selected neighbour. This is sometimes referred to as the "annealed network" assumption, which is different from the "quenched network", where nodes never change their neighbours [57].

SIS heterogeneous mean-field at the single level

$$[\dot{S}_k]_f = \gamma[I_k]_f - \tau k[S_k]_f \pi_{I,f}, \tag{5.10a}$$

$$[\dot{I}_k]_f = \tau k[S_k]_f \pi_{I,f} - \gamma[I_k]_f, \tag{5.10b}$$

$$\pi_{I,f} = \sum_{l=1}^{M} l[I_l]_f \bigg/ \sum_{l=1}^{M} lN_l, \tag{5.10c}$$

for $k = 1, 2, \ldots, M$.

Here, $[S_k]_f$ is an approximation of $[S_k]$ and $[I_k]_f$ is an approximation of $[I_k]$. Similarly to the homogeneous case, the subscript "f" refers to "first" since this can be considered as the first approximation when the number of pairs is expressed in terms of the number of singles. Note that $[S_k]_f + [I_k]_f$ is constant, so we can eliminate any of their differential equations, for example, by setting $[I_k]_f = N_k - [S_k]_f$.

In the case of the SIR epidemic, the term $\gamma[I_k]_f$ does not appear in the differential equation of $[S_k]_f$; hence, the closed system becomes

$$[\dot{S}_k]_f = -\tau k[S_k]_f \pi_{I,f},$$
$$[\dot{I}_k]_f = \tau k[S_k]_f \pi_{I,f} - \gamma[I_k]_f,$$
$$[\dot{R}_k]_f = \gamma[I_k]_f,$$
$$\pi_{I,f} = \sum_{l=1}^{M} l[I_l]_f \bigg/ \sum_{l=1}^{M} lN_l,$$

where $k = 1, 2, \ldots, M$. Note that $[S_k]_f + [I_k]_f + [R_k]_f$ is constant, so we can eliminate any one of the differential equations, for example, by setting $[I_k]_f = N_k - [S_k]_f - [R_k]_f$.

The size of this system can be reduced as follows. We can integrate each of the $[S_k]_f$ equations using the integrating factor $\exp\left(k\tau \int_0^t \pi_{I,f}(\hat{t})\,d\hat{t}\right)$. Multiplying the differential equation $[\dot{S}_k]_f + \tau k \pi_{I,f}[S_k]_f = 0$ by this integrating factor, we obtain

$$\frac{d}{dt}\left([S_k]_f(t)\left(e^{\tau \int_0^t \pi_{I,f}(\hat{t})\,d\hat{t}}\right)^k\right) = 0.$$

If we set

$$\theta_f(t) = e^{-\tau \int_0^t \pi_{I,f}(\hat{t})\,d\hat{t}},$$

then it follows that $[S_k]_f(t) = S_k(0)\theta_f^k(t)$ and $\dot{\theta}_f = -\tau \pi_{I,f}\theta_f$, where $S_k(0)$ is the original number of susceptible nodes with degree k and $\theta_f(0) = 1$. Here, θ_f represents the per-edge probability that a node has not yet received transmission. If we also use $[S_k]_f + [I_k]_f + [R_k]_f = [N_k]_f$, we can rewrite the system as below.

SIR heterogeneous mean-field at the single level

$$[S_k]_f = S_k(0)\theta_f^k, \tag{5.11a}$$

$$[I_k]_f = [N_k]_f - [S_k]_f - [R_k]_f, \tag{5.11b}$$

$$[\dot{R}_k]_f = \gamma[I_k]_f, \tag{5.11c}$$

$$\pi_{I,f} = \sum_{l=1}^{M} l[I_l]_f \bigg/ \sum_{l=1}^{M} lN_l, \tag{5.11d}$$

$$\dot{\theta}_f = -\tau\pi_{I,f}\theta_f, \tag{5.11e}$$

with initial conditions $\theta_f(0) = 1$, $[R_k]_f(0) = R_k(0)$ and for $k = 1, 2, \ldots, M$.

If we are only interested in the total number of susceptible, infected or recovered nodes, we can define

$$\hat{\psi}(x) = \frac{1}{N}\sum_{k=1}^{M} S_k(0)x^k \tag{5.12}$$

and then

$$[S]_f = \sum_{k=1}^{M} S_k(0)\theta_f^k = N\hat{\psi}(\theta_f),$$

with $[I]_f = N - [S]_f - [I]_f$ and $[\dot{R}]_f = \gamma[I]_f$. Note that if we are interested in the proportion infected, rather than the number, N can be scaled out of the system.

The following exercise shows that we can reduce the SIR heterogeneous mean-field model further by expressing $\pi_{I,f}$ in terms of θ_f. The resulting system was derived in [222], and an equivalent system appears in [4].

Exercise 5.1. For the SIR heterogeneous mean-field model, define

$$\pi_{S,f} = \sum_{l=1}^{M} l[S_l]_f \bigg/ \sum_{l=1}^{M} lN_l \quad \text{and} \quad \pi_{R,f} = \sum_{l=1}^{M} l[R_l]_f \bigg/ \sum_{l=1}^{M} lN_l.$$

a. Explain the meaning of $\pi_{S,f}$ and $\pi_{R,f}$ and show that $\pi_{S,f} + \pi_{I,f} + \pi_{R,f} = 1$.

b. Show that $\dot{\pi}_R = \gamma\pi_{I,f}$, and so solve for $\dot{\pi}_R$ in terms of θ_f and $\dot{\theta}_f$.

c. Use the resulting differential equation for $\dot{\pi}_{R,f}$ to show that

$$\pi_{R,f} = \pi_{R,f}(0) - \frac{\gamma}{\tau}\ln(\theta_f).$$

d. Show directly from the definition of $\pi_{S,f}$ that $\pi_{S,f} = \theta_f\frac{\hat{\psi}'(\theta_f)}{\langle K\rangle}$.

e. Using $\pi_{S,f} + \pi_{I,f} + \pi_{R,f} = 1$, show that

$$\dot{\theta}_f = -\tau\theta_f + \tau\theta_f^2\frac{\hat{\psi}'(\theta_f)}{\langle K\rangle} - \gamma\theta_f\ln(\theta_f) + \tau\theta_f\pi_{R,f}(0).$$

We turn now to the closure of the pairwise models at the level of triples. Using the triple closure (5.9) in the SIS heterogenous pairwise system (5.5) leads to the closed system below.

SIS heterogeneous pairwise

$$[\dot{S}_k]_s = \gamma[I_k]_s - \tau[S_kI]_s, \tag{5.13a}$$

$$[\dot{I}_k]_s = \tau[S_kI]_s - \gamma[I_k]_s, \tag{5.13b}$$

$$[\dot{S}_kI_l]_s = \gamma([I_kI_l]_s - [S_kI_l]_s) + \tau([S_kS_lI]_s - [IS_kI_l]_s - [S_kI_l]_s), \tag{5.13c}$$

$$[\dot{S}_kS_l]_s = \gamma([S_kI_l]_s + [I_kS_l]_s) - \tau([S_kS_lI]_s + [IS_kS_l]_s), \tag{5.13d}$$

$$[\dot{I}_kI_l]_s = \tau([S_kI_l]_s + [I_kS_l]_s) - 2\gamma[I_kI_l]_s + \tau([IS_kI_l]_s + [I_kS_lI]_s), \tag{5.13e}$$

where $[S_kI]$ is given in (5.4) and

$$[A_lS_kI]_s = \frac{k-1}{k}\frac{[A_lS_k]_s[S_kI]_s}{[S_k]_s}, \quad [IS_kA_l]_s = \frac{k-1}{k}\frac{[IS_k]_s[S_kA_l]_s}{[S_k]_s}. \tag{5.14}$$

The subscript "s" refers to the fact that this is the second approximation. The functions $[A_k]_s$ and $[A_kB_l]_s$ are defined as the solutions of this system and can be considered as approximations of the exact expected values $[A_k]$ and $[A_kB_l]$, respectively. Note that $[S_kI]_s = [IS_k]_s$ and $[S_kA_l]_s = [A_lS_k]_s$. The closed system contains $2M$ equations for the singles and $M^2 + M(M+1)$ equations for the pairs, yielding $2M^2 + 3M$ equations altogether. For networks with high-degree nodes, the system can be difficult to handle even numerically. The approximation

$$[A_kB_l] \approx \frac{[A_kB][AB_l]}{[AB]}.$$

led [90] to an approximating system containing only order M equations. The variables of the approximating system are $[S_k]$, $[I_k]$, $[S_kI]$, $[SI_l]$, $[S_kS]$ and $[II_l]$, and in the closures the above approximation is used where it is needed. (This closure was formulated for the more general case of preferential mixing in [90].)

In the SIR case, the term $\gamma[I_k]_s$ does not appear in the $[S_k]_s$ differential equation, $\gamma[I_kI_l]_s$ does not appear in the $[S_kI_l]_s$ differential equation and $\gamma([S_kI_l]_s + [I_kS_l]_s)$ does not appear in the $[S_kS_l]_s$ differential equation. The closed system is given below.

SIR heterogeneous pairwise

$$[\dot{S_k}]_s = -\tau[S_kI]_s, \tag{5.15a}$$

$$[\dot{I_k}]_s = \tau[S_kI]_s - \gamma[I_k]_s, \tag{5.15b}$$

$$[\dot{R_k}]_s = \gamma[I_k]_s, \tag{5.15c}$$

$$[\dot{S_kI_l}]_s = -\gamma[S_kI_l]_s + \tau([S_kS_lI]_s - [IS_kI_l]_s - [S_kI_l]_s), \tag{5.15d}$$

$$[\dot{S_kS_l}]_s = -\tau([S_kS_lI]_s + [IS_kS_l]_s), \tag{5.15e}$$

with the same closures as in the SIS case:

$$[A_lS_kI]_s = \frac{k-1}{k}\frac{[A_lS_k]_s[S_kI]_s}{[S_k]_s}, \quad [IS_kA_l]_s = \frac{k-1}{k}\frac{[IS_k]_s[S_kA_l]_s}{[S_k]_s}. \tag{5.16}$$

5.2.3 Compact pairwise model

The heterogeneous pairwise model has $\mathcal{O}(M^2)$ equations. For practical reasons, we need a simpler model. To address this, House and Keeling [151] proposed the approximation

$$\frac{[S_kI]}{k[S_k]} \approx \frac{[SI]}{\sum_{l=1}^{M} l[S_l]} \quad \text{which implies} \quad [S_kI] \approx [SI]\frac{k[S_k]}{\sum_{l=1}^{M} l[S_l]}. \tag{5.17}$$

This assumes neighbours of all susceptible nodes are interchangeable, or in other words a degree k susceptible node's neighbour is as likely to be infected as any other susceptible node's neighbour, while it may be different from the probability an infected node's neighbour is infected. Using this approximation, the triple closure (5.9) takes the form

$$[A_lS_kI] \approx (k-1)[A_lS_k]\frac{[SI]}{[SX]},$$

where $[SX] = \sum_{l=1}^{M} l[S_l]$ is the total number of stubs, starting from susceptible nodes. Now using $\sum_{l=1}^{M}[A_lS_k] = [AS_k]$ and applying approximation (5.17) yields

$$\sum_{l=1}^{M}[A_lS_kI] \approx (k-1)[AS_k]\frac{[SI]}{[SX]} \approx (k-1)k[S_k]\frac{[AS][SI]}{[SX]^2},$$

leading to

$$\sum_{k=1}^{M}\sum_{l=1}^{M}[A_lS_kI] \approx \frac{[AS][SI]}{[SX]^2}\sum_{k=1}^{M}(k-1)k[S_k].$$

We now consider the exact SIS heterogeneous pairwise system (5.5) and sum equations of pairs for k and l from 1 to M. Using $[AB] = \sum_{k=1}^{M} \sum_{l=1}^{M} [A_l B_k]$ and applying the above approximation allows us to recast (5.5c) as

$$[\dot{SI}] \approx \gamma([II] - [SI]) + \tau \frac{([SS] - [SI])[SI]}{[SX]^2} \sum_{k=1}^{M} (k-1)k[S_k] - \tau[SI].$$

Similarly, the sum of equations (5.5d) for $k = 1, 2, \ldots, M$ yields

$$[\dot{SS}] \approx 2\gamma[SI] - 2\tau \frac{[SS][SI]}{[SX]^2} \sum_{k=1}^{M} (k-1)k[S_k],$$

and the sum of equations (5.5e) leads to

$$[\dot{II}] \approx 2\tau[SI] - 2\gamma[II] + 2\tau \frac{[SI]^2}{[SX]^2} \sum_{k=1}^{M} (k-1)k[S_k].$$

The variables in the compact pairwise model will be denoted by $[S_k]_c$, $[I_k]_c$, $[SI]_c$, $[SS]_c$, $[II]_c$, and $[SX]_c$. Using the approximation (5.17) in (5.5a) and (5.5b), the compact pairwise approximation gives the equations below.

SIS compact pairwise

$$[\dot{S_k}]_c = \gamma[I_k]_c - \tau k[S_k]_c \frac{[SI]_c}{[SX]_c}, \tag{5.18a}$$

$$[\dot{I_k}]_c = \tau k[S_k]_c \frac{[SI]_c}{[SX]_c} - \gamma[I_k]_c, \tag{5.18b}$$

$$[\dot{SI}]_c = \gamma([II]_c - [SI]_c) + \tau([SS]_c - [SI]_c)[SI]_c Q_c - \tau[SI]_c, \tag{5.18c}$$

$$[\dot{SS}]_c = 2\gamma[SI]_c - 2\tau[SS]_c[SI]_c Q_c, \tag{5.18d}$$

$$[\dot{II}]_c = 2\tau[SI]_c - 2\gamma[II]_c + 2\tau[SI]_c^2 Q_c, \tag{5.18e}$$

$$[SX]_c = \sum_{k=1}^{M} k[S_k]_c, \tag{5.18f}$$

$$Q_c = \frac{1}{[SX]_c^2} \sum_{k=1}^{M} (k-1)k[S_k]_c. \tag{5.18g}$$

Observe that the compact pairwise model consists of only $2M + 3$ differential equations. This number decreases even further if not all degrees between 1 and M occur. For example, in the case of a bimodal graph with only two different degrees d_1 and d_2, the variables of the system are $[S_{d_1}]_c$, $[S_{d_2}]_c$, $[I_{d_1}]_c$, $[I_{d_2}]_c$, $[SI]_c$, $[SS]_c$ and $[II]_c$, giving rise to only 7 equations.

In the case of SIR epidemics, the term $\gamma[I_k]_c$ does not appear in the differential equation of $[S_k]_c$, the term $\gamma[II]_c$ does not appear in the differential equation of $[SI]_c$

and the term $\gamma([SI]_c)$ does not appear in the differential equation of $[SS]_c$; hence, the closed system takes the form

$$[\dot{S_k}]_c = -\tau k[S_k]_c \frac{[SI]_c}{[SX]_c},$$

$$[\dot{I_k}]_c = \tau k[S_k]_c \frac{[SI]_c}{[SX]_c} - \gamma[I_k]_c,$$

$$[\dot{R_k}]_c = \gamma[I_k]_c,$$

$$[\dot{SI}]_c = -\gamma[SI]_c + \tau([SS]_c - [SI]_c)[SI]_c Q_c - \tau[SI]_c,$$

$$[\dot{SS}]_c = -2\tau[SS]_c[SI]_c Q_c.$$

This system can be reduced significantly by observing that adding the first three equations for all $k = 1, 2, \ldots, M$, we get $[S]_c + [I]_c + [R]_c = N$, and adding the $[R_k]_c$ equations, we obtain $[\dot{R}]_c = \gamma[I]_c$. Thus, if it is sufficient to know $[S]_c$, $[I]_c$ and $[R]_c$, the $[I_k]_c$ and $[R_k]_c$ variables can be eliminated, leading to the equations below.

SIR compact pairwise

$$[\dot{S_k}]_c = -\tau k[S_k]_c \frac{[SI]_c}{[SX]_c}, \tag{5.19a}$$

$$[\dot{SI}]_c = -\gamma[SI]_c + \tau([SS]_c - [SI]_c)[SI]_c Q_c - \tau[SI]_c, \tag{5.19b}$$

$$[\dot{SS}]_c = -2\tau[SS]_c[SI]_c Q_c, \tag{5.19c}$$

$$[SX]_c = \sum_{k=1}^{M} k[S_k]_c, \tag{5.19d}$$

$$Q_c = \frac{1}{[SX]_c^2} \sum_{k=1}^{M} (k-1)k[S_k]_c, \tag{5.19e}$$

$$[S]_c = \sum_{k=1}^{M} [S_k]_c, \quad [I]_c = N - [S]_c - [R]_c, \quad [\dot{R}]_c = \gamma[I]_c. \tag{5.19f}$$

5.2.4 Super-compact pairwise model

The number of equations in the compact pairwise model is $\mathcal{O}(M)$. However, this can be reduced further by noting that we are generally interested only in the total number of infected nodes $[I]$ rather than the number infected of each degree $[I_k]$. If we sum the $[\dot{I_k}]_c$ equations, we arrive at an equation for $[\dot{I}]_c$,

$$[\dot{I}]_c = \tau[SI]_c - \gamma[I]_c.$$

In the SIS case, this is enough to determine $[S]_c$ because $[I] + [S] = N$; thus, we no longer need $[SX]_c$ except in the expression for Q_c. We can eliminate $[SX]_c$ by approximating Q_c, which will yield the "super-compact pairwise SIS model". In the SIR case, we will see that $[S_k]_c$ takes a simple form which allows us to eliminate $[SX]_c$ and express $[S]_c$ and Q_c in terms of a single unknown variable $\theta(t)$, for which we can write a single differential equation. Thus, with no further approximation, we arrive at the "super-compact pairwise SIR model".

For the SIS case, we write $s_k = [S_k]_c / [S]_c$. Then, $\sum s_k = 1$ and we define

$$n_S := \sum_{k=1}^M k s_k = \frac{[SI]_c + [SS]_c}{[S]_c}.$$

To find an approximation for s_k, we use the numerical solutions of the compact pairwise model and compare $[S_k]_c / [S]_c$ to $p_k = N_k / N$. Numerical results show that these appear linearly related, that is s_k / p_k is a linear function of the degree k. Assuming a linear relation, we arrive at an approximate Q_c (derived in [287]):

$$Q_c \approx \frac{1}{n_S^2 [S]} \left(\frac{\langle K^2 \rangle (\langle K^2 \rangle - n_S \langle K \rangle) + \langle K^3 \rangle (n_S - \langle K \rangle)}{\langle K^2 \rangle - \langle K \rangle^2} - n_S \right) =: Q_{sc},$$

where $\langle K^i \rangle = \sum k^i p_k$ is the ith moment of the degree distribution. This yields the model below.

SIS super-compact pairwise

$$[\dot{I}]_{sc} = \tau [SI]_{sc} - \gamma [I]_{sc}, \tag{5.20a}$$

$$[\dot{SI}]_{sc} = \gamma ([II]_{sc} - [SI]_{sc}) + \tau [SI]_{sc} ([SS]_{sc} - [SI]_{sc}) Q_{sc} - \tau [SI]_{sc}, \tag{5.20b}$$

$$[\dot{SS}]_{sc} = 2\gamma [SI]_{sc} - 2\tau [SI]_{sc} [SS]_{sc} Q_{sc}, \tag{5.20c}$$

$$[\dot{II}]_{sc} = -2\gamma [II]_{sc} + 2\tau [SI]_{sc}^2 Q_{sc} + 2\tau [SI]_{sc}, \tag{5.20d}$$

$$[S]_{sc} = N - [I]_{sc} \tag{5.20e}$$

$$Q_{sc} = \frac{1}{n_S [S]_{sc}} \left(\frac{\langle K^2 \rangle (\langle K^2 \rangle - n_S \langle K \rangle) + \langle K^3 \rangle (n_S - \langle K \rangle)}{n_S \left(\langle K^2 \rangle - \langle K \rangle^2 \right)} - 1 \right), \tag{5.20f}$$

$$n_S = \frac{[SI]_{sc} + [SS]_{sc}}{[S]_{sc}}. \tag{5.20g}$$

Numerical comparison shows that the prevalence obtained from the super-compact pairwise model is in excellent agreement with that obtained from the compact pairwise model for several network types (see [287]).

For the SIR case, we can solve Eq. (5.19a) by using the integrating factor

$$\exp \left(\tau k \int_0^t \frac{[SI]_c(\hat{t})}{[SX]_c(\hat{t})} \, d\hat{t} \right),$$

yielding

$$[S_k]_c(t) = [S_k]_c(0)\theta(t)^k,$$

where

$$\theta(t) = \exp\left(-\tau \int_0^t \frac{[SI]_c(\hat{t})}{[SX]_c(\hat{t})} \, d\hat{t}\right).$$

The initial condition $\theta(0) = 1$ holds, and differentiation shows that θ satisfies the differential equation $\dot{\theta} = -\tau [SI]_c \theta / [SX]_c$. Using again that $\hat{\psi}(x) = \frac{1}{N}\sum_{k=1}^M S_k(0)x^k$ and the initial condition $[S_k]_c(0) = S_k(0)$ leads to

$$[S]_c(t) = \sum_{k=1}^M [S_k]_c(t) = N\hat{\psi}(\theta(t)).$$

Furthermore,

$$[SX]_c(t) = \sum_{k=1}^M k[S_k]_c(t) = \sum_{k=1}^M kS_k(0)\theta^k(t) = N\theta(t)\hat{\psi}'(\theta(t))$$

and

$$\begin{aligned}
Q_c(t) &= \sum_{k=1}^M (k-1)k[S_k]_c(t) \left/ \left(\sum_{k=1}^M k[S_k]_c(t)\right)^2\right. \\
&= \sum_{k=1}^M (k-1)k[S_k]_c(0)\theta^k(t) \left/ \left(\sum_{k=1}^M k[S_k]_c(t)\right)^2\right. \\
&= \frac{\theta^2(t)\hat{\psi}''(\theta(t))}{N(\theta(t)\hat{\psi}'(\theta(t)))^2} = \frac{\hat{\psi}''(\theta(t))}{N(\hat{\psi}'(\theta(t)))^2}.
\end{aligned} \tag{5.21}$$

We define this to be Q_{sc}. In the SIR case, no approximation is needed to derive the super-compact pairwise model (below) from the compact pairwise model.

SIR super-compact pairwise

$$\dot{\theta}_{sc} = -\tau \frac{[SI]_{sc}}{N\hat{\psi}'(\theta_{sc})}, \tag{5.22a}$$

$$[\dot{SI}]_{sc} = -\gamma[SI]_{sc} + \tau([SS]_{sc} - [SI]_{sc})[SI]_{sc}Q_{sc} - \tau[SI]_{sc}, \tag{5.22b}$$

$$[\dot{SS}]_{sc} = -2\tau[SS]_{sc}[SI]_{sc}Q_{sc}, \tag{5.22c}$$

$$[\dot{R}]_{sc} = \gamma[I]_{sc}, \qquad [S]_{sc} = N\hat{\psi}(\theta_{sc}), \qquad [I]_{sc} = N - [S]_{sc} - [R]_{sc}, \tag{5.22d}$$

$$Q_{sc} = \frac{\hat{\psi}''(\theta_{sc})}{N(\hat{\psi}'(\theta_{sc}))^2}, \tag{5.22e}$$

with $\theta_{sc}(0) = 1$.

Thus, the degree heterogeneity can be captured by a system containing only four differential equations. This system was derived using a different approach in [151]. It is important to note that this simple system exactly recovers the solution of the compact pairwise system (5.19), i.e. $[S]_{sc} = [S]_c$, $[I]_{sc} = [I]_c$ and $[R]_{sc} = [R]_c$. Although we do not do it here, we could eliminate N from the system by dividing all singles and pairs by N and multiplying Q_{sc} by N.

Exercise 5.2. Based on systems (5.13) and (5.15), write down the differential equations of the heterogeneous pairwise model for a bimodal Configuration Model network with two different degrees d_1 and d_2 and with $N_1 = N_2 = N/2$ for both SIS and SIR epidemics. Plot the prediction $I(t)$ for $N = 1000$, $d_1 = 3$, $d_2 = 6$ and $\tau = \gamma = 1$, starting with a fraction 0.01 infected at random.

Exercise 5.3. Based on systems (5.18) and (5.19), write down the differential equations of the compact pairwise model for a bimodal graph with two different degrees d_1 and d_2 for both SIS and SIR epidemics. Compare the solution to that obtained in Exercise 5.2 by using the same parameter values.

Exercise 5.4. Based on systems (5.20) and (5.22), write down the differential equations of the super-compact pairwise model for a bimodal graph with two different degrees d_1 and d_2 for both SIS and SIR epidemics. Compare the solution to that obtained in Exercise 5.2 by using the same parameter values.

5.3 Analysis of the closed systems

In this section, we analyse some of the mean-field models developed above using the tools of the qualitative theory of ODEs. We postpone the more technical parts of the derivations to Section 5.5.

5.3.1 SIS heterogeneous mean-field model at the single level

We consider SIS epidemics closed at the level of pairs, system (5.10). Adding the equations corresponding to $[S_k]$ and $[I_k]$, we get that $[S_k]_f(t) + [I_k]_f(t)$ is constant in time. If the initial condition satisfies $[S_k]_f(0) + [I_k]_f(0) = N_k$, then $[S_k]_f(t) + [I_k]_f(t) = N_k$. By using $[S_k]_f(t) = N_k - [I_k]_f(t)$, the system can be reduced to equations for $[I_k]$ only,

$$[\dot{I_k}]_f = \tau k(N - [I_k]_f)\pi_{I,f} - \gamma[I_k]_f, \quad k = 1, 2, \ldots, M, \qquad (5.23)$$

where $\pi_{I,f}$ is given in (5.10c). This system has a disease-free steady state where $[I_k]_f = 0$ for all k. In order to find the endemic steady state, we express $[I_k]_f$ from the steady state equation in terms of $\pi_{I,f}$ as

$$[I_k]_f = \frac{\tau k N_k \pi_{I,f}}{\gamma + \tau k \pi_{I,f}}. \qquad (5.24)$$

Multiplying this by k, summing over k and using the definition of $\pi_{I,f}$ yields

$$\pi_{I,f} \sum_{k=1}^{M} k N_k = \sum_{k=1}^{M} \frac{\tau k^2 N_k \pi_{I,f}}{\gamma + \tau k \pi_{I,f}}.$$

This determines the value of $\pi_{I,f}$ at steady state and (5.24) gives the steady state value of $[I_k]_f$. This has the trivial solution $\pi_{I,f} = 0$ corresponding to the disease-free steady state. Dividing by $\pi_{I,f}$ eliminates this trivial solution and yields

$$\sum_{k=1}^{M} k N_k = \sum_{k=1}^{M} \frac{\tau k^2 N_k}{\gamma + \tau k \pi_{I,f}},$$

which determines the endemic steady state. We find the condition for the transcritical bifurcation, when the endemic and the disease-free steady states coincide, by setting $\pi_{I,f} = 0$. We get

$$\frac{\tau}{\gamma} = \sum_{k=1}^{M} k N_k \bigg/ \sum_{k=1}^{M} k^2 N_k = \frac{\langle K \rangle}{\langle K^2 \rangle}.$$

So the critical value of τ predicted by the heterogeneous mean-field approximation at single level is

$$\tau_c^{HMF} = \gamma \frac{\langle K \rangle}{\langle K^2 \rangle}.$$

Section 5.5.1 proves that for $\tau < \tau_c^{HMF}$, the heterogeneous mean-field approximation at single level predicts that the disease-free steady state is stable, and for $\tau > \tau_c^{HMF}$ it is unstable and there is a unique endemic steady state. We can also find the stability of the disease-free steady state as follows (see [177]). Multiply Eq. (5.23) by k and sum over k. Introducing $J_f = \sum_{k=1}^{M} k[I_k]_f$ and $L_f = \sum_{k=1}^{M} k^2 [I_k]_f$, we have

$$\dot{J}_f = J_f \left(\tau \frac{\langle K^2 \rangle}{\langle K \rangle} - \gamma - \frac{\tau L_f}{N \langle K \rangle} \right).$$

This implies that for $\tau < \tau_c^{HMF}$, we have $\dot{J}_f < -c J_f$ with some positive constant c; hence, J_f is strictly decreasing and converges to zero. Therefore, $[I_k]_f$ also tends to zero for all k, implying the stability of the disease-free steady state.

The stability of the endemic steady state is studied in the book [104], which has a detailed analytical investigation of several mean-field models. Theorem 10.35 of [104] shows that the endemic steady state is globally stable.

Section 5.5.1 shows that the largest eigenvalue of the Jacobian at the disease-free steady state is $\tau \frac{\langle K^2 \rangle}{\langle K \rangle} - \gamma$. Hence, according to (4.18) the growth-based basic reproductive ratio in the heterogeneous SIS case is

$$\overline{R}_0^{HMF} = \frac{\tau \langle K^2 \rangle}{\gamma \langle K \rangle}, \tag{5.25}$$

showing that for power law networks $\overline{R}_0^{\text{HMF}}$ may be unbounded (if $\langle K^2 \rangle$ is unbounded), enabling epidemic invasion even for arbitrarily small transmission rates. We summarise results for the closed system (5.10) below.

Analytical results for SIS heterogeneous mean-field at single level

- Conservation of singles: $[S_k]_f(t) + [I_k]_f(t) = N_k$.
- Reducibility of equations: $[\dot{I}_k]_f = \tau k (N - [I_k]_f) \pi_{I,f} - \gamma [I_k]_f$.
- Transcritical bifurcation at $\tau = \tau_c^{\text{HMF}} = \gamma \langle K \rangle / \langle K^2 \rangle$: the disease-free steady state loses its stability and an endemic steady state appears.
- Global behaviour for $\tau < \tau_c^{\text{HMF}}$: all solutions converge to the disease-free steady state $[I_k]_f = 0$.
- Uniqueness of the endemic steady state for $\tau > \tau_c^{\text{HMF}}$.
- Global behaviour for $\tau > \tau_c^{\text{HMF}}$: all solutions converge to the endemic steady state.
- The growth-based basic reproductive ratio is $\overline{R}_0^{\text{HMF}} = \tau \langle K^2 \rangle / \gamma \langle K \rangle$.

5.3.2 SIR heterogeneous mean-field model at the single level

We consider SIR epidemics closed at the level of pairs, system (5.11). This system has infinitely many steady states, namely $[I_k]_f = 0$ and $[S_k]_f = S_k^*$ with any $S_k^* \in [0, N_k]$. At any non-equilibrium value, we expect that $[S_k]_f$ is decreasing and $[R_k]_f$ is increasing, with both bounded in $[0, N_k]$. This implies that these functions approach a limit as time goes to infinity, which we denote by $[S_k]_f^\infty$ and $[R_k]_f^\infty$. Because $[\dot{R}_k]_f = \gamma [I_k]_f$, we have $[I_k]^\infty = 0$. The system converges to a steady state, which depends on the initial condition. In Section 5.5.2, this is determined analytically in terms of the initial conditions, and the final epidemic size $[R_k]_f^\infty$ is

$$[R_k]_f^\infty = N_k - S_k(0)\theta_f(\infty)^k, \tag{5.26}$$

where $S_k(0) = [S_k]_f(0)$ is the initial condition and $\theta_f(\infty)$ is given implicitly by

$$\theta_f(\infty) = \exp\left[-\frac{\tau}{\gamma}\left(1 - \frac{\theta_f(\infty)\hat{\psi}'(\theta_f(\infty))}{\langle K \rangle} - \frac{\sum_l lR_l(0)}{N \langle K \rangle}\right)\right]. \tag{5.27}$$

Section 5.5.2 shows that there is a unique solution for $\theta_f(\infty)$ in $[0,1]$, which can be found by iteration. We summarise results for the closed system (5.11) below.

Analytical results for SIR heterogeneous mean-field at the single level

- Conservation of singles: $[S_k]_f(t) + [I_k]_f(t) + [R_k]_f(t) = N_k$.
- Global behaviour: all solutions converge to a steady state, which depends on the initial conditions.
- The steady state is determined by the final epidemic size $[R_k]_f^\infty$, which is given analytically by Eqs. (5.26) and (5.27).

5.3.3 SIS compact pairwise model

We consider the compact pairwise model for an SIS epidemic (5.18). Note that summing equations corresponding to $[S_k]$ leads to $[\dot{S}]_c = \gamma(N - [S]_c) - \tau[SI]_c$, where $[S]_c = \sum_{k=1}^{M}[S_k]_c$ is the expected number of susceptible nodes. Thus, the solution of the compact pairwise model satisfies the exact equation $[\dot{S}] = \gamma[I] - \tau[SI]$. Adding the differential equations of the pairs, we get that the total number of pairs $[SS]_c + 2[SI]_c + [II]_c$ is constant in time. If this is equal to nN at the initial time instant, then pair conservation $[SS]_c + 2[SI]_c + [II]_c = nN$ holds for all time, where $n = \langle K \rangle$ is the average degree. Section 5.5.3 shows that the stub conservation relation $\sum_{k=1}^{M} k[S_k]_c = [SS]_c + [SI]_c$ also holds for this model, i.e. the total number of stubs starting from susceptible nodes equals the total number of SS and SI pairs. Using the pair and stub conservation relations, the system can be reduced to the differential equation of the singles and to the differential equation of the SI pairs; the remaining two differential equations for the SS and II pairs are not independent.

The system has a trivial steady state called the disease-free steady state $[S_k]_c = N_k$, $[SI]_c = 0$, $[SS]_c = nN$ and $[II]_c = 0$. Section 5.5.3 proves that an eigenvalue of the Jacobian at this steady state passes through zero when

$$\frac{\tau}{\gamma} = \frac{\langle K \rangle}{\langle K^2 \rangle - \langle K \rangle}.$$

This relation defines the critical value of τ in the compact pairwise model as

$$\tau_c^{\mathrm{CPW}} = \gamma \frac{\langle K \rangle}{\langle K^2 \rangle - \langle K \rangle}.$$

Section 5.5.3 shows that if $\tau < \tau_c^{\mathrm{CPW}}$, the disease-free steady state is stable. For $\tau > \tau_c^{\mathrm{CPW}}$, it loses stability and there is an endemic steady state. Moreover, we find an expression for the eigenvalues at the disease-free steady state. Using the largest eigenvalue and Eq. (4.18), the growth-based basic reproductive ratio in the compact pairwise model for an SIS epidemic is

$$\overline{R}_0^{CPW} = \frac{a + \sqrt{(2\gamma - a)^2 + 8\gamma(a + \tau)}}{2\gamma},$$ (5.28)

where $a = \tau \frac{\langle K^2 \rangle - 2\langle K \rangle}{\langle K \rangle} - \gamma$. We note that for a regular network, $\langle K \rangle = n$ and $\langle K^2 \rangle = n^2$; hence, $a = \tau(n - 2) - \gamma$. We can then reduce this formula to the one obtained by using the homogeneous pairwise model in (4.19), i.e. $\overline{R}_0^{CPW} = \overline{R}_0^{PW}$ holds for homogeneous networks. Summarising, the closed system (5.18) leads to the following.

Analytical results for SIS compact pairwise

- Conservation of pairs: $[SS]_c + 2[SI]_c + [II]_c = \langle K \rangle N$.
- Conservation of stubs: $\sum_{k=1}^{M} k[S_k]_c = [SS]_c + [SI]_c$.
- Transcritical bifurcation at $\tau = \tau_c^{CPW} = \gamma \frac{\langle K \rangle}{\langle K^2 \rangle - \langle K \rangle}$: the disease-free steady state loses stability and an endemic steady state appears.
- The growth-based basic reproductive ratio is $\overline{R}_0^{CPW}$ given in (5.28).

The following open questions remain and lack a rigorous mathematical proof:

1. Characterisation of the global behaviour for $\tau < \tau_c^{CPW}$: do all solutions converge to the disease-free steady state?
2. Is the endemic steady state unique for $\tau > \tau_c^{CPW}$?
3. Is the endemic steady state stable (locally or globally) for $\tau > \tau_c^{CPW}$?

5.4 Comparison of models to simulation

We now study the performance of mean-field models for heterogeneous networks in a similar way as in the case of homogeneous mean-field models. Consider SIS and SIR epidemics on a Configuration Model network with N nodes. The number of infected nodes is determined from stochastic simulation. The average of several simulations, approximating the expected value of the prevalence, is denoted by $[I](t)$. This is compared numerically to the prevalence given based on the heterogeneous mean-field system closed at the level of pairs, (5.10), the compact pairwise model closed at the level of triples, (5.18), and the homogeneous pairwise model, (4.10). Hence, the following four functions are compared: $[I](t)$, $[I]_f(t)$, $[I]_c(t)$ and $[I]_s(t)$. Note that $[I]_s(t)$ corresponds to the homogeneous pairwise model, while the other two approximations are for heterogeneous networks. The homogeneous approximation $[I]_s(t)$ is used here only to show that a mean-field model based on the homogeneity assumption, and having only a few differential equations, does not perform well for networks with significant degree heterogeneity. As before, we consider the following parameters:

- infection rate τ and recovery rate γ,
- average degree $\langle K \rangle$,
- variance of the degree distribution $\langle K^2 \rangle - \langle K \rangle^2$.

This numerical analysis will be carried out for the following three network types, as defined in Section 1.2.3:

- bimodal random graphs,
- Erdős–Rényi random graphs,
- random graphs with a power law degree distribution.

First, we study the dependence on the disease parameters, τ and γ. By scaling time, we can take $\gamma = 1$ and just vary τ. We focus on the performance near the critical value $\tau_c^{HMF} = \gamma \langle K \rangle / \langle K^2 \rangle$. We take bimodal Configuration Model networks with degrees $d_1 = 5$ and $d_2 = 35$, and use N_i to be the number of degree d_i nodes.

We take four values of τ: $\tau = 0.9\tau_c^{HMF}$, $\tau = \tau_c^{HMF}$, $\tau = 1.1\tau_c^{HMF}$ and $\tau = 1.5\tau_c^{HMF}$ and plot $[I](t)$, $[I]_f(t)$, $[I]_c(t)$ and $[I]_s(t)$ in Fig. 5.2 for a fixed network size $N = 1000$. Close to the critical value, the agreement is poorer and improves away from it. Further testing shows that for $\tau > 2\tau_c^{HMF}$ and $\tau < 0.8\tau_c^{HMF}$, simulations are effectively indistinguishable from the predictions of the compact pairwise approximation. Moreover, the compact pairwise model yields significantly better agreement than the mean-field model closed at the level of pairs. The homogeneous pairwise model completely fails to predict the prevalence for this parameter range.

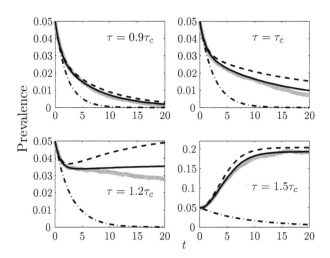

Fig. 5.2: The time dependence of the number of infected nodes for an SIS epidemic for a bimodal random network based on individual-based stochastic simulation, $[I](t)$ (grey thick lines), heterogeneous mean-field model closed at the level of pairs, $[I]_f(t)$ (dashed lines), the compact pairwise model, $[I]_c(t)$ (solid lines), and homogeneous pairwise model, $[I]_s(t)$ (dashed-dotted lines). The parameter values are $N = 1000$, $N_1 = N_2 = 500$, $d_1 = 5$, $d_2 = 35$, $\langle K \rangle = (d_1 + d_2)/2 = 20$ and $\gamma = 1$; the τ values are shown in each subfigure.

We study the effect of the average degree on the performance of the ODE models using Erdős–Rényi random networks. Figure 5.3 shows $[I](t)$, $[I]_f(t)$, $[I]_c(t)$ and $[I]_s(t)$ for $\langle K \rangle = 50$ and for $\langle K \rangle = 10$. Each edge exists with probability $p = \langle K \rangle / (N-1)$ to give the desired average degree. The value of τ in both cases is twice the predicted critical value; $\tau = 2\tau_c^{\text{HMF}}$. We see better agreement for denser graphs. It is worth noting that the heterogeneous mean-field model does not perform better than the homogeneous pairwise model. This can be explained in part by the fact that the Erdős–Rényi random network is not significantly heterogeneous. When there is little degree heterogeneity, it is more important for closures to capture pair correlations (i.e. the fact that neighbours of infected and susceptible nodes are not interchangeable). Thus, for Erdős–Rényi networks the homogeneous pairwise model is a reasonable choice as a mean-field ODE approximation, especially because apart from yielding accurate results, the system has only a few differential equations.

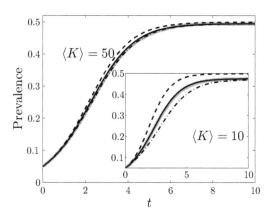

Fig. 5.3: The time dependence of the expected number of infected nodes for an SIS epidemic for Erdős–Rényi random networks based on individual-based stochastic simulation, $[I](t)$ (grey thick lines), heterogeneous mean-field model closed at the level of pairs, $[I]_f(t)$ (dashed lines), compact pairwise model, $[I]_c(t)$ (solid lines), and homogeneous pairwise model, $[I]_s(t)$ (dashed-dotted lines), for a denser graph with $\langle K \rangle = 50$ and for a sparser graph with $\langle K \rangle = 10$ (inset). The parameter values are $N = 1000$, $\gamma = 1$ and $\tau = 2\tau_c^{\text{HMF}}$.

We finally consider the effect of the average degree on the performance of the ODE models for Configuration Model random networks with a truncated power law degree distribution. These random networks are given by a minimal degree, k_{min}, a maximal degree, k_{max}, and exponent, α, of the degree distribution. The degree distribution of the graph is $p_k = Ck^{-\alpha}$ for $k = k_{\text{min}}, k_{\text{min}}+1, \ldots, k_{\text{max}}$ with the normalisation constant C given by $\frac{1}{C} = \sum_{k=k_{\text{min}}}^{k_{\text{max}}} k^{-\alpha}$. In Fig. 5.4, $[I](t)$, $[I]_f(t)$, $[I]_c(t)$ and $[I]_s(t)$ are plotted for a sparser and a denser power law Configuration Model random network with power $\alpha = 2$. For the sparser network, the minimal degree is

$k_{min} = 1$ and the maximal degree is $k_{max} = 40$. The average degree is $\langle K \rangle = 2.7$, and
the standard deviation of the degree distribution is $\sqrt{\langle K^2 \rangle - \langle K \rangle^2} = 4.2$. (We use
standard deviation to show spread around the average degree.) For the denser net-
work, the minimal degree is $k_{min} = 10$ and the maximal degree is $k_{max} = 150$. The
average degree is $\langle K \rangle = 28.7$, and the standard deviation of the degree distribution is
$\sqrt{\langle K^2 \rangle - \langle K \rangle^2} = 26.5$. The value of τ in both cases is chosen based on the average
degree of the network as $\tau = 1.5\gamma/\langle K \rangle$. The figure shows that the ODEs perform
better for denser graphs despite the fact that the heterogeneity is much stronger
(the standard deviation of the degree distribution is 6 times larger for the denser
network). As expected, the performance of the heterogeneous mean-field model is
much better than that of the homogeneous pairwise model due to high heterogeneity of the degree distribution. Thus, while the homogeneous pairwise model works
for networks with limited degree heterogeneity, it is inappropriate for graphs with
significant heterogeneity.

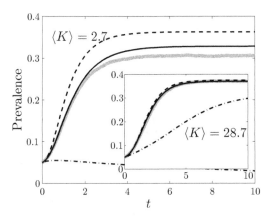

Fig. 5.4: The time dependence of the number of infected nodes for an SIS epidemic
for power law Configuration Model networks based on individual-based stochastic
simulation, $[I](t)$ (grey thick lines), heterogeneous mean-field model closed at the
level of pairs, $[I]_f(t)$ (dashed lines), compact pairwise model, $[I]_c(t)$ (solid lines),
and homogeneous pairwise model, $[I]_s(t)$ (dashed-dotted lines), for a sparser net-
work with $\langle K \rangle = 2.7$ and for a denser network with $\langle K \rangle = 28.7$ (inset). The parameter values are $N = 1000$, $\gamma = 1$ and $\tau = 1.5\gamma/\langle K \rangle$.

We turn now to SIR epidemics. We first investigate the performance of the heterogeneous mean-field model at single level and the compact pairwise model for a
bimodal random network, in which half of the nodes have degree d_1 and the other
half have degree d_2. Figure 5.5 shows the functions $[I](t)$, $[I]_f(t)$, $[I]_c(t)$ and $[I]_s(t)$
for a denser network with $\langle K \rangle = 50$ and for a sparser network with $\langle K \rangle = 10$ (in the
inset). In the denser one, the degrees are $d_1 = 30$ and $d_2 = 70$; in the sparser one,

they are $d_1 = 5$ and $d_2 = 15$. The value of τ in both cases is double that of the critical value, i.e. $\tau = 2\tau_c^{\text{HMF}}$. We observe that the agreement is better for the denser network and the compact pairwise model captures the pair correlations; hence, it yields excellent agreement. It is worth noting that the performance of the heterogeneous mean-field model at single level for the sparser network is not better than that of the homogeneous pairwise model.

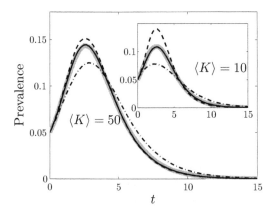

Fig. 5.5: The time dependence of the expected number of infected nodes for an SIR epidemic for bimodal random networks based on individual-based stochastic simulation, $[I](t)$ (grey thick lines), heterogeneous mean-field model closed at the level of pairs, $[I]_f(t)$ (dashed lines), compact pairwise model, $[I]_c(t)$ (solid lines) and homogeneous pairwise model, $[I]_s(t)$ (dashed-dotted lines), for a denser network with $\langle K \rangle = 50$ ($d_1 = 30$ and $d_2 = 70$) and for a sparser network with $\langle K \rangle = 10$ ($d_1 = 5$ and $d_2 = 15$) (inset). The parameter values are $N = 1000$, $N_1 = N_2 = 500$, $\gamma = 1$ and $\tau = 2\tau_c^{\text{HMF}}$.

The systematic comparison of simulation results to output from the heterogeneous mean-field and compact pairwise models can be complemented by attempting the following exercises. For algorithm descriptions and ready-to-run source codes accompanying the book, see Appendix A.1 and the following link:

https://springer-math.github.io/Mathematics-of-Epidemics-on-Networks/

Exercise 5.5. Investigate the effect of system size N on the performance of mean-field models for bimodal random networks. Create bimodal random networks of size $N = 100$, $N = 200$, $N = 400$ and $N = 800$, with $N_1 = N_2 = N/2$ and degrees $d_1 = 5$ and $d_2 = 15$. Use the SIS and the SIR dynamics and determine $[I](t)$ by averaging over 100 simulations. Compare $[I](t)$ to $[I]_f(t)$ (heterogeneous mean-field model at single level) and $[I]_c(t)$ (compact pairwise model) for each system size separately.

Exercise 5.6. Create two 1000-node bimodal random networks. Take $N_1 = N_2 = N/2$ and choose the degrees such that one has average degree $\langle K \rangle = 5$ and the other has average degree $\langle K \rangle = 25$, but both have the same variance $\langle K^2 \rangle - \langle K \rangle^2$. The predicted critical value of τ is $\tau_c^{\text{HMF}} = \gamma \langle K \rangle / \langle K^2 \rangle$. Take $\tau = 2\tau_c^{\text{HMF}}$. Use the SIS and the SIR dynamics and determine $[I](t)$ by averaging over 100 simulations. Compare $[I](t)$ to $[I]_f(t)$ (heterogeneous mean-field model at single level) and $[I]_c(t)$ (compact pairwise model) for both networks.

Exercise 5.7. Create a 1000-node bimodal and a power law random network with the same average degree and same variance in the degree distribution. Let $N_1 = N_2 = N/2$. The predicted critical value of τ is $\tau_c^{\text{HMF}} = \gamma \langle K \rangle / \langle K^2 \rangle$. Take $\tau = 2\tau_c^{\text{HMF}} = 2\gamma \langle K \rangle / \langle K^2 \rangle$. Use the SIS and the SIR dynamics and determine $[I](t)$ by averaging over 100 simulations. Compare $[I](t)$ to $[I]_f(t)$ (heterogeneous mean-field model at single level) and $[I]_c(t)$ (compact pairwise model) for both networks.

Exercise 5.8. Create 1000-node power law random graphs with two different α values. The predicted critical value of τ is $\tau_c^{\text{HMF}} = \gamma \langle K \rangle / \langle K^2 \rangle$. Take $\tau = 2\tau_c^{\text{HMF}}$. Use the SIS and the SIR dynamics and determine $[I](t)$ by averaging over 100 simulations. Compare $[I](t)$ to $[I]_f(t)$ (heterogeneous mean-field model at single level) and $[I]_c(t)$ (compact pairwise model) for both networks.

Exercise 5.9. Create two 1000-node bimodal random networks, one with $N_1 = N_2 = N/2$, the other one with $N_1 = N/4$, $N_2 = 3N/4$, keeping the average degree and variance at the same value. The predicted critical value of τ is $\tau_c^{\text{HMF}} = \gamma \langle K \rangle / \langle K^2 \rangle$. Take $\tau = 2\tau_c^{\text{HMF}}$. Use the SIS and the SIR dynamics and determine $[I](t)$ by averaging over 100 simulations. Compare $[I](t)$ to $[I]_f(t)$ (heterogeneous mean-field model at single level) and $[I]_c(t)$ (compact pairwise model) for both networks.

This concludes our overview of mean-field models and summary of analytical and numerical results. The next section performs a rigorous mathematical derivation of some of the results presented thus far.

5.5 Detailed analytical study of the mean-field models

The technical details of the analytical investigations presented in Section 5.3 are given below.

5.5.1 Heterogeneous SIS mean-field model at the single level

We now complete the analysis of system (5.23). We study the local stability of the disease-free steady state by explicitly determining the eigenvalues of the Jacobian.

We then verify the existence and uniqueness of the endemic steady state for τ values above the transcritical bifurcation value. Recall that the disease-free steady state of the system is given by $[I_k]_f = 0$ for all k and the critical value of τ is

$$\tau_c^{HMF} = \gamma \frac{\langle K \rangle}{\langle K^2 \rangle}.$$

We first investigate the stability of this equilibrium.

Proposition 5.1 *The disease-free steady state of system* (5.23) *is asymptotically stable for* $\tau < \tau_c^{HMF}$ *and unstable for* $\tau > \tau_c^{HMF}$.

Proof. We prove this by direct investigation of the eigenvalues of the Jacobian of system (5.23) at the disease-free steady state $[I_k]_f = 0$. The Jacobian takes the form $\tau P / \langle K \rangle - \gamma I$, where

$$P = \begin{pmatrix} p_1 & 2p_1 & \cdots & M p_1 \\ \vdots & \vdots & \ddots & \vdots \\ k p_k & 2k p_k & \cdots & M k p_k \\ \vdots & \vdots & \ddots & \vdots \\ M p_M & 2M p_M & \cdots & M^2 p_M \end{pmatrix}.$$

The eigenvalues of $\tau P / \langle K \rangle - \gamma I$ take the form $\lambda_i = \tau \mu_i / \langle K \rangle - \gamma$, where μ_i is the ith eigenvalue of P. To find eigenvalues of P, we solve $\det(P - \mu I) = 0$ using some simplifying row and column manipulations. For each $j = 1, 2, \ldots, M-1$, we replace column j of $P - \mu I$ by the result of subtracting j/M times the final column. The new matrix has $-\mu$ in the main diagonal (except in the last column), $j\mu/M$ at the jth entry in last row and zeros everywhere else (except in the last column). Now for $j = 1, \ldots, M-1$, we add M/j times the jth row to the final row. The result is an upper triangular matrix, with all but the final diagonal entry being $-\mu$. The final entry is $-\mu + \sum_{k=1}^{M} k^2 p_k$. Setting the determinant to zero yields

$$(-\mu)^{M-1} \left(-\mu + \sum_{k=1}^{M} k^2 p_k \right) = 0.$$

By properties of row and column operations, the determinant of $P - \mu I$ is zero exactly when the determinant of this new matrix is zero. Thus, the eigenvalues of P are $\mu_1 = \mu_2 = \cdots = \mu_{M-1} = 0$ and $\mu_M = \sum_{k=1}^{M} k^2 p_k = \langle K^2 \rangle$.

Therefore, the eigenvalues of the Jacobian are $\lambda_1 = \lambda_2 = \cdots = \lambda_{M-1} = -\gamma$ and $\lambda_M = \tau \frac{\langle K^2 \rangle}{\langle K \rangle} - \gamma$. Thus, all eigenvalues have negative real part if and only if $\tau \frac{\langle K^2 \rangle}{\langle K \rangle} - \gamma < 0$, i.e. $\tau < \tau_c^{HMF}$. On the other hand, if $\tau > \tau_c^{HMF}$, then there is a positive eigenvalue of the Jacobian; hence, the disease-free steady state is unstable. $\qquad\square$

Concerning the endemic steady state, the following statement holds.

Proposition 5.2 *If $\tau < \tau_c^{HMF}$, then system (5.23) does not have an endemic steady state. If $\tau > \tau_c^{HMF}$, then there is a unique endemic steady state.*

Proof. In Section 5.3, we showed that the endemic steady state is

$$[I_k]_f = \frac{\tau k N_k \pi_{I,f}}{\gamma + \tau k \pi_{I,f}}, \qquad k = 1, \ldots, M,$$

where $\pi_{I,f}$ solves

$$\sum_{k=1}^{M} k N_k = \sum_{k=1}^{M} \frac{\tau k^2 N_k}{\gamma + \tau k \pi_{I,f}}.$$

Here, we show that this equation has a unique solution if $\tau > \tau_c^{HMF}$ and no solution if $\tau < \tau_c^{HMF}$. The equation takes the form $f(\pi_{I,f}) = 1$, where

$$f(\pi_{I,f}) = \frac{\tau}{N \langle K \rangle} \sum_{k=1}^{M} \frac{k^2 N_k}{\gamma + \tau k \pi_{I,f}}.$$

We seek a solution $\pi_{I,f}$ in the interval $[0, 1]$. Observe that f is strictly decreasing between 0 and 1 with

$$f(0) = \frac{\tau}{\gamma} \frac{\langle K^2 \rangle}{\langle K \rangle}$$

$$f(1) = \frac{\tau}{N \langle K \rangle} \sum_{k=1}^{M} \frac{k^2 N_k}{\gamma + \tau k} < \frac{\tau}{N \langle K \rangle} \sum_{k=1}^{M} \frac{k^2 N_k}{\tau k} = 1.$$

Hence, in the case $f(0) < 1$ there is no $\pi_{I,f} \in [0, 1]$ satisfying $f(\pi_{I,f}) = 1$. In the case $f(0) > 1$, there is a unique $\pi_{I,f} \in [0, 1]$ satisfying $f(\pi_{I,f}) = 1$. The condition $f(0) = 1$, is exactly $\tau = \tau_c^{HMF}$; thus, the statement is proved. $\square$

5.5.2 Heterogeneous SIR mean-field model at the single level

Consider the heterogeneous SIR model closed at the level of pairs, system (5.11), with initial conditions $[S_k]_f(0) = S_k(0)$, $[I_k]_f(0) = I_k(0)$, $[R_k]_f(0) = R_k(0)$ and $\theta_f(0) = 1$:

$$[S_k]_f = S_k(0) \theta_f^k,$$
$$[I_k]_f = [N_k]_f - [S_k]_f - [R_k]_f,$$
$$[\dot{R}_k]_f = \gamma [I_k]_f,$$
$$\pi_{I,f} = \sum_{l=1}^{M} l[I_l]_f \bigg/ \sum_{l=1}^{M} l N_l,$$
$$\dot{\theta}_f = -\tau \pi_{I,f} \theta_f.$$

We can show that this must converge to an equilibrium for biologically relevant initial conditions. Note that $[I_k]_f(0) \geq 0$ for all k. Thus, $\pi_{I,f}(0) \geq 0$. As long as $\pi_{I,f}$ remains non-negative, θ_f decreases monotonically, but remains positive. Thus, $[S_k]_f$ remains non-negative. Looking at $[\dot{I}_k]_f$, we have

$$[\dot{I}_k]_f = \tau \pi_{I,f} [S_k]_f - \gamma [I_k]_f.$$

While $\pi_{I,f}$ and $[S_k]_f$ are non-negative, $[I_k]_f$ cannot become negative. Further, $\pi_{I,f}$ remains positive because it comes from a weighted average of $[I_k]_f$. Hence, $[R_k]_f$ increases monotonically. Since $[S_k]_f$ and $[I_k]_f$ remain non-negative, $[R_k]_f$ is bounded from above by N_k. Thus, θ_f, $[S_k]_f$ and $[R_k]_f$ must each approach a limit, which we denote by $\theta_f(\infty)$, $[S_k]_f^\infty$ and $[R_k]_f^\infty$, respectively. From the $[\dot{R}_k]_f$ equation, it is clear that as such a limit is approached $[I_k]_f \to 0$, so $[I_k]_f^\infty = 0$ and $\pi_{I,f}^\infty = 0$ as well.

We now derive Eqs. (5.26) and (5.27) for the final epidemic size.

Proposition 5.3 *For system* (5.11) *with initial conditions* $[S_k]_f(0) = S_k(0)$, $[I_k]_f(0) = I_k(0)$, $[R_k]_f(0) = R_k(0)$ *and* $\theta_f(0) = 1$, *the final epidemic size* $[R_k]_f^\infty$ *can be given as*

$$[R_k]_f^\infty = N_k - S_k(0)\theta(\infty)^k, \tag{5.29}$$

where $\theta(\infty)$ *is given by the implicit equation*

$$\theta_f(\infty) = \exp\left[-\frac{\tau}{\gamma}\left(1 - \frac{\theta_f(\infty)\hat{\psi}'(\theta_f(\infty))}{\langle K \rangle} - \frac{\sum l R_l(0)}{N \langle K \rangle}\right)\right]. \tag{5.30}$$

Proof. As noted above, $[I_k]_f^\infty = 0$ for all k and thus $\pi_{I,f}^\infty = 0$ as well. Hence, Eq. (5.29) follows from $[R_k]_f^\infty + [S_k]_f^\infty = N_k$ and from $[S_k]_f(t) = S_k(0)\theta(t)^k$. We can solve the equation for $\dot{\theta}_f$ with an integrating factor so that

$$\theta_f(\infty) = \exp\left(-\tau \int_0^\infty \pi_{I,f}(\hat{t})d\hat{t}\right).$$

Expanding $\pi_{I,f}$ in the integral, we see that

$$\theta_f(\infty) = \exp\left(-\tau \sum_{l=1}^M \int_0^\infty l[I_l]_f(\hat{t})d\hat{t} \Big/ \sum_{l=1}^M lN_l\right).$$

However, because $[\dot{R}_l]_f = \gamma[I_l]_f$, we have $\int_0^\infty l[I_l]_f(\hat{t})\,d\hat{t} = l([R_l]_f^\infty - [R_l]_f(0))/\gamma = l(N_l - [S_l]_f^\infty - [R_l]_f(0))/\gamma$. Thus,

$$\theta_f(\infty) = \exp\left(-\tau \frac{\sum_{l=1}^M l(N_l - [S_l]_f^\infty - [R_l]_f(0))}{\gamma N \langle K \rangle}\right).$$

Some manipulation using $\sum_{l=1}^{M} lN_l = N \langle K \rangle$ and $\sum_{l=1}^{M} l[S_l]_f = N\theta_f \hat{\psi}'(\theta_f)$ gives

$$\theta_f(\infty) = \exp\left[-\frac{\tau}{\gamma}\left(1 - \frac{\theta_f(\infty)\hat{\psi}'(\theta_f(\infty))}{\langle K \rangle} - \frac{\sum lR_l(0)}{N\langle K \rangle}\right)\right],$$

which gives the desired implicit equation for $\theta_f(\infty)$. $\square$

The final epidemic size can be determined by computing the value of $\theta_f(\infty)$. It is natural to ask if Eq. (5.30) has a positive solution for $\theta_f(\infty)$ and if it is unique.

Proposition 5.4 *The implicit equation Eq. (5.30) has a unique solution in* $[0, 1]$.

Proof. We define

$$f(x) = \exp\left[-\frac{\tau}{\gamma}\left(1 - \frac{x\hat{\psi}'(x)}{\langle K \rangle} - \frac{\sum lR_l(0)}{N\langle K \rangle}\right)\right]$$

to be the function on the right-hand side of Eq. (5.30). We note that f is a convex function, having $f(0) > 0$ and $f(1) < 1$. Thus, it must intersect the function $g(x) = x$ in exactly one value between 0 and 1. Consequently, there is a unique solution $\theta_f(\infty)$ for Eq. (5.30). $\square$

5.5.3 Compact pairwise SIS model

The analytical results for the compact pairwise model (5.18) are verified here.

Proposition 5.5 *In the compact pairwise SIS model of system* (5.18)*, the stub conservation* $\sum_{k=1}^{M} k[S_k]_c = [SS]_c + [SI]_c$ *holds.*

Proof. Introducing $F = [SS]_c + [SI]_c - [SX]_c$ and differentiating with respect to time, the differential equations in the compact pairwise model yield

$$\dot{F} = \gamma([II]_c + [SI]_c - \langle K \rangle N - [SX]_c) + \tau[SI]_c\left(\frac{1}{[SX]_c}\sum_{k=1}^{M} k^2[S_k]_c - 1 - ([SS]_c + [SI]_c)Q_c\right).$$

Using the pair conservation $[SS]_c + 2[SI]_c + [II]_c = \langle K \rangle N$ and the equation $[SX]_c Q_c + 1 = \frac{1}{[SX]_c}\sum_{k=1}^{M} k^2[S_k]_c$ leads to

$$\dot{F} = -(\gamma + \tau Q_c[SI]_c)F.$$

The constant zero function is a trivial solution of this differential equation. Hence, assuming that $F(0) = 0$, that is the stub conservation holds at the initial time, we get the desired relation $F(t) = 0$ for all t by the uniqueness of the solution of this simple linear differential equation. $\square$

Proposition 5.6 *The disease-free steady state in the SIS compact pairwise model of system (5.18) is* $[S_k]_c = N_k$, $[SI]_c = 0$, $[SS]_c = nN$ *and* $[II]_c = 0$. *Its stability changes at*

$$\tau_c^{CPW} = \gamma \frac{\langle K \rangle}{\langle K^2 \rangle - \langle K \rangle}.$$

For $\tau < \tau_c^{CPW}$, *the disease-free steady state is stable, and for* $\tau > \tau_c^{CPW}$ *it is unstable.*

Proof. Simple calculation shows that the values given in the statement determine a steady state. The investigation of its stability is based on the linearisation of the right-hand side of the differential equations at the disease-free steady state. Differentiating the right-hand side of the differential equation of $[S_k]_c$ with respect to $[S_l]_c$ (with $l \neq k$) yields zero since $[SI]_c = 0$ in this equilibrium. Differentiating with respect to $[S_k]_c$ yields $-\gamma$ for the same reason. Differentiating the right-hand sides of the differential equations of $[SI]_c$, $[SS]_c$ and $[II]_c$ with respect to $[S_k]_c$ yields again zero. Differentiating the right-hand sides of the differential equations of $[SI]_c$, $[SS]_c$ and $[II]_c$ with respect to these variables yields the matrix

$$P = \begin{pmatrix} a & 0 & \gamma \\ b & 0 & 0 \\ 2\tau & 0 & -2\gamma \end{pmatrix},$$

where

$$a = \tau \frac{\langle K^2 \rangle - \langle K \rangle}{\langle K \rangle} - (\tau + \gamma), \qquad b = 2\gamma - 2\tau \frac{\langle K^2 \rangle - \langle K \rangle}{\langle K \rangle}.$$

Thus, the Jacobian at the disease-free steady state takes the form

$$J = \begin{pmatrix} -\gamma \hat{I} & V \\ 0 & P \end{pmatrix},$$

with some $M \times 3$ matrix V, $\hat{I}$ denoting the $M \times M$ identity matrix, and 0 denoting the $3 \times M$ zero matrix. This matrix is block triangular, and so its eigenvalues are the eigenvalues of the diagonal block matrices. These are $-\gamma$ with multiplicity M (from $-\gamma \hat{I}$) and the eigenvalues of P, one of which is zero and the other two are the roots of the quadratic polynomial

$$\lambda^2 + \lambda(2\gamma - a) - 2\gamma(a + \tau).$$

The zero eigenvalue appears due to the fact that $[SS] + [II] + 2[SI]$ is constant (and so the equations are not independent). This eigenvalue is always zero and its eigenvector corresponds to the observation that a small initial change in the number of edges will persist. It does not affect stability. The roots of the quadratic polynomial have negative real part if and only if $a + \tau < 0$ since in this case the coefficients of the quadratic are positive. The relation $a + \tau = 0$ leads to the critical value of τ given in the statement and $a + \tau < 0$ holds if and only if $\tau < \tau_c^{CPW}$. □

Finally, we prove that a transcritical bifurcation occurs at the critical value, i.e. when the disease-free steady state loses its stability, then an endemic steady state appears.

Proposition 5.7 *If $\tau > \tau_c^{CPW}$, then the compact pairwise model of system (5.18) has an endemic steady state with $[S_k]_c < N_k$.*

Proof. For simpler notation, let X_k, Z, U and V denote the endemic steady state values of $[S_k]_c$, $[SI]_c$, $[SS]_c$ and $[II]_c$, respectively. Then, from the first M equations we get

$$\gamma N_k = X_k \left(\gamma + \tau k \frac{Z}{Z+U} \right).$$

Expressing X_k in terms of the other variables, multiplying the equation by $k(k-1)$ and summing them for $k = 1, 2, \ldots, M$ yields

$$\gamma \sum_{k=1}^{M} \frac{k(k-1)N_k}{\gamma + \tau k \frac{Z}{Z+U}} = \sum_{k=1}^{M} k(k-1)X_k = Q(U+Z)^2.$$

The steady state equation of the variable $[SS]_c$ yields $\tau QU = \gamma$. Hence, the above equation can be written in the form

$$1 = \frac{\tau U}{U+Z} \sum_{k=1}^{M} \frac{k(k-1)N_k}{\gamma(U+Z) + \tau kZ}.$$

The steady state equation of the variable $[II]_c$ yields $\gamma V - \tau Z = \tau Z^2 Q$. Combining this equation with the equation $\tau QU = \gamma$ and with the pair conservation $[SS]_c + 2[SI]_c + [II]_c = \langle K \rangle N$, we get the following relation between U and Z:

$$\gamma \langle K \rangle NU = \gamma Z^2 + ZU(\tau + 2\gamma) + \gamma U^2. \tag{5.31}$$

This can be solved for Z in terms of U, that is for any $U \in [0, \langle K \rangle N]$ the equation above has a unique non-negative solution for Z. Let this solution be denoted by $Z = g(U)$. It is straightforward to check that $g(0) = 0$ and $g(\langle K \rangle N) = 0$.
    Introducing

$$f(U) = \frac{\tau U}{U + g(U)} \sum_{k=1}^{M} \frac{k(k-1)N_k}{\gamma(U + g(U)) + \tau kg(U)},$$

the existence of the endemic steady state is proved if one can show that there is a $U \in (0, \langle K \rangle N)$, for which $f(U) = 1$. To show this, we investigate the values of f at the endpoints of the interval. We first determine the limit of f as $U \to 0$ (note that $f(0)$ is not defined). Equation (5.31) leads to

$$\frac{U}{(U+Z)^2} = \frac{\gamma}{\gamma \langle K \rangle N - \tau Z}.$$

Hence,

$$\lim_{U \to 0} \frac{U}{(U + g(U))^2} = \frac{1}{\langle K \rangle N}.$$

Dividing (5.31) by Z yields

$$\lim_{U \to 0} \frac{U}{g(U)} = 0.$$

Using these limits, one obtains

$$\lim_{U \to 0} f(U) = \frac{\tau}{\langle K \rangle N} \sum_{k=1}^{M} \frac{k(k-1)N_k}{\gamma + \tau k} < \frac{\tau}{\langle K \rangle N} \sum_{k=1}^{M} \frac{k(k-1)N_k}{\tau k} = \frac{N(\langle K \rangle - 1)}{\langle K \rangle N} < 1.$$

We determine the value of f at the other endpoint of the interval.

$$f(\langle K \rangle N) = \frac{\tau}{\langle K \rangle N} \sum_{k=1}^{M} \frac{k(k-1)N_k}{\gamma} = \frac{\tau}{\gamma} \frac{\langle K^2 \rangle - \langle K \rangle}{\langle K \rangle}$$

shows that for $\tau > \tau_c^{\text{CPW}}$, $f(\langle K \rangle N) > 1$. Hence, due to $\lim_{U \to 0} f(U) < 1$ and by the continuity of f, the equation $f(U) = 1$ has a solution $U \in (0, \langle K \rangle N)$, i.e. there exists an endemic steady state. $\square$

5.6 Effective degree models

The closures presented above were formulated in terms of pairs or triples of nodes. To extend this approach further, we would consider possible arrangements of sets of four nodes, then five nodes, etc. At each stage, we increase the number of nodes to be considered as a unit. At the four-node level, we would consider paths containing four nodes as well as central nodes with three neighbours, and at the five-node level, a wide variety of structures emerge. The number of equations required explodes. In this section, we explore an alternative approach to building up a sequence of closures, known as "effective degree models" [197] or "approximate master equations" [115, 116].

Instead of focusing on collections of particular numbers of nodes, we can look at collections of nodes within a certain distance of a central node. The number of variables we must track grows rapidly with the distance we consider, so we will focus just on the first non-trivial level: we track possible arrangements of star-like motifs. For the SIS model, we define $S_{s,i}$ to be the number of susceptible nodes with s susceptible and i infected neighbours and $I_{s,i}$ to be the number of infected nodes with s susceptible and i infected neighbours. For the SIR model, we use $S_{s,i,r}$ and $I_{s,i,r}$ to also count the number of recovered neighbours. An advantage of this approach in the SIS case is that it allows us to better track the number of infected neighbours a node has, and thus, when a node recovers we have a better estimate for its risk of reinfection [115, 116, 197].

Figure 5.6 shows the transitions that define the SIR version of the effective degree model, in the special case of a node with degree 3. Below, we use this to derive the governing equations, which can be simplified further.

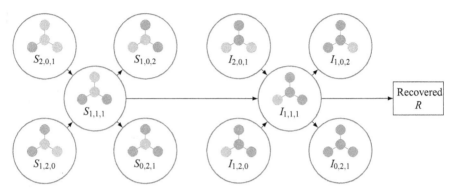

Fig. 5.6: Example flows in and out of $S_{1,1,1}$ and $I_{1,1,1}$ relevant to the SIR system, showing S (○), I (◉) and R (◉) nodes. Other paths exist into and out of the other compartments, but they are not shown. The equations that arise naturally from this diagram can be simplified by appropriate choice of new variables.

5.6.1 Basic effective degree model

We begin by looking at the SIS model. The transitions are shown in Fig. 5.7.

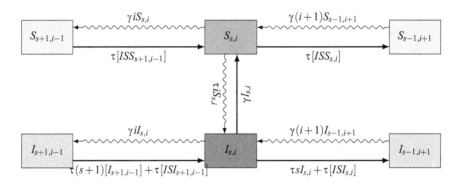

Fig. 5.7: A flow diagram demonstrating the unclosed effective degree equations for the SIS model. Those terms with γ correspond to a recovery and those with τ correspond to transmission. Only transitions involving $I_{s,i}$ and $S_{s,i}$ are shown.

The value of $S_{s,i}$ can change in three different ways: infection of the central node, infection of one of its neighbours or the recovery of one of its infected neighbours. Similarly, the value of $I_{s,i}$ can change in the following three different ways: recovery of the central node, infection of one of its susceptible neighbours or the recovery of one of its infected neighbours. For the SIS epidemic, these satisfy the system of differential equations given below.

$$\dot{S}_{s,i} = -\tau i S_{s,i} + \gamma I_{s,i} + \gamma\big((i+1)S_{s-1,i+1} - iS_{s,i}\big)$$
$$+ \tau[ISS_{s+1,i-1}] - \tau[ISS_{s,i}], \tag{5.32a}$$

$$\dot{I}_{s,i} = \tau i S_{s,i} - \gamma I_{s,i} + \gamma\big((i+1)I_{s-1,i+1} - iI_{s,i}\big) + \tau[ISI_{s+1,i-1}]$$
$$- \tau[ISI_{s,i}] - \tau s I_{s,i} + \tau(s+1)I_{s+1,i-1}, \tag{5.32b}$$

where $[ISS_{s,i}]$ denotes the expected number of ways to select three nodes such that the "focal" third node has s susceptible neighbours, i infected neighbours, and is susceptible; the second node is one of its susceptible neighbours; and the first node is an infected neighbour of that neighbour. There is a similar definition for $[ISI_{s,i}]$. The indices in the system run through the set $\{(s,i) : s \geq 0,\ i \geq 0,\ 1 \leq s+i \leq M\}$, where M is the maximum degree in the network. This exact system was introduced in [298] and needs closure for the triples. The closed system was developed by Lindquist et al. [197], and closely related systems by [115, 205]. The closure is based on the assumption that the susceptible neighbours of any susceptible central node are interchangeable, and similarly susceptible neighbours of any infected central node are interchangeable. That is, given a central susceptible node, we can assume that the rate at which any of its susceptible neighbours become infected is the same as for any other susceptible node's susceptible neighbours, and a similar result holds for infected nodes

$$[ISS_{s,i}] \approx \frac{[ISS]}{[SS]} s S_{s,i} \quad \text{and} \quad [ISI_{s,i}] \approx \frac{[ISI]}{[SI]} s I_{s,i}. \tag{5.33}$$

The number of SI links can be expressed in terms of the star-like motifs in two different ways, leading to the consistency condition

$$\sum_{s,i} i S_{s,i} = [SI] = \sum_{s,i} s I_{s,i}, \tag{5.34}$$

where the sums are taken over all possible pairs s and i, i.e. the summation $\sum_{s,i}$ stands for the double summation over all possible s and i: $\sum_{\kappa=1}^{M} \sum_{s+i=\kappa}$. The number of triples and SS pairs can be also expressed as

$$[ISS] = \sum_{s,i} s i S_{s,i}, \quad [ISI] = \sum_{s,i} i(i-1) S_{s,i} \quad \text{and} \quad [SS] = \sum_{s,i} s S_{s,i}, \tag{5.35}$$

where the same summation convention is used. Using the triple closures (5.33) in system (5.32) leads to the effective degree model below for SIS dynamics.

SIS effective degree model

$$\dot{S}_{s,i} = -\tau i S_{s,i} + \gamma I_{s,i} + \gamma((i+1)S_{s-1,i+1} - iS_{s,i})$$

$$+ \tau \frac{[ISS]}{[SS]}((s+1)S_{s+1,i-1} - sS_{s,i}), \tag{5.36a}$$

$$\dot{I}_{s,i} = \tau i S_{s,i} - \gamma I_{s,i} + \gamma((i+1)I_{s-1,i+1} - iI_{s,i})$$

$$+ \tau \left(\frac{[ISI]}{[SI]} + 1\right)((s+1)I_{s+1,i-1} - sI_{s,i}), \tag{5.36b}$$

$$S = \sum_{s,i} S_{s,i}, \qquad I = \sum_{s,i} I_{s,i}, \tag{5.36c}$$

with

$$[SS] = \sum_{s,i} sS_{s,i}, \qquad [SI] = \sum_{s,i} iS_{s,i}, \qquad [ISS] = \sum_{s,i} isS_{s,i}, \qquad [ISI] = \sum_{s,i} i(i-1)S_{s,i}.$$

We now look at the SIR dynamics. We first look at the $S_{s,i,r}$ and $I_{s,i,r}$ equations and show how these can be grouped to eliminate r. We start from

$$\dot{S}_{s,i,r} = -\tau i S_{s,i,r} + \gamma((i+1)S_{s,i+1,r-1} - iS_{s,i,r}) + \tau([ISS_{s+1,i-1,r}] - [ISS_{s,i,r}]) \tag{5.37a}$$

$$\dot{I}_{s,i,r} = \tau i S_{s,i,r} + \gamma((i+1)I_{s,i+1,r-1} - iI_{s,i,r}) + \tau([ISI_{s+1,i-1,r}] - [ISI_{s,i,r}]),$$
$$+ \tau((s+1)I_{s+1,i-1,r} - sI_{s,i,r}) - \gamma I_{s,i,r}. \tag{5.37b}$$

The first term in the $\dot{S}$ equation represents infection of the central node. The terms with γ represent recovery of an infected neighbour. The final terms represent infection of a neighbour from outside. The corresponding terms of the $\dot{I}$ equation represent the same actions, with additional terms representing infection of neighbours by the central node and recovery of the central node.

We can simplify these equations by combining terms. The value of r does not influence the dynamics so we define $S_{s,i} = \sum_r S_{s,i,r}$, and similarly for the triples. So the "effective degree" of a node is $i + s$. We sum the S equations over r to get

$$\dot{S}_{s,i} = -\tau i S_{s,i} + \gamma[(i+1)S_{s,i+1} - iS_{s,i}] + \tau([ISS_{s+1,i-1}] - [ISS_{s,i}]).$$

We could perform a similar operation with $I_{s,i,r}$, but we go further. No term in the $\dot{S}_{s,i}$ equation depends on the $I_{*,*}$ variables. Thus, we can jump directly to $I = \sum_{s,i,r} I_{s,i,r}$ without affecting any $S_{*,*}$ equation. When we sum the equations, the terms representing recovery of a neighbour cancel, terms representing infection of a neighbour from an outside source cancel and terms representing infection of neighbours by the central node also cancel. This represents the fact that once the central node is infected, the statuses of its neighbours have no influence on its status. Combining these, we are left with

$$\dot{I} = -\gamma I + \tau \sum i S_{s,i}.$$

Defining $S = \sum_{s,i} S_{s,i}$, we similarly find $\dot{S} = -\tau \sum i S_{s,i}$. These do not significantly simplify our equations because we still have the sums involving $S_{s,i}$. However, because $R = N - S - I$, one can see that $\dot{R} = 0 - \dot{S} - \dot{I} = \gamma I$. Using this, we finally have

$$I = N - S - R, \qquad \dot{R} = \gamma I,$$

and then we just calculate S.

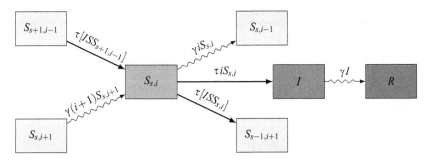

Fig. 5.8: The flow diagram for the SIR effective degree model. The I compartment has fluxes from all of the $S_{s,i}$ compartments for which $i \neq 0$.

If we now substitute for the triples, we arrive at a model which is equivalent to that of [197] in a simplified form.

SIR effective degree model

$$\dot{S}_{s,i} = -\tau i S_{s,i} + \gamma \big((i+1) S_{s,i+1} - i S_{s,i} \big)$$
$$+ \tau \frac{[ISS]}{[SS]} \big((s+1) S_{s+1,i-1} - s S_{s,i} \big), \tag{5.38a}$$

$$\dot{R} = \gamma I, \qquad S = \sum_{s,i} S_{s,i}, \qquad I = N - S - R, \tag{5.38b}$$

where the pair and the triple are given as for the SIS system.

The flow diagram corresponding to the SIR effective degree model is shown in Fig. 5.8. Assuming a network with maximum degree M, the number of equations for SIS epidemics will be $2 \sum_{\kappa=1}^{M} (\kappa + 1) = M(M+3)$. This is because for each degree

$\kappa \leq M$ there are $\kappa + 1$ equations due to $(s,i) \in \{(s,i) : s \geq 0, i \geq 0, s+i \leq \kappa\}$, with the factor two accounting for both S and I nodes with each neighbourhood combination. For SIR epidemics, the additional simplifications reduce it to about half of this total.

For both models, it is possible to compute the threshold between the endemic and disease-free behaviour, numerically for the SIS case and analytically for the SIR case (see [197]). Using a next-generation matrix approach, [197] found $R_0 = \frac{\tau}{\tau+\gamma} \frac{\langle K^2 - K \rangle}{\langle K \rangle}$. In Chapter 6, we will use percolation-based techniques to derive this relation. For the growth-based R_0, we will find

$$\overline{R}_0 = \frac{\tau}{\gamma} \left(\frac{\langle K^2 - K \rangle}{\langle K \rangle} - 1 \right).$$

That is, early in the epidemic, for every individual that recovers in a short time interval, $\overline{R}_0$ are infected.

Exercise 5.10. Show that if $\frac{\tau}{\tau+\gamma} \frac{\langle K^2-K \rangle}{\langle K \rangle} = 1$, then $\frac{\tau}{\gamma} \left(\frac{\langle K^2-K \rangle}{\langle K \rangle} - 1 \right) = 1$ as well.

5.6.2 Compact effective degree model

As we saw in the previous subsection, the number of equations in the effective degree model is $\mathcal{O}(M^2)$, where M is the maximum degree. We look for a reduced system. This reduced system will be the compact effective degree model, where instead of the $S_{s,i}$ variables we use $S_\kappa = \sum_{s+i=\kappa} S_{s,i}$, the number of susceptible nodes with κ non-recovered neighbours ($\kappa = 1, 2, \ldots, M$), and $[SI]$, the number of SI edges. Thus, we need an *a priori* assumption relating $S_{s,i}$ to S_κ. Following [13], we assume that non-recovered neighbours of susceptible nodes are interchangeable. That is, given a susceptible neighbour having κ non-recovered neighbours, the probability that i are infected and $s = \kappa - i$ are susceptible is

$$\binom{\kappa}{i} \langle I \rangle^i \langle S \rangle^s,$$

where $\langle I \rangle$ is the probability a neighbour of a susceptible node is infected (given that the neighbour is not recovered), and $\langle S \rangle$ is the probability a neighbour of a susceptible node is susceptible (given that it is not recovered), i.e.

$$\langle I \rangle = \frac{[SI]}{[SI] + [SS]} \quad \text{and} \quad \langle S \rangle = \frac{[SS]}{[SI] + [SS]}.$$

If $s + i = \kappa$, then using $\langle I \rangle + \langle S \rangle = 1$, we can approximate $S_{s,i}$ as

$$S_{s,i} \approx \binom{\kappa}{i} \langle I \rangle^i (1 - \langle I \rangle)^{\kappa-i} S_\kappa. \tag{5.39}$$

This leads to the compact effective degree model. The variables of the approximating system should be denoted by new symbols. Thus, we would need $S_{s,i}^c$ instead of $S_{s,i}$, S_κ^c instead of S_κ, $[SI]^c$ instead of $[SI]$, $[SS]^c$ instead of $[SS]$ and $\langle I \rangle^c$ instead of $\langle I \rangle$. However, in order to keep the notation simple, we omit the superscript "c" in what follows. By using the binomial theorem, it is straightforward to check that Eq. (5.39) leads to $S_\kappa = \sum_{s+i=\kappa} S_{s,i}$.

Our aim is to derive differential equations for the new variables S_κ, $\kappa = 1, 2, \ldots, M$ and $[SI]$. For reference, we derive a few identities that will be needed.

Lemma 5.5 *For any positive integer r, the following identities hold:*

$$\sum_{a=0}^{r} a \binom{r}{a} p^a q^{r-a} = rp(p+q)^{r-1},$$

$$\sum_{a=0}^{r} a(a-1)\binom{r}{a} p^a q^{r-a} = r(r-1)p^2(p+q)^{r-2},$$

$$\sum_{a=0}^{r} a(r-a)\binom{r}{a} p^a q^{r-a} = r(r-1)pq(p+q)^{r-2}.$$

Proof. We have

$$\sum_{a=0}^{r} a \binom{r}{a} p^a q^{r-a} = p\frac{\partial}{\partial p}\sum_{a=0}^{r} \binom{r}{a} p^a q^{r-a} = p\frac{\partial}{\partial p}[p+q]^r = rp(p+q)^{r-1}$$

$$\sum_{a=0}^{r} a(a-1)\binom{r}{a} p^a q^{r-a} = p^2\frac{\partial}{\partial p}\left(\frac{\partial}{\partial p}\sum_{a=0}^{r} \binom{r}{a} p^a q^{r-a}\right) = p^2\frac{\partial}{\partial p}\left(\frac{\partial}{\partial p}[p+q]^r\right)$$
$$= r(r-1)p^2(p+q)^{r-2}$$

and

$$\sum_{a=0}^{r} a^2 \binom{r}{a} p^a q^{r-a} = q\frac{\partial}{\partial q}\left(p\frac{\partial}{\partial p}\sum_{a=0}^{r} \binom{r}{a} p^a q^{r-a}\right) = q\frac{\partial}{\partial q}\left(p\frac{\partial}{\partial p}[p+q]^r\right)$$
$$= r(r-1)pq(p+q)^{r-2}$$

□

Proposition 5.8 *Assuming the closure in Eq. (5.39), the following identities hold:*

$$\sum_{s+i=\kappa} iS_{s,i} = \kappa S_\kappa \langle I \rangle, \tag{5.40}$$

$$\sum_{s+i=\kappa} i(i-1)S_{s,i} = \kappa(\kappa-1)S_\kappa \langle I \rangle^2, \tag{5.41}$$

$$\sum_{s+i=\kappa} siS_{s,i} = \kappa(\kappa-1)S_\kappa \langle I \rangle (1-\langle I \rangle). \tag{5.42}$$

Proof. For the first identity, we use the approximation in Eq. (5.39) to get $\sum_{s+i=\kappa} iS_{s,i} = S_\kappa \sum_{s+i=\kappa} i\binom{\kappa}{i} \langle I \rangle^i (1 - \langle I \rangle)^s$, where the summations are taken over all pairs s and i that sum to a particular value of κ. Taking $s = \kappa - i$ leads to $S_\kappa \sum_{i=0}^{\kappa} i\binom{\kappa}{i} \langle I \rangle^i (1 - \langle I \rangle)^{\kappa-i}$. Based on Lemma 5.5, this becomes $S_\kappa \kappa \langle I \rangle (\langle I \rangle + 1 - \langle I \rangle)^{\kappa-1} = S_\kappa \kappa \langle I \rangle$, and this completes the first equation. The second and third equations follow similarly. □

As mentioned above, the number of edges between susceptible and infected neighbours is $[SI] = \sum_{s,i} iS_{s,i}$. The first statement of Proposition 5.8 gives

$$[SI] = \sum_{s,i} iS_{s,i} = \sum_\kappa \sum_{s+i=\kappa} iS_{s,i} = \sum_\kappa \kappa S_\kappa \langle I \rangle = \langle I \rangle \sum_\kappa \kappa S_\kappa = \langle I \rangle ([SI] + [SS]).$$

Note that this equation is not the same as the definition of $\langle I \rangle$ since this is for the approximating variables with the previously mentioned upper index "c". Thus, we conclude that $\langle I \rangle = [SI]/\sum_\kappa \kappa S_\kappa$. This makes intuitive sense because it states that the probability a neighbour of a susceptible node is infected is equal to the number of SI edges divided by the total number of edges from S to either S or I nodes (counting SS edges once in each direction).

We can now derive a simplified model. Let us start with the SIR case. We seek equations for S_κ and $[SI]$. We have

$$\dot{S}_\kappa = \sum_{s+i=\kappa} \dot{S}_{s,i}$$

$$= \sum_{s+i=\kappa} -\tau iS_{s,i} + \gamma((i+1)S_{s,i+1} - iS_{s,i}) + \tau\frac{[ISS]}{[SS]}((s+1)S_{s+1,i-1} - sS_{s,i})$$

$$= -\tau \sum_{s+i=\kappa} iS_{s,i} + \gamma \sum_{s+i=\kappa} (i+1)S_{s,i+1} - \gamma \sum_{s+i=\kappa} iS_{s,i}$$

$$+ \tau\frac{[ISS]}{[SS]} \left(\sum_{s+i=\kappa} (s+1)S_{s+1,i-1} - \sum_{s+i=\kappa} sS_{s,i} \right)$$

$$= -\tau\kappa S_\kappa \langle I \rangle + \gamma(\kappa+1)S_{\kappa+1}\langle I \rangle - \gamma\kappa S_\kappa \langle I \rangle,$$

where we have used Proposition 5.8 and the fact that $\sum_{s+i=\kappa}(s+1)S_{s+1,i-1} = \sum_{s+i=\kappa} sS_{s,i}$.

We now look for an equation for $[\dot{S}I]$. By Theorem 4.4, $[SI]$ satisfies the exact differential equation

$$[\dot{S}I] = -\gamma[SI] + \tau([ISS] - [ISI] - [SI]).$$

Hence, we need only to express the triples in terms of our variables S_κ and $[SI]$. The triples are given in (5.35), so Proposition 5.8 yields

$$[ISS] = \sum_{s,i} siS_{s,i} = \sum_\kappa \sum_{s+i=\kappa} siS_{s,i} = \langle I \rangle (1 - \langle I \rangle) \sum_\kappa \kappa(\kappa-1)S_\kappa$$

and

$$[ISI] = \sum_{s,i} i(i-1)S_{s,i} = \sum_{\kappa} \sum_{s+i=\kappa} i(i-1)S_{s,i} = \langle I \rangle^2 \sum_{\kappa} \kappa(\kappa-1)S_\kappa.$$

Thus, we arrive at the system below.

SIR compact effective degree model

$$\dot{S}_\kappa = \langle I \rangle \left(-(\tau+\gamma)\kappa S_\kappa + \gamma(\kappa+1)S_{\kappa+1} \right), \tag{5.43a}$$

$$[\dot{S}I] = -(\tau+\gamma)[SI] + \tau\left(\langle I \rangle - 2\langle I \rangle^2 \right)\sum_\kappa \kappa(\kappa-1)S_\kappa, \tag{5.43b}$$

$$\langle I \rangle = \frac{[SI]}{\sum_\kappa \kappa S_\kappa}, \qquad \dot{R} = \gamma I, \qquad S = \sum_\kappa S_\kappa, \qquad I = N - S - R. \tag{5.43c}$$

This model is equivalent to that of [13], but takes a simpler form.

We now turn to the SIS case, seeking equations for S_κ, $[SI]$ and $[II]$. We have

$$\dot{S}_\kappa = \sum_{s+i=\kappa} \dot{S}_{s,i}$$

$$= \sum_{s+i=\kappa} -\tau i S_{s,i} + \gamma I_{s,i} + \gamma((i+1)S_{s-1,i+1} - i S_{s,i}) + \tau \frac{[ISS]}{[SS]}((s+1)S_{s+1,i-1} - s S_{s,i})$$

$$= -\tau \sum_{s+i=\kappa} i S_{s,i} + \gamma \sum_{s+i=\kappa} I_{s,i} + \gamma \sum_{s+i=\kappa} (i+1)S_{s-1,i+1} - \gamma \sum_{s+i=\kappa} i S_{s,i}$$

$$+ \tau \frac{[ISS]}{[SS]} \left(\sum_{s+i=\kappa} (s+1)S_{s+1,i-1} - \sum_{s+i=\kappa} s S_{s,i} \right)$$

$$= -\tau \kappa S_\kappa \langle I \rangle + \gamma(N_\kappa - S_\kappa),$$

where N_κ is the number of nodes of degree κ, $I_\kappa = N_\kappa - S_\kappa$ and we used that $\sum_{s+i=\kappa}(s+1)S_{s+1,i-1} = \sum_{s+i=\kappa} s S_{s,i}$ and $\sum_{s+i=\kappa}(i+1)S_{s-1,i+1} = \sum_{s+i=\kappa} i S_{s,i}$.

We now look for differential equations for $[SI]$ and $[II]$. According to Theorem 4.3, they satisfy the exact differential equation

$$[\dot{S}I] = \gamma([II] - [SI]) + \tau([ISS] - [ISI] - [SI]), \qquad [\dot{I}I] = -2\gamma[II] + 2\tau([ISI] + [SI]).$$

Hence, we need only to express the triples in terms of our variables S_κ, $[SI]$ and $[II]$. The triples are given in (5.35); hence, by using Proposition 5.8, we get

$$[ISS] = \sum_{s,i} s i S_{s,i} = \sum_{\kappa} \sum_{s+i=\kappa} s i S_{s,i} = \langle I \rangle (1 - \langle I \rangle) \sum_\kappa \kappa(\kappa-1)S_\kappa$$

and

$$[ISI] = \sum_{s,i} i(i-1)S_{s,i} = \sum_{\kappa} \sum_{s+i=\kappa} i(i-1)S_{s,i} = \langle I \rangle^2 \sum_\kappa \kappa(\kappa-1)S_\kappa.$$

The system we arrive at is the same as the SIS compact pairwise model (5.18).

SIS compact effective degree model

$$\dot{S}_\kappa = -\tau\kappa S_\kappa \langle I \rangle + \gamma(N_\kappa - S_\kappa), \tag{5.44a}$$

$$[\dot{SI}] = \gamma[II] - (\tau + \gamma)[SI] + \tau\left(\langle I \rangle - 2\langle I \rangle^2\right) \sum_\kappa \kappa(\kappa - 1)S_\kappa, \tag{5.44b}$$

$$[\dot{II}] = -2\gamma[II] + 2\tau[SI] + 2\tau\langle I \rangle^2 \sum_\kappa \kappa(\kappa - 1)S_\kappa, \tag{5.44c}$$

$$\langle I \rangle = \frac{[SI]}{\sum_\kappa \kappa S_\kappa}, \qquad S = \sum_\kappa S_\kappa, \qquad I = N - S, \tag{5.44d}$$

for $\kappa = 1, 2, \ldots, M$.

5.7 Conclusions and outlook

The heterogeneous mean-field model at single level has appeared independently several times under different names in multiple equivalent formulations. It appears that it was first introduced by Hethcote, Yorke and Nold [140, 141, 241] (where it was called the proportional mixing model), studied in [4] (where it was called social heterogeneity) and used by Pastor-Satorras and Vespignani for power law networks in [243] (where it was called dynamical mean-field reaction rate equation). It is frequently referred to as the heterogeneous mean-field [57] or degree-based mean-field model [247]. This ODE system was also introduced in a formally different but equivalent form by Eames and Keeling in [90], and yet another equivalent version in the SIR case (mean-field social heterogeneity) appears in [222]. The heterogeneous pairwise model and its reduced approximate form were introduced also in [90]. The compact pairwise model is developed in [151]. This model is generalised for a system with five node statuses in [138].

The heterogeneous models have been analysed by using ODE techniques in several papers. Based on the continuous approximation of the degree distribution for power law networks, the endemic steady state of the heterogeneous mean-field model at single level is determined analytically and compared to simulation in [243]. It was shown first in the same paper that there is no critical value of the transmission rate for these networks with exponents between 2 and 3. These results were extended to power law graphs with exponential cutoff and finite cutoff in [245]. It was later shown by [63] that in fact this absence of a threshold exists for any exponent (without a cutoff) because the disease would persist around high-degree nodes. This observation shows that for SIS disease, models that do not account for partnership duration can have qualitatively different outcomes.

The theory developed in [243] for the SIS case is extended to the SIR epidemic in [226], where the final epidemic size is also determined from the heterogeneous mean-field system at single level. The same theory is extended also to the case of preferential mixing in [39], where the heterogeneous mean-field system at single level is written down for this case and the critical value of τ is shown to be $1/\Lambda_m$, where Λ_m is the largest eigenvalue of the connectivity matrix. In [40], the conditions under which Λ_m tends to infinity as $N \rightarrow \infty$ is investigated, i.e. when does the critical value of the transmission rate disappear for the case of preferential mixing? The stability of the disease-free steady state of the heterogeneous mean-field model at single level is studied by the eigenvalues of the Jacobian in [27], where also the average degree of the newly infected nodes is investigated, and it is observed that the hubs are infected first and lower degree nodes are infected later. The distinction between quenched networks (networks with adjacency matrix fixed in time) and annealed networks (constant rewiring with fixed degree distribution) is studied in [57]. The heterogeneous mean-field system at single level is claimed to be exact on annealed networks. The threshold behaviour observed in simulations is compared to thresholds in ODE models in [41, 57, 98].

A significant development, identified as a "next step" in the study of the complexity of networked systems, is that of multilayer or multiplex networks (see [37, 183]) for extensive reviews. As the name suggests, networks are generalised to the case where either multiple networks over different nodes are interconnected or nodes of the same network are connected via different and multiple layers of connectivity, which can overlap. The main premise of such models is the idea that networks themselves do not fully capture the connectivity pattern observed in real-world systems [30, 37, 188]. For example, networks of sexual contacts are made up of different connectivity layers based on nodes' sexual orientations but with links between such different layers. Such a setup is referred to as interconnected networks [274, 320]. Multilayer networks have been applied to model the co-evolution of the transmission of epidemics and the spread of awareness about the disease, where typically the two processes spread on different networks over the same nodes [2, 106, 120, 136, 159]. Other fruitful areas where multilayer networks have been used successfully are (a) to model competing pathogens which spread within the same population but do so via different networks of contacts [105, 206], and (b) social contagion [72]. This area remains extremely active with many generalisations of results from the simple/single network setup emerging or, more importantly, novel phenomena being discovered as a result of the complex interconnectedness.

Chapter 6
Percolation-based approaches for disease modelling

The methods introduced thus far are applicable to both SIS and SIR diseases. This chapter focuses primarily on SIR disease. Once a node u becomes infected with an SIR disease, no other node affects the timing of any other action of u. This permits simplifications leading to a more complete theory of SIR disease.

We begin developing this theory by simulating SIR outbreaks in finite networks and extrapolating our observations to the infinite network limit. We use this to analytically calculate the epidemic probability in large Configuration Model networks. We then investigate the relationship between "bond percolation" and SIR disease in arbitrary networks. We use percolation to study an equivalence between the epidemic probability $\mathcal{P}$ and the attack rate (fraction infected) $\mathcal{A}$ in large networks. Next, we calculate $\mathcal{A}$ in the case of Configuration Model networks. Finally (Section 6.5), we adapt this calculation to derive the edge-based compartmental model (EBCM), which gives the time dependence of the number of susceptible, infected, and recovered nodes in Configuration Model networks. The resulting model has dramatically fewer equations than the pairwise and effective degree models, and Chapter 7 will show that, excluding pathological cases, the predictions are identical. We end the chapter with a brief discussion of percolation-like approaches to SIS disease.

The derivations in this chapter are based on percolation concepts, rather than on master equations. The presentation thus takes a different flavour from our earlier chapters and also builds up significant background on the relation between percolation and SIR disease. Although this background is helpful in understanding Section 6.5, a reader interested in just the EBCM methods may skip to there.

6.1 Typical SIR outbreaks

We begin by observing outbreaks for finite networks and then consider the infinite network limit. For simplicity, we temporarily depart from the continuous-time Markovian model where nodes transmit with rate τ and recover with rate γ. We use a

© Springer International Publishing AG 2017

I.Z. Kiss et al., *Mathematics of Epidemics on Networks*, Interdisciplinary Applied Mathematics 46, DOI 10.1007/978-3-319-50806-1_6

discrete-time Markovian model where nodes transmit with probability p and recover in a single time step. We introduce a definition.

Definition 6.1 *Let G and p be given. Define $x_{G,p}(m)$ to be the probability that exactly m nodes are infected for an SIR disease spreading with probability p per edge, starting from a randomly chosen node in the network G.*

If the context is clear, we sometimes simplify this to $x(m)$.

6.1.1 Dependence on network size

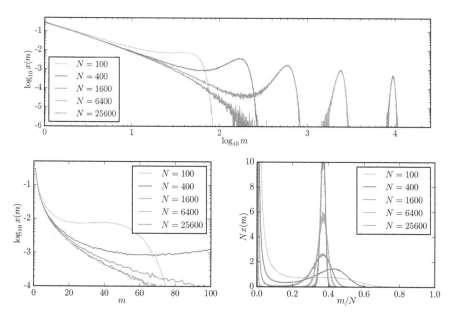

Fig. 6.1: The frequency of different final outbreak sizes in 10^6 stochastic simulations initialised with a single randomly chosen infection for five Erdős–Rényi random networks with different N, and edge probability $5/(N-1)$. Transmission probability is $p = 0.25$. (Top) Log-log plot showing that different networks have similar small outbreak behaviour, while large outbreaks grow with population size. (Bottom left) The absolute sizes of small outbreaks. For fixed m, as $N \to \infty$, the value of $x_{G,p}(m)$ converges. (Bottom right) The proportional sizes of outbreaks. As the number of nodes increases, the proportion infected in small outbreaks approaches 0, while epidemics (large outbreaks) become tightly peaked around a single value.

Figure 6.1 shows simulated results for discrete-time Markovian epidemics in Erdős–Rényi random networks of expected degree 5. For each N, a single network is used. The outbreaks are initialised with a single randomly chosen infected node, and transmissions occur with probability $p = 0.25$. As N increases, the proportion infected in small outbreaks approaches 0, while the epidemics (large outbreaks) are tightly peaked around a single value. If the network is large enough, outbreaks typically appear in one of two types: the number infected is small compared to the population size or it is proportional to the population size. There are few intermediate outbreaks. For smaller networks, this dichotomy is less clear [240].

For larger N, there are two clear peaks [210], one corresponding to a proportion of approximately 0 infected, and the other having a proportion just under 0.4 infected. We refer to outbreaks close to 0 as "small outbreaks" and those corresponding to the other peak as "epidemic outbreaks", or simply "epidemics".

Focusing on small outbreaks we see that, although the proportion infected goes to zero as N increases, the frequency of a given *number* infected m approaches a limit. This suggests an underlying distinction between small outbreaks and epidemics. Small outbreaks are not affected by network size and go extinct without "seeing" the network size. Epidemics do "see" the network size. In fact, a small outbreak goes extinct because not enough transmission happens for the disease to establish itself, while an epidemic goes extinct because after it is established the number of nodes still available to infect is reduced.

6.1.2 Dependence on transmission probability

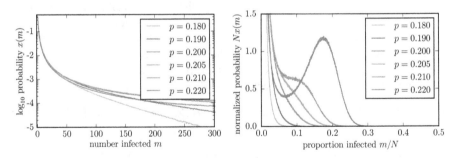

Fig. 6.2: Final outbreak size frequency in 5×10^7 stochastic simulations for varying per-edge transmission probability p. All simulations use the same 6400 node Erdős–Rényi random network having average degree 5, but each simulation has a randomly selected index node. For $p < 0.2$, there are no "large" outbreaks. At $p = 0.2$, there is no peak corresponding to epidemics, but the frequency of sizes decays very slowly. For $p > 0.2$, the distinction begins to emerge and would be clearer for larger N.

Figure 6.2 investigates how outbreaks depend on the transmission probability p, holding G fixed, with $N = 6400$. For small p, there is only a small outbreak peak in $x_{G,p}(m)$. As p increases, the peak grows wider. As p crosses the *epidemic threshold*

$p_c = 0.2$, a new peak emerges that corresponds to epidemic outbreaks. Initially it is obscured by the small outbreak peak. As p increases it becomes more distinct.

Combining these observations with the effect of varying N, we infer that for any fixed $p > p_c$, if N is large enough small outbreaks and epidemic outbreaks are clearly distinct. However, for any fixed N, as $p \to p_c$ from above, small outbreaks and epidemic outbreaks are eventually indistinguishable.

6.1.3 Epidemic definitions for finite networks

To develop a mathematical theory, we require clear definitions of what is and is not an epidemic. There are many reasonable criteria we could use to distinguish small outbreaks from epidemics. Heuristically, an epidemic is an outbreak that is associated with the larger peak and a small outbreak is an outbreak associated with the smaller peak. If N is large enough and the transmission probability is sufficiently far above the threshold, it is typically obvious whether an outbreak is an epidemic or not. While in practice the distinction is generally clear, our definition must account for the (possibly rare) outbreaks of intermediate size.

We will eventually define epidemics in the limit $N \to \infty$, but for now, we will assume that for large values of N there are two peaks in $x(m)$: one at M_0, assumed to be close to zero, and another at M_L assumed to be much larger. We take M^* to be the location where $x(m)$ takes its minimum between them. Those outbreaks with size smaller than M^* are called *small outbreaks*, while those with size at least M^* are called *epidemics*. Figure 6.3 demonstrates the regions attributed to epidemics and small outbreaks in our definition, using the same simulations as in Fig. 6.1.

Close to the epidemic threshold, it may be hard to distinguish epidemics from small outbreaks. Indeed, if N is not large, there may not even be two local maxima. If there is only a single maximum close to 0, the picture we have described does not exist and we say that there are no epidemics.

Definition 6.2 *Assuming M_0, M^* and M_L are as described for a network G with N nodes, we define*

$$\mathcal{P}(G, p) = P(m \geq M^*) = \sum_{m=M^*}^{N} x_{G,p}(m)$$

to be the probability of an epidemic *and*

$$\mathcal{A}(G, p) = \mathbb{E}\left(\frac{m}{N} \Big| m \geq M^*\right) = \frac{1}{\mathcal{P}(G, p)} \sum_{m=M^*}^{N} \frac{m}{N} x_{G,p}(m)$$

to be the average fraction infected in epidemics, *that is, the* attack rate.

If the context is clear, we often omit the arguments of $\mathcal{P}$ and $\mathcal{A}$.

Our definition implicitly assumes there is only one typical epidemic size. This is usually a valid assumption for large networks, but it breaks down, for example, in

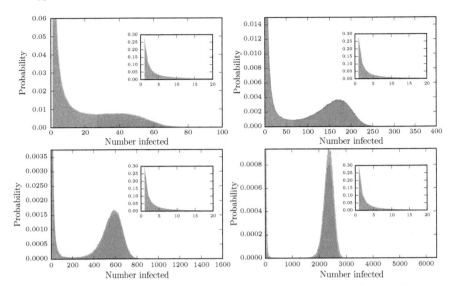

Fig. 6.3: The distinction between epidemics and small outbreaks for four networks of Fig. 6.1. The regions above and below M^* are shaded in different colours. The inset shows small outbreaks more clearly. Note the vertical scale of the insets does not change, but it does for the main plots. As the network grows, epidemics and small outbreaks become more distinct. The network size is (top left) $N = 100$, (top right) $N = 400$, (bottom left) $N = 1600$, and (bottom right) $N = 6400$.

networks with multiple large components with a few paths joining them [144]. It is relatively straightforward to adapt our definition to these, but we do not do so here.

In Table 6.1 we give $\mathcal{P}(G, p)$ and $\mathcal{A}(G, p)$ based on Fig. 6.3. Remarkably, $\mathcal{P}$ and $\mathcal{A}$ appear to coincide as N increases. Later, we will show that this is not a coincidence and does not rely on special properties of Erdős–Rényi random networks.

N	$\mathcal{P}$	$\mathcal{A}$
100	0.237	0.423
400	0.340	0.387
1600	0.339	0.350
6400	0.365	0.366
25600	0.368	0.368

Table 6.1: Values of $\mathcal{P}$ and $\mathcal{A}$ based on simulations of Fig. 6.1. As N increases, these approach the same limit. The mismatch for smaller networks is primarily due to the difficulty in distinguishing epidemic outbreaks from small outbreaks.

6.1.4 Epidemic definition in the infinite network limit

Our observations above suggest that as N increases, small outbreaks affect the same *number* of nodes, while epidemics affect the same *proportion* of the nodes. It will be easier to study epidemics in the limit of infinite networks for which "finite-size effects" (such as intermediate-sized outbreaks or peak width) may be ignored.

We focus on outbreaks in large networks chosen from some random graph model $\mathcal{G}$, where each graph in $\mathcal{G}$ has an associated probability. For example, $\mathcal{G}$ might be Configuration Model networks (see Section 1.2.3) having a particular degree distribution or Erdős–Rényi random networks with a particular edge probability.

Our interest is in what happens to $\mathcal{P}(G,p)$ and $\mathcal{A}(G,p)$ for $G \in \mathcal{G}$ as $N \to \infty$. We anticipate that as N increases, $\mathcal{P}$ and $\mathcal{A}$ approach limits. Although we can construct graph classes for which this is not true (see Exercise 6.1), we expect that most important classes will satisfy this. Ideally, we would like to consider the behaviour of individual networks $G \in \mathcal{G}$ as $N \to \infty$. However, rare networks that deviate significantly may exist, which complicates proofs.[1] It will be simpler to consider all networks $G \in \mathcal{G}$ with a given number of nodes N, observe the average behaviour and then consider this average as $N \to \infty$. We will assume that a single large network closely approximates the expected behaviour.

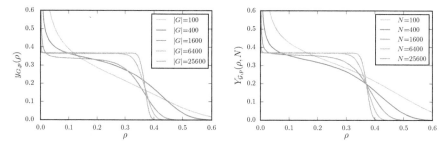

Fig. 6.4: The probability that the proportion infected exceeds ρ for networks of different size. (Left) $y_{G,p}(\rho)$ using 10^6 simulations from Fig. 6.1, using a single network for each size. (Right) $Y_{\mathcal{G},p}(\rho,N)$ using 10^6 simulations, with each simulation using a different network. In both cases, there is a threshold ρ_c such that if $0 < \rho < \rho_c$, then $y(\rho)$ or $Y(\rho)$ converges to a non-zero value (which appears to be approximately ρ_c), and if $\rho_c < \rho$, then it converges to 0.

---

[1] For such cases, we should use *convergence in probability*, which means, in some sense, that almost all networks behave as expected and the few that do not become rarer as $N \to \infty$.

Definition 6.3 *Define* $y_{G,p}(\rho)$ *to be the probability that a random outbreak in the N-node network G with transmission probability p infects at least a proportion* ρ *of the population,*

$$y_{G,p}(\rho) = \sum_{m \geq \rho N} x_{G,p}(m).$$

For the network class $\mathcal{G}$, *define* $Y_{\mathcal{G},p}(\rho,N)$ *to be the expected value of* $y_{G,p}(\rho)$ *for N-node networks* $G \in \mathcal{G}$:

$$Y_{\mathcal{G},p}(\rho,N) = \mathbb{E}\big(y_{G,p}(\rho)\big||G| = N\big),$$

where $|G|$ *denotes the number of nodes in G. Finally, if the limit exists, define*

$$Y_{\mathcal{G},p}(\rho) = \lim_{N \to \infty} Y_{\mathcal{G},p}(\rho,N).$$

In Fig. 6.1, for each N the same Erdős–Rényi random network having expected degree 5 was used in all simulations. In Fig. 6.4, we show $y_{G,p}(\rho)$ for these simulations as well as $Y_{\mathcal{G},p}(\rho,N)$ for those same values of N, but with each simulation using a new network. We see similar convergence to $Y_{\mathcal{G},p}(\rho)$ for both as the network size tends to infinity, but $Y_{\mathcal{G},p}(\rho,N)$ converges faster.

The figure suggests that the large network limit $Y_{\mathcal{G},p}(\rho)$ is a positive constant just below 0.4 for small positive ρ, and at some ρ_c slightly less than 0.4, it abruptly switches to 0. So as $N \to \infty$, outbreaks either infect a negligible proportion (occurring about 60% of the time) or almost exactly $\rho_c N$ nodes (about 40% of the time). This motivates the following definitions.

Definition 6.4 *Assume that* $Y_{\mathcal{G},p}(\rho)$ *is positive and constant for* $0 < \rho < \rho_c$ *and that it is zero for* $\rho > \rho_c$. *Then, we define* $\mathcal{P}(\mathcal{G},p)$ *to be the value* $Y_{\mathcal{G},p}(\rho)$ *takes for* $0 < \rho < \rho_c$ *and* $\mathcal{A}(G,p) = \rho_c$. *If* $Y_{\mathcal{G},p}(\rho) = 0$ *for all positive* ρ, *then we define* $\mathcal{P}(\mathcal{G},p) = 0$ *and* $\mathcal{A}(\mathcal{G},p) = 0$.

So $\mathcal{P}$ can be thought of as the probability that a node in an infinite network would spark an outbreak that infects an infinite number of nodes.

Exercise 6.1. Consider a class of networks $\mathcal{G}$ (all with ≥ 11 nodes) such that if $N = |G|$ is odd, every node in G has degree 10, while if N is even, every node has degree 6. Explain why Def. 6.4 does not apply.

Exercise 6.2. Consider a class of networks $\mathcal{G}$ (all with ≥ 11 nodes) such that if $G \in \mathcal{G}$, then with probability 0.5 every node has degree 10, and otherwise every node has degree 6, independently of N. Explain why Def. 6.4 does not apply. [Hint: What would Fig. 6.4 look like if $\mathcal{G}$ were made up of just degree 10 or just degree 6 networks?]

Exercise 6.3. This exercise shows that for some network classes, "pathological" networks may exist. Consider the class $\mathcal{G}$ of Erdős–Rényi random networks formed by placing edges between any two nodes independently with probability $q = 5/(N-1)$.

  a. Show that the expected degree of nodes in graphs in $\mathcal{G}$ is 5.
  b. Take the transmission probability $p = 0.25$. What is $y_{G_0,p}(\rho)$ for any $\rho > 0$ for G_0 if G_0 has no edges?
  c. How is this different from what we saw in Fig 6.4?
  d. Calculate the probability $G \in \mathcal{G}$ has no edges as a function of N if $N > 6$.
More generally, it can be proven that as $N \to \infty$, the probability that a random network in $\mathcal{G}$ gives a result far from expected goes to zero.

Exercises 6.1–6.3 give some insight into why a formal theory needs careful attention. We want to focus on network classes in which $\mathcal{P}$ and $\mathcal{A}$ are well defined, and for which the spread in a random large network is well predicted by the average behaviour. Proving this holds for a given $\mathcal{G}$ can be challenging, and we simply assume it holds. One necessary property of $\mathcal{G}$ that should be highlighted is that as the network size increases, the density of edges must remain bounded. That is, given a randomly chosen node u in a random network $G \in \mathcal{G}$, the expected number of nodes at any distance d from u converges to a finite value as $N \to \infty$. This assumption excludes networks with degree distributions having $\langle K^2 \rangle = \infty$. Configuration Model networks having a degree distribution with finite second moment $(\langle K^2 \rangle < \infty)$ have the properties we want [156].

Using the fact that the expected number of nodes at distance d from the index case remains bounded allows us to say that if an epidemic occurs, it must travel further than distance d from the initial node. To make this more precise, we define the following.

Definition 6.5 *Let $q(G, p, d)$ be the probability for a given network G that an outbreak of a disease with transmission probability p will travel at least distance d from the initial randomly chosen node. Define*

$$Q(\mathcal{G}, p, d, N) = \mathbb{E}\big(q(G, p, d)\big| |G| = N\big)$$

that is, the expected value of $q(G, p, d)$ over all N-node networks in $\mathcal{G}$. Then take

$$Q(\mathcal{G}, p, d) = \lim_{N \to \infty} Q(\mathcal{G}, p, d, N)$$

to be the large network limit. Finally, set

$$Q(\mathcal{G}, p) = \lim_{d \to \infty} Q(\mathcal{G}, p, d).$$

In less rigorous terms, $Q(\mathcal{G}, p)$ can be thought of as the probability that an index node in an infinite network starts an infinite chain of infections. We have the following theorem.

Theorem 6.6 *Let a network class $\mathcal{G}$ and transmission probability p for which Def. 6.4 applies be given. Assume that for any $d > 0$, the expected number of nodes* *at distance d from a randomly chosen node $u \in G$ for $G \in \mathcal{G}$ remains bounded as $N \to \infty$. Then,*

$$\mathcal{P}(\mathcal{G}, p) = Q(\mathcal{G}, p).$$

So to calculate epidemic probability $\mathcal{P}(\mathcal{G}, p)$, it suffices to calculate, for all fixed finite d, the probability that the disease spreads at least distance d as $N \to \infty$. We sketch the proof. The fundamental observation is that small outbreaks cannot travel far. Conversely, epidemic outbreaks must travel far.

Let us consider a fixed d. As $N \to \infty$, epidemic outbreaks infect close to ρ_c of the nodes, which is eventually so large that $\rho_c N$ is larger than the expected number of nodes within distance d of the index case. This guarantees that epidemic outbreaks travel farther than d from the initial node. So $\mathcal{P}(\mathcal{G}, p) \le Q(\mathcal{G}, p, d)$ for any d, and thus $\mathcal{P}(\mathcal{G}, p) \le \lim_{d \to \infty} Q(\mathcal{G}, p, d) = Q(\mathcal{G}, p)$.

Conversely, the size distribution of small outbreaks becomes constant as $N \to \infty$, decaying as the number of infected m nodes increases. For $m < d$, we know that an outbreak affecting m nodes cannot travel a distance d. As d increases, the probability that a small outbreak travels further than d must go to zero. So $1 - \mathcal{P}(\mathcal{G}, p) \le 1 - \lim_{d \to \infty} Q(\mathcal{G}, p, d)$. This establishes that $\mathcal{P}(\mathcal{G}, p) \ge \lim_{d \to \infty} Q(\mathcal{G}, p, d) = Q(\mathcal{G}, p)$. Combined with our previous result, we conclude $\mathcal{P}(\mathcal{G}, p) = Q(\mathcal{G}, p)$.

This gives us a convenient way to calculate epidemic probability. We simply calculate the probability that an outbreak spreads arbitrarily far in the $N \to \infty$ limit.

6.2 Epidemic probability in Configuration Model networks

We now derive equations giving the epidemic probability in Configuration Model networks. The equations are very similar for discrete-time, continuous-time and even non-Markovian epidemics.

Let $\mathcal{G}$ be the Configuration Model network class with a given degree distribution. Let $P(k)$ be the probability that a randomly chosen node's degree is k and let

$$\psi(x) = \sum_k P(k) x^k$$

be the *probability-generating function* of the degree distribution. The average degree of the network is $\langle K \rangle = \sum_k k P(k)$. As noted in the introduction, the probability that a node reached by choosing a random edge has degree k is $P_n(k) = k P(k) / \langle K \rangle$, representing the fact that nodes are then chosen proportionally to their degree.

A few observations are in order: $\psi(1) = \sum P(k) = 1$, $\psi'(1) = \sum k P(k) = \langle K \rangle$ and $\psi''(1) = \sum k(k-1) P(k) = \langle K^2 - K \rangle$. Note that $\sum_k P_n(k) x^{k-1}$ becomes $\psi'(x) / \langle K \rangle$. We could rewrite the denominator as $\psi'(1)$, but for later consistency, we do not.

Our goal will be to take a given fixed d and calculate the probability that the outbreak has a chain of at least d transmissions. We can assume that the network is sufficiently large that the finite size cannot be detected within d steps from the initial node. Once we find this probability, the limit as $d \to \infty$ gives the probability that the outbreak would spread infinitely far in an infinite network. We will assume that $\langle K^2 \rangle$ is finite. The number of nodes a given distance d from a node scales like $\langle K^2 \rangle^d / \langle K \rangle^d$, which remains finite as N increases.

There are a few observations that will be useful for this calculation:

- We can neglect cycles because d is fixed (in the $N \to \infty$ limit for Configuration Model networks, with high probability the initial node is in no cycle of length less than $2d$).
- The initial node has degree k with probability $P(k)$ (it is chosen uniformly from the network).
- Every other node within distance d from the initial node has degree k with probability $P_n(k) = kP(k)/\langle K \rangle$ (their stubs are chosen uniformly).
- Aside from the index case, a newly infected node whose degree is k has $(k-1)$ susceptible neighbours (it cannot infect its infector, and since there are no cycles no other neighbours are already infected).

So for our calculations, each newly infected node has its degree k chosen from the distribution defined by $P_n(k)$ and has $(k-1)$ susceptible neighbours. We will consider three versions of the SIR model. In the first, we use the discrete-time Markovian model considered so far in this chapter: each transmission occurs independently with probability p. In the second, we use the continuous-time Markovian model considered in previous chapters: infection durations are exponentially distributed with rate γ and time to transmission is exponentially distributed with rate τ. The global transmission probability is $\tau/(\tau+\gamma)$, but transmissions from a fixed node to its neighbours all depend on the same random variable, its infection duration, and thus they are not independent. The final model generalises this, with the transmission probability from u to v depending on other properties of u and v.

6.2.1 Discrete-time Markovian model

For the discrete-time Markovian model, each transmission occurs independently with probability p. This becomes a *Galton–Watson* process[2], with a small modification because the index case is different from all later cases.

We define α_d in the limit $N \to \infty$ to be the probability that an edge from an infected node u to a random neighbour v (of as yet unknown degree) is *not* the start of a chain of infections that will reach a distance of at least d from u. Then the probability that none of the edges from u start a chain of infections of distance at least d is α_d^k, where k is the degree of u. Averaging this over all possible degrees

---

[2] The Galton–Watson process was initially studied to understand the probability that a family name would go extinct. The analysis investigates the probability that a family name will last at least g generations by expressing the probability that it lasts g generations from a father in terms of the number of sons and the probability that it lasts $(g-1)$ generations from at least one of them.

of u, we find that the probability that if we infect a randomly chosen node u, no edges from u start a chain of distance at least d is $\sum_k P(k)\alpha_d^k = \psi(\alpha_d)$. Setting $\alpha = \lim_{d\to\infty}\alpha_d$, Theorem 6.6 leads to

$$\mathcal{P} = 1 - \psi(\alpha).$$

We can find each α_d iteratively, as shown in Fig. 6.5. We begin with $\alpha_1 = 1 - p$, the probability that the first transmission to v does not happen. Then, α_2 is the probability that the edge either does not transmit to v or it does, but v causes no further transmissions. The degree of v is chosen using $P_n(k)$, so

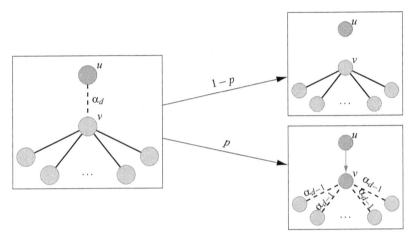

Fig. 6.5: The probability that an edge from an infected node u to a susceptible node v (of unknown, random degree k_v) does not spark a chain of transmissions of length d or more is α_d. This equals the probability that the edge does not transmit $(1-p)$, or that it does but none of the remaining $(k_v - 1)$ edges from v starts a transmission chain of length $(d-1)$ or more $\left(p\sum_{k_v} P_n(k_v)\alpha_{d-1}^{k_v-1}\right)$. So $\alpha_d = (1-p) + p\psi'(\alpha_{d-1})/\langle K\rangle$.

$$\alpha_2 = (1-p) + p\sum_k P_n(k)\alpha_1^{k-1} = (1-p) + p\frac{\psi'(\alpha_1)}{\langle K\rangle}.$$

More generally, for $d > 1$, α_d is the probability that the edge does not transmit to v or it does but none of the other edges from v result in a chain of length at least $d-1$:

$$\alpha_d = 1 - p + p\frac{\psi'(\alpha_{d-1})}{\langle K\rangle} = f(\alpha_{d-1})$$

where $f(x) = 1 - p + p\psi'(x)/\langle K\rangle$.

We represent this iteration in the cobweb diagrams in Fig. 6.6, showing convergence to $\alpha = \lim_{d \to \infty} \alpha_d$. A cobweb diagram for a function $g(x)$ is generated by plotting the diagonal and $g(x)$ [272]. Then, by drawing vertical lines from the diagonal to the function followed by horizontal lines back to the diagonal, we find the successive values solving $x_n = g(x_{n-1})$.

It is straightforward to show α_d increases monotonically and its limit solves

$$\alpha = f(\alpha). \tag{6.1}$$

So α is a *fixed point* of f. By direct observation, $f(1) = 1$ while $f(0) > 0$. Because every coefficient in the expansion of $\psi(x)$ is non-negative, f is a convex function, and it follows that if $f'(1) \le 1$, the only solution to $x = f(x)$ with $0 \le x \le 1$ is $x = 1$.

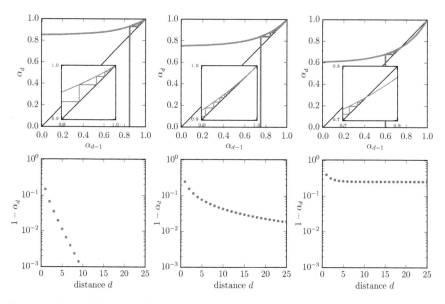

Fig. 6.6: (Top) "Cobweb diagrams" showing α_d below $(p = 0.15)$, at $(p = 0.25)$ and above $(p = 0.4)$ threshold. The degree distribution is Poisson with mean 4, so $\psi(x) = e^{4(x-1)}$. The green curve is $f(\alpha_{d-1}) = (1 - p) + p\psi'(\alpha_{d-1})/\langle K \rangle$. We take $\alpha_1 = 1 - p$ for the initial value. Insets enlarge regions about the attracting point. (Bottom) The value of $(1 - \alpha_d)$ decays rapidly below threshold, very slowly at threshold, and quickly approaches a non-zero equilibrium above threshold.

Thus, in this case $\alpha = 1$. If $\alpha = 1$, then $\mathcal{P} = 1 - \psi(\alpha)$ is zero. So we know that if $f'(1) \le 1$, then $\mathcal{P} = 0$.

In contrast, if $f'(1) > 1$, then there is exactly one other solution to $x = f(x)$, with $0 < x < 1$. Standard techniques show that because $f'(1) > 1$, the iteration converges to the smaller solution (where $f' < 1$). Then, $\mathcal{P} > 0$. We find that $f'(1) > 1$ exactly when $p\langle K^2 - K \rangle / \langle K \rangle > 1$. In this case, $\alpha < 1$ and $\mathcal{P} = 1 - \psi(\alpha) > 0$. Thus, epidemics are possible if $p\langle K^2 - K \rangle / \langle K \rangle > 1$.

We now summarise.

$\mathcal{P}$ for discrete-time Markovian Model

For the discrete-time Markovian model, the probability of an epidemic in the limit of large Configuration Model networks is

$$\mathcal{P} = 1 - \psi(\alpha), \tag{6.2a}$$

where α is the limit as $d \to \infty$ for the iteration

$$\alpha_d = 1 - p + p\frac{\psi'(\alpha_{d-1})}{\langle K \rangle} \qquad d = 2,3,\dots \tag{6.2b}$$

with the initial value $\alpha_1 = 1 - p$. The critical value of p at which epidemics become possible is $p_c = \langle K \rangle / \langle K^2 - K \rangle$. If $p \leq p_c$, then $\alpha = 1$ and $\mathcal{P} = 0$. If $p > p_c$, then $\alpha < 1$ and $\mathcal{P} > 0$.

Exercise 6.4. Redefine α to be the probability that a given edge from an infected node does not "result in an epidemic". Using a circular definition, we say that an edge results in an epidemic if the disease crosses it and then crosses another edge which results in an epidemic. Using this definition, derive equation (6.1).

Comment: This only shows that α solves equation (6.1). It does not show which solution to choose when there are two solutions. However, the underlying process naturally corresponds to the iterative process we previously used to derive α. Thus, when there are two solutions, the solution that is attracting under iteration is the correct choice.

Exercise 6.5. A regular graph is a graph in which every node has the same degree.
   a. Determine the critical value p_c for a regular random graph with each node having degree k_0.
   b. Using system 6.2, determine $\mathcal{P}$ when $k_0 = 3$ and $p > p_c$.
   c. Using system 6.2, find $\mathcal{P}$ when $k_0 = 1$ or $k_0 = 2$.
   d. Explain why we should be able to anticipate the results when $k_0 = 1$ or $k_0 = 2$ without performing the calculations.

Exercise 6.6. Consider Erdős–Rényi random networks for which an edge exists between two nodes with probability $q = 5/(N-1)$. The expected degree of nodes is 5. In the limit of large networks, the degree distribution is Poissonian, so the probability of having degree k is $5^k e^{-5}/k!$.
   a. Show that $\psi(x) = e^{-5(1-x)}$.
   b. Find $\mathcal{P}$ for transmission probability $p = 0.25$ and compare with Table 6.1.

Exercise 6.7. We develop an alternate derivation of the epidemic threshold. Early in an outbreak, the probability that newly infected nodes have degree k is $P_n(k) = kP(k)/\langle K \rangle$. Each newly infected node has $(k-1)$ susceptible neighbours. Recall that R_0 is the expected number of infections caused by an infected node early in the outbreak (other than the index case). Use this to explain the epidemic threshold.

6.2.2 Continuous-time Markovian model

We now consider the continuous-time Markovian model with infection transmitting along an edge at rate τ and recovery happening at rate γ. The following form of the law of total probability plays an important role in the subsequent calculations.

Proposition 6.1 *Let X be an exponentially distributed random variable with parameter γ. Let A be an event for which the conditional probability $P(A|X = T)$ is known. Then, the probability of A can be given as*

$$P(A) = \int_0^\infty P(A|X = T)\gamma e^{-\gamma T}\, dT.$$

If the duration of infection for u is known to be T, u will transmit to each neighbour independently with probability (Exercise 6.8) given by

$$p(T) = 1 - e^{-\tau T}.$$

However, if the duration is unknown, we cannot treat transmission to each neighbour as independent events: each transmission depends on the same random variable, the duration of u's infection.

We modify α_d to be a function of infection duration: $\alpha_d(T)$ is the probability that an edge from a node whose infection duration is T is not the start of a transmission chain of length at least d. If we integrate over possible recovery times of the possibly infected neighbours, using Proposition 6.1 and $p(T)$, then

$$\alpha_d(T) = 1 - p(T) + p(T)\int_0^\infty \sum_k P_n(k)\alpha_{d-1}(\hat{T})^{k-1}\gamma e^{-\gamma \hat{T}}\, d\hat{T}$$

$$= 1 - p(T) + p(T)\int_0^\infty \frac{\psi'(\alpha_{d-1}(\hat{T}))}{\langle K\rangle}\gamma e^{-\gamma \hat{T}}\, d\hat{T}.$$

Taking $\alpha(T) = \lim_{d\to\infty}\alpha_d(T)$, we conclude that the probability that the index case starts an epidemic is $\mathcal{P} = 1 - \int_0^\infty \psi(\alpha(T))\gamma e^{-\gamma T}\, dT$.

If we try to find conditions under which $\mathcal{P} > 0$, as we did for the discrete-time Markovian case, it is a little more difficult to prove results about the limit of the functions $\alpha_d(T)$. Instead, we introduce a new variable $\chi_d = \int_0^\infty \psi'(\alpha_d(T))\gamma e^{-\gamma T}\, dT/\langle K\rangle$, which does not depend on T. This is the probability that if u infects its neighbour v, then v is not the first node in a chain of length at least $(d-1)$. For $d = 1$, we find $\alpha_1(T) = 1 - p(T) = e^{-\tau T}$ and so

$$\chi_1 = \int_0^\infty \psi'(e^{-\tau T})\gamma e^{-\gamma T}\, dT/\langle K\rangle,$$

and for $d > 1$ we get $\alpha_d(T) = 1 - p(T) + p(T)\chi_d$ so χ_d is found by iteration:

$$\chi_d = \int_0^\infty \frac{\psi'(1 - p(T) + p(T)\chi_{d-1})}{\langle K \rangle} \gamma e^{-\gamma T} \, dT.$$

For τ below a threshold value the limit is $\chi = 1$ for which $\alpha(T) = 1$ and so $\mathcal{P}$ evaluates to 0. Above the threshold, $\chi < 1$ and epidemics are possible. The threshold occurs when $\frac{d}{d\chi} \int_0^\infty \psi'(1 - p(T) + p(T)\chi)\gamma e^{-\gamma T} \, dT / \langle K \rangle \big|_{\chi=1} = 1$, that is, the derivative with respect to χ evaluated at $\chi = 1$ is 1, which becomes

$$\int_0^\infty \psi''(1)p(T)\gamma e^{-\gamma T} \, dT / \langle K \rangle = 1.$$

Using $\psi''(1) = \langle K^2 - K \rangle$, the threshold becomes $\int_0^\infty p(T)\gamma e^{-\gamma T} \, dT = \langle K \rangle / \langle K^2 - K \rangle$. Exercise 6.8 shows that the integral is $\tau/(\gamma + \tau)$. Epidemics are possible if

$$\frac{\tau}{\gamma + \tau} \frac{\langle K^2 - K \rangle}{\langle K \rangle} > 1.$$

$\mathcal{P}$ for continuous-time Markovian model

For the continuous-time Markovian model, the probability of an epidemic is

$$\mathcal{P} = 1 - \int_0^\infty \psi(\alpha(T))\gamma e^{-\gamma T} \, dT, \tag{6.3a}$$

where $\alpha(T)$ is the limit as $d \to \infty$ of the iteration

$$\alpha_d(T) = 1 - p(T) + p(T) \int_0^\infty \frac{\psi'(\alpha_{d-1}(\hat{T}))}{\langle K \rangle} \gamma e^{-\gamma \hat{T}} \, d\hat{T}, \tag{6.3b}$$

with $p(T) = 1 - e^{-\tau T}$ and the initial value $\alpha_1(T) = e^{-\tau T}$. If

$$\frac{\tau}{\tau + \gamma} \frac{\langle K^2 - K \rangle}{\langle K \rangle} > 1 \tag{6.4}$$

epidemics are possible. Otherwise, they are not.

Exercise 6.8. In this exercise, we investigate two ways to find the probability that a random infected node u transmits to a random neighbour v.

a. (First method):

    i. Show that if a node has known infection duration T, the probability it transmits to a neighbour is $p(T) = 1 - e^{-\tau T}$. [Hint: Let $q(T) = 1 - p(T)$ and find an equation for $q'(T)$.]

    ii. By directly integrating

$$\int_0^\infty p(T)\gamma e^{-\gamma T}\, dT\,,$$

show that the probability that an edge from an infected node u (having unknown duration) would transmit to the neighbour v is $\tau/(\tau + \gamma)$.

b. (Second method): In a short-enough time interval Δt, we can ignore the possibility that u would both transmit and recover. The probability that u transmits to v is approximately $\tau \Delta t$ and the probability that u recovers is $\gamma \Delta t$. Without calculating the integral above, consider the time interval in which the first event happens. Show that the probability the event is a transmission is $\tau/(\tau + \gamma)$.

c. Find the probability that u recovers without transmitting to v.

Exercise 6.9. Consider a node v infected by a neighbour early in the epidemic.

a. Explain why the degree k of v is chosen using the distribution $P_n(k)$.

b. Explain why the expected number of additional infections caused by v, given its degree k is $(k-1)\tau/(\tau + \gamma)$.

c. Show that the expected number of infections caused by v averaged over all k is $R_0 = [\tau/(\tau + \gamma)][\langle K^2 - K \rangle / \langle K \rangle]$.

6.2.3 Non-Markovian model

Most research into infectious disease spread assumes the discrete- or continuous-time Markovian frameworks. However, in reality, the details are more complicated. For example, the infection duration distribution may be more complicated or nodes may become more or less infectious as their infection progresses. The probability that u transmits to v may depend on other intrinsic properties of u and v such as age or background immunity. Dynamic models of non-Markovian spread are discussed further in Chapter 9, but we briefly investigate non-Markovian spread here.

    We assume that if u and v are neighbours and u becomes infected, then all factors that influence the probability that u transmits to v can be expressed as properties of u or v. We use $\xi(u)$ and $\zeta(v)$ to denote all properties of u and all properties of v that might affect the transmission from u to v.

    We assume that $\xi(u)$ and $\zeta(u)$ are assigned independently of one another and of other nodes. We say that infectiousness and susceptibility are independent. With some abuse of notation, we use $P(\xi)$ and $P(\zeta)$ to denote their probabilities. If ξ is known for u and ζ for v and u becomes infected, then transmission from u to its neighbour v occurs with probability $p(\xi(u), \zeta(v))$ independently of any other events in the epidemic. Although we refer to these models as non-Markovian, once ξ and ζ are assigned to all nodes, we can treat the spread as a Markov process.

Definition 6.7 *Given* $\xi(u)$, *set*

$$p_o(\xi) = \int p(\xi, \zeta) P(\zeta) \, d\zeta$$

to be the expected value of the transmission probability $p(\xi, \zeta)$ *over all possible* ζs. *For a given node u, we will occasionally write* $p_o(u)$ *to mean* $p_o(\xi(u))$. *This is the* out-transmission *probability of u, the probability that u would transmit to a random neighbour if infected conditional on u having* ξ. *Similarly, set*

$$p_i(\zeta) = \int p(\xi, \zeta) P(\xi) \, d\xi$$

to be the expected value of $p(\xi, \zeta)$ *over all possible* ξs. *This is the* in-transmission *probability of a node v, the probability that v would receive a transmission from a random infected neighbour conditional on v having* ζ.

So p_o and p_i are the transmission probabilities conditional on knowing the properties of the infector or the infectee, respectively. We now define the average of p.

Definition 6.8 *We set*

$$\langle p \rangle = \iint p(\xi, \zeta) P(\xi) P(\zeta) \, d\xi d\zeta$$

to be the average transmission probability. *This is the expected value of* $p(\xi, \zeta)$.

We note that if $\xi(u)$ and $\zeta(u)$ were not assigned to u independently of one another, then the observed transmission probabilities may be higher or lower than $\langle p \rangle$. For example, if more infectious nodes were also more susceptible, the observed transmission probabilities would be higher because the infected nodes are sampled from the nodes with higher infectiousness.

For calculating $\mathcal{P}$, we get an equation very similar to the continuous-time Markovian model. We will derive an expression for α_d, writing it as a function of ξ:

$$\alpha_d(\xi) = \iint \left(1 - p(\xi, \zeta) + p(\xi, \zeta) \sum_k P_n(k) \alpha_{d-1}(\hat{\xi})^{k-1} \right) P(\hat{\xi}) P(\zeta) \, d\hat{\xi} d\zeta$$

$$= 1 - p_o(\xi) + p_o(\xi) \int \frac{\psi'(\alpha_{d-1}(\hat{\xi}))}{\langle K \rangle} P(\hat{\xi}) \, d\hat{\xi}.$$

The calculation of the threshold condition follows similarly to the continuous-time case by introducing the variable $\chi_d = \int \psi'(\alpha_d(\xi))P(\xi)\,d\xi\,/\,\langle K\rangle$ and showing that it satisfies the iteration

$$\chi_d = \int \frac{\psi'(1 - p_o(\xi) + p_o(\xi)\chi_{d-1})}{\langle K\rangle} P(\xi)\,d\xi.$$

$\mathcal{P}$ for non-Markovian model

If infectiousness and susceptibility are independent, the probability of an epidemic is

$$\mathcal{P} = 1 - \int \psi(\alpha(\xi))P(\xi)\,d\xi, \tag{6.5a}$$

where $\alpha(\xi)$ is the limit of the iteration

$$\alpha_d(\xi) = 1 - p_o(\xi) + p_o(\xi) \int \frac{\psi'(\alpha_{d-1}(\hat{\xi}))}{\langle K\rangle} P(\hat{\xi})\,d\hat{\xi}, \tag{6.5b}$$

starting with $\alpha_1(\xi) = 1 - p_o(\xi)$ for every ξ. An epidemic is possible when

$$\langle p\rangle \frac{\langle K^2 - K\rangle}{\langle K\rangle} > 1.$$

So the epidemic probability depends entirely on how p_o is distributed, and not p_i.

Exercise 6.10. Following the derivation in the continuous-time Markovian case, derive the condition that epidemics can occur when $\langle p\rangle\langle K^2 - K\rangle/\langle K\rangle > 1$ for non-Markovian spread.

Exercise 6.11. In this exercise, we see that the discrete and continuous-time Markovian models may be thought of as a special case of the non-Markovian model with independent infectiousness and susceptibility.
  a. Derive system (6.2) directly from system (6.5).
  b. Similarly, derive system (6.3).

Homogeneous infectiousness or susceptibility

There are two important cases which are worth particular attention. Sometimes, the probability of transmission along any edge does not depend on the properties of the node receiving infection, in which case we say susceptibility is homogeneous. Sometimes, the probability of transmission does not depend on the source of the infection, in which case we say infectiousness is homogeneous.

Definition 6.9 *If $p(\xi,\zeta) = p_o(\xi)$ for all ξ and ζ (so p is independent of ζ), then we say that the network has* homogeneous susceptibility. *If $p(\xi,\zeta) = p_i(\zeta)$ (so p is* *independent of* ξ), *then we say that the network has* homogeneous infectiousness. *Otherwise, we say that it has* heterogeneous susceptibility *or* infectiousness.

Example 6.1. The discrete-time Markovian model has both homogeneous suscepti- bility and infectiousness. This follows because the probability that u transmits to v is simply p. This is independent of any property of u or v.

Example 6.2. The continuous-time Markovian model has heterogeneous infectious- ness, but homogeneous susceptibility. This follows because the transmission rate from u to v is τ. So if the duration of infection of u is T_u, then the probability of transmission is $1 - e^{-\tau T_u}$. This depends on u's duration, which varies from node to node, but it does not depend on any property of v. Note that if T_u is an unknown variable, then transmission from u to v and u to w are not independent.

We can think of heterogeneity in infectiousness as stating that if u is infected and we have no information about u's infectiousness, then transmissions from u to a neighbour v and to another neighbour w are dependent events. They both depend on the same random variable $\xi(u)$. Heterogeneity in susceptibility can be thought of symmetrically.

Exercise 6.12. Consider the following examples. Determine whether the network has homogeneous infectiousness, homogeneous susceptibility, both, or neither.
   a. The discrete-time Markovian model.
   b. The continuous-time Markovian model.
   c. A network in which transmission occurs at rate τ, and all infection durations are the fixed value T.
   d. A network in which transmission occurs at rate τ, and duration of infections comes from some other distribution.
   e. A network in which the transmission rate depends on the degrees of both nodes, and all infection durations are the fixed value T.
   f. A network in which the transmission rate depends on only the degree of the receiving node, and all infection durations are the fixed value T.

Exercise 6.13. Prove that if infectiousness is homogeneous, then $\mathcal{P}$ is the same as for the discrete-time Markovian model.

Upper and lower bounds

The iteration for χ_d becomes $\chi_d = h(\chi_{d-1})$, where h is convex. Jensen's inequality lets us prove the following theorem [173, 211].

Theorem 6.10 *For Configuration Model networks with a given average transmis- sion probability* $\langle p \rangle$ *and independent infectiousness and susceptibility:*

- *Epidemic probability is maximised with homogeneous infectiousness.*
- *Epidemic probability is minimised when a fraction* $\langle p \rangle$ *of the population trans- mit to all neighbours and a fraction* $(1 - \langle p \rangle)$ *transmit to none (that is, infec- tiousness is as heterogeneous as possible).*

As a general rule, if $\langle p \rangle$ is fixed, increased heterogeneity in infectiousness reduces the probability of an epidemic. This can be understood crudely by recognising that if multiple paths from a single node lead to epidemics, then some of these paths are "wasted". At a first approximation, increasing heterogeneity in infectiousness has little effect on the total number of paths leading to epidemics from all nodes. However, more of them come from the more infectious nodes, and fewer from the less infectious nodes, so a higher fraction is wasted. The probability a random node has at least one such path leading from it goes down, decreasing $\mathcal{P}$.

6.3 Percolation and SIR disease

Our observation for Erdős–Rényi random networks in Table 6.1 suggests that for the discrete time SIR model, the epidemic probability $\mathcal{P}$ equals the attack rate $\mathcal{A}$. In this section, we use percolation theory to show that this is generally true across a much wider range of networks. We will see that this result is specific to the discrete-time Markovian model. However, the tools we develop will allow us to investigate the dynamic spread of the continuous-time model in much more detail, and even make progress with non-Markovian disease processes.

Percolation is a large branch of network theory concerned with the outcome of deleting nodes or edges from networks. We will show a close relation between SIR epidemics and "bond percolation", where edges are deleted with some probability, and discuss some properties of bond percolation which are relevant to our understanding of SIR disease spread. These observations will form the basis for the remainder of the chapter.

We begin with some properties of bond percolation. We then demonstrate the relationship with the discrete-time SIR model we have considered here by showing that simulating an epidemic with a given p starting from a given node is equivalent to performing bond percolation with that given p and then following edges in the remaining network from the initial node. We then use this fact to demonstrate that $\mathcal{P} = \mathcal{A}$, and finally, we adapt the bond percolation analogy to the usual continuous-time epidemic model and our non-Markovian model.

6.3.1 Properties of bond percolation

To investigate bond percolation, we require two simple definitions.

Definition 6.11 *Given a network G and probability p, we define the* percolated network *H to be the new network formed by taking a copy of G and deleting each edge independently with probability* $(1 - p)$.

When $p = 0$, H has no edges. When $p = 1$, H and G are identical.

Definition 6.12 *A set of nodes B in an undirected network forms a connected component if there is a path between any two nodes in B, and no node in B has an edge to any node outside of B.*

If we delete many edges in a network, it eventually breaks into multiple connected components. Typically, if G and p are large enough, then H has a single large "giant" connected component and possibly many other much smaller connected components. The precise definition of a giant component is somewhat vague for a fixed finite network, as was the definition of an epidemic. However, as the network grows larger and/or p grows, the distinction between giant components and smaller components becomes clearer.

Figure 6.7 shows the result of percolation for different values of p and different network sizes in Erdős–Rényi random networks of expected degree 5. As we increase the network size for small p, the largest component grows slowly and is not significantly larger than the second largest component. As $N \to \infty$, the proportion of the nodes in the largest component goes to zero. This changes abruptly at some $p = p_c$ (here $p_c = 0.2$), known as the percolation threshold. Above p_c, the largest component is significantly larger than the second largest component. As $N \to \infty$, the proportion of nodes in the largest component goes to a non-zero limit. Figure 6.7 shows that the transition becomes sharper as N increases.

For Configuration Model networks, a well-developed theory can predict the small component size distribution [235] as well as the proportion of nodes in the giant component [173, 211, 224, 225, 234].

Proposition 6.2 *Consider a Configuration Model network G created with degree distribution given by $P(k)$ with finite $\langle K^2 \rangle$. Let H be a network with the same nodes as G generated by keeping edges of G with probability p. As $N \to \infty$, if $p \langle K^2 - K \rangle / \langle K \rangle > 1$, then with high probability H has a unique giant component whose size approaches a fixed fraction of N as $N \to \infty$. If $p \langle K^2 - K \rangle / \langle K \rangle < 1$, then with high probability H does not have a giant component.*

This was proven in [224, 225]. The fraction of nodes in the giant component can be calculated by an argument similar to that used to calculate $\mathcal{P}$.

6.3.2 Bond percolation and discrete-time SIR disease

There are strong similarities between the size and existence of giant components in networks and the size and existence of epidemics in the discrete-time Markovian version of the SIR model. We will now establish the relationship. Later, we will extend this to *directed* bond percolation and apply it to the continuous-time Markovian SIR model and more general non-Markovian models.

To show the relation between SIR disease and bond percolation, we first give an algorithm to simulate the spread of SIR disease in a given network. We can carefully modify the algorithm so that the results are not changed but the approach becomes equivalent to bond percolation.

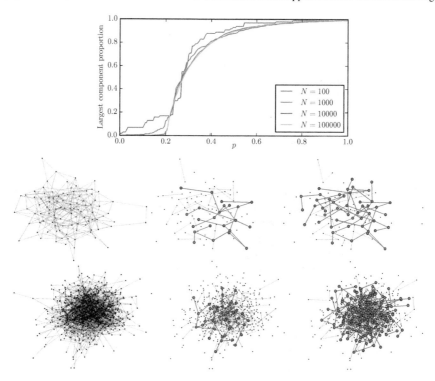

Fig. 6.7: Percolation demonstration. (Top) The proportion of nodes in the largest component as a function of p for Erdős–Rényi random networks with expected degree 5. A transition occurs at $p_c = 0.2$. It sharpens as N increases. (Middle and bottom) 100-node and 500-node networks with degree 0 nodes removed. Each edge is assigned a random weight between 0 and 1. (Left) The original networks. (Centre) Edges with weight less than $0.15 < p_c$ are kept. The largest component is coloured red and the second largest blue. The red components are of similar size in both networks and are not much larger than the blue components. (Right) Edges with weight less than $0.25 > p_c$ are kept. The red component in the larger network is about 5 times as large as in the smaller network, and both are much larger than the blue components.

Figure 6.8 provides pseudocode for basic_discrete_SIR_epidemic. This simulates SIR disease through a static network G, assuming infection lasts a single time step and transmits with probability p. Starting with a set of initial infected nodes, it iterates over all edges of infected nodes, checking to see if transmission will occur.

The majority of the work is done in discrete_SIR_epidemic, which requires an input function test_transmission that determines whether an edge will transmit. In this case, it simply returns True with probability p. More sophisticated models can be designed with different choices of test_transmission.

Figure 6.9 shows all possible paths that can occur in a simple four-node line network, starting from one infected node. The arrow labels give the probabilities to go from one state to the next. The process terminates when no infections remain. The probability of following a particular path to a final state is given by multiplying the probabilities of each arrow in Fig. 6.9. There are six possible terminal states, and their probabilities are shown. Note that in the simulation, each edge that ever had an opportunity to transmit had a single random number generated for it. With some thought, we can see that the same outcomes occur whether we generate the random number for an edge immediately before using it or at the very start of the algorithm prior to tracking the disease.

This observation helps lead to bond percolation. We can initially generate a random number between 0 and 1 for every edge prior to tracing the disease spread, and allow the disease to spread over those edges with an assigned number less than p. The outcome is the same as our existing approach. If we delete those edges assigned a value greater than p, then we are keeping the edges that would successfully transmit and deleting those that would not. However, we are also performing bond percolation. If we call this new network H, then H is the percolated network, and we can define a new algorithm for tracing the disease: when testing for transmission, the algorithm simply checks if the edge is in H. This algorithm is shown in Fig. 6.10, and relies on discrete_SIR_epidemic from Fig. 6.8. An important observation is that the disease reaches exactly the nodes that are in the same connected component in H as the initially infected node(s).

For comparison, we show the possible networks H resulting from the same network as in Fig. 6.9. The probability of each outcome is shown in Fig. 6.11. The probability, a set of nodes ends up infected in Fig. 6.9 is exactly equal to the probability, the set of nodes is the connected component containing the initial node.

The argument we have given establishes a theorem.

Theorem 6.13 *Consider a network G and a disease that transmits independently for each edge with probability p. Let X be a set of nodes in G, with $u \in X$. The probability that an outbreak initialised with only u infected will infect exactly the nodes in X is the same as the probability that the connected component containing u following bond percolation is made up of exactly the nodes in X.*

Consequences of Theorem 6.13

The percolation model is a valuable conceptual tool because it yields a single static object (the percolated network H) to study. The disease spreads to fill the connected component of H containing the index node. An epidemic occurs if H has a giant component and the initial node is in that giant component. Below the threshold for a giant component, epidemics are impossible. We demonstrate the power of these observations through the result suggested in Table 6.1.

Corollary 6.14 *Consider a class of networks G for which percolation results in a single giant component and the proportion of nodes that are in the giant component*

Input: Network G, index node(s) initial_infecteds, and transmission probability p.
Output: Lists t, S, I, and R giving number in each status at each time.

```
function discrete_SIR_epidemic(G, test_transmission, parameters, initial_infecteds)
    infecteds ← initial_infecteds
    t, S, I, R ← [0], [N-length(Infecteds)], [length(Infecteds)], [0]
    for u in G.nodes do
        u.susceptible ← True
    for u in infecteds do
        u.susceptible ← False
    while infecteds is not Empty do
        new_infecteds ← []
        for u in infecteds do
            for v in G.neighbours(u) do
                if v.susceptible and test_transmission(u, v, parameters) then
                    new_infecteds.append(v)
                    v.susceptible ← False
        infecteds ← new_infecteds
        R.append(R.last + I.last)
        I.append(length(infecteds))
        S.append(S.last - I.last)
        t.append(t.last+1)
    return t, S, I, R
function simple_test_transmission  (u, v, [p])
    return random(0,1)< p                              ▷ Returns True with probability p.
function basic _discrete _SIR_epidemic (G, p, initial_infecteds)
    return discrete_SIR_epidemic(G, simple_test_transmission, [p], initial_infecteds)
```

Fig. 6.8: An algorithm to calculate the number of infected at each time, assuming infections last a single time step. The function discrete_SIR_epidemic calculates the events occurring in the outbreak. When the algorithm considers transmission from u to v, it calls an auxiliary function test_transmission. The function basic_discrete_SIR_epidemic implements the discrete-time Markovian model by calling discrete_SIR_epidemic with appropriate inputs. In this case, simple_test_transmission returns True with probability p.

converges to a non-zero limit as $N \to \infty$. Then, this limit is $\mathcal{P}(\mathcal{G}, p)$ and

$$\mathcal{P}(\mathcal{G}, p) = \mathcal{A}(\mathcal{G}, p)$$

for all p, where $\mathcal{P}$ and $\mathcal{A}$ are as given in Def. 6.4.

The proof relies on the fact that for $N \to \infty$, after percolation in one of these networks, the giant component always occupies the same proportion. The probability of an epidemic is the probability that the initial node is in the giant component, which equals this fraction. If an epidemic occurs, the entire component is infected, so the proportion infected is also this fraction. A more rigorous proof cleans up the details of what it means to say that we can treat the giant components as occupying

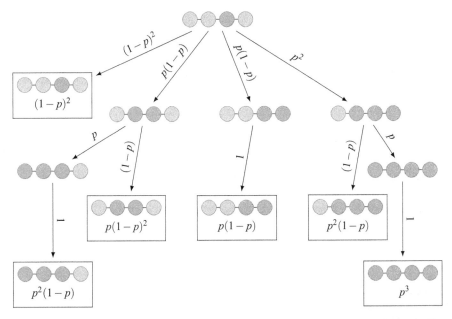

Fig. 6.9: Transmissions in a path of 4 nodes, showing susceptible nodes as (○), infected nodes as (◐) and recovered nodes as (◑). Infection starts in one of the two middle nodes. At each step, transmission occurs with probability p. All possible transmission chains are shown. The probability that the final state is one of the boxed terminal states equals the product of the probabilities of the intermediate transitions.

the same fraction of the network, and how we handle rare cases where percolation might not result in a single component.

Without using the underlying percolated network, this would be much harder to prove. We would have to study the disease as a dynamic, spreading process and somehow relate the possible final states to possible early behaviours.

This relationship between SIR disease spread and percolation has been widely used [56, 121, 212, 234]. In Fig. 6.12, we present an algorithm that can be used to approximate the size and probability of epidemics in large networks for the discrete-time Markovian model. The algorithm finds the largest component of the network following percolation and returns the fraction of nodes in that component.

6.3.3 Directed percolation and continuous time SIR disease

We now return to the continuous-time Markovian SIR model. Infected nodes transmit to each neighbour with rate τ and recover with rate γ. Unlike the discrete-time Markovian SIR model, we cannot assume that nodes transmit independently with the same probability for all nodes because transmissions from a given node to each

Input: Network G, index node(s) initial_infecteds, and transmission probability p.
Output: Lists t, S, I, and R giving number in each status at each time.

```
function percolate _network (G,p)
    H ← empty_graph
    H.nodes ← G.nodes
    for edge in G.edges do
        if random(0,1)< p then
            H.add_edge(edge)
    return H

function edge _exists (u, v, [H])
    return H.has_edge(u,v)                        ▷ Returns True if H has the edge.

function percolation_based_discrete_SIR_epidemic(G, p, initial_infecteds)
    H ← percolate_network(G, p)
    return discrete_SIR_epidemic(H, edge_exists, [H], initial_infecteds)
```

Fig. 6.10: Algorithm using bond percolation to produce the same outcomes as basic_discrete_SIR_epidemic in Fig. 6.8. The implementation is different: percolation_based_discrete_SIR_epidemic first generates the percolated network H. Then, to determine if transmission occurs, it checks whether the edge exists in H. If so, transmission succeeds. If not, it fails. By only looking at the results of edge_exists here or simple_test_transmission in Fig. 6.8, it is impossible to tell which one is being used.

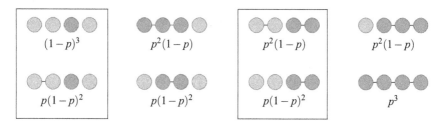

Fig. 6.11: All possible networks resulting from percolation on the line graph of Fig. 6.9, along with their probabilities. In each case, the nodes in the same component as the index node in Fig. 6.9 are shown as (⬤) and the nodes not in that component are shown as (◯). Different arrangements that result in the same colour scheme are joined together. The probability of a particular colouring is the same as in Fig. 6.9.

of its neighbours depend on the same random variable: the duration of its infection [87, 135, 172, 173, 211, 301]. Nodes with long duration infect a larger proportion of their neighbours. Consequently, the continuous-time model cannot be equated with the usual bond percolation.

Input: Network G, transmission probability p. Assumes existence of percolate_network from Fig. 6.10, and an algorithm connected_components which returns the nodes in each connected component of a graph.
Output: $\mathcal{P}$: an estimate of epidemic probability (and final size).

   function estimate_SIR_prob_size(G, p)
      $H \leftarrow$ percolate_network(G,p)
      Components $\leftarrow$ connected_components(H)
      $L \leftarrow$ largest(Components)
      return $|L|/|G|, |L|/|G|$ ▷ Be careful about integer division

Fig. 6.12: An algorithm to estimate $\mathcal{P}$ and $\mathcal{A}$ for the discrete-time Markovian model. In large populations, the fraction of the network in the largest component approximates the attack rate and the epidemic probability.

Instead, we adapt the process to a directed version of percolation. Prior to the introduction of the infection, we can take a fatalistic view of the transmission process and create a new network H. For each node u, we add it to H and we determine its infection duration if infected (chosen from an exponential distribution with rate γ).[3] Once that is known, all transmissions from u to its neighbours are independent. For each neighbour v we independently calculate the waiting time until u transmits to v (chosen from an exponential distribution of rate τ). If it is shorter than u's duration, we place a *directed* edge from u to v into H with a weight associated with the waiting time before transmission. Note that placing an edge from v to u must be considered as well. We refer to this as *directed percolation*.

Thus a directed edge from u to v in H is equivalent to the statement that u will transmit to v if u is ever infected. The algorithm, leading to a new percolated network H with directed edges, is demonstrated in Fig. 6.13.

Given H and the index nodes, we can identify the nodes infected in an outbreak by following edges in H from the index infections, respecting edge direction. This is described in the algorithm in Fig. 6.15.

This process gives all infected nodes, but without information about the timing. In fact, if there are multiple transmission paths from the index node to a given "target" node, the path that has the fewest transmissions may not be the one that is the first to complete [202]. If we distinguish the nodes by "generation" (the transmission chain with the fewest steps from an index node to a given node), we can adapt the function percolation_based_discrete_epidemic from Fig. 6.10. However, to determine the actual times of infection, when we add the directed edges, we can assign a time representing the waiting time until transmission. Then, we can use Dijkstra's algorithm to calculate the infection time of each node. The algorithm presented in Appendix A.1.2 can be thought of as a modification of this idea.

---

[3] Most programming languages have standard tools to generate a number x from an exponential distribution with rate r. If these are not available, another option is to generate a number s uniformly between 0 and 1, and take $x = (-\ln s)/r$.

Input: Network G, transmission rate per edge τ, and recovery rate γ.
Output: The percolated network H.

```
function directed_percolate_network(G, τ, γ)
    H ← empty_directed_graph
    for u in G.nodes do
        H.add_node(u)
        duration ← exponential_variate(γ)
        for v in G.neighbours(u) do
            time_to_infect ← exponential_variate(τ)
            if time_to_infect<duration then
                H.add_directed_edge(u, v)
    return H
```

Fig. 6.13: Algorithm for creating a directed percolation network corresponding to the continuous-time SIR model.

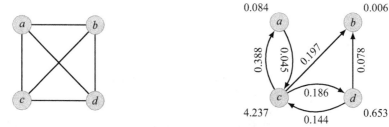

Fig. 6.14: (Left) The original network G. (Right) H, a percolated network found for $\tau = 1$, $\gamma = 0.5$ generated by the algorithm in Fig. 6.13. Nodes are labelled by the infection duration, and edges by the waiting time until transmission.

Exercise 6.14. Consider the percolated network H in Fig. 6.14.
  a. If node a is initially infected at $t = 0$, determine which other nodes become infected, when they become infected and when they recover.
  b. Repeat this for nodes b, c and d.

As before, we can deduce properties of the epidemic from the directed network H. To do this, we need the definition below.

Definition 6.15 *Given a directed network H, the* out-component *of a node u is the set of nodes including u that can be reached from u following the direction of the edges in H. Similarly, the* in-component *of u is the set of nodes including u from which u is reachable following the direction of edges.*

If u is initially infected, then every node in its out-component is eventually infected. Further, a node is eventually infected if and only if at least one node in its in-component is initially infected. So the final set of infections corresponds exactly to the combined out-components of the index nodes in the percolated network H (Fig. 6.15).

Input: Network G, set of index node(s) initial_infecteds, transmission rate per edge τ, and recovery rate γ.
Output: Full set of infected nodes (no time information).

```
function out_component(G, source)
    unprocessed ← []
    for node in H do
        found[node] ← False
    for node in source do
        found[node] ← True
        unprocessed.append(node)
    while unprocessed is not empty do
        u ← unprocessed.pop
        for v in H.out_neighbours(u) do
            if not found[v] then
                found[v] ← True
                unprocessed.append(v)
    return nodes u such that found[u] is True

function get_infected_nodes(G, τ, γ, initial_infecteds)
    H ← directed_percolate_network(G, τ, γ)
    infected_nodes ← out_component(G, initial_infecteds)
    return infected_nodes
```

Fig. 6.15: A directed percolation-based algorithm for simulating the final result of continuous-time epidemics. The function get_infected_nodes first creates the network H using directed_percolate_network and then finds the nodes reachable from the initial infections using out_component, which includes the source nodes. This algorithm will give only the total number of infections to occur, not the exact times. Note that many network packages contain built-in algorithms to find the out-component of a node or set of nodes, usually using a depth-first or breadth-first search in which case there is no need to implement out_component. An equivalent approach can be used to find the in-component, following edges in reverse.

Let H_{SCC} denote the largest strongly connected component of the directed network H. For given τ and γ, we typically find that $|H_{\text{SCC}}|/N$ approaches a constant as $N \to \infty$. If epidemics are possible, then there is a critical value p_c such that if $\tau/(\gamma + \tau) > p_c$, then $|H_{\text{SCC}}|/N$ approaches a positive value as $N \to \infty$. If so, then H_{SCC} is called a "giant strongly connected component" [84]. We use H_{IN} to denote the "giant in-component", that is all nodes from which H_{SCC} can be reached but which are not in H_{SCC}. We use H_{OUT} to denote the "giant out-component", that is all nodes reachable from H_{SCC} but which are not in H_{SCC}. This is sometimes represented with a "bow-tie" diagram [52], as in Fig. 6.16. An epidemic occurs if the infection reaches H_{SCC}.

Let us consider a single index node. Assuming that it is in H_{IN} or H_{SCC}, the outbreak will infect all nodes of H_{SCC} and H_{OUT}. If it happens to be in H_{IN}, it may infect a few additional nodes, but this represents a vanishing fraction in the limit $N \to \infty$. Thus, the probability that the index case leads to an epidemic is equal to the

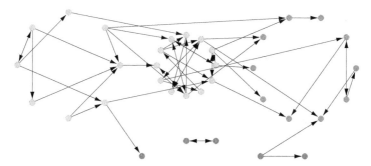

Fig. 6.16: The "bow-tie" structure of a network following directed percolation above the threshold [52]. The "knot" or central nodes (○) are H_{SCC}. Every node in H_{SCC} is reachable from every other. The left "wing" (○) is H_{IN}. From each node in H_{IN}, we can reach H_{SCC} following the edges. The right "wing" is H_{out} (◉). Every node in H_{OUT} is reachable from H_{SCC}. Some nodes (●) are not in any of these. Note $|H_{IN}| \neq |H_{OUT}|$.

proportion of nodes in $H_{IN} \cup H_{SCC}$, while in the $N \to \infty$ limit the fraction infected is equal to the proportion of nodes in $H_{SCC} \cup H_{OUT}$.

We can define epidemic probability and size for a given network G analogously to what we did in the discrete-time model, with $x_{G,\tau,\gamma}(m)$ playing the role of $x_{G,p}(m)$ before. We similarly use $\mathcal{P}(\mathcal{G},\tau,\gamma)$ and $\mathcal{A}(\mathcal{G},\tau,\gamma)$. Although they are equal for the discrete-time Markovian model, in general $\mathcal{P} \neq \mathcal{A}$ because $|H_{IN}| \neq |H_{OUT}|$.

Theorem 6.16 *For a class of networks $\mathcal{G}$ and given transmission rate τ and recovery rate γ, if H is generated by the process in Fig. 6.13, then $\mathcal{P}$ and $\mathcal{A}$ satisfy*

$$\mathcal{P}(\mathcal{G},\tau,\gamma) = \lim_{|G| \to \infty} \frac{\mathbb{E}(|H_{IN}| + |H_{SCC}|)}{|G|},$$

$$\mathcal{A}(\mathcal{G},\tau,\gamma) = \lim_{|G| \to \infty} \frac{\mathbb{E}(|H_{SCC}| + |H_{OUT}|)}{|G|}.$$

As before, we anticipate that an individual large network $G \in \mathcal{G}$ will give results that are very close to $\mathcal{P}(\mathcal{G},\tau,\gamma)$ and $\mathcal{A}(\mathcal{G},\tau,\gamma)$. For a large network G, we can thus approximate the epidemic size and probability using the algorithms in Fig. 6.13 and Fig. 6.17. This generates a percolated network H, finds H_{SCC} and then finds one node u in H_{SCC}. The in- and out-components of u give $H_{IN} \cup H_{SCC}$ and $H_{OUT} \cup H_{SCC}$, respectively. So their sizes give an approximation to $\mathcal{P}$ and $\mathcal{A}$.

6.3.4 Non-Markovian disease spread

The directed percolation interpretation applies naturally to non-Markovian SIR disease spread, as long as the probability that u would transmit to v if infected depends only on u and v and no external factors [173].

Figure 6.18 shows how to generate the directed percolation version for a non-Markovian epidemic model. With some modification of notation, we can use $\mathcal{P}(\mathcal{G})$ and $\mathcal{A}(\mathcal{G})$ to be the obvious generalisations of $\mathcal{P}$ and $\mathcal{A}$ to the given disease process.

Input: Network G, transmission rate per edge τ and recovery rate γ.
Output: Estimates of $\mathcal{P}$ and $\mathcal{A}$.

    function estimate_SIR_prob_size_from_dir_perc(H)
        SccList ← strongly_connected_components(H)
        Hscc ← Component in SccList with largest size
        u ← random node from Hscc
        In ← in_component(H, [u])
        Out ← out_component(H, [u])
        $\mathcal{P}$ ← |In|/|H| ▷ Be careful about integer division
        $\mathcal{A}$ ← |Out|/|H|
        return $\mathcal{P}$, $\mathcal{A}$
    function estimate_directed_SIR_prob_size(G, τ, γ)
        H ← directed_percolate_network(G, τ, γ)
        $\mathcal{P}$, $\mathcal{A}$ ← estimate_SIR_prob_size_from_dir_perc(H)
        return $\mathcal{P}$, $\mathcal{A}$

Fig. 6.17: Algorithm to estimate $\mathcal{P}$ and $\mathcal{A}$ in a network using directed percolation. It finds the largest strongly connected component and chooses a random node in that component. Assuming that the system is above the percolation threshold, then the component Out will consist of H_{SCC} and H_{OUT} while the component In consists of H_{SCC} and H_{IN}. We assume the existence of an algorithm that finds all strongly connected components (a common example is Tarjan's algorithm [297]). We assume a method in_component with similar definition to out_component. Below the threshold, the returned values will be close to zero.

Then, the following statements hold.

Theorem 6.17 *For a class of networks $\mathcal{G}$ and some transmission rule, if H is the result of directed percolation, then $\mathcal{P}$ and $\mathcal{A}$ satisfy*

$$\mathcal{P}(\mathcal{G}) = \lim_{|G| \to \infty} \frac{\mathbb{E}(|H_{IN}| + |H_{OUT}|)}{|G|},$$

$$\mathcal{A}(\mathcal{G}) = \lim_{|G| \to \infty} \frac{\mathbb{E}(|H_{SCC}| + |H_{OUT}|)}{|G|}.$$

The algorithms in Fig. 6.17 use a percolated network to approximate $\mathcal{P}$ and $\mathcal{A}$.

An important, powerful result is that switching the edge directions in the bow-tie diagram interchanges $\mathcal{P}$ and $\mathcal{A}$. Thus, we can focus our attention just on studying

Input: Network G, ξ, ζ for each node, and transmission: a function determining if transmission happens.

Output: The percolated network H.

```
function nonMarkov_directed_percolate_network(G, ξ, ζ, transmission)
    H ← empty_directed_graph
    for u in G.nodes do
        H.add_node(u))
        for v in G.neighbours(u) do
            transmit ← transmission(ξ[u], ζ[v])
            if transmit then
                H.add_directed_edge(u, v)
    return H
```

Fig. 6.18: Algorithm for creating directed percolation network corresponding to a non-Markovian SIR model. Then, estimate_SIR_prob_size_from_dir_perc in Fig. 6.17 estimates $\mathcal{P}$ and $\mathcal{A}$.

the general impact of heterogeneity on $\mathcal{P}$, and our results immediately apply to $\mathcal{A}$. This has played an important role in a number of studies [1, 211, 212].

It can be shown that regardless of the network class under consideration, for a given average transmission probability, increased heterogeneity in infectiousness or susceptibility for the disease tends to decrease the size and probability of an epidemic [212, 301]. So for a given average transmission probability, the highest $\mathcal{P}$ and $\mathcal{A}$ occur for the discrete time model.

Theorem 6.18 *Assume that there is a distribution of infectiousness such that each node u is given a probability $p_o(u)$ of a transmission going out (independently) to a random neighbour. For, given average transmission probability p, $\mathcal{P}$ takes its maximum value when every node u has $p_o(u) = p$, and it takes its minimum value when a fraction p of the nodes have $p_o(u) = 1$ and the remainder have $p_o(u) = 0$.*

The proofs of this are based on techniques of [190, 191], showing that the probability of having a long directed path towards or away from a random node in the directed percolation network is reduced by increased heterogeneity. Using the equivalence between $\mathcal{P}$ and $\mathcal{A}$, we immediately have the following.

Corollary 6.19 *Assume that there is a distribution of susceptibility such that each node u is independently assigned a probability $p_i(u)$ of a transmission coming in from a random infected neighbour. For a given average $\langle p \rangle$, $\mathcal{A}$ takes its maximum value when every node u has $p_i(u) = \langle p \rangle$, and it takes its minimum when a fraction $\langle p \rangle$ of the nodes have $p_i(u) = 1$ and the remainder have $p_i(u) = 0$.*

Theorem 6.20 *For any unclustered network class and average transmission probability $\langle p \rangle$, heterogeneity in infectiousness affects $\mathcal{P}$ but not $\mathcal{A}$, while heterogeneity in susceptibility affects $\mathcal{A}$ but not $\mathcal{P}$.*

This was proven for Configuration Model networks by [173, 211] and for more general network classes with few short cycles by [212]. If there are many short cycles, then heterogeneity in infectiousness also affects $\mathcal{A}$ and heterogeneity in susceptibility also affects $\mathcal{P}$.

Exercise 6.15. The concept of undirected bond percolation was introduced for comparison with the discrete-time Markovian SIR model. However, for the other models, the equivalence was shown for a directed network. By thinking of undirected percolation as saying what will happen the first time the disease attempts to cross an edge, explain why directed percolation using probability p in both directions will result in the same predictions for $\mathcal{P}$ and $\mathcal{A}$ as undirected percolation.

6.4 Epidemic size in Configuration Model networks

In Section 6.2, we focused on the epidemic probability, which we found by calculating the probability that a random node is the source of at least one long chain of infections. In this section, we look at the probability that a given node is infected, meaning it is the target of a chain of infections starting from an infected node. A number of authors have used the idea that we can learn much about epidemics by studying the transmission chains that reach a random node, or equivalently, by following a reverse transmission process [12, 24, 80, 110].

We begin with the limit in which a negligibly small proportion is initially infected. In the next section, we look at the case in which a non-negligible fraction is initially infected. Our focus will be on calculating the probability that a random node u remains susceptible if an epidemic happens. We will calculate this in terms of the probability that a neighbour v of u would transmit to u if v is ever infected.

We emphasise a technical point. We are looking for a path of transmissions to u through v. This can have no cycles. So if a path reaches u from w, follows an edge from u to v and there is an edge back to u from v, we do not count this as a path through v to u. It is counted as a path through w. In what follows, we will repeatedly calculate the probability of a length $(d-1)$ path to a neighbour v of a random node u, but we exclude any path that reaches v through u. To make the arguments more concise, we introduce the *test node* assumption.

Definition 6.21 A test node *is a single randomly chosen node u which is then prevented from transmitting to its neighbours.*

So in the case of directed percolation, if u is a test node, we remove all edges that point from u to a neighbour in the percolated network. In the case of undirected percolation, we replace all edges involving u with directed edges pointing towards u. The concept of a test node is effectively identical to the idea of the cavity state, which is used in a "message-passing" formulation [163].

We will calculate the proportion infected in the $N \to \infty$ limit if an epidemic occurs with a single index node. If an epidemic occurs, then the disease spreads from the index node to all of H_{SCC}, with at most a vanishingly small number of infections in H_{IN}. From there it will spread to all of H_{OUT}. As $N \to \infty$, a node is in either $H_{SCC} \cup H_{OUT}$ if and only if it has an arbitrarily long path of transmissions to it. So we calculate the proportion infected by calculating the probability a random node has a long transmission chain to it.

6.4.1 Discrete-time Markovian model

We proceed similarly to the calculation of α. We consider the original network G and the percolated network H. We choose a random test node u and define ω_d to be the probability that a u–v edge in G does *not* produce a v–u edge in H, which is the final edge of a transmission chain of length at least d. Then, $\omega_1 = 1 - p$ and we can express ω_d in terms of ω_{d-1} (Fig. 6.19):

$$\omega_d = (1 - p) + p \frac{\psi'(\omega_{d-1})}{\langle K \rangle}.$$

Taking $\omega = \lim_{d \to \infty} \omega_d$, we find that the probability that a randomly chosen node u is never infected is $\psi(\omega)$ (in the $N \to \infty$ limit assuming an epidemic happens). This is identical to the steps that lead to the probability of an epidemic, which is as we expect given Corollary 6.14. As before, the critical value is $p_c = \langle K \rangle / \langle K^2 - K \rangle$.

$\mathcal{A}$ for small initial condition in discrete-time Markovian model

For the discrete-time Markovian model, the fraction infected in an epidemic assuming a small initial condition and a large Configuration Model network is

$$\mathcal{A} = 1 - \psi(\omega), \tag{6.6a}$$

where ω is the limit as $d \to \infty$ for the iteration

$$\omega_d = 1 - p + p \frac{\psi'(\omega_{d-1})}{\langle K \rangle} \qquad d = 1,2,3,\ldots \tag{6.6b}$$

with the initial value $\omega_1 = 1 - p$. If $p < p_c = \langle K \rangle / \langle K^2 - K \rangle$, then $\omega = 1$ and $\mathcal{A} = 0$, while if $p > p_c$, then $\omega < 1$ and $\mathcal{A} > 0$.

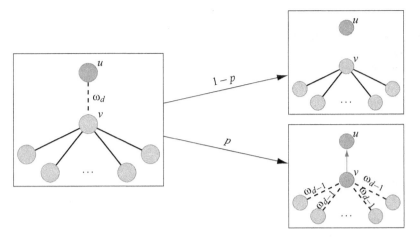

Fig. 6.19: The probability that an edge to a node u from node v (of unknown, random degree k_v) is not the final edge of a chain of potential transmissions to u of length d or more is ω_d. This equals the probability that the edge does not transmit $(1-p)$, or that it does but none of the remaining (k_v-1) edges of v are the final edge of a transmission chain to v of length $(d-1)$ or more $\left(p\sum_{k_v}P_n(k_v)\omega_{d-1}^{k_v-1}\right)$. So $\omega_d = (1-p)+p\psi'(\omega_{d-1})/\langle K\rangle$.

6.4.2 Continuous-time Markovian model

Remarkably, the continuous-time Markovian model calculation of $\mathcal{A}$ reduces to the same equations as for the discrete-time Markovian model. This contrasts with the calculation for $\mathcal{P}$, which involved an integral over the duration of u's infection because infectiousness is heterogeneous. Transmissions from different neighbours of the test node u to u are independent events because the durations of infection of neighbours of u are independent. Thus, susceptibility is homogeneous, and so $\mathcal{A}$ should be the same. Going through the steps in detail, we have

$$\omega_d = \int_0^\infty \left[1 - p(T_v) + p(T_v)\sum_k P_n(k)\omega_{d-1}^{k-1}\right]\gamma e^{-\gamma T_v}\,\mathrm{d}T_v,$$

where $p(T_v) = 1 - e^{-\tau T_v}$ is the transmission probability given the duration of v's infection. Following Exercise 6.8, we can write

$$\omega_d = \frac{\gamma}{\gamma+\tau} + \frac{\tau}{\gamma+\tau}\frac{\psi'(\omega_{d-1})}{\langle K\rangle}.$$

$\mathcal{A}$ for small initial condition in continuous-time Markovian model

For the continuous-time Markovian model with a single introduced infection, we have

$$\mathcal{A} = 1 - \psi(\omega), \tag{6.7a}$$

where ω is the limit of the iteration

$$\omega_d = \frac{\gamma}{\gamma + \tau} + \frac{\tau}{\gamma + \tau} \frac{\psi'(\omega_{d-1})}{\langle K \rangle}, \tag{6.7b}$$

with $\omega_1 = \gamma/(\gamma + \tau)$.

If we replace $\tau/(\gamma + \tau)$ (the expected transmission probability) by p, we recover the discrete-time Markovian model equations. Thus, although the probability of an epidemic is different for the discrete- and continuous-time Markovian models, the fraction infected is the same.

6.4.3 Non-Markovian model

When we calculate $\mathcal{A}$ for the non-Markovian model, we will continue our earlier assumptions when we calculated $\mathcal{P}$ [equations (6.5a) and (6.5b)] and assume that infectiousness and susceptibility are independent. The approach is symmetric to the calculation of $\mathcal{P}$, with p_i replacing p_o, and ω replacing α [211].

$\mathcal{A}$ for small initial condition in non-Markovian model

For a non-Markovian SIR disease spreading in a Configuration Model network, the attack rate of an epidemic is

$$\mathcal{A} = 1 - \int \psi(\omega(\zeta)) P(\xi) \, d\xi, \tag{6.8a}$$

where $\omega(\zeta)$ is the limit of the iteration

$$\omega_d(\zeta) = 1 - p_i(\zeta) + p_i(\zeta) \int \frac{\psi'(\omega_{d-1}(\hat{\zeta}))}{\langle K \rangle} P(\hat{\zeta}) \, d\hat{\zeta}, \tag{6.8b}$$

with $\omega_1(\zeta) = 1 - p_i(\zeta)$.

Example 6.3. Consider an Erdős–Rényi random network with expected degree 5 and take the discrete-time Markovian SIR model with $p = 1$, that is, the disease transmits to all neighbours. Assume that two vaccines are under consideration. Both appear to prevent 70% of transmissions. One does so by giving complete protection to 70%

of the population and no protection to the remaining 30%. The other gives equal protection to all nodes, by reducing their susceptibility per partnership by 70%.

In both cases $p_o(u)$, the probability that u would transmit to its neighbour if infected is 0.7. The infectiousness is homogeneous in both cases. However, the case where the disease gives complete protection to 70% of the population has very heterogeneous susceptibility, while the other case has homogeneous susceptibility.

For the heterogeneous case, we use $\zeta = 1$ if the vaccination of a node was unsuccessful and $\zeta = 0$ if the node is successfully vaccinated. We have $\psi(\omega) = e^{-5(1-\omega)}$, $p_i(0) = 0$, $p_i(1) = 1$ and

$$A = 1 - 0.7\psi(\omega(0)) - 0.3\psi(\omega(1)),$$

$$\omega(\xi) = 1 - p_i(\xi) + p_i(\xi)\left[\frac{0.7\psi'(\omega(0)) + 0.3\psi'(\omega(1))}{5}\right], \quad \xi = 0,1.$$

It is quickly clear that $\omega(0) = 1$ and it follows that $\omega(1) = [0.7\psi'(1) + 0.3\psi'(\omega(1))]/5 = 0.7 + 0.3\psi'(\omega(1))/5$. This can be solved iteratively, giving $\omega(1) \approx 0.825$. Then,

$$A \approx 1 - 0.7\psi(1) - 0.3\psi(0.825) \approx 0.175$$

Exercise 6.16. Explain why for the discrete-time and continuous-time Markovian models susceptibility is homogeneous. Use this to show that system (6.8) reduces to systems (6.6) and (6.7) in these cases.

Exercise 6.17. Consider Erdős–Rényi random networks of expected degree 10. To each node assign a pair of integers (ξ, ζ). With probability $1/2$ $\xi = 0$, otherwise $\xi = 1$. Similarly, ζ is 0 or 1 with probability $1/2$, independently of ξ. Let the v to u transmission probability be 1 if $\xi = 1$ and $\zeta = 0$, $1/2$ if $\xi = \zeta = 1$, and 0 if $\xi = 0$.

  a. Calculate $\mathcal{P}$ and A in the large network limit.

  b. Generate a percolated network H for a network of 50000 nodes and use the algorithms in Fig. 6.17 to estimate $\mathcal{P}$ and A.

Exercise 6.18. This problem revisits Theorem 6.10.

  a. Explain the implications of Theorem 6.10 for A.

  b. Let $\langle p \rangle$ be fixed. Consider a population that minimises $\mathcal{P}$. What can we say about p_i for this population, and what conclusions does that lead to about A?

  c. Using Erdős–Rényi random networks of 10^5 nodes and expected degree 10, use the algorithms in Figs. 6.18 and 6.17 for the population minimizing $\mathcal{P}$ and plot $\mathcal{P}$ and A against $\langle p \rangle$.

6.5 Edge-based compartmental modelling: epidemic dynamics

In this section, we modify the derivation for A to predict the dynamics of an epidemic, assuming either the discrete-time or the continuous-time Markovian model. We will calculate the number of nodes that are susceptible, infected or recovered at any time t. We proceed in the large network limit, assuming some non-zero (but possibly small) proportion of nodes initially infected. In practice, we expect this to accurately model a finite network as long as the initial amount of infection is large enough that the dynamics proceed deterministically.

6.5.1 Epidemic size with many initially infected nodes

We begin by calculating the final epidemic size in the discrete-time model, assuming a nonzero initial proportion infected. As much as possible, we follow the derivation we had for the final size assuming a vanishing proportion initially infected, but some changes are required. Once this derivation is complete, it is a small step to calculating the dynamics of the infection rather than just the final epidemic size.

In the context of the percolated network H, the index nodes that are not in H_{SCC} or H_{OUT} will cause some infections that lie outside H_{SCC} and H_{OUT}. For a single index node, this amount is insignificant, but when the number of such nodes is proportional to the population size, the combined effect can be large. Thus, u can be infected even if there is not a long path that reaches u, as long as a short path through susceptible nodes starting from one of the index nodes reaches u. So we must now include the possibility of short paths in our calculations.

We will perform this calculation only for the discrete-time Markovian model. As with the single initial infection, the calculation will be identical for the continuous-time Markovian case. It can be adapted for the non-Markovian case.

We assume that some proportion $(1-\rho)$ of the population is initially susceptible, so a proportion ρ are either infected or recovered. This may be because the disease has been spreading for some time before we initialise the equations (in which case some nodes will be recovered and infected nodes will disproportionately connect to infected or recovered nodes). Alternately, we allow the possibility that the disease is introduced by choosing index infections and instantaneously infecting them. We set $S(k,0)$ to be the probability that a degree k node is susceptible at $t=0$. Then,

$$\sum_k S(k,0)P(k)x^k = \frac{1}{N}\sum_k S_k(0)x^k = \hat{\psi}(x)$$

following Eq. (5.12), where $S_k(0)$ is the number of susceptible degree k nodes at time 0. If $S(k,0) \to 1$, then $\hat{\psi}(x) \to \psi(x)$. If the initial condition has a proportion ρ of the nodes randomly infected, then $\hat{\psi}(x) = (1-\rho)\psi(x)$.

Definition 6.22 *We say that a path from w to u in G transmits by generation d if w is initially infected, the length of the path is at most d and it exists in the directed percolated network H.*

Given a randomly chosen initially susceptible node u, we define $\theta(d)$ to be the probability that a random u–v edge in G is not the final edge of a path that transmits by generation d. Since u is initially susceptible, $\theta(0) = 1$

(contrast with $\alpha_1 = \omega_1 = 1 - p$ due to differences in the definitions). Thus, $\sum_k S(k,0)P(k)\theta(d)^k = \hat{\psi}(\theta(d))$ is the probability that a random node is still susceptible after d transmission generations, and the probability that a random, initially susceptible neighbour of an initially susceptible test node is still susceptible is $\sum_k S(k,0)P_n(k)\theta(d)^{k-1}/\sum_k S(k,0)P_n(k) = \hat{\psi}'(\theta(d))/\hat{\psi}'(1)$.

We assume that if we look at a susceptible node at $t = 0$, the statuses of its neighbours are independent.[4] We define $\phi_S(0)$ to be the probability that an edge of an initially susceptible node connects to another initially susceptible node. Similarly, $\phi_I(0)$ and $\phi_R(0)$ are the probabilities that an edge of an initially susceptible node connects to an initially infected or recovered node, respectively. These must sum to 1. We take a test node u chosen randomly from the population, and let v be a random neighbour of u. Then, $\theta(d)$ is (see Exercise 6.19)

$$\theta(d) = (1-p) + p\left[\phi_R(0) + \phi_S(0)\frac{\hat{\psi}'(\theta(d-1))}{\hat{\psi}'(1)}\right]. \tag{6.9}$$

Iterating this leads to the results below.

$\mathcal{A}$ for large initial condition in discrete-time Markovian model

The fraction eventually infected for the discrete-time Markovian model in a large Configuration Model network is given by

$$\mathcal{A} = 1 - \hat{\psi}(\theta), \tag{6.10a}$$

where θ is the limit of the iteration

$$\theta(d) = (1-p) + p\left[\phi_R(0) + \phi_S(0)\frac{\hat{\psi}'(\theta(d-1))}{\hat{\psi}'(1)}\right], \tag{6.10b}$$

with $\theta(0) = 1$ and $\hat{\psi}(x) = \sum_k S(k,0)P(k)x^k$.

If the disease is introduced at time $t = 0$ by infecting a proportion ρ of the nodes uniformly at random, then $\phi_S(0) = 1 - \rho$, $\hat{\psi}(x) = (1-\rho)\psi(x)$ and $\phi_R(0) = 0$.

Exercise 6.19. In this exercise, we derive two interpretations of equation (6.9).

a. Explain why $\phi_R(0) + \phi_S(0)\hat{\psi}'(\theta(d-1))/\hat{\psi}'(1)$ is the probability that a random neighbour has never been infected after $(d-1)$ steps. Using this, explain (6.9).

b. Rewrite the right-hand side of (6.9) as $1 - p\left[1 - \phi_R(0) - \phi_S(0)\frac{\hat{\psi}'(\theta(d-1))}{\hat{\psi}'(1)}\right]$. What is the meaning of the term in brackets? Using this, explain (6.9).

---

[4] This assumption fails in a Configuration Model network if, for example, at $t = 0$ we select some nodes and preferentially infect their neighbours.

6.5.2 Predicting time evolution of the epidemic

Our basic approach to deriving the epidemic dynamics is similar to what we have just used to calculate $\mathcal{A}$ with a non-zero initial condition. Rather than calculating the probability that a node is eventually infected, we calculate the probability that it is infected by time t. Effectively, we are considering chains of infections approaching a node u that have distance less than t and start from an initially infected node. As time increases, the chains we consider grow, and we will be able to find a closed system of equations that we can use to calculate the relevant growth properties.

Conceptually, when the percolated network H is created, we assign each edge a weight equal to the waiting time before transmission. In the discrete-time model this weight is simply 1. In the continuous-time model, this is taken from the exponential distribution with rate τ truncated at the infection duration.

We modify the definition of "transmits by generation d" to "transmits by time t".

Definition 6.23 *We say that a path from w to u in G transmits by time t if w is initially infected, the path exists in H and the sum of weights along the path is less than t.*

We define $\theta(t)$ to be the probability that for a randomly chosen initially susceptible node u and random neighbour v, there is no path to u through v that is transmitting by time t. That is, $\theta(t)$ is the probability that v has not transmitted to u by time t. Thus, a test node u is susceptible at time t if it is initially susceptible and there is no path to it that is transmitting by time t.

We again use $S(k,0)$ to denote the probability that a random degree k node is initially susceptible. We define $\phi_S(t)$, $\phi_I(t)$, and $\phi_R(t)$ to be the probabilities that a random neighbour v of a random, initially susceptible test node u is susceptible, infected or recovered at time t and the edge from v to u is not transmitting by time t. Note that $\theta = \phi_S + \phi_I + \phi_R$.

Following the steps we took for $\mathcal{A}$ with many initial infections, we can solve for S and ϕ_S at a given time t in terms of θ. The method is modified slightly to account for S, I, and R being numbers of nodes rather than proportions.[5] We have

$$S = N\sum_k S(k,0)P(k)\theta^k = N\hat{\psi}(\theta),$$

$$\phi_S = \phi_S(0)\frac{\sum_k S(k,0)P_n(k)\theta^{k-1}}{\sum_k S(k,0)P_n(k)} = \phi_S(0)\frac{\hat{\psi}'(\theta)}{\hat{\psi}'(1)},$$

and by noting that $S+I+R=N$ and $\phi_S + \phi_I + \phi_R = \theta$, we have

$$I = N - S - R,$$

$$\phi_I = \theta - \phi_S - \phi_R.$$

---

[5] We do this for consistency with other parts of the book. Most previous studies using this approach take S, I and R to be proportions

Our challenge now is to find how θ changes in time and to determine I, R, ϕ_I, and ϕ_R. This will depend on the specific details of the model. Figures 6.20 and 6.21 summarise what we know so far.

Fig. 6.20: A flow diagram showing the fluxes between S, I, and R. We know S in terms of θ, and can express I in terms of S and R. The I to R flow is model dependent.

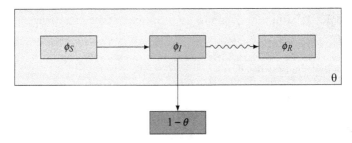

Fig. 6.21: A flow diagram for the basic edge-based compartmental model. We set ϕ_S, ϕ_I and ϕ_R to be the probabilities that a random neighbour v of a random test node u has not yet transmitted to u and is susceptible, infected or recovered. We know ϕ_S in terms of θ, and we know these sum up to θ, so we can express ϕ_I in terms of θ, ϕ_S and ϕ_R. The compartment $(1 - \theta)$ represents the probability that the path is transmitting by time t.

We begin with the discrete-time Markovian model and then the continuous-time Markovian model. Chapter 9, system (9.36), will present non-Markovian dynamics.

Discrete-time Markovian model

Our calculation here closely resembles the final epidemic size calculation, noting that if a node first becomes infected after t steps, then it must be infectious at time t.

For the discrete-time Markovian model, any infected node recovers after a single time step. Thus, we know that the flow from I to R at time step t is simply $I(t)$. Similarly, an infected neighbour v will either transmit (with probability p) in the next time step or recover. So we know the flow from ϕ_I to $(1 - \theta)$ is $p\phi_I$ and the flow to ϕ_R is $(1 - p)\phi_I$.

Because $\theta(0) = 1$, the initial value of $(1 - \theta)$ is 0. The flow from ϕ_I to ϕ_R is always proportional to the flow from ϕ_I to $(1 - \theta)$ with proportionality constant $(1 - p)/p$. So the total change to ϕ_R is $(1 - p)/p$ times $(1 - \theta)$. Thus,

$$\phi_R = \phi_R(0) + \frac{1 - p}{p}(1 - \theta).$$

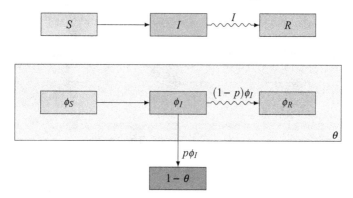

Fig. 6.22: The full flow diagrams for the discrete-time Markovian model.

We put this into the flow diagrams in Fig. 6.22. We get

$$\theta(t) = \theta(t-1) - p\phi_I(t-1)$$
$$= \theta(t-1) - p\left[\theta(t-1) - \phi_S(0)\frac{\hat{\psi}'(\theta(t-1))}{\hat{\psi}'(1)} - \phi_R(0) - \frac{1-p}{p}(1-\theta(t-1))\right].$$

Simplifying, the discrete-time edge-based compartmental model takes the form below.

Discrete-time EBCM model

$$\theta(t) = (1-p) + p\left[\phi_R(0) + \phi_S(0)\frac{\hat{\psi}'(\theta(t-1))}{\hat{\psi}'(1)}\right], \qquad (6.11a)$$
$$R(t) = R(t-1) + I(t-1), \quad S(t) = N\hat{\psi}(\theta(t)), \quad I(t) = N - S - R, \quad (6.11b)$$

where $\hat{\psi}(x) = \sum_k P(k)S(k,0)x^k$ and $\theta(0) = 1$.

In the limit $S(0) \to N$ and $\phi_S(0) \to 1$, where almost all nodes are initially suscep-tible, this system is equivalent to that of [305]. Compare this with system (6.10).

Continuous-time Markovian model

The analysis for the continuous-time Markovian model is similar. Infected nodes recover at rate γ rather than after a single time step. Thus, the flow from I to R is γI. Similarly, an infected neighbour v transmits with rate τ and recovers with rate γ. So the flow from ϕ_I to ϕ_R is $\gamma\phi_I$ and the flow from ϕ_I to $(1-\theta)$ is $\tau\phi_I$. As before, these are proportional to one another, so $\phi_R = \phi_R(0) + (1-\theta)\gamma/\tau$ (see Fig. 6.23).

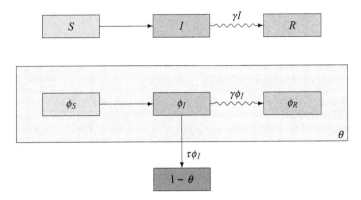

Fig. 6.23: The full flow diagrams for the continuous-time Markovian model.

We then have $\dot{\theta} = -\tau\phi_I$ and $\phi_I = \theta - \phi_S - \phi_R$. Substituting for ϕ_S and ϕ_R gives the following form for the continuous-time edge-based compartmental model.

Continuous-time EBCM model

$$\dot{\theta} = -\tau\theta + \tau\phi_S(0)\frac{\hat{\psi}'(\theta)}{\hat{\psi}'(1)} + \gamma(1-\theta) + \tau\phi_R(0), \qquad (6.12a)$$

$$\dot{R} = \gamma I, \qquad S = N\hat{\psi}(\theta), \qquad I = N - S - R, \qquad (6.12b)$$

where $\hat{\psi}(x) = \sum_k P(k)S(k,0)x^k$ and $\theta(0) = 1$.

Remarkably, this system is governed by a single differential equation for θ, which depends only on disease parameters, the degree distribution, and θ. We need additional equations to find S, I and R, but they do not feed back into the dynamics.

Exercise 6.20. By noting that at the final state $\dot{\theta} = 0$, derive an equation for the final epidemic size and compare it with the final epidemic size equation we found for many introduced infections in (6.10).

Example 6.4. We consider the spread of an SIR disease in four populations, all having average degree 5. In the first population, the degree distribution is of the form $P(k) \propto k^{-\alpha}e^{-k/40}$. So this appears like a power law distribution, but the tail is truncated around $k = 40$. So that the average is about 5, we take $\alpha \approx 1.42$. In the second, we take a bimodal distribution, with $P(8) = P(2) = 0.5$. In the third, we use an Erdős–Rényi random network of average degree 5, which is equivalent to saying that the degree distribution is Poisson with mean 5. Finally, in the fourth, we take $P(5) = 1$ so all nodes have the same degree.

Figure 6.24 compares the predicted and observed spread of the disease taking $\tau = 0.4$ and $\gamma = 1$. The agreement is excellent. We have already seen that the level of heterogeneity in the degree distribution affects the epidemic threshold, but here we see that the distributions with highest heterogeneity show the greatest initial growth as the high degree nodes are infected early and transmit to their neighbours. The epidemics in these populations quickly die off once the highest degree nodes are depleted and the remaining network has typical degree well below 5. Generally epidemics in more homogeneous networks grow slower but persist longer.

The following exercise provides equations that account for degree correlations.

Exercise 6.21. Consider networks that have degree correlations. The probability that v, a random neighbour of a degree k node u, has degree $\hat{k}$ is $P_n(\hat{k}|k)$.

a. Define θ_k to be the probability that a neighbour of a degree k test node u has not yet transmitted to u. Find the number of susceptible nodes in terms of all of the θ_k.

b. Define $\phi_{S|k}$ to be the probability a neighbour of a degree k test node u is susceptible. Similarly, generalise $\phi_{I|k}$ and $\phi_{R|k}$. Find an equation for $\dot{\theta}_k$ in terms of the other $\theta_{\hat{k}}$.

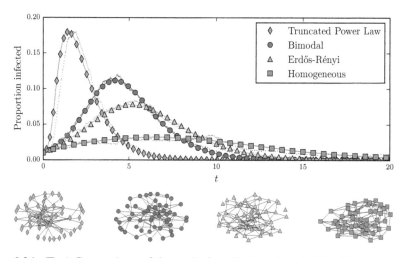

Fig. 6.24: (Top) Comparison of theoretical predictions (symbols) with simulations for populations having 5000 nodes (dotted) and 500000 nodes (solid) and different degree distributions with average degree 5. In all cases, $\tau = 0.4$ and $\gamma = 1$. We begin with 1% of the population infected instantaneously at $t = 0$. The symbols show analytical predictions. (Bottom) Sample networks with the given degree distribution, but only 50 nodes.

Rigorous proofs

The EBCM derivations we have given are heuristic. Our assumptions that neighbours of u are independent of one another could fail if a few nodes have very large degree. For example, if one node has degree $(N-1)$, then the status of u's neighbours are all likely to depend heavily on when (or if) the high degree node is infected.

More rigorous proofs of these equations have been derived. The earliest of these was [76] which showed that if the fifth moment is bounded, then the edge-based-compartmental model equations are correct for the continuous-time Markovian model. Later, [156] improved on this result and showed that if the second moment of the degree distribution is bounded, then the EBCM equations accurately predict large-network dynamics for the continuous-time Markovian model. If the second moment is not bounded, then the equations break down.

6.6 SIS disease

We finish by briefly discussing a percolation-like approach for SIS disease [63, 123, 134, 145, 326]. We assume a given transmission rate τ and recovery rate γ. We assign a sequence of potential transmission times to each edge and a sequence of potential recovery times to each node generated from Poisson processes with rates τ and γ, respectively. We can trace infection forward from an index node with infected nodes transmitting at potential transmission times and recovering at potential recovery times. The nodes infected at a later time T are the nodes which are the target of a path which moves upwards from an initially infected node that respects the transmissions and recoveries, shown in Fig. 6.25.

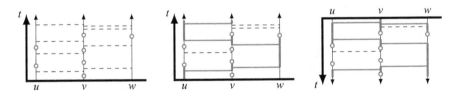

Fig. 6.25: A symmetry of the Markovian SIS process. We consider a line network with v connected to u and w. (Left) We assign transmission times (dashed lines) and recovery times (small circles) as Poisson processes. (Middle) Then, we trace a path upwards from the initially infected node u. (Right) When we invert the process, we see that any path in the original process from u to v exists in the inverted process from v to u. Thus, infection of u at time 0 is as likely to lead to the infection of v at time T as infection of v at time 0 is to lead to the infection of u at time T.

A symmetry emerges: a Poisson process is as likely to assign an event to time $T - t_0$ as it is to assign to time t_0. This implies that we can study an inverted process where we take the diagram shown on the right in Fig. 6.25, starting with $t = 0$ at the top and t increasing downwards. The resulting disease spread is also a legitimate outbreak and occurs with equal probability to the original. A chain from node u to node v in the original exists iff a chain from v to u exists in this new direction.

A consequence of this observation is that if we start with a single infected node u at time 0, then the expected number of infections at a later time T equals the probability that if a single random node is infected at time 0, then u is infected at time T. Pushing this farther, we can show that the probability u leads to a chain of infections causing an epidemic equals the proportion of time that u is infected in the endemic equilibrium. Unfortunately, this approach depends on the properties of Poisson processes and does not generalise to non-Markovian transmission and recovery processes.

6.7 Conclusions and outlook

This chapter has focused on studying disease spread from a percolation-based perspective. We used this, combined with a Galton–Watson process approach, in order to develop simple equations that predict many properties of SIR disease spread in Configuration Model networks. We ultimately derived a system of equations, the EBCM equations, that predict the dynamics of SIR disease. For the continuous-time model, the resulting system is governed by a single differential equation for θ in terms of only the parameters, the initial conditions and θ. Given the solution to this equation, we can calculate S, I and R with a few additional equations.

Our approach is similar to "susceptibility set" methods that others have used for calculating final sizes [9, 12, 12, 24]. The susceptibility set of a node u is the set of nodes whose infection would lead to the infection of node u. In our context, this is the in-component of u in H. The EBCM approach can be interpreted in terms of a susceptibility set by thinking of susceptibility sets growing in time; that is, we consider the set of nodes whose infection would result in, infection of u within t units of time. The proportion infected by time t is the probability a random node's susceptibility set contains an initially infected node by time t.

Compared to our earlier systems, the EBCM equations are remarkably frugal. For networks with an unbounded number of distinct degrees, the previous models would all require an unbounded number of equations. For the EBCM equations, the total number of equations is small and does not change as the degree distribution changes. In the next chapter, we will show that in fact the EBCM equations are equivalent to the previous models if we make (very) mild assumptions on the initial conditions.

The EBCM method we have used is based on an approach introduced in [316]. This derivation and equations were simplified in [215], and further refined in [220–222], allowing for dynamic networks and a variety of other assumptions about the disease or network structure. These approaches assumed an infinitesimally small initial proportion of infected nodes, and the simplifications limited the appropriateness of the model below the epidemic threshold. This was resolved in [217].

A closely related approach [163] derives an integral equation by using a cavity state in a message passing formulation, which in this context is equivalent to the assumption of a test node not transmitting. The EBCM approach has been used in many contexts, including disease modelling [184, 200, 258, 305, 318, 321] and study of the spread of complex contagions, processes (such as rumours) in which previous exposures can lead to more successful transmission [218, 322]. In Section 9.3, we apply this approach to non-Markovian SIR spread.

A major challenge of disease modelling is understanding the role of network clustering. The best available tools to do this analysis appear to be based in percolation theory. In many networks, clustering has a surprisingly small impact on the final size of an epidemic [209, 214], but it does impact the dynamics [151, 214]. In special cases, it can be studied analytically for SIR disease [114, 213, 265, 318]. We use numerical simulations to explore this in more detail in Chapter 11. At present however, a general theory for clustered networks is lacking. In part, this is due to computational challenges when neighbours' statuses are correlated, and in part this is because no clear consensus exists on which clustered networks to consider. This remains an open problem without a satisfactory conclusion.

Chapter 7
Hierarchies of SIR models

This chapter focuses on the relationships between the continuous-time SIR models we have previously derived and identifying conditions under which they are appropriate. Unless otherwise noted, the models discussed in this chapter are SIR models. Each of these models involves some assumptions, and to understand their limitations, we need to understand whether the true spread "respects" these assumptions.

It is generally not a surprise that the heterogeneous mean-field model at the single level (HetMF) and the homogeneous mean-field model at the single level (HomMF) appear as limiting cases of the more complex models. These simpler models assume that when a transmission occurs, the recipient is selected randomly from the population (perhaps weighted by degree). The more complex models assume that the neighbours are chosen randomly from the population *a priori* and then the transmissions occur within this subset. In the low transmission limit in which edges are unlikely to transmit twice, these models should have the same prediction.

However, it may be surprising that for Configuration Model networks, the pairwise (PW) model , the effective degree (ED) model , and the edge-based compartmental model (EBCM) are equivalent if the infection is introduced uniformly at random into the population. These models were derived through very different approaches, but their underlying assumptions are similar, and we can show that they are equivalent except for unusual initial conditions. Even then, we anticipate that after a short time, information about the initial condition will decay, after which all three models will be equivalent and the simplest model becomes preferable.

7.1 The models and their assumptions

We have introduced several models of SIR disease spread, based on different assumptions. Each model creates a reduced state space by grouping similar nodes (and/or edges) together and treating them as interchangeable. Based on the current expected state, the model continuously updates the expected network state as time

© Springer International Publishing AG 2017

I.Z. Kiss et al., *Mathematics of Epidemics on Networks*, Interdisciplinary Applied Mathematics 46, DOI 10.1007/978-3-319-50806-1_7

moves forward. So for example, the HetMF model groups nodes by degree, and assumes that all neighbours are interchangeable. The assumption that neighbours are interchangeable means that if we choose a random node of degree k, the number of infected neighbours is binomially distributed, with a probability that is independent of the node's status or degree.

A model's quality depends on how accurately it predicts future states. This depends strongly on how the state space is reduced. By grouping a larger set of nodes or edges together, we anticipate a lower dimensional system of equations. However, this may mask heterogeneities or correlations that develop due to the transmission process. For example, due to how the disease transmits, susceptible nodes will tend to have fewer neighbours that have been infected than infected nodes. If these masked heterogeneities are important to the disease dynamics, this limits the model's accuracy.

7.1.1 Mean-field models

We first consider the HomMF and HetMF models, systems (4.9) and (5.11), respectively. The HomMF model assumes that all nodes with the same status are interchangeable. The HetMF model assumes that all nodes with the same status and degree are interchangeable. So in the HomMF model, when a node becomes infected, we move an "average" node to the infected class, and when a node recovers, we move an "average" node to the recovered class. In contrast, the HetMF model assumes that when a node is infected, we move an average node of the given degree to the infected class, and similarly when recovery happens we move an average node of the given degree to the recovered class.

Both models assume that all neighbours are interchangeable. This means that the probability that a random neighbour of a susceptible node has a given status is equal to the probability that a random neighbour of an infected node has that given status. This assumption is equivalent to assuming that nodes are continuously changing their neighbours at very high rate.

As the disease spreads, it does not respect these assumptions. Treating all nodes as average results in inaccuracies in the HomMF model: a higher degree node is more likely to become infected early in the epidemic, and in turn it will cause more infections. We can improve the accuracy by turning to the HetMF model, but even this is insufficient. In reality, infected nodes have received transmission from (at least) one neighbour, and may have transmitted to other neighbours. Thus, neighbours of infected nodes are more likely to be infected or recovered than neighbours of susceptible nodes. Treating all neighbours as interchangeable ignores this.

7.1.2 Improved models

We have considered three model classes that are more careful about tracking edges, namely the EBCM, PW and ED models. These models reduce the state space in different ways. The EBCM approach groups nodes by their degree and status, and it assumes that neighbours of susceptible nodes are interchangeable (but not with neighbours of infected nodes). The PW model treats susceptible nodes of a given degree as interchangeable, and it assumes that neighbours of each group of susceptible nodes are interchangeable. The ED model groups nodes together if they have the same number of susceptible and infected neighbours, and it assumes that susceptible neighbours of susceptible nodes are interchangeable.

It turns out that for SIR disease on a Configuration Model network, the spread "respects" the assumptions of all three of these models. This means that if the assumptions of the model hold at time $t = 0$, they continue to hold later. We do not address the question of "stability": if the initial condition does not exactly satisfy the assumptions or if random events cause the true dynamics to vary slightly from the predicted solution, do the true dynamics remain close to the prediction? Simulations show that unless the number of infections is sufficiently small that it can go extinct stochastically, the predictions are typically excellent approximations.

It should be highlighted that the PW model distinguishes between edges starting from degree k_1 nodes and edges starting from degree k_2 nodes. This allows the PW model to account for random networks with assortative or disassortative mixing. The EBCM approach can be modified for this through introducing θ_k (Exercise 6.21). The ED model can be modified for this by grouping nodes together if they have the same number of susceptible and infected neighbours and the same degree, and then treating the susceptible neighbours of susceptible nodes of each degree k_0 as interchangeable, system (5.37). This increases the number of equations in the two models roughly by a factor of the total number of distinct degrees.

7.2 Hierarchy

We now develop a hierarchy of the models, shown in Fig. 7.1, assuming Configuration Model networks following [219, 221]. We will allow for arbitrary placement of the initial infected nodes, but assume that the number of infections is large enough that the dynamics are deterministic.

Perhaps surprisingly, we will see that the PW model and ED models are distinct: each can apply in situations where the other does not. The EBCM model is a special case of both. If we incorporate additional assumptions into the PW and ED models (namely that the neighbours of all susceptible nodes are interchangeable), then they can be simplified to the compact PW system (5.19) and compact ED system (5.43), which we will see are equivalent to the EBCM system (6.12).

We saw already in Section 5.6.2 that the compact ED model can be derived from the ED model with an additional assumption. Similarly, Section 5.2.3 showed that the compact PW model can be derived from the PW model under an additional assumption. In both cases, the additional assumption corresponds to stating that neighbours of all susceptible nodes are interchangeable, which is the assumption of the EBCM approach. We will show that

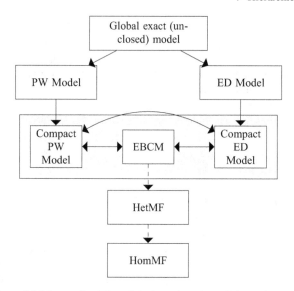

Fig. 7.1: **SIR model hierarchy**. The global unclosed model can be reduced to either the pairwise (PW) model or the effective degree (ED) model through specific simplifying assumptions: the PW model results from assuming that for all degrees k, neighbours of susceptible degree k nodes are indistinguishable, while the ED model results from assuming that susceptible neighbours of susceptible nodes are indistinguishable. These systems are not equivalent, but with additional assumptions, both reduce to the EBCM equations. All of these simplifying assumptions are consistent with the disease dynamics in the sense that if they hold initially, they hold for later times. Dashed lines denote reductions that hold only in particular limits. If the edge transmission probability tends to zero (while the degrees tend to infinity), these models converge to the HetMF model. In turn, this converges to the HomMF model if the degrees are similar enough.

- the PW and ED models are not equivalent and neither is a special case of the other;
- the compact ED model is equivalent to the EBCM model;
- the compact PW model is equivalent to the EBCM model;
- the EBCM model reduces to the HetMF model in certain limits; and
- the HetMF model reduces to the HomMF model in certain limits.

7.2.1 Non-equivalence of PW, ED and EBCM models

To illustrate the distinctions between the models, we will consider cases where their predictions differ due to differences in the initial conditions.

Consider a Configuration Model network G and assume that to begin the disease spread, we choose some nodes and infect all their neighbours (but not them). These central nodes are initially in many SI edges, and when they get infected, rather than new SI edges being formed, many SI edges become II edges. By treating neighbours as interchangeable, the closures for the PW model give the wrong values for $[ISI]$ and $[ISS]$, and the EBCM model assumption of neighbour independence fails. However, the ED model can account for this.

Example 7.1. For a specific example, we begin with a random regular network built using the Configuration Model: all nodes have degree 4. As an initial condition, we choose a proportion r of the nodes uniformly at random and infect their neighbours (but not the chosen nodes). The probability a random node is initially infected is the probability at least one of its 4 neighbours was initially chosen, $\rho = 1 - (1-r)^4$.

Those nodes which were initially chosen have all of their neighbours infected, so there are significantly more susceptible nodes with all neighbours infected than would be predicted assuming independent neighbour statuses. Figure 7.2 compares the simulated epidemic spread with predictions from the three models for $N = 10^6$, $\tau = \gamma = 1$ and $r = 0.01$. Although the difference is small, we see that the ED model accurately predicts the dynamics, but the PW and EBCM models overestimate the early growth because they do not recognise the increased early depletion of SI edges.

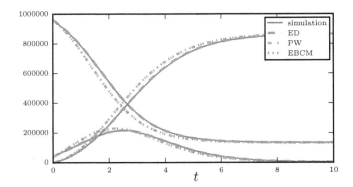

Fig. 7.2: A comparison of simulations and model predictions of S, I and R for an epidemic spreading in a random regular network having degree 4, $N = 10^6$ and $\tau = \gamma = 1$. A proportion 0.01 of the nodes are chosen at random, and all of their *neighbours* are infected. The ED model accurately reproduces the dynamics, but the other models do not.

We can also find cases where the PW model applies but the ED and EBCM models fail. Assume that we select some edges, and if the edge happens to be between two nodes of the same degree, we infect one with some probability. A susceptible neighbour of a susceptible degree k_0 node is likely to have a different degree (and therefore a different risk) compared to a susceptible neighbour of a degree k_1 node.

Because the ED model assumes that susceptible neighbours of susceptible degree k_0 nodes are at the same risk of infection as susceptible neighbours of susceptible degree k_1 nodes, it will incorrectly calculate the rate of infection of those neighbours, which will translate into an incorrect calculation of the risk to the central susceptible nodes. The EBCM model will similarly fail because it assumes θ is degree independent. The PW model can account for this however.

Example 7.2. For a specific example, we consider Configuration Model networks having two degrees, 10 and 2, distributed with $P(2) = 5/6$ and $P(10) = 1/6$. With this distribution, $P_n(2) = P_n(10) = 1/2$. For any edge between two nodes of the same degree, we select one node of the edge and infect it with probability 0.3. This unusual method of choosing the neighbours ensures that the susceptible neighbours of susceptible degree 2 nodes are more likely to have degree 10 than the susceptible neighbours of susceptible degree 10 nodes. Thus, they are not interchangeable. The PW model is appropriate in this case, but not the ED or EBCM models.

Figure 7.3 compares the simulated epidemic spread with predictions from the three models for $N = 10^6$ and $\tau = \gamma = 1$. Although the difference is small, the PW model accurately predicts the dynamics, but the ED and EBCM models do not.

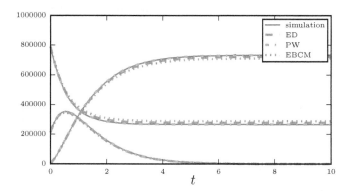

Fig. 7.3: A comparison of simulations and model predictions of S, I and R in a 10^6 node network having $P(2) = 5/6$ and $P(10) = 1/6$. If an edge contains two nodes of the same degree, with probability 0.3, one node is chosen and infected. The PW model accurately reproduces the dynamics, but the other models do not.

In one of the most commonly studied cases, all three models apply. Consider a Configuration Model network initialised by infecting a proportion of the nodes randomly. All three models apply.

Example 7.3. As our final example, we take an Erdős–Rényi random network with edge probability $5/(N-1)$ and take $N = 10^6$. We take a proportion 0.1 of the nodes uniformly at random and infect them. We again take $\tau = \gamma = 1$. Figure 7.4 shows that all three models successfully predict the dynamics.

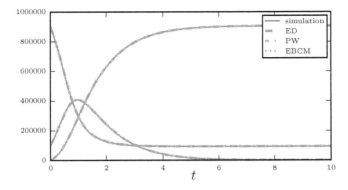

Fig. 7.4: A comparison of simulations and model predictions of S, I and R in a 10^6 node Erdős–Rényi random network of average degree 5. A proportion 0.1 of the nodes are chosen uniformly at random and infected. All three models accurately reproduce the dynamics.

We must conclude that the models are not equivalent as they can give different predictions for the same initial conditions. Further, which model is appropriate depends on the initial condition. Although the EBCM equations are slightly more specialised than the other approaches, the initial conditions we must choose to make it fail are likely unrealistic, and even in these cases it provides reasonable agreement with simulation.

7.3 Equivalence of compact ED, compact PW and EBCM models

The compact ED model was derived from the ED model and the compact PW model was derived from the PW model by adding the assumption that neighbours of any susceptible node are interchangeable. In particular, the status of one neighbour of a susceptible node is independent of the status of any other. This assumption matches the EBCM assumptions so we would expect all three models to be equivalent.

In this section, we show that these models are mathematically equivalent. We begin by showing that the compact ED and EBCM models are equivalent by showing that each can be derived from the other. Then, we do the same for the compact PW and EBCM models. It then follows that the compact ED and compact PW models are equivalent. Throughout, we will assume that at time $t = 0$, a fraction ρ of the nodes are chosen uniformly at random and then infected. Under this assumption, $S = N(1 - \rho)\psi(\theta)$, $\hat{\psi}(x) = (1 - \rho)\psi(x)$, $\phi_S(0) = 1 - \rho$ and $\phi_R(0) = 0$. In particular, this gives $\phi_S(0)\hat{\psi}'(\theta)/\hat{\psi}'(1) = (1 - \rho)\psi(\theta)/\langle K \rangle$. This assumption simplifies the algebra, but the same results hold without it. The important detail is that all three models require that in the initial condition the neighbours of susceptible nodes are interchangeable.

7.3.1 Equivalence of compact ED and EBCM models

Theorem 7.1 *The compact effective degree SIR model, system (5.43), and the EBCM model, system (6.12), are equivalent.*

This subsection proves this result by deriving each model from the other.

Deriving the compact ED model from the EBCM model

We first show that the compact effective degree model can be derived from the EBCM approach. We begin with the EBCM equations, system (6.12), assuming random introduction at $t = 0$

$$\dot{\theta} = -\tau\theta(t) + \tau(1-\rho)\frac{\psi'(\theta(t))}{\langle K \rangle} + \gamma(1-\theta),$$

$$\dot{R} = \gamma I, \qquad S = N(1-\rho)\psi(\theta), \qquad I = N - S - R.$$

where $\psi(x) = \sum_k P(k)x^k$. We augment these equations with the variables $\phi_S = (1 - \rho)\psi'(\theta)/\langle K \rangle$, $\phi_R = \gamma(1-\theta)/\tau$ and $\phi_I = \theta - \phi_S - \phi_R$. We note that $\dot{\theta} = -\tau\phi_I$, $\frac{d}{dt}(\phi_S + \phi_I) = -(\tau + \gamma)\phi_I$ and $\dot{\phi}_R = \gamma\phi_I$.

  We now write down some new variables in terms of the EBCM variables, with the definitions chosen so that each new variable should correspond to an ED variable. We then derive a system of differential equations for these new variables using the EBCM equations and show that the final system is the same as the ED system. We use K rather than k to denote the actual degree of a node in order to avoid confusion between k and κ. We define

$$X_\kappa = N(1-\rho)\sum_{K \geq \kappa} P(K)\binom{K}{\kappa}(\phi_S + \phi_I)^\kappa \phi_R^{K-\kappa},$$

$$[XY] = N\phi_I(1-\rho)\psi'(\theta),$$

$$\langle Y \rangle = \phi_I/(\phi_S + \phi_I),$$

$$X = S, \qquad Y = I, \qquad Z = R.$$

Exercise 7.1. By explaining the biological meaning of each variable, explain why we anticipate $X_\kappa = S_\kappa$, $\langle Y \rangle = \langle I \rangle$ and $[XY] = [SI]$.

  Our challenge is to find $\dot{X}_\kappa$, $\langle Y \rangle$ and $[XY]$ in terms of themselves rather than the EBCM variables and show that the system matches system (5.43) (it is obvious that the equations for Y and Z match those for I and R). We have

$$\dot{X}_\kappa = N(1-\rho)\sum_{K \geq \kappa} P(K)\binom{K}{\kappa}\frac{d}{dt}\left((\phi_S + \phi_I)^\kappa \phi_R^{K-\kappa}\right)$$

$$= N(1-\rho)\sum_{K \geq \kappa} P(K)\binom{K}{\kappa}\left[-\kappa(\tau+\gamma)\phi_I(\phi_S + \phi_I)^{\kappa-1}\phi_R^{K-\kappa} + \gamma\phi_I(K-\kappa)(\phi_S + \phi_I)^\kappa \phi_R^{K-\kappa-1}\right]$$

$$= N\frac{(1-\rho)\phi_I}{\phi_S+\phi_I}\sum_{K\geq\kappa}P(K)\left[(\tau+\gamma)\kappa\binom{K}{\kappa}(\phi_S+\phi_I)^\kappa\phi_R^{K-\kappa}+\gamma(K-\kappa)\binom{K}{\kappa}(\phi_S+\phi_I)^{\kappa+1}\phi_R^{K-\kappa-1}\right]$$

$$= \langle Y\rangle\left[-(\tau+\gamma)\kappa X_\kappa+\gamma(\kappa+1)X_{\kappa+1}\right],$$

where in the last step we used Exercise 7.2. This matches the equation for $\dot{S}_\kappa$ in the original compact ED model.

Exercise 7.2.
  a. Show that $(K-\kappa)\binom{K}{\kappa}=(\kappa+1)\binom{K}{\kappa+1}$.
  b. Use this to show that $N(1-\rho)\sum_{K\geq\kappa}P(K)(K-\kappa)\binom{K}{\kappa}(\phi_S+\phi_I)^{\kappa+1}\phi_R^{K-\kappa-1}=$
  $(\kappa+1)X_{\kappa+1}$.

We now look for an equation for $[\dot{X}Y]$.

$$[\dot{X}Y] = N(1-\rho)\frac{d}{dt}\left(\phi_I\psi'(\theta)\right)$$

$$= N(1-\rho)\psi'(\theta)\left(\frac{d}{dt}\left((\phi_S+\phi_I)-\phi_S\right)\right)+N(1-\rho)\phi_I(-\tau\phi_I)\psi''(\theta)$$

$$= N(1-\rho)\psi'(\theta)\left(-(\tau+\gamma)\phi_I+\frac{(1-\rho)}{\langle K\rangle}\tau\phi_I\psi''(\theta)\right)-N\tau(1-\rho)\phi_I^2\psi''(\theta)$$

$$= -(\tau+\gamma)[XY]+N\tau(1-\rho)\phi_S\phi_I\psi''(\theta)-N\tau(1-\rho)\phi_I^2\psi''(\theta)$$

$$= -(\tau+\gamma)[XY]+N\tau(1-\rho)\phi_I\psi''(\theta)(\phi_S-\phi_I)$$

$$= -(\tau+\gamma)[XY]+N\tau(1-\rho)\phi_I\frac{\phi_S+\phi_I}{\phi_S+\phi_I}\psi''(\theta)[\phi_S+(\phi_I-\phi_I)-\phi_I]\frac{\phi_S+\phi_I}{\phi_S+\phi_I}$$

$$= -(\tau+\gamma)[XY]+N\tau(1-\rho)\langle Y\rangle(\phi_S+\phi_I)^2\psi''(\theta)(1-2\langle Y\rangle)$$

$$= -(\tau+\gamma)[XY]+N\tau\langle Y\rangle(1-2\langle Y\rangle)(\phi_S+\phi_I)^2(1-\rho)\sum_K K(K-1)P(K)\theta^{K-2}.$$

Exercise 7.3. Using $\theta=\phi_S+\phi_I+\phi_R$:
  a. Show that $P(K)K(K-1)=\sum_\kappa\binom{K}{\kappa}\kappa(\kappa-1)P(K)(\phi_S+\phi_I)^\kappa\phi_R^{K-\kappa}$.
  b. From this, show that $N(\phi_S+\phi_I)(1-\rho)\sum_K K(K-1)P(K)\theta^{K-2}=\sum_\kappa\kappa(\kappa-1)X_\kappa$.

From Exercise 7.3, we can conclude that

$$[\dot{X}Y] = -(\tau+\gamma)[XY]+\tau\langle Y\rangle(1-2\langle Y\rangle)\sum_\kappa\kappa(\kappa-1)X_\kappa,$$

This matches the equation for $[SI]$. Now, we find the equation for $\langle Y\rangle$.

Exercise 7.4. Consider $\langle Y\rangle=\phi_I/(\phi_S+\phi_I)$. Show that $\langle Y\rangle=[XY]/\sum_\kappa S_\kappa$.

This completes the derivation of the compact ED model from the EBCM model.

Deriving the EBCM model from the compact ED model

We now go in the opposite direction, beginning with the compact ED equations, system (5.43),

$$\dot{S}_\kappa = \langle I \rangle \left[-(\tau + \gamma)\kappa S_\kappa + \gamma(\kappa + 1)S_{\kappa+1} \right],$$

$$[\dot{SI}] = -(\tau + \gamma)[SI] + \tau \left(\langle I \rangle - 2\langle I \rangle^2 \right) \sum_\kappa \kappa(\kappa - 1)S_\kappa,$$

$$\langle I \rangle = \frac{[SI]}{\sum_\kappa \kappa S_\kappa},$$

$$\dot{R} = \gamma I, \qquad S = \sum_\kappa S_\kappa, \qquad I = N - S - R.$$

For convenience, we will not introduce new variable names for the remainder of these derivations. Conceptually, we are starting with one system, defining new variables in terms of the system and then showing that the newly defined variables satisfy the appropriate system of equations.

We will start deriving the EBCM equations by finding an expression we will define to be ϕ_I. We take the equation for $[SI]$ and modify it to remove the term corresponding to infection of the susceptible node in an edge from a source outside the edge. To help this, we write

$$[\dot{SI}] = -(\tau + \gamma)[SI] + \tau \left(\langle I \rangle - 2\langle I \rangle^2 \right) \sum_\kappa \kappa(\kappa - 1)S_\kappa$$

$$= (-\tau + \gamma)[SI] + \tau \langle I \rangle (1 - \langle I \rangle) \sum_\kappa \kappa(\kappa - 1)S_\kappa - \tau \langle I \rangle^2 \sum_\kappa \kappa(\kappa - 1)S_\kappa$$

$$= (-\tau + \gamma)[SI] + \tau \langle I \rangle (1 - \langle I \rangle) \sum_\kappa \kappa(\kappa - 1)S_\kappa - \tau \langle I \rangle \frac{\sum_\kappa \kappa(\kappa - 1)S_\kappa}{\sum_\kappa \kappa S_\kappa}[SI].$$

$$(7.1)$$

The first term represents recovery of the infected node and transmission within the edge. The second term represents new $[SI]$ edges created when the second node in an $[SS]$ pair is infected. The final term represents infection of the susceptible node of an $[SI]$ pair from a different neighbour. It is this final term we want to eliminate. We will do this by using an integrating factor. Define $F(t)$ such that $F'(t) = \tau \langle I \rangle \frac{\sum_\kappa \kappa(\kappa-1)S_\kappa}{\sum_\kappa \kappa S_\kappa}$. There is an arbitrary constant of integration to be determined later. We define $\phi_I = e^{F(t)}[SI]$. We note that the natural definition of $[SS]$ would give $\sum_\kappa \kappa S_\kappa - [SI]$, so we define $\phi_S = e^{F(t)} \left(\sum_\kappa \kappa S_\kappa - [SI] \right)$. Note that $\langle I \rangle = \phi_I / (\phi_S + \phi_I)$ and $\phi_I + \phi_S = e^{F(t)} \sum_\kappa \kappa S_\kappa$.

Multiplying equation (7.1) by $e^{F(t)}$ and using $\dot{\phi}_I = F'e^{F(t)}[SI] + e^{F(t)}[\dot{SI}]$ yields

$$\dot{\phi}_I = -(\tau + \gamma)\phi_I + \tau \langle I \rangle \phi_S \frac{\sum_\kappa \kappa(\kappa - 1)S_\kappa}{\sum_\kappa \kappa S_\kappa}$$

$$= -(\tau + \gamma)\phi_I + \phi_S F'(t).$$

We have a single equation, but with two variables ϕ_I and ϕ_S. We now solve for $\dot{\phi}_I + \dot{\phi}_S$, which yields ϕ_S. We start with $\sum_\kappa \kappa \dot{S}_\kappa$

$$\frac{d}{dt} \sum_\kappa \kappa S_\kappa = \sum_\kappa \kappa \langle I \rangle \left[-(\tau + \gamma)\kappa S_\kappa + \gamma(\kappa + 1)S_{\kappa+1} \right] \tag{7.2a}$$

$$= \gamma \langle I \rangle \sum_\kappa \left((\kappa + 1)\kappa S_{\kappa+1} - \kappa^2 S_\kappa \right)$$

$$- \tau \langle I \rangle \frac{\sum_\kappa \kappa(\kappa - 1)S_\kappa}{\sum_\kappa \kappa S_\kappa} \sum_\kappa \kappa S_\kappa - \tau \langle I \rangle \sum_\kappa \kappa S_\kappa . \tag{7.2b}$$

Exercise 7.5.
  a. Fill in the steps between equation (7.2a) and (7.2b).
  b. Show also that $\sum_\kappa \left((\kappa + 1)\kappa S_{\kappa+1} - \kappa^2 S_\kappa \right) = -\sum_\kappa (\kappa + 1)S_{\kappa+1}$.

Now again using the integrating factor $e^{F(t)}$, we can absorb the penultimate term on the right-hand side into the left-hand side, leading to

$$\frac{d}{dt}(\phi_I + \phi_S) = \gamma \langle I \rangle \sum_\kappa (\kappa + 1)S_{\kappa+1}e^{F(t)} - \tau \sum_\kappa \kappa S_\kappa e^{F(t)}$$

$$= -\gamma \langle I \rangle (\phi_I + \phi_S) - \tau \langle I \rangle (\phi_I + \phi_S) .$$

Because $\langle I \rangle = \phi_I / (\phi_I + \phi_S)$, this gives $\frac{d}{dt}(\phi_I + \phi_S) = -(\gamma + \tau)\phi_I$. Combining this with our expression for $\dot{\phi}_I$, we can conclude that

$$\dot{\phi}_S = -\phi_S F'(t) ,$$

$$\dot{\phi}_I = -(\tau + \gamma)\phi_I + \phi_S F'(t) .$$

We define the variables θ and ϕ_R, where $\theta(0) = 1$, $\dot{\theta} = -\tau \phi_I$ and $\phi_R = \theta - \phi_S - \phi_I$. It follows that $\dot{\phi}_R = \gamma \phi_I$. So far, ϕ_S and ϕ_I are known up to a multiplicative constant due to the constant of integration in F. We can arbitrarily set $\phi_S(0) + \phi_I(0) = 1$ so that $\phi_R(0) = 0$, but any value in $[0, 1]$ is acceptable to model the presence of already recovered neighbours of random nodes at $t = 0$.

Exercise 7.6. This exercise continues the derivation of the EBCM equations. We need to show that $S = N(1 - \rho)\psi(\theta)$, where θ satisfies the equations we have derived. Because we assume that a fraction ρ are infected at time 0, $\phi_R(0) = 0$, and so F is chosen so that $\phi_S(0) + \phi_I(0) = 1$. Define $X_\kappa = N(1 - \rho)\sum_{K \geq \kappa} \binom{K}{\kappa}(\phi_S + \phi_I)^\kappa \phi_R^{K-\kappa}$.
  a. Explain why $S_\kappa(0) = N(1 - \rho)P(\kappa)$.
  b. Show that $\dot{X}_\kappa = \langle I \rangle \left[-(\tau + \gamma)\kappa X_\kappa + \gamma(\kappa + 1)X_{\kappa+1} \right]$.
  c. Show that $X_\kappa(0) = S_\kappa(0)$ and thus prove that $X_\kappa = S_\kappa$.

If $\phi_R(0) \neq 0$, the same result holds, but we do not show the steps. We now take $\psi(\theta) = \sum_K P(K)\theta^K$. Then,

$$\sum_\kappa S_\kappa = N \sum_K \sum_\kappa P(K)(1-\rho)\binom{K}{\kappa}\phi_R^{K-\kappa}(\phi_S + \phi_I)^\kappa$$
$$= N(1-\rho)\sum_K P(K)(\phi_S + \phi_I + \phi_R)^K$$
$$= N(1-\rho)\psi(\theta).$$

As a final technical detail, we find $F'(t)$ in terms of the EBCM variables. This is straightforward using the identities of Lemma 5.5.

Exercise 7.7. Write

$$F'(t) = \tau\langle I\rangle \frac{N(1-\rho)\sum_K \sum_\kappa \kappa(\kappa-1)P(K)\binom{K}{\kappa}(\phi_S + \phi_I)^\kappa \phi_R^{K-\kappa}}{N(1-\rho)\sum_K \sum_\kappa \kappa P(K)\binom{K}{\kappa}(\phi_S + \phi_I)^\kappa \phi_R^{K-\kappa}}.$$

Using the identities of Lemma 5.5, show

$$F'(t) = \tau\phi_I \frac{\psi''(\theta)}{\psi'(\theta)}.$$

Comment: We can interpret $\psi''(\theta)/\psi'(\theta)$ as the expected "excess degree", or the expected number of edges of a susceptible neighbour ignoring the edge from which it was reached.

Thus, we finally have

$$\dot{\phi}_S = -\phi_S F'(t) = -\phi_S \tau\phi_I \frac{\psi''(\theta)}{\psi'(\theta)},$$

and the ansatz $\phi_S(t) = (1-\rho)\psi'(\theta)/\langle K\rangle$ is a solution to this equation with appropriate initial condition. This completes the derivation of the EBCM equations starting from the compact ED model.

7.3.2 Equivalence of compact PW and EBCM model

Theorem 7.2 *The compact pairwise SIR model, system (5.19), and the EBCM model, system (6.12), are equivalent.*

Deriving the compact PW model from the EBCM model

We assume that the EBCM equations hold and define some quantities in terms of the EBCM variables.

$$[S_k] = N(1-\rho)P(k)\theta^k,$$
$$[SS] = N(1-\rho)\psi'(\theta)\phi_S,$$
$$[SI] = N(1-\rho)\psi'(\theta)\phi_I.$$

We expect these new quantities to match the corresponding compact PW model variables of system (5.19), and we will show that they satisfy the same system of equations.

| **Exercise 7.8.** Explain why these relationships for $[S_k]$, $[SS]$, and $[SI]$ should hold.

We define $[SX] = \sum_k k[S_k]$ and solve for $[SX]$ in terms of θ.

$$[SX] = \sum_{k=1}^{M} k[S_k] = N(1-\rho)\sum_{k=1}^{M} kP(k)\theta^k = N(1-\rho)\theta\psi'(\theta),$$

$$Q = \frac{1}{[SX]^2}\sum_{k=1}^{M}(k-1)k[S_k] = \frac{N(1-\rho)\sum_{k=1}^{M}k(k-1)P(k)\theta^k}{(N(1-\rho)\theta\psi'(\theta))^2}$$

$$= \frac{\psi''(\theta)}{N(1-\rho)\psi'(\theta)^2}.$$

It is straightforward to see that

$$[\dot{S}_k] = N(1-\rho)P(k)k\dot{\theta}\theta^{k-1} = -N(1-\rho)P(k)k\tau\phi_I\theta^{k-1}$$

$$= -N(1-\rho)P(k)\tau k\frac{\theta^k}{\theta}\phi_I\frac{N(1-\rho)\psi'(\theta)}{N(1-\rho)\psi'(\theta)} = -\tau k[S_k]\frac{[SI]}{[SX]}.$$

Similarly, taking $\dot{\phi}_I = -(\tau+\gamma)\phi_I - \dot{\phi}_S$ and using $\dot{\phi}_S = -\tau\phi_S\phi_I\psi''(\theta)/\psi'(\theta)$ we get

$$[\dot{SI}] = N(1-\rho)\dot{\theta}\psi''(\theta)\phi_I + N(1-\rho)\psi'(\theta)\dot{\phi}_I$$

$$= -N(1-\rho)\tau\phi_I^2\psi''(\theta) + N(1-\rho)\psi'(\theta)\left(-(\tau+\gamma)\phi_I + \tau\phi_S\phi_I\frac{\psi''(\theta)}{\psi'(\theta)}\right)$$

$$= -N(1-\rho)(\tau+\gamma)\phi_I\psi'(\theta) - \frac{N^2(1-\rho)^2}{N(1-\rho)}\tau\phi_I^2\psi''(\theta)\left(\frac{\psi'(\theta)}{\psi'(\theta)}\right)^2$$

$$+ \frac{N^2(1-\rho)^2}{N(1-\rho)}\tau\phi_I\phi_S\psi''(\theta)\left(\frac{\psi'(\theta)}{\psi'(\theta)}\right)^2$$

$$= -(\tau+\gamma)[SI] - \tau[SI]^2 Q + \tau[SI][SS]Q,$$

which matches the earlier expression.

| **Exercise 7.9.** Show that the derivative of $[SS]$ when expressed in terms of the EBCM variables is equivalent to the compact PW equations.

Deriving EBCM model from compact PW model

We have shown that the super-compact PW model, system (5.22), can be derived from the compact PW model. This looks similar to the EBCM equations.

Exercise 7.10. By writing $\phi_S = [SS]/N(1-\rho)\psi'(\theta)$ and $\phi_I = [SI]/N(1-\rho)\psi'(\theta)$, derive the EBCM equations from the super-compact PW model.

This completes the task of showing that the EBCM equations and compact PW equations are equivalent. Combined with our earlier result, the compact ED equations are also equivalent.

7.4 Limiting approximations

We consider two important limits of the EBCM equations. In the first, $\langle K \rangle$ increases, holding $\tau \langle K \rangle$ fixed. The number of edges transmitting twice goes down as τ decreases, but the increase in $\langle K \rangle$ keeps the total number of transmissions per node at a similar level. We anticipate that if the disease never crosses the same edge twice, the disease cannot know that the network is static rather than rapidly changing. So the HetMF model should apply even if the network is static.

In the second limit, we additionally want to study what happens if the degrees are all similar. We expect the HomMF model to arise.

7.4.1 Reduction of EBCM model to HetMF model

Theorem 7.3 *Given a sequence of degree distributions and transmission rates for which $\langle K \rangle \to \infty$, with $\hat{\tau} = \tau \langle K \rangle$ fixed, the EBCM model of system (6.12) reduces to the HetMF model (5.11).*

We assume that we have a sequence of networks with $\langle K \rangle \to \infty$ and $\hat{\tau} = \tau \langle K \rangle$ fixed, so that infectiousness scales inversely with average degree. We have

$$\dot{\theta}_c = -\tau \theta_c + \tau (1 - \rho) \frac{\psi'(\theta_c)}{\langle K \rangle} + \gamma (1 - \theta_c),$$

while for the HetMF model we use Exercise 5.1 and replace $\hat{\psi}'(\theta)$ by $(1-\rho)\psi'(\theta)$ to get

$$\dot{\theta}_f = -\tau \theta_f + \tau (1 - \rho) \theta_f^2 \frac{\psi'(\theta_f)}{\langle K \rangle} - \gamma \theta_f \ln(\theta_f).$$

These equations are similar, and we will show that under appropriate conditions they have the same limiting behaviour. As a first observation, we note that both θ_f and θ_c lie between 0 and 1. In fact, further analysis shows that $(1-\theta_f)$ and $(1-\theta_c)$ are both $\mathcal{O}(1/\langle K \rangle)$. We "add zero" to the EBCM equation for θ_c to get

$$\dot{\theta}_c = -\tau\theta_c + \tau(1-\rho)\theta_c^2\frac{\psi'(\theta_c)}{\langle K \rangle} - \gamma\theta_c\ln(\theta_c)$$

$$+ \tau(1-\rho)(1-\theta_c^2)\frac{\psi'(\theta_c)}{\langle K \rangle} + \gamma(1-\theta_c+\theta_c\ln(\theta_c)).$$

We want to show that the last two terms are $\mathcal{O}\left(1/\langle K \rangle^2\right)$. We Taylor expand $\theta_c\ln(\theta_c)$ about 1, and then in the final two terms we write $\theta_c = 1 - r/\langle K \rangle$. We find

$$\dot{\theta}_c = -\tau\theta_c + \tau(1-\rho)\theta_c^2\frac{\psi'(\theta_c)}{\langle K \rangle} - \gamma\theta_c\ln(\theta_c) + \tau(1-\rho)(1-\theta_c^2)\frac{\psi'(\theta_c)}{\langle K \rangle}$$

$$+ \gamma\left(1-\theta_c+1\ln 1+(\theta_c-1)(1/1+\ln 1)+\frac{(\theta_c-1)^2}{2}+\cdots\right)$$

$$= -\tau\theta_c + \tau(1-\rho)\theta_c^2\frac{\psi'(\theta_c)}{\langle K \rangle} - \gamma\theta_c\ln(\theta_c)$$

$$+ \frac{\hat{\tau}}{\langle K \rangle}(1-\rho)\left(\frac{2r}{\langle K \rangle}-\frac{r^2}{\langle K \rangle^2}\right)\frac{\psi'(\theta_c)}{\langle K \rangle} + \gamma\left(\frac{r^2}{2\langle K \rangle^2}\right)+\cdots.$$

Because $\psi'(\theta_c)/\langle K \rangle < 1$, we conclude that

$$\dot{\theta}_c = -\tau\theta_c + \tau(1-\rho)\theta_c^2\frac{\psi'(\theta_c)}{\langle K \rangle} - \gamma\theta_c\ln(\theta_c) + \mathcal{O}\left(1/\langle K \rangle^2\right),$$

and so up to (and including) order $1/\langle K \rangle$, we recover the HetMF equations, with deviation occurring only at order $1/\langle K \rangle^2$. This means that we can reasonably use the HetMF model if $\langle K \rangle$ is large and τ is small.

7.4.2 Reduction of HetMF model to HomMF model

In this section, we show that the HetMF model reduces to the HomMF model as the population becomes homogeneous.

Exercise 7.11. We can explore the homogeneous SIR equations in more detail.

a. Consider system (4.9), assuming that a proportion ρ is instantaneously infected at $t = 0$. Following the steps in the derivation of system (5.11) and Exercise 5.1, write $[S] = S(0)\theta_h^n$ and derive the equation

$$\dot{\theta}_h = -\tau\theta_h + \tau(1-\rho)\theta_h^{n+1} - \gamma\theta_h\ln(\theta_h),$$

where n is the degree.

b. Alternately, write $\xi(t) = \tau(n/N)\int_0^t [I]_f(\hat{t})\,d\hat{t}$ and consider the equation for $[S]_h$, using an integrating factor $e^{-\xi(t)}$.

    i. Write $[\dot{R}]_h$ as $c\dot{\xi}$ (find c).

    ii. Write $[S]_h$ as a function of $S(0)$ and $\xi(t)$.

    iii. Taking $[I]_h = N - [S]_h - [R]_h$ and differentiating $\xi(t)$, write down a complete system of equations for $[S]_h$, $[I]_h$, $[R]_h$ and ξ that involves just a single differential equation.

    iv. What is the mathematical relationship between ξ and θ_h?

We have two similar theorems, depending on whether $\langle K \rangle \to \langle K \rangle_0$ or $\langle K \rangle \to \infty$.

Theorem 7.4 *Given a sequence of degree distributions and transmission rates such that*

- $\langle K \rangle$ *approaches a constant* $\langle K \rangle_0$
- $\langle K^2 \rangle$ *approaches* $\langle K \rangle_0^2$ *(the variance is zero), and*
- $\langle K \rangle \tau$ *remains constant,*

the HetMF model approaches the HomMF model.

Theorem 7.5 *Given a sequence of degree distributions and transmission rates such that*

- $\langle K \rangle \to \infty$,
- $\langle K^2 \rangle / \langle K \rangle^2 \to 1 + \mathcal{O}(1/\langle K \rangle)$, *while*
- $\langle K \rangle \tau$ *remains constant, and*
- $\langle K^n \rangle / \langle K \rangle^n$ *are uniformly bounded,*

the HetMF model approaches the HomMF model.

To prove each, we start from the HetMF model.

$$\dot{\theta}_f = -\tau\theta_f + \tau(1-\rho)\theta_f^2\frac{\psi'(\theta_f)}{\langle K \rangle} - \gamma\theta_f\ln(\theta_f).$$

The HomMF equations are the same equations with $\psi(x) = x^{\langle K \rangle}$:

$$\dot{\theta}_h = -\tau\theta_h + \tau(1-\rho)\theta_h^{\langle K \rangle+1} - \gamma\theta_h\ln(\theta_h),$$

where we use h rather than f as a subscript to avoid repetition. Our challenge is to show that $\tau(1-\rho)\theta_f^2\frac{\psi'(\theta_f)}{\langle K \rangle}$ is well approximated by $\tau(1-\rho)\theta_h^{\langle K \rangle+1}$.

For our first theorem, where $\langle K \rangle_0$ is finite, we note that $\psi(x) \to x^{\langle K \rangle_0}$. This is because if a distribution has finite mean $\langle K \rangle_0$ and zero variance, then the only value in the distribution is $\langle K \rangle_0$. In this case, the result is straightforward.

It is more difficult if $\langle K \rangle \to \infty$. Here, the distribution does not have to take only a single value, but rather it is concentrated around a growing average. We have

$$\dot{\theta}_f = -\tau\theta_f + \tau(1-\rho)\theta_f^{\langle K \rangle + 1} - \gamma\theta_f\ln(\theta_f)$$
$$+ \tau(1-\rho)\theta_f\left(\theta_f\frac{\psi'(\theta_f)}{\langle K \rangle} - \theta_f^{\langle K \rangle}\right).$$

As $\langle K \rangle \to \infty$, we find that $\dot{\theta}_f \to 0$ like $1/\langle K \rangle$, so we write $\theta_f(t) = 1 - r(t)/\langle K \rangle + \mathcal{O}\left(1/\langle K \rangle^2\right)$. Our challenge is to show that as $\langle K \rangle \to \infty$, the final term $\tau(1 - \rho)\theta_f\left(\theta_f\psi'(\theta_f)/\langle K \rangle - \theta_f\right)$ goes to zero like $1/\langle K \rangle^2$, and we have written it in a way to make this easier.

We note that $\tau = \hat{\tau}/\langle K \rangle$, where $\hat{\tau}$ is constant. So we must show that

$$\left(\theta_f\frac{\psi'(\theta_f)}{\langle K \rangle} - \theta_f^{\langle K \rangle}\right)$$

goes to 0 like $1/\langle K \rangle$. We Taylor expand this about $\theta = 1$, getting

$$\left[1\frac{\psi'(1)}{\langle K \rangle} - \frac{r}{\langle K \rangle}\left(1\frac{\psi'(1)}{\langle K \rangle} + 1\frac{\psi''(1)}{\langle K \rangle}\right) + \cdots\right] - \left[1 - \frac{r}{\langle K \rangle}\langle K \rangle + \cdots\right].$$

Noting that $\psi'(1) = \langle K \rangle$ and by assumption $\psi''(1)/\langle K \rangle^2 = 1 + \mathcal{O}(1/\langle K \rangle)$, we find that this is $\mathcal{O}(1/\langle K \rangle)$. This establishes that up to $\mathcal{O}\left(1/\langle K \rangle^2\right)$, θ_f and θ_h coincide. Similar steps show that S, I and R are the same for both models up to $\mathcal{O}(N/\langle K \rangle)$. A consequence of this result is that the EBCM equations for a degree distribution with $\langle K^2 \rangle/\langle K \rangle^2 \approx 1$ and $\langle K \rangle$ large are effectively identical to the HomMF model.

7.5 Conclusions and outlook

We have seen that the spread of SIR disease in Configuration Model networks is well-described by the EBCM, ED, and PW models, with the EBCM model containing significantly fewer equations. The distinction between the models comes in differences in the initial conditions. As time progresses, we expect the epidemic to "forget" the specific details of its initial conditions, so we anticipate that for realistic systems shortly after introduction, any of these models would be equally appropriate. It would be interesting to study the "information content" of the PW and ED models to see how information about the initial conditions decays. Once the information about the initial condition is lost, the EBCM model would apply.

The HetMF and HomMF models make assumptions that neighbours of all nodes are interchangeable. This is equivalent to assuming that all edges are extremely short-lived. A node connects to a new neighbour, interacts and then replaces the neighbour in an almost negligibly short period of time. This assumption is sometimes referred to as the "annealed" network limit, which is contrasted with the "quenched" (or frozen, unchanging) network. We have shown that in fact the annealed network arises as a natural limit of the quenched network when the probability that an edge would ever transmit twice goes to zero. This can be understood by imagining the disease learning about the network as it spreads. If it never attempts to transmit across an edge twice, it never has the ability to learn whether the network is "quenched" or not.

In fact, it is possible to interpolate between the quenched and the annealed network to allow finite, non-zero edge duration, which appears in Chapter 8. Understanding the spread of disease on dynamic networks produces a number of new challenges and opens up the possibility of people altering their interactions in response to the epidemic.

We end with a comment on why we do not have models for SIS spread which are exact in the limit of large networks. To do this, we would need to be able to create classes of nodes for which it is acceptable to interchange neighbours, and the fact that an infected node alters its neighbourhood in a way that affects its future susceptibility appears to prevent this. The neighbours of nodes which are currently susceptible but were previously infected cannot be interchanged with neighbours of susceptible nodes that have never been infected. Their neighbours may have different statuses and are at different risks of future infection. This effect grows as nodes are reinfected multiple times.

This caused one of the most significant failures of simple mean-field models. Using the SIS version of the HetMF model for a network with $P(k) \sim k^{-\alpha}$ for $2 < \alpha < 3$, we would predict that $\langle K^2 \rangle = \infty$ but $\langle K \rangle < \infty$. Then, for any positive τ, we have $\langle K^2 \rangle \tau / \langle K \rangle \gamma = \infty$. So $\tau_c = 0$: any positive value of τ leads to a possible epidemic. If $\alpha > 3$, however, $\langle K^2 \rangle < \infty$, and so epidemics are only possible for τ above a positive threshold $\tau_c > 0$. The prediction based on the SIS HetMF approximation turns out to be wrong. It can be proven rigorously that for a Configuration Model network with $P(k) \sim k^{-\alpha}$, even if $\alpha > 3$, epidemics are possible for any τ. The error is because once a high-degree node is infected, it will infect many neighbours, which will reinfect it once it recovers. This leads to an island of infection which persists long enough for infection to spread from the island to other high-degree nodes where the process repeats [63, 88].

We expect the ED model to outperform the others for SIS disease because it captures the fact that when an infected node with many infected neighbours recovers, it becomes a susceptible node with many infected neighbours. However, even in the ED case, this does not fully resolve the problem.

Chapter 8
Dynamic and adaptive networks

An important feature of many real-world networks is the transient nature of some interactions. Thus far, our models have explicitly assumed that the network is static. That is, we assume that the rate of partner turnover is so slow that we can ignore its impact on epidemic dynamics.

Over the past decade, there has been tremendous progress in modelling and analysing disease spread in non-static networks, i.e. networks whose structure changes due to endogenous or exogenous factors, or because of the disease dynamics unfolding on the network. The terminology used to describe such networks is not standardised. We summarise some common terminology and link it to the relation between the time scales of the *dynamics on the network* and the *dynamics of the network*. Ordered from static to networks that change quickly, we have:

1. "Static" or "quenched" networks: links in the network do not change;
2. "Evolving" networks: the network may change appreciably over the time scale of the epidemic, but little change occurs over the time scale of a single infection. Examples often include Internet and friendship networks;
3. "Dynamic" or "adaptive" networks: the individual connections of an infected node may change considerably during the node's infectious period. Often, these changes may be in response to observed infection status;
4. "Annealed" or "fast" networks: individuals change partners so quickly that we can ignore the possibility of a single edge transmitting twice. These are well modelled by the heterogeneous mean-field models at node or single level.

Further refinement of this terminology is motivated by data-driven analysis and modelling. For example, "temporal" or "time-varying" networks [146] are used to describe cases where the order of interactions matter, often because networks have significant structural variations at different times. For example: a virus spreading through mobile networks will be affected by the time of day based on whether individuals are at home or at work, or the spread of livestock diseases is affected by seasonal animal movements [178].

This chapter focuses on dynamic and adaptive networks, where time scales of individual partnerships and individual transmissions are comparable. This leads to the

© Springer International Publishing AG 2017

I.Z. Kiss et al., *Mathematics of Epidemics on Networks*, Interdisciplinary Applied Mathematics 46, DOI 10.1007/978-3-319-50806-1_8

richest collection of possible behaviours, including bistability and oscillations. The body of research focusing on this area has grown extensively over the last decade [22, 125, 126, 128, 205, 262, 273]. The main driver of such research is provided by empirical observations, which clearly show that many real-world networks, ranging from biological and social to technological, are best described by non-static networks. The overarching observation is that in many cases, dynamics happening on networks and changes in networks are interlinked with a clear feedback loop. In many cases, the structure of networks affects unfolding of processes on networks but, equally, the evolution of networks is partly determined by the statuses of the nodes, which are driven by the dynamics on networks (see Fig. 8.1).

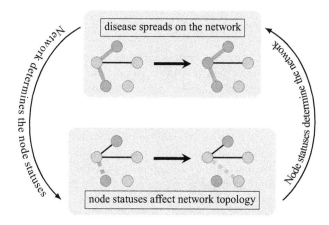

Fig. 8.1: Feedback loop showing the interaction of the dynamics on and of the network. A susceptible node (○) becomes infected (◐), as shown by the transition from the top-left to the top-right network. Changes in network structure are shown by breaking an SI link and creating an SS link instead, shown by the change from the bottom-left to the bottom-right network.

The chapter is divided according to the type of link dynamics, and it focuses on SIS dynamics with (a) contact-conserving rewiring, (b) random link activation and deletion, (c) link-status-dependent activation and deletion, and (d) link deactivation and activation on a fixed network. For these scenarios, we offer and analyse pairwise models, effective degree models or both, with an emphasis on the stability of the disease-free steady state. Finally, we look at SIR disease in a dynamic network in which individuals preserve their degree, and we use an edge-based compartmental modelling approach. This is a rapidly evolving field, so compared to previous chapters, we provide less detail, and instead try to give an overview of different modelling options.

8.1 Link-conserving rewiring models

One of the very first dynamic network models [128] used the pairwise modelling framework for adaptive networks. In this model, S nodes break their links to I nodes and reconnect to a randomly chosen susceptible node from the population (see Fig. 8.2). As a result, the number of edges and the average degree remain constant.

Fig. 8.2: Illustration of link-conserving rewiring: susceptible nodes break links to infectious nodes followed by their instantaneous reconnection to randomly chosen susceptible nodes.

8.1.1 Pairwise model

For the SIS epidemic with per-contact transmission rate τ, recovery rate γ and a rewiring rate ω, the pairwise model (4.3) can be extended to the system below.

SIS pairwise model with contact-conserving rewiring

$$[\dot{S}] = \gamma[I] - \tau[SI], \tag{8.1a}$$

$$[\dot{I}] = \tau[SI] - \gamma[I], \tag{8.1b}$$

$$[\dot{SI}] = -(\tau + \gamma)[SI] + \tau([SSI] - [ISI]) + \gamma[II] \underbrace{-\omega[SI]}_{\text{loss due to rewiring}}, \tag{8.1c}$$

$$[\dot{II}] = -2\gamma[II] + 2\tau([ISI] + [SI]), \tag{8.1d}$$

$$[\dot{SS}] = 2\gamma[SI] - 2\tau[SSI] \underbrace{+2\omega[SI]}_{\text{gain due to rewiring}}. \tag{8.1e}$$

Following [128], we take the closures $[SSI] = \frac{[SS][SI]}{[S]}$ and $[ISI] = \frac{[SI][SI]}{[S]}$.

We note that the closure above, as used in [128], does not involve the average number of stubs starting from a susceptible node, n_S, as typically used for closing pairwise models. However, here the original model is followed and the potential discrepancies of this approach are not discussed. To help simplify our analysis, we can eliminate the equations for $[S]$ and $[II]$ using conservation of the number of nodes and edges. We have $[S] = N - [I]$ and $[II] = N\langle K \rangle - [SS] - 2[SI]$. So we can consider the three-dimensional system with $[I]$, $[SI]$, and $[SS]$. The governing equations are

$$[\dot{I}] = \tau[SI] - \gamma[I],$$

$$[\dot{SI}] = -(\tau + \gamma)[SI] + \tau\left(\frac{[SS][SI]}{N - [I]} - \frac{[SI]^2}{N - [I]}\right) + \gamma(N\langle K \rangle - [SS] - 2[SI]) - \omega[SI],$$

$$[\dot{SS}] = 2\gamma[SI] - 2\tau\frac{[SS][SI]}{N - [I]} + 2\omega[SI].$$

The disease-free steady state is $([I], [SI], [SS]) = (0, 0, N\langle K \rangle)$. We will find the endemic equilibrium in terms of $[I]_{\mathrm{eq}}$. Because $[\dot{I}] = 0$ at equilibrium, we have

$$[SI]_{\mathrm{eq}} = \frac{\gamma[I]_{\mathrm{eq}}}{\tau}.$$

From $[\dot{SS}] = 0$, we conclude that

$$[SS]_{\mathrm{eq}} = \frac{(\gamma + \omega)(N - [I]_{\mathrm{eq}})}{\tau}.$$

Plugging these into $[\dot{SI}] = 0$, we have

$$b[I]_{\mathrm{eq}}^2 - (a + b)N[I]_{\mathrm{eq}} + (a - \gamma)N^2 = 0, \tag{8.2}$$

where $a = \tau\langle K \rangle - \omega$ and $b = \tau - \omega$. A careful analysis of this quadratic equation leads to the proposition below.

Proposition 8.1

1. If $\tau\langle K \rangle > \omega + \gamma$, then a single endemic steady state exists;
2. If $\tau\langle K \rangle < \omega + \gamma$, $\tau < \omega$, $\tau(\langle K \rangle + 1) < 2\omega$ and $\tau^2(\langle K \rangle - 1)^2 + 4(\tau - \omega)\gamma > 0$, then two endemic steady states exist;
3. If $\tau\langle K \rangle < \omega + \gamma$ but one of the three additional conditions at item 2 does not hold, then no endemic steady state exists.

The critical line $\omega = \tau\langle K \rangle - \gamma$ *(transcritical bifurcation) and parabola* $\omega = \frac{\tau^2(\langle K \rangle - 1)^2 + 4\tau\gamma}{4\gamma}$ *(saddle-node bifurcation) are represented in Fig. 8.3a by the dash-dotted and dotted line, respectively.*

The Jacobian at an arbitrary point evaluates to

$$J = \begin{pmatrix} -\gamma & \tau & 0 \\ \tau\frac{([SS] - [SI])[SI]}{(N - [I])^2} & -\tau - 3\gamma + \tau\frac{[SS]}{N-[I]} - 2\tau\frac{[SI]}{N-[I]} - \omega & \tau\frac{[SI]}{N-[I]} - \gamma \\ -2\tau\frac{[SS][SI]}{(N-[I])^2} & 2\gamma - 2\tau\frac{[SS]}{N-[I]} + 2\omega & -2\tau\frac{[SI]}{N-[I]} \end{pmatrix}. \tag{8.3}$$

In the disease-free steady state, $[I] = [SI] = 0$ and $[SS] = N\langle K \rangle$, so J simplifies significantly. One eigenvalue is $\lambda_1 = -\gamma$, and the other two solve

$$\lambda^2 + (3\gamma + \omega - (\langle K \rangle - 1)\tau)\lambda + 2\gamma(\gamma + \omega - \tau\langle K \rangle) = 0. \qquad (8.4)$$

If $\gamma + \omega - \tau\langle K \rangle > 0$, then λ_2 and λ_3 are both negative with the disease-free state being stable. On the other hand, if $\gamma + \omega - \tau\langle K \rangle < 0$, then $\lambda_2\lambda_3 < 0$, which implies that this steady state loses stability.

The coupling between disease and network dynamics leads to a richer spectrum of behaviour, including "backward" transcritical bifurcation with bistability and oscillations (see Fig. 8.3). Furthermore, concerning the steady states, the following scenarios are possible: (a) the disease-free steady state is stable and no other plausible steady states exist, (b) the disease-free steady state loses stability and the unique endemic steady state is stable and, finally, (c) two non-trivial steady states exist and the parameter space corresponding to this regime divides further into two, according to their stability. The analysis of the original model by Gross has been refined in different ways by either a more detailed analysis of the bifurcation picture [158] or by considering alternative rewiring mechanisms [262]. It is worth noting that in this original model, oscillations are observed over a very narrow region, and further studies reveal that numerical closures (involving counts taken from the simulation), rather than the current ones, give rise to more robust oscillations [127]. Oscillatory behaviour is also studied in [269, 295]. In [269], an SIR model with smart rewiring, node birth and death, and link cutting and creation is considered, and oscillatory regimes are mapped out by identifying the peak spectrum and peak prevalence within the signal given by the stochastic simulation of the full network model. Simon et al. [295] consider a pairwise SIS model with link-status-dependent link activation and deletion, with full details given later in the chapter.

Exercise 8.1. Use the Jacobian at a general steady state, as given by Eq. (8.3), to work out the stability of the disease-free steady state. In particular, find all eigenvalues of $J|_{\text{DFSS}}$ and show that these have the properties stated.

8.1.2 Effective degree model

Another prominent modelling framework for epidemics on networks is based on *effective degree* models, see system (5.32), where the model focuses on evolution equations for star-like structures, denoting nodes and their neighbours in all possible configurations. Hence, if the degree distribution of the network, $P(k) = p_k$, spans degrees $1, 2, \ldots, M$, then for SIS epidemics, the system needs to account for

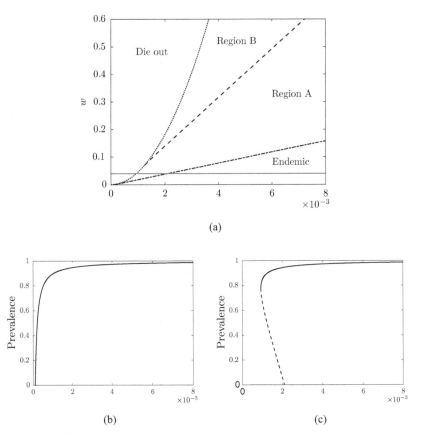

(a)

(b)   (c)

Fig. 8.3: (a) Global bifurcation diagram for $N = 10^5$ nodes, $\langle K \rangle = 20$ and $\gamma = 0.002$. The solid line corresponds to the cross section for $w = 0.04$, which is plotted in detail in panel (c). In both regions A and B, two non-zero steady states exist. In region A, one is stable and the other is not. In region B, both are unstable. Examples of a transcritical bifurcation (b) and bistability (c) are given for $w = 0$ and $w = 0.04$, respectively. The dashed line is the numerically computed Hopf bifurcation curve. For more details, see [128].

variables such as $S_{s,i}$ and $I_{s,i}$, which denote the expected number of susceptible and infectious nodes with s susceptible and i infectious neighbours, respectively. If the degree of the node is k, then $(s,i) \in \{(a, k-a) : 0 \le a \le k\}$, which gives rise to $2(k+1)$ variables for central susceptible and infected nodes. Summing over all ks leads to $2\sum_{k=1}^{M}(k+1)$ equations, which yields $M(M+3)$ different variables. For dynamic networks, where the degree of nodes changes, the number of equations could be higher, but different ways to control this number exist. As will be seen later, for random link activation deletion, it is possible to impose a limit on the maximum value that degrees can achieve, and this helps control the dimension of the system.

An important improvement on the model of system (8.1) is the corresponding effective degree model [205]. This offers a higher degree of accuracy at a finer scale. The effective degree model is presented below.

SIS effective degree model with contact-conserving rewiring

$$\dot{S}_{si} = \tau \frac{\sum_{k=0}^{M} \sum_{j+l=k} jl S_{jl}}{\sum_{k=0}^{M} \sum_{j+l=k} jS_{jl}} [(s+1)S_{s+1,i-1} - sS_{si}]$$

$$+ \omega \frac{\sum_{k=0}^{M} \sum_{j+l=k} l S_{jl}}{\sum_{k=0}^{M} \sum_{j+l=k} S_{jl}} (S_{s-1,i} - S_{s,i}) \qquad (8.5a)$$

$$+ \gamma I_{si} - \tau i S_{si} + \gamma[(i+1)S_{s-1,i+1} - iS_{si}] + \omega(i+1)S_{s,i+1} - \omega i S_{si},$$

$$\dot{I}_{si} = \tau \frac{\sum_{k=1}^{M} \sum_{j+l=k} l^2 S_{jl}}{\sum_{k=1}^{M} \sum_{j+l=k} jl I_{jl}} [(s+1)I_{s+1,i-1} - sI_{si}] \qquad (8.5b)$$

$$- \gamma I_{si} + \tau i S_{si} + \gamma[(i+1)I_{s-1,i+1} - iI_{si}] + \omega[(s+1)I_{s+1,i} - sI_{si}].$$

This system extends the effective degree SIS model on static networks, system (5.36). Rewiring is accounted for by redistributing the total rate of cutting SI links, i.e. $-\omega \sum_{k=0}^{M} \sum_{j+l=k} l S_{jl}$, towards the creation of new SS links. This is done by uniformly spreading the total rate of cutting over all susceptible nodes that will subsequently gain new susceptible neighbours. This is accounted for by the link creation rate

$$\omega \frac{\sum_{k=0}^{M} \sum_{j+l=k} l S_{jl}}{\sum_{k=0}^{M} \sum_{j+l=k} S_{jl}} = \omega \frac{[SI]}{[S]},$$

which when summed across all susceptible nodes yields $\omega[SI]$, and thus the number of edges is conserved.

The major outcome of this model is a more accurate characterisation of the system behaviour, such as being able to track the number of nodes of different degrees, how the neighbourhood of a node evolves or how correlation between different node types develops. Details are provided in [205], where the effective degree model shows excellent agreement with simulation. An added benefit of this more complex model is that it can handle networks with a relatively low average degree, where the original pairwise model fails to accurately capture the endemic steady state. While analytical results from such a model are difficult to obtain, it provides a valuable tool to better understand the interplay of the two processes.

8.2 Random link activation and deletion models

We continue with another simple dynamic network. Any existing edge is deleted at random according to a Poisson process with rate ω. Any pair of nodes without an edge is joined following a Poisson process with rate α. These rates are independent of the nodes' statuses. Such networks naturally couple with SIS dynamics, and the resulting pairwise and effective degree models are presented below.

However, focusing on the expected number of edges alone, $E(t)$ (counting each edge in both directions), we predict $E(t)$ to evolve as

$$\dot{E}(t) = \alpha(N(N-1) - E(t)) - \omega E(t), \tag{8.6}$$

with the steady state given by $E_{\text{eq}} = \frac{\alpha N(N-1)}{\alpha + \omega}$, where the equilibrium average degree is then $\langle K \rangle_{\text{eq}} = \frac{\alpha(N-1)}{\alpha + \omega}$. Alternatively, we can derive this by noting that at equilibrium a random pair of nodes is joined with probability $\alpha/(\alpha + \omega)$.

8.2.1 Pairwise model

We can extend the pairwise models in system (4.10) to account for a range of link dynamics combined with disease dynamics. We begin with random link activation and deletion. The corresponding pairwise equations for the SIS model are as follows.

SIS pairwise model with random link activation and deletion

$$[\dot{S}] = \gamma[I] - \tau[SI], \tag{8.7a}$$

$$[\dot{I}] = \tau[SI] - \gamma[I], \tag{8.7b}$$

$$[\dot{SI}] = -(\tau + \gamma)[SI] + \tau([SSI] - [ISI]) + \gamma[II]$$
$$+ \underbrace{\alpha([S][I] - [SI])}_{\text{link activation}} \underbrace{-\omega[SI]}_{\text{link deletion}}, \tag{8.7c}$$

$$[\dot{II}] = -2\gamma[II] + 2\tau([ISI] + [SI]) + \underbrace{\alpha([I]([I]-1) - [II])}_{\text{link activation}} \underbrace{-\omega[II]}_{\text{link deletion}}, \tag{8.7d}$$

$$[\dot{SS}] = 2\gamma[SI] - 2\tau[SSI] + \underbrace{\alpha([S]([S]-1) - [SS])}_{\text{link activation}} \underbrace{-\omega[SS]}_{\text{link deletion}}, \tag{8.7e}$$

with the closures

$$[SSI] = \frac{(n_S - 1)}{n_S} \frac{[SS][SI]}{[S]} \quad \text{and} \quad [ISI] = \frac{(n_S - 1)}{n_S} \frac{[SI][SI]}{[S]}, \tag{8.8}$$

employed to generate a solvable self-consistent system, where $n_S(t) = ([SS] + [SI])/[S]$ is the average degree of susceptible nodes.

As before, τ and γ represent the per-contact transmission and recovery rates, respectively. Here, α and ω denote the link activation and deletion rates, respectively.

Terms such as $\alpha([S][I] - [SI])$ and $\omega[SI]$ account for the activation and deletion of SI links, where $([S][I] - [SI])$ is the number of potential SI pairs that are not yet connected. We can eliminate the equation for $[S]$ using conservation of nodes: $[S] = N - [I]$.

Steady states and their stability

This compact model lends itself to a rigorous bifurcation analysis, including the determination of the steady states and their stability, as well as mapping out the full spectrum of possible system behaviours. The disease-free steady state is $([I], [SI], [II], [SS]) = (0, 0, 0, N \langle K \rangle_{eq})$. The Jacobian matrix corresponding to this steady state is

$$
J = \begin{pmatrix}
-\gamma & \tau & 0 & 0 \\
\alpha N & \tau(\langle K \rangle_{eq} - 2) - \gamma - \omega - \alpha & \gamma & 0 \\
-\alpha & 2\tau & -2\gamma - \omega - \alpha & 0 \\
\alpha(1 - 2N) & 2\gamma - 2\tau(\langle K \rangle_{eq} - 1) & 0 & -\omega - \alpha
\end{pmatrix}. \tag{8.9}
$$

The equilibrium loses stability at a transcritical bifurcation when an eigenvalue passes through zero, meaning that the determinant goes through zero. Taking $\langle K \rangle_{eq} = \alpha(N-1)/(\omega + \alpha)$ and solving $\det(J) = 0$ for τ, we find that the transcritical bifurcation occurs at

$$
\tau_c = \frac{\gamma(2\gamma + \alpha + \omega)}{\alpha N + 2\gamma(\langle K \rangle_{eq} - 1)}. \tag{8.10}
$$

As expected, numerical investigation shows that for $\tau < \tau_c$ the solutions of the system tend to the disease-free steady state, while for $\tau > \tau_c$ the solutions converge (without oscillation) to the endemic steady state. These local predictions based on the Jacobian can be made rigorous.

Theorem 8.1 *Let* $\langle K \rangle_{eq} = \frac{\alpha(N-1)}{\alpha + \omega}$ *and assume that* $[SS](0) + 2[SI](0) + [II](0) = \langle K \rangle_{eq} N$ *and* $[SS](0) + [SI](0) = \langle K \rangle_{eq} [S](0)$. *In this case, system (8.7), with closures in Eq. (8.8), reduces to*

$$
[\dot{S}] = \gamma N - (\gamma + \tau \langle K \rangle_{eq})[S] + \tau[SS], \tag{8.11a}
$$

$$
[\dot{SS}] = 2(\langle K \rangle_{eq}[S] - [SS]) \left(\gamma - \frac{\tau(\langle K \rangle_{eq} - 1)}{\langle K \rangle_{eq}} \frac{[SS]}{[S]} \right)
$$
$$
+ \alpha[S]([S] - 1) - (\alpha + \omega)[SS]. \tag{8.11b}
$$

For this system, the following statements hold:

1. *If* $\tau < \tau_c$, *then the disease-free steady state is globally stable;*
2. *If* $\tau > \tau_c$, *then the disease-free steady state is unstable but an endemic steady state exists and is globally stable;*
3. *The system cannot exhibit oscillations.*

| **Exercise 8.2.** Derive Eq. (8.6) directly from system (8.7).

| **Exercise 8.3.** Derive the Jacobian given in Eq. (8.9) and the expression for the critical τ value given in Eq. (8.10).

| **Exercise 8.4.** Carry out the phase plane analysis of the two-dimensional system given in Theorem 8.1 and complete the proof of the theorem.

Limiting behaviour for fast network dynamics

It is useful to consider the behaviour of system (8.7) in the limit that the time scale of the network dynamics is fast relative to that of the epidemic. In this limit, the prevalence resulting from system (8.7) converges to the infectious prevalence resulting from the simple compartmental system

$$\dot{I}(t) = \frac{\tau \langle K \rangle_{eq} I(t)(N - I(t))}{N} - \gamma I(t). \tag{8.12}$$

We can think of $\tau \langle K \rangle_{eq}$ as the transmission rate β used widely for compartmental models [see also Eq. (1.1b)]. Figure 8.4 shows that the agreement improves with faster network dynamics.

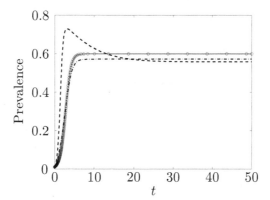

Fig. 8.4: Numerical solutions of system (8.7) for $\alpha = 0.001$ (dashed line), $\alpha = 0.01$ (dash-dotted line) and $\alpha = 1$ (∘), respectively, with $\omega = \frac{((N-1)-\langle K \rangle_{eq})\alpha}{\langle K \rangle_{eq}}$, where in this case $\langle K \rangle_{eq} = 5$. The numerical solution of Eq. (8.12) is given by the solid line. The other parameters are $\tau = 0.5$, $\gamma = 1$, the average degree at time zero is equal to 10, $[I](0) = 10$ and $N = 1000$. The initial conditions for the pairwise model are $[SI](0) = \frac{10[I](0)(N-[I](0))}{N}$, $[II](0) = \frac{10([I](0))^2}{N}$, and $[SS](0) = \frac{10(N-[I](0))^2}{N}$.

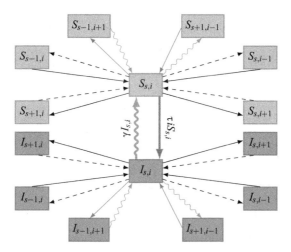

Fig. 8.5: Flow diagram showing transitions required for the SIS effective degree model. Thick coloured lines represent infection/recovery of the focal node, thin coloured lines represent infection/recovery of a neighbour; dashed lines represent removal of an edge, and solid uncoloured lines denote addition of an edge.

8.2.2 Effective degree model

The effective degree framework can be extended to this case as well, but it is important to be able to control the range of degrees that nodes can achieve since this will determine the number of equations. This can be achieved by imposing a local constraint such that no node can have more than M links. To accomplish this, we assign each node M stubs, and those stubs may break and reform into edges. This automatically bounds the number of equations from above by $2\sum_{k=0}^{M}(k+1) = (M+1)(M+2)$, which is slightly higher than for the static case, because we must allow for nodes to have no edges. In Fig. 8.5, a diagram of all possible transitions is shown, differentiating between transitions due to infection, recovery, link creation and link deletion. Based on this, the effective degree equations for the SIS model coupled with random link activation and deletion, see also [300], are as follows.

SIS effective degree model with random link activation and deletion

$$\dot{S}_{si} = -\tau i S_{si} + \gamma I_{si} + \gamma[(i+1)S_{s-1,i+1} - iS_{si}] \tag{8.13a}$$

$$+ \tau \frac{\sum_{k=0}^{M} \sum_{j+l=k} jl S_{jl}}{\sum_{k=0}^{M} \sum_{j+l=k} j S_{jl}} [(s+1)S_{s+1,i-1} - sS_{si}] \tag{8.13b}$$

$$\underbrace{-\omega[(s+i)S_{si}]}_{\text{loss via link deletion}} + \underbrace{\omega[(i+1)S_{s,i+1} + (s+1)S_{s+1,i}]}_{\text{gain via link deletion}} \tag{8.13c}$$

$$\underbrace{-\alpha(M-(s+i))S_{si}}_{\text{loss via link activation}} + \underbrace{\alpha(M-(s-1+i))P_S S_{s-1,i}}_{\text{gain via link activation}} \tag{8.13d}$$

$$\underbrace{+\alpha(M-(s+i-1))P_I S_{s,i-1}}_{\text{gain via link activation}}, \tag{8.13e}$$

$$\dot{I}_{si} = \tau i S_{si} - \gamma I_{si} + \gamma[(i+1)I_{s-1,i+1} - iI_{si}] \tag{8.13f}$$

$$+ \tau \frac{\sum_{k=1}^{M} \sum_{j+l=k} l^2 S_{jl}}{\sum_{k=1}^{M} \sum_{j+l=k} j I_{jl}} [(s+1)I_{s+1,i-1} - sI_{si}] \tag{8.13g}$$

$$\underbrace{-\omega[(s+i)I_{si}]}_{\text{loss via link deletion}} + \underbrace{\omega[(i+1)I_{s,i+1} + (s+1)I_{s+1,i}]}_{\text{gain via link deletion}} \tag{8.13h}$$

$$\underbrace{-\alpha(M-(s+i))I_{si}}_{\text{loss via link activation}} + \underbrace{\alpha(M-(s-1+i))P_S I_{s-1,i}}_{\text{gain via link activation}} \tag{8.13i}$$

$$\underbrace{+\alpha(M-(s+i-1))P_I I_{s,i-1}}_{\text{gain via link activation}}, \tag{8.13j}$$

where for $X \in \{S,I\}$, $P_X = \frac{\sum_{k=0}^{M} \sum_{j+l=k} (M-(j+l)) X_{jl}}{\sum_{k=0}^{M} \sum_{j+l=k} (M-(j+l))(S_{jl}+I_{jl})}$ is the probability of picking an available stub belonging to a node of type X.

In more detail, the denominator of P_X is the number of free stubs in the whole network which are not yet connected. The numerator, on the other hand, stands for the number of such stubs originating from nodes of type X. Hence, their ratio represents the probability that a randomly chosen free stub originates from a node of type X. Removing the link activation and deletion terms leads to the original effective degree model [197]. The average degree at equilibrium is known *a priori*, and the distribution can be treated as binomial with $(N-1)$ trials and probability of success $\alpha/(\alpha+\omega)$. Hence, even if we set $M = (N-1)$, we can generally truncate the degrees at some $k_{\max}$ for which the expected number of nodes with degree at least $k_{\max}$ is less than one.

The total number of stubs that are part of active links at time t, $L_A(t)$, and stubs that are dormant, $L_D(t)$, can be calculated from the effective degree formulation as

$$L_A(t) = \sum_{k=0}^{M} \sum_{j+l=k} (j+l)(S_{jl}+I_{jl}),$$

$$L_D(t) = \sum_{k=0}^{M} \sum_{j+l=k} (M-(j+l))(S_{jl}+I_{jl}),$$

with the average degree given by $\langle k \rangle(t) = \frac{L_A(t)}{N}$. At equilibrium, $\alpha L_D = \omega L_A$, which leads to the average degree at equilibrium

$$\langle K \rangle_{eq} = \frac{\alpha M}{\alpha + \omega}, \tag{8.14}$$

which depends on the maximal degree M rather than the number of nodes N. This means that in the dynamic network case, M itself can be regarded as a parameter which controls the level of network connectedness.

The effective degree model shows excellent agreement with simulation, as illustrated in Figs. 8.6 & 8.7. The crucial role of M is illustrated in Fig. 8.6. Since M controls the average degree at equilibrium, as per Eq. (8.14), depending on the initial network configuration and values of α and ω, the network can either undergo a link thinning process or could gain links. Hence, even if the initial network can sustain an epidemic, the disease can die out deterministically if a small enough value of M leads to significant link thinning (see the left column of Fig. 8.6).

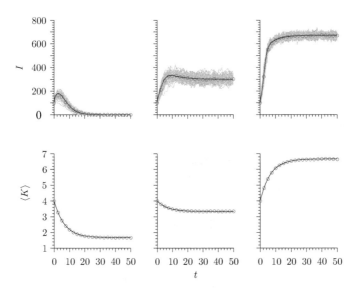

Fig. 8.6: $I(t) = \sum_{k=0}^{M} \sum_{j+l=k} I_{jl}(t)$ and $\langle K \rangle(t) = \frac{L_A(t)}{N}$ plotted for three different values of M, i.e. the left, middle and right columns correspond to $M = 5$, $M = 10$ and $M = 20$, respectively. Results based on the effective degree model are given by solid lines and those based on simulation by circles. In all cases $N = 1000$, $I_0 = 100$, $\alpha = 0.05$, $\omega = 0.1$, $\tau = 0.5$ and $\gamma = 1$. The initial network is a regular random graph with $k_i = 4$ for all nodes $i = 1, 2, \ldots, N$. In each case, simulation results are averaged over 100 realisations, with the individual ones plotted in grey.

The converse can occur. If M is sufficiently large to allow the network to gain links fast, then the epidemic can become established even when the initial network cannot support epidemics. Of course, all this is dependent on the time scale of the network and epidemic dynamics, but nevertheless controlling the value of M could be thought of as a mechanism or measure through which the impact of the epidemic can be mitigated, and it can be linked to local properties of nodes/individuals, which can be inferred more readily. The creation and deletion of links happens at random, and as such the network evolves independently of the disease, although the evolution of the disease is dependent upon the network. Indeed, this leads to observing dynamic networks with degree distributions close to Poisson with the mean given by Eq. (8.14), and this is confirmed by results from both the effective degree model and simulation.

The epidemic thresholds based on the effective degree model and simulation are compared in Fig. 8.7. The agreement demonstrates that the dynamic effective degree model accurately captures the evolution of the epidemic. For a given starting network, the critical τ value to sustain an epidemic on a static network can be computed. For the setup considered in Fig. 8.7, this is $\tau_c \approx 1/3$, upon assuming that $R_0 = \frac{\tau}{\tau+\gamma} \left(\frac{\langle K^2 \rangle}{\langle K \rangle} \right)$. So for a fixed value of α, and for infection rates below the threshold, the cutting rate still needs to be high enough to overcome the creation of links. Here, the time scales of the two processes are crucial, since creating links quickly can lead to the network becoming sufficiently connected to support an epidemic even if the original network and the value of τ could not lead to an epidemic. However, high values of τ may require large values of ω in order to stop an epidemic. On the other hand, such high values of the cutting rate may lead to a disconnected network. In between the two regimes, and given an initial network which is favourable for the disease to spread, epidemics may become established but then be curtailed by progressive link removal while the network remains well connected.

Fig. 8.7 highlights how network dynamics can be exploited as a control measure. A critical cutting rate can be determined based on disease parameters, above which epidemics die out. The following scenarios are possible: (a) starting with a network and disease parameters that can give rise to an initial growth in prevalence, a large cutting rate can lead to extinction, and (b) starting with a sparse network or disease parameters leading to extinction, a large link activation rate can lead to persistence.

8.3 Link-status-dependent link activation and deletion models

We have explored the possibility that all edges are created and removed at a rate independent of the nodes' statuses, and a case in which only SI edges may be deleted and are immediately replaced by SS edges. However, it seems likely that some intermediate regime exists, where any edges may be created or destroyed, but at a rate

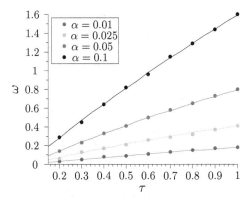

Fig. 8.7: Epidemic threshold in the (τ, ω) space for four distinct values of α. Results based on the effective degree model are given by solid lines and those from simulation by solid points. In each case, $N = 1000$, $I_0 = 10$, $M = 20$ and $\gamma = 1$. The initial network is a regular random graph with $k_i = 4$ for all nodes $i = 1, 2, \dots, N$.

that depends on the statuses of the associated nodes. We now present and investigate both simulation and pairwise models for various scenarios of this general setup.

8.3.1 Oscillating epidemics in a dynamic network model

The pairwise model given in system (8.7) can be generalised to make link activation and deletion link-status dependent. Note that this model does not conserve the number of edges. The resulting model is as follows.

SIS pairwise model with link-status-dependent activation and deletion

$$[\dot{S}] = \gamma[I] - \tau[SI] \tag{8.15a}$$

$$[\dot{I}] = \tau[SI] - \gamma[I], \tag{8.15b}$$

$$[\dot{SI}] = -(\tau + \gamma)[SI] + \tau([SSI] - [ISI]) + \gamma([II]) \\ + \alpha_{SI}([S][I] - [SI]) - \omega_{SI}[SI], \tag{8.15c}$$

$$[\dot{II}] = -2\gamma[II] + 2\tau([ISI] + [SI]) + \alpha_{II}([I]([I] - 1) - [II]) - \omega_{II}[II], \tag{8.15d}$$

$$[\dot{SS}] = 2\gamma[SI] - 2\tau[SSI] + \alpha_{SS}([S]([S] - 1) - [SS]) - \omega_{SS}[SS]. \tag{8.15e}$$

The activation and deletion rates are now link-status dependent. Due to the high dimensionality of the parameter space, a full study is difficult. Instead, researchers have focused on exploring parts of parameter space with simpler systems, mainly by setting a subset of the parameters to zero. In [180], the authors studied a model with closures involving the average degree of the network, which itself changes in time, as well as a number of simple dynamic network models with no epidemics, or on networks where nodes are labelled but still without an epidemic on the network.

Including network dynamics produces a richer spectrum of behaviour compared to the static network case, with the following possible outcomes: (a) epidemic extinction, (b) endemic equilibrium and (c) oscillations (see Fig. 8.8). Moreover, network dynamics when tuned to the right regime can be thought of as a potential epidemic control mechanism, and this is framed as a classical control problem in [277].

8.3.2 Bifurcation analysis of the pairwise model

In [294], the authors performed a detailed bifurcation analysis of the pairwise model assuming that the average degree remains constant, and carried out the analysis for a wide range of parameter combinations, with [295] focusing solely on characterising the oscillatory regime when all network dynamic parameters are set to zero except α_{SS} and ω_{SI}. So edges are deleted only if the two nodes have a different status and new edges are created only between susceptible nodes. We will focus on this case, which is in fact similar to the assumptions of system (8.1), but without requiring that the removal of SI edges and creation of SS is perfectly balanced. Indeed, if there is no disease, then the model will saturate the network with all edges being present. We will see that under these assumptions, oscillations can occur because as the disease spreads the network becomes sparse, but as it fades the network becomes densely connected again. This sets the stage for disease resurgence.

Results for $\alpha_{SS} \neq 0$, $\omega_{SI} \neq 0$ and $\alpha_{SI} = \alpha_{II} = \omega_{SS} = \omega_{II} = 0$ are summarised below. Using the closures given in equation (8.8), system (8.15) becomes self-consistent and can be analysed using bifurcation theory. First, $[S]$'s equation is eliminated using conservation of nodes: $[S] = N - [I]$. The disease-free steady state is given by $([I],[SI],[II],[SS]) = (0,0,0,N(N-1))$ and the following statement holds.

Proposition 8.2 *Assume $\alpha_{SS} \neq 0$, $\omega_{SI} \neq 0$ and $\alpha_{SI} = \alpha_{II} = \omega_{SS} = \omega_{II} = 0$. Then, the disease-free steady state is stable if and only if $\omega_{SI} > \tau(N-2) - \gamma$.*

Proof. If we order our variables $([SI],[II],[SS],[I])$, the Jacobian of the system at the disease-free steady state is lower block triangular:

$$
J = \begin{pmatrix}
-\gamma + \tau(N-2) - \tau - \omega_{SI} & \gamma & 0 & 0 \\
2\tau & -2\gamma & 0 & 0 \\
2\gamma - 2\tau(N-2) & 0 & -\alpha_{SS} & \alpha_{SS}(1-2N) \\
\tau & 0 & 0 & -\gamma
\end{pmatrix}.
$$

A block triangular matrix's eigenvalues are the eigenvalues of the diagonal blocks:

$$\begin{pmatrix} -\gamma + \tau(N-2) - \tau - \omega_{SI} & \gamma \\ 2\tau & -2\gamma \end{pmatrix}, \quad \begin{pmatrix} -\alpha_{SS} & \alpha_{SS}(1-2N) \\ 0 & -\gamma \end{pmatrix}.$$

The disease-free steady state is stable if and only if all the eigenvalues have negative real part. The second matrix has negative eigenvalues $-\alpha_{SS}$ and $-\gamma$. We must check the first matrix. For a 2×2 matrix, both eigenvalues have negative real part if and only if its determinant is positive and its trace is negative. The determinant is positive if $\gamma + \omega_{SI} - \tau(N-2) > 0$. The trace is negative if $3\gamma + \omega_{SI} - \tau(N-3) > 0$. The first condition implies the second one and, hence, Proposition 8.2 holds. □

The endemic steady state is the solution of a quartic equation and its analytical representation is likely to be of little practical use. Numerical investigations suggest that only a single biologically plausible endemic steady state exists. The stability analysis of this, which is detailed in [295], reveals the bifurcation diagram in Fig. 8.8a with three distinct outcomes in the (τ, ω_{SI}) plane. These are: (a) a stable disease-free steady state with no other plausible steady states, (b) a unique stable endemic equilibrium and, finally, (c) a unique stable limit cycle.

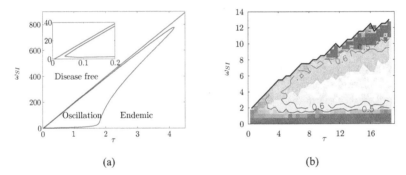

(a)   (b)

Fig. 8.8: (a) Bifurcation diagram for the pairwise ODE model (8.15) in the (τ, ω_{SI}) parameter space for $N = 200$, $\gamma = 1$ and $\alpha_{SS} = 0.04$. The transcritical bifurcation occurs along the straight red line, and a Hopf bifurcation occurs along the perimeter of the island. (b) Identification of the oscillatory regime relies on the values of the frequency of peak power. The equivalent of the bifurcation diagram in the left panel, but based on simulation, showing two potential boundaries provided by the iso-lines with the values of the frequency of peak power being 0.5 and 0.6. Peak frequency was ≈ 0.75 (yellow colour). Near-zero frequencies are shown in dark red. The thick black line shows one boundary for the disease-free regime determined as the value of ω_{SI} above which all realisations die out. The bottom boundary of the grey-shaded area represents an alternative boundary determined as the value of ω_{SI} under which no realisations die out.

8.3.3 Simulation-based bifurcation analysis and tracking of the oscillatory cycle

Network-based stochastic simulation shows that the system exhibits the range of behaviours predicted by the pair-based mean-field model. The typical oscillatory and endemic behaviours are illustrated in Fig. 8.9. As sharp bifurcation boundaries between these regimes in the stochastic model cannot be expected, empirical definitions for the boundaries between the different regimes have to be given. We now describe the bifurcations shown in Fig. 8.8b.

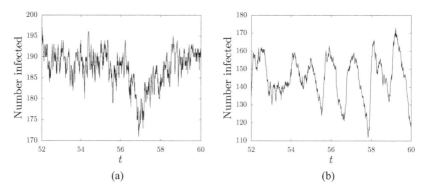

(a)   (b)

Fig. 8.9: Typical time series in the (a) endemic and (b) oscillatory regimes with $\omega_{SI} = 0.5$ and 4.0, respectively. The samples were chosen at random from multiple realisations with parameters: $N = 200$, $\tau = 12$, $\gamma = 1$ and $\alpha_{SS} = 0.04$.

Boundary of the disease-free regime: For each parameter pair (τ, ω_{SI}) and taking a maximum simulation time $T_{max} = 1320$, the proportion of realisations in which the epidemic died out is determined. It turns out that there is sharp increase in the proportion of realisations in which epidemics die out. Hence, two potential boundaries can be defined by focusing on the critical cutting rate up to which no epidemic dies out, and on the critical cutting rate above which all epidemics die out. These two boundaries define a region or band that can be considered as the bifurcation boundary which separates the disease-free regime from other possible states. The width of this region is shown by the grey-shaded area in Fig. 8.8b.

Boundary of the oscillating regime: For epidemics that have survived, evidence for oscillatory behaviour can be assessed based on the power spectrum of the time series of the number of infected nodes. In essence, the presence of a non-trivial oscillatory behaviour can be assessed by the existence of statistically significant power at a non-zero frequency peak in the power spectrum (see [132]). Excluding zero frequency is required because time series in which the spectrum is dominated by the zero mode can be difficult to distinguish from pure white noise [269]. Hence, to identify the boundaries of the oscillatory regime, in [295] the power and frequency of the main peak of the power spectrum were recorded for a number of different parameter configurations. As illustrated in Fig. 8.8b, for large values of τ, the peak frequency can decrease to close to zero levels and a stable endemic equilibrium is observed (the dark red area in the vicinity of the disease-free area). Qualitatively, this is consistent with the picture provided by the theoretical bifurcation diagram, which shows a narrow region where the endemic regime exists between the oscillatory and the disease-free regime (e.g. see $\tau = 4$ and $\omega_{SI} \approx 780$ in Fig. 8.8a).

Bifurcation diagram in the (τ, ω_{SI}) **plane:** The equivalent of Fig. 8.8a, but in terms of simulation, is shown in Fig. 8.8b, with the bifurcation "curves" and the regions of the three different behaviours. The plot qualitatively confirms the prediction of the theoretical model where a bounded oscillatory domain appears to be curved out of the endemic regime but close to the disease-free domain delimited by the strict boundary where all epidemics die out. For larger values of the transmission rate, the oscillatory and the disease-free regimes separate more readily due to the reappearance of the endemic regime. When considering the situation where some epidemics survive but some die out, the disease-free and the oscillatory regimes overlap. In terms of simulation, in this regime the evolution of prevalence is characterised by large seesaw patterns of a long continuous decrease followed by a rapid and significant increase.

The oscillatory cycle in terms of the network: Oscillations are notoriously difficult to map out from adaptive network models, especially directly from simulation. Oscillations usually arise as a combination of a positive and negative feedback with a suitable time delay. Following this idea, it is possible to heuristically explain and characterise the oscillations in the model. In order to do this, we focus on the prevalence $[I]$ and the average degree $\langle K \rangle$, for which the exact differential equations are

$$[\dot{I}] = \tau[SI] - \gamma[I],$$
$$\langle \dot{K} \rangle = \alpha_{SS}([S]([S]-1) - [SS]) - 2\omega_{SI}[SI].$$

A close inspection of these reveals that the direction of change of these quantities is determined by the relative magnitude of the following four processes:

- **A**: Infection with rate $\tau[SI]$,
- **B**: Recovery of I nodes with rate $\gamma[I]$,
- **C**: Creation of SS links with rate $\alpha_{SS}([S]([S]-1) - [SS])$, and
- **D**: SI link deletion with rate $\omega_{SI}[SI]$.

The stages of the oscillatory cycle are determined by the strength of process **A** relative to **B**, and that of **C** compared to **D**. Consider an epidemic that is well-established and is capable of sustaining oscillations, and start observing the dynamics when the epidemic is about to take off, i.e. when process **A** dominates **B**, and **C** is stronger than **D**. In this case, the growth of the epidemic goes hand in hand with the network becoming more connected. This is due to the fact that the total rate of link cutting is smaller than the total rate of link creation, i.e. $2\omega_{SI}[SI] < \alpha_{SS}([S]([S]-1)-[SS])$. This is a direct result of the epidemic having been just recovered from an excursion where the connectivity of the network was low and the number of susceptible nodes was large.

As the epidemic expands, the balance of processes **C** and **D** changes. Namely, as the prevalence grows, the number of susceptible nodes decreases while link cutting acts on an increasing number of SI links. This tilts the balance in favour of edge cutting, meaning that $2\omega_{SI}[SI] > \alpha_{SS}([S]([S]-1)-[SS])$. In Fig. 8.10a, a snapshot close to peak connectivity is shown. At this stage, the epidemic continues to grow, i.e. **A** still dominates **B**.

Close to the highest level of infection, shown in Fig. 8.10b, the recovery and the cutting of occasional SI links will outcompete the spread of infection, that is, process **B** will outcompete process **A**. The ongoing loss of links leads to "segregation" of the network, as shown in Fig. 8.10c, meaning that clusters/clumps of either susceptible or infected nodes will form, with occasional between-cluster links.

Due to the increasing number of susceptible nodes, link creation takes over and the susceptible clusters/clumps will become more densely connected. At this crucial point, the survival of the epidemic relies on the few inter-clump links which are the only ways in which the epidemic can reseed itself in the densely connected susceptible parts of the network. If successful, this process is followed by a sudden epidemic expansion, i.e. the dominance of **A** over **B** (see Fig. 8.10d). This finishes the cycle and the four stages of the oscillation can be characterised as follows:

1. $\mathbf{A} > \mathbf{B}, \mathbf{C} > \mathbf{D}, [I]$ increasing, $\langle K \rangle$ increasing,
2. $\mathbf{A} > \mathbf{B}, \mathbf{C} < \mathbf{D}, [I]$ increasing, $\langle K \rangle$ decreasing,
3. $\mathbf{A} < \mathbf{B}, \mathbf{C} < \mathbf{D}, [I]$ decreasing, $\langle K \rangle$ decreasing,
4. $\mathbf{A} < \mathbf{B}, \mathbf{C} > \mathbf{D}, [I]$ decreasing, $\langle K \rangle$ increasing.

8.4 Link deactivation and activation on fixed networks

In [285, 304], the authors propose a dynamic network model with a fixed underlying network topology. Susceptible individuals can temporarily deactivate links to infectious individuals with the potential of re activation once both are susceptible. In contrast with the previous dynamic network models, changes in the network play out on a fixed structure, where fixed links are switched on and off, as dictated by the model. The pairwise evolution equations are as follows.

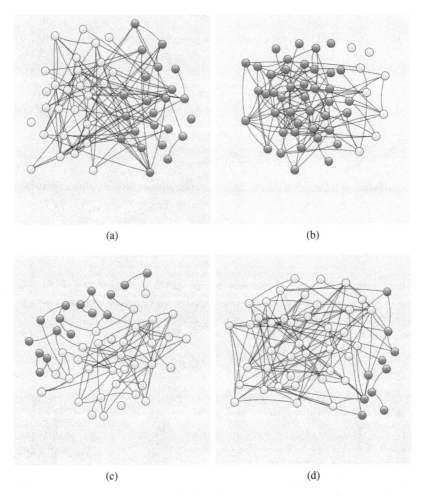

Fig. 8.10: Snapshots of the network during the main phases of the oscillatory cycle: (a) the growing phase of the epidemic with $\langle K \rangle$ close to its maximum, (b) close to the maximum prevalence and a decreasing average degree, (c) decreasing prevalence with $\langle K \rangle$ close to its minimum and, finally, (d) minimal prevalence but with growing average degree. Parameter values are $N = 50$, $\tau = \gamma = 1$, $\omega_{SI} = 1.3$ and $\alpha_{SS} = 0.04$, with all the other activation and deletion rates being equal to zero. Susceptible and infected nodes are denoted by red and yellow filled circles, respectively.

SIS pairwise model with link deactivation-activation on fixed networks

$$[\dot{S}] = \gamma[I] - \tau[SI], \tag{8.16a}$$

$$[\dot{I}] = \tau[SI] - \gamma[I], \tag{8.16b}$$

$$[\dot{SI}] = -(\tau+\gamma)[SI] + \tau([SSI] - [ISI]) + \gamma([II]) - d[SI], \tag{8.16c}$$

$$[\dot{II}] = -2\gamma[II] + 2\tau([ISI] + [SI]), \tag{8.16d}$$

$$[\dot{SS}] = 2\gamma[SI] - 2\tau[SSI] + a[\widehat{SS}], \tag{8.16e}$$

$$[\dot{\widehat{SI}}] = -\gamma[\widehat{SI}] + \tau([\widehat{SSI}] - [I\widehat{SI}]) + \gamma[\widehat{II}] + d[SI], \tag{8.16f}$$

$$[\dot{\widehat{II}}] = -2\gamma[\widehat{II}] + 2\tau[I\widehat{SI}], \tag{8.16g}$$

$$[\dot{\widehat{SS}}] = 2\gamma[\widehat{SI}] - 2\tau[\widehat{SSI}] - a[\widehat{SS}], \tag{8.16h}$$

where $[\widehat{AB}]$, with $A, B \in \{S, I\}$, denotes deactivated links. The $[SI]$ links can transmit infection, but $[\widehat{SI}]$ links cannot. The transmission and recovery rates are τ and γ, as before, and d and a are the deactivation and activation rates, respectively.

We use the closures

$$[ABC] = \frac{[AB][BC]}{[B]}, \qquad [\widehat{ABC}] = \frac{[\widehat{AB}][BC]}{[B]},$$

where $A, B, C \in \{S, I\}$, where again we stick with the choices of the original paper and do not use n_S explicitly. Here, we note the identities: $[\widehat{SI}] = [\widehat{IS}]$, $[\widehat{SSI}] = [I\widehat{SS}]$ and $[\widehat{ISI}] = [I\widehat{SI}]$. We can eliminate $[S]$ from our differential equations through conservation of nodes: $[S] = N - [I]$. We can eliminate $[\widehat{II}]$ from our equations through conservation of edges: $[\widehat{II}] = N\langle K\rangle - [SS] - [II] - 2[SI] - [\widehat{SS}] - 2[\widehat{SI}]$.

In the disease-free steady state, all variables with I are zero and all edges are active. The equilibrium is $([SI], [II], [SS], [\widehat{SI}], [\widehat{SS}], [I]) = (0, 0, N\langle K\rangle, 0, 0, 0)$. With this ordering of variables, the Jacobian is block lower triangular:

$$J = \begin{pmatrix} \tau\langle K\rangle - (\tau+\gamma+d) & \gamma & 0 & 0 & 0 & 0 \\ 2\tau & -2\gamma & 0 & 0 & 0 & 0 \\ 2\gamma - 2\tau\langle K\rangle & 0 & 0 & 0 & a & 0 \\ -2\gamma & -\gamma & -\gamma & -3\gamma & -\gamma & 0 \\ 0 & 0 & 0 & 2\gamma & -a & 0 \\ \tau & 0 & 0 & 0 & 0 & -\gamma \end{pmatrix}.$$

The eigenvalues are given by the eigenvalues of the diagonal blocks:

$$\begin{pmatrix} \tau\langle K\rangle - (\tau+\gamma+d) & \gamma \\ 2\tau & -2\gamma \end{pmatrix}, \quad \begin{pmatrix} 0 & 0 & a \\ -\gamma & -3\gamma & -\gamma \\ 0 & 2\gamma & -a \end{pmatrix}, \quad (-\gamma).$$

From the second and third matrices, we have $\lambda_1 = -a$, $\lambda_2 = \lambda_3 = -\gamma$ and $\lambda_4 = -2\gamma$. The first matrix shows that λ_5 and λ_6 are given by the quadratic equation

$$\lambda^2 + ((\gamma + d - \tau \langle K \rangle) + \tau + 2\gamma)\lambda + 2\gamma(\gamma + d - \tau \langle K \rangle) = 0. \qquad (8.17)$$

If $\gamma + d - \tau \langle K \rangle > 0$, then the trace is negative and the determinant positive. If this holds, the disease-free steady state is stable. If, however, $\gamma + d - \tau \langle K \rangle < 0$, then the determinant is negative, so one eigenvalue is positive and the other is negative. Hence, the disease-free steady state is unstable. The bifurcation point is given by

$$\tau_c = \frac{d + \gamma}{\langle K \rangle},$$

which does not depend on the activation rate a.

In [304], the authors showed that while the threshold does not depend on a, larger values of a lead to a steady state with a higher fraction of infected nodes, where the activation effectively negates the positive impact of deactivation. Furthermore, in contrast to the link-conserving rewiring model, which can display a backward transcritical bifurcation with bistability, this model exhibits only a forward transcritical bifurcation separating the disease-free and endemic regimes.

This model can also be considered in the slow and fast network dynamics regimes with some important consequences. For slow network dynamics, it turns out that the endemic steady state is well approximated by an equation which is functionally equivalent to that of the full model, see [304], by simply replacing the average degree with the effective degree, which can be obtained as $a, d \to 0$ and $d/a = c$ remains constant. This means that for slow network dynamics, a good approximation is provided by static networks with an effective mean degree whose value is slightly smaller than that of the initial starting network.

The authors also consider the case when the dynamic of the network is comparable to or faster than disease transmission. In this case, the interplay is more subtle and susceptible nodes experience a strong reduction in infectious pressure. Furthermore, pair correlations will be strong and model reductions are unlikely. These make it less likely that a static network analogue exists. This work has been further extended to characterise the active part of the network when the process plays out on highly heterogenous networks [285]. The authors have concentrated on mapping out and understanding the behaviour of the active part of the network during an epidemic. The topology of the subnetwork consisting of active links can be significantly different from the original network. They also proposed a closure for the traditional pairwise model by taking into account degree heterogeneity, but without increasing the number of equations in the mean-field model. This is an approach that has also been used in [287] and is based on the following simple argument, which is presented in terms of the notation used for pairwise models:

$$[ASI] = \sum_k [AS_kI] \approx \sum_k \frac{k-1}{k} \frac{[AS_k][S_kI]}{[S_k]}$$

$$= \sum_k \frac{k-1}{k[S_k]} \underbrace{[AS] \frac{k[S_k]}{\sum_l l[S_l]}}_{[AS_k]} \underbrace{[SI] \frac{k[S_k]}{\sum_l l[S_l]}}_{[S_kI]}$$

$$= \frac{[AS][SI]}{(\sum_l l[S_l])^2} \sum_k (k-1)k[S_k]$$

$$= [AS][SI] \frac{\sum_k (k^2[S_k] - k[S_k])}{(\sum_l l[S_l])^2} = [AS][SI] \frac{S_2 - S_1}{S_1^2}, \qquad (8.18)$$

where $S_1 = \sum_k k[S_k]$ and $S_2 = \sum_k k^2[S_k]$ are the time-evolving first and second moment of stubs emanating from susceptible nodes. This in turn allows us to keep the original equations and have a self-consistent closure apart from some uncertainty about the precise values of S_1 and S_2. For a simple SIS model, $[SS] + [SI] = n_s[S]$, and this is equivalent to $S_1 = \sum_k kS_k$. However, to work out the second moment some heuristic arguments are needed. These can be situation or model dependent, and for SIS dynamics, in [287] it has been shown that S_2 can be approximated by assuming a favourable linear relation between the original degree distribution of the network and the distribution over susceptible nodes in time (see Section 5.2.4). In the deactivation and activation model, the authors have estimated this quantity in two different limits: near bifurcation and in the limit of large values of the transmission rate. Other approximations, for example, based on some intuition gained from simulation, are possible, but these make it unlikely to derive analytical results.

8.5 EBCM-based approach

A potential problem with many of the dynamic network models considered above is that as edges are broken and formed, there is no constraint which forces nodes to preserve their degrees. This question was addressed in part by [317] and simplified by [222], both under the assumption of a vanishingly small initial condition. We now adapt the EBCM approach to networks in which edges change over time, allowing for a finite fraction to be infected initially. Related approaches are being developed for understanding disease spread in populations with entry and exit [195, 196].

8.5.1 SIR disease in dynamic degree-preserving networks

We now look at how the edge-based compartmental models can be used to study SIR disease spread in dynamic networks. We consider a different model for how individuals change their partners. Following the Configuration Model assumptions, each node's degree is chosen from a distribution and node i is given k_i stubs. Each edge breaks according to a Poisson process with rate η, freeing the stubs. The stubs form new edges with other stubs that are newly freed. We neglect the between-partner period and assume that a node replaces a neighbour instantly. In simulation, this can be done by choosing multiple edges, breaking them and reconnecting the stubs so that the nodes have new neighbours.

We can adapt the EBCM approach presented for static networks in Section 6.5 to this case. We have a new parameter, η, which represents the rate at which an edge breaks. Our definitions of the ϕ and θ variables changes slightly. Given a test node u, these previously represented the probability a neighbour v had never transmitted to u and had a given status. Because the neighbour the stub connects to can change, we make our definitions more precise. We have

- ϕ_S: The probability a random stub belonging to the test node u has never been involved in a transmission of infection to u and is *currently* connected to a susceptible neighbour,
- ϕ_I: The same definition as ϕ_S except that the current neighbour is infected,
- ϕ_R: The same definition as ϕ_S except that the current neighbour is recovered,
- $\theta = \phi_S + \phi_I + \phi_R$: The probability a random stub belonging to the test node u has never been involved in a transmission of infection to u.

We assume a proportion ρ is selected uniformly from the population and infected at time $t = 0$. All other individuals are initially susceptible. It is straightforward to relax this assumption and allow the initial infections to be degree dependent or some to be initially recovered. We use

$$\psi(x) = \sum_k P(k)x^k,$$

and conclude that $S(t) = N(1-\rho)\psi(\theta(t))$. Our goal is to find $\theta(t)$.

Previously, ϕ_S could be calculated explicitly by finding the probability that a random neighbour is susceptible. However, a stub that currently joins u to a susceptible node may have previously connected to an infected neighbour and brought infection to u. The previous method no longer applies. Similarly, we cannot find ϕ_R explicitly. We will instead find differential equations for ϕ_S and ϕ_R. To do this, we will need to know the probability that a newly formed edge connects to a susceptible, infected or recovered individual. We call these probabilities π_S, π_I and π_R and note that they are the probabilities that a random stub belongs to a node of each type. The flow diagram of Fig. 8.11 demonstrates the fluxes between the various compartments of interest. As before, we have

$$S = N(1-\rho)\psi(\theta),$$
$$I = N - S - R,$$
$$\dot{R} = \gamma I.$$

We can find π_S, π_I and π_R similarly. The probability that a random stub belongs to a degree k node is $P_n(k) = kP(k)/\langle K\rangle$. The probability that a node was initially susceptible is $1 - \rho$, and the probability it has not been infected since is θ^k. Thus, $\pi_S = \sum_k kP(k)(1-\rho)\theta^k/\langle K\rangle$. This can be condensed using ψ to give

$$\pi_S = (1-\rho)\frac{\theta\psi'(\theta)}{\langle K\rangle},$$
$$\pi_I = 1 - \pi_S - \pi_R,$$
$$\dot{\pi}_R = \gamma\pi_I.$$

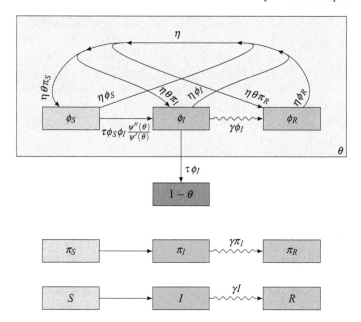

Fig. 8.11: (Top) Flow diagram for the ϕ and θ variables for the EBCM model with edge turnover. Compared to Fig. 6.23, new paths exist due to edge rewiring. These result in flows out of each ϕ compartment at rate equal to η times the compartment size, giving a total flux from all three of $\eta\theta$s. This is redistributed into each of the ϕ compartments at rate $\eta\theta$ times the corresponding π variable. (Middle) Flow diagram for π variables. (Bottom) Flow diagram for the individual-level variables S, I and R.

All that remains is the calculation of the ϕ and θ variables. The fluxes along most arrows in Fig. 8.11 are straightforward. The fluxes from ϕ_I to $1-\theta$ and to ϕ_R are as before. The fluxes out of each ϕ compartment due to edge breaking are $\eta\phi_S$, $\eta\phi_I$ and $\eta\phi_R$. These give a total of $\eta\theta$, which is distributed back into the ϕ compartments at rates $\eta\theta\pi_S$, $\eta\theta\pi_I$ and $\eta\theta\pi_R$, that is according to the probability that the newly found stub belongs to a node of the given status. From this, we have

$$\dot{\phi}_R = \gamma\phi_I - \eta\phi_R + \eta\theta\pi_R,$$
$$\dot{\theta} = -\tau\phi_I,$$
$$\phi_I = \theta - \phi_S - \phi_R.$$

We still need a differential equation for ϕ_S. We show two derivations. With some effort (not given here) it can be shown that if v is a susceptible neighbour of a test node u, then the expected number of neighbours of v (other than u) is $\psi''(\theta)/\psi'(\theta)$. This is commonly called the excess degree. Thus, the expected number of infected neighbours of v, conditional on v being susceptible, is $\phi_I\psi''(\theta)/\psi'(\theta)$. Each such edge transmits at rate τ. So the ϕ_S to ϕ_I transition due to infection of v is $\tau\phi_S\phi_I\psi''(\theta)/\psi'(\theta)$ (the ϕ_S accounts for the probability the stub satisfies the definition for ϕ_S). Combined with other fluxes, this gives

$$\dot{\phi}_S = \eta\theta\pi_S - \eta\phi_S - \tau\phi_S\phi_I\frac{\psi''(\theta)}{\psi'(\theta)}.$$

An alternate derivation comes from attempting to reproduce the explicit formula for ϕ_S from the static case. Consider a random test node u and a random neighbour v. Let $\sigma(t)$ be the probability the two stubs joining v and u had not previously been involved in transmitting to v or u prior to edge forming. The probability that v is susceptible at time t and that u's stub had not previously been involved in transmitting to u is $\sigma(t)\theta(t)^{k-1}$, where k is the degree of v. Averaging over k gives $\phi_S(t) = \sum_k(1-\rho)P_n(k)\sigma(t)\theta^{k-1} = (1-\rho)\sigma(t)\psi'(\theta)/\langle K\rangle$. The derivative of $\sigma(t)$ is given by the rate at which such edges form, $\eta\theta^2$ (one θ is for v's stub and the other is for u's stub), minus the rate at which such edges break, $\eta\sigma$:

$$\dot{\sigma} = \eta\theta^2 - \eta\sigma.$$

Thus, the derivative $\dot{\phi}_S(t)$ is as follows.

$$\begin{aligned}
\dot{\phi}_S(t) &= \dot{\sigma}(1-\rho)\frac{\psi'(\theta)}{\langle K\rangle} + \sigma(1-\rho)\frac{\dot{\theta}\psi''(\theta)}{\langle K\rangle}\\
&= (1-\rho)\eta\theta^2\frac{\psi'(\theta)}{\langle K\rangle} - (1-\rho)\eta\sigma\frac{\psi'(\theta)}{\langle K\rangle} - \sigma(1-\rho)\tau\phi_I\frac{\psi''(\theta)}{\langle K\rangle}\\
&= \eta\theta\pi_S - \eta\phi_S - \tau\phi_S\phi_I\frac{\psi''(\theta)}{\psi'(\theta)}.
\end{aligned}$$

The final step used $\pi_S = (1-\rho)\theta\psi'(\theta)/\langle K\rangle$ and eliminated σ using $\phi_S = \sigma(1-\rho)\psi'(\theta)/\langle K\rangle$. Our final system is

EBCM model for SIR and dynamic network

$$\dot{\theta} = -\tau\phi_I, \tag{8.19a}$$

$$\dot{\phi}_S = \eta\theta\pi_S - \eta\phi_S - \tau\phi_S\phi_I\frac{\psi''(\theta)}{\psi'(\theta)}, \tag{8.19b}$$

$$\dot{\phi}_R = \eta\theta\pi_R - \eta\phi_R + \gamma\phi_I, \tag{8.19c}$$

$$\phi_I = \theta - \phi_S - \phi_R, \tag{8.19d}$$

$$\dot{\pi}_R = \gamma\pi_I, \qquad \pi_S = (1-\rho)\theta\frac{\psi'(\theta)}{\langle K\rangle}, \qquad \pi_I = 1 - \pi_S - \pi_I, \tag{8.19e}$$

$$\dot{R} = \gamma I, \qquad S = N(1-\rho)\psi(\theta), \qquad I = N - S - R. \tag{8.19f}$$

Our initial conditions are

$$\theta(0) = 1, \quad \phi_S(0) = 1-\rho, \quad \phi_R(0) = 0, \quad \pi_R(0) = 0, \quad R(0) = 0.$$

An equivalent system of equations replaces the differential equation for ϕ_S with

$$\dot{\sigma} = \eta \theta^2 - \eta \sigma, \quad \phi_S = \frac{\sigma \pi_S}{\theta},$$

taking $\sigma(0) = 1$. Other equivalent variations are possible.

To demonstrate the accuracy of the model, we consider the same disease and populations as in Example 6.4. That example corresponds to $\eta = 0$ in this model. We now consider $\eta = 1$ and $\eta = 2$, and Fig. 8.12 shows that the agreement is very good. All nodes (including in the Erdős–Rényi network) preserve their degree throughout the simulations.

Increasing η results in increased spread. This is because multiple transmissions from an infected individual along the same stub may now go to different individuals. Because the time scale of the epidemic on a truncated power law network is very short, the impact of edge breaking and forming is smaller than for the other populations.

Exercise 8.5. Find a differential equation for ϕ_I and show that using this, we can modify system (8.19) to find a complete system of equations without including ϕ_R.

Exercise 8.6. Consider an SIR disease in a network in which half the individuals have degree 2 and half have degree 5. Find $\psi(x)$. Plot some numerical solutions for $\rho = 0.01$, $\tau = 1$, $\gamma = 2$ and $\eta = 0.1, 1, 10$.

Exercise 8.7. Repeat the derivation of system (8.19), but allowing for the initial infections to be chosen according to degree, with $\hat{\psi}(x) = \sum_k P(k)S(k,0)x^k$. Then, give the full equations for an SIR disease in a network in which half the individuals have degree 2 and half have degree 5 (as in Exercise 8.6), but none of the degree 2 individuals are initially infected and 2% of the degree 5 individuals are. Plot some numerical solutions for $\tau = 1$, $\gamma = 2$ and $\eta = 0.1, 1, 10$. [Hint: Consider the derivation of system (6.12).]

8.6 Conclusions and outlook

In this chapter, we have focused on dynamic or adaptive networks, where coupling of the network dynamics to the epidemic process on the network is key. The focus was on the development of mean-field models, be it pairwise, effective degree or edge-based compartmental type, and on the subsequent analysis of these in order to elucidate the role of concurrently considering the dynamics on and of the network. A particularly promising direction, but yet unexplored, relates to the rigorous study of the emerging networks. It is apparent that the coupled process in the right parameter regimes can either make the network densely connected or make it fall apart into disjoint components. Moreover, we hypothesise that such changes extend to other network characteristics, such as degree distribution, clustering, etc. Mapping out such changes in a systematic and rigorous way will lead to the equivalent

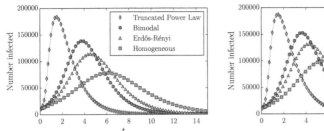
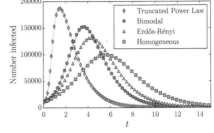

Fig. 8.12: The spread of the same disease as in Fig. 6.24, but with $\eta = 1$ (left) and $\eta = 2$ (right). Increased η increases the amount of growth by reducing the probability that a transmission is repeated to the same neighbour. The populations in the simulations have 10^6 nodes. Solid curves show simulation and symbols show theoretical predictions.

of a bifurcation analysis, but for the network. This could open up a new direction focusing on a more organic way of generating synthetic networks. Such networks will emerge naturally from the interplay of various processes, as is likely to happen in real life. In order to do this, models of higher resolution are needed, and they should be used in conjunction with simulation.

The edge-based compartmental approach can be used to implement a series of other rewiring or dynamic network scenarios. For example, it is possible to generalise it to allow stubs to remain "dormant" for a period following an edge-breaking event, or for creation and breaking of edges to be completely independent events. Details of this and related models in the limit of a vanishingly small number of initially infected nodes appear in [222]. Pairwise and effective degree models can also be applied for different scenarios of network dynamics, for example, the link-status-dependent activation and deletion model has only been studied for a limited number of parameter combinations. Moreover, dynamic and adaptive networks have been used recently in the context of control theory, where it is feasible to formulate optimal control problems in the context of dynamic networks [277, 331].

The research literature on non-static networks is vast and is often divided based on the time scale of changes of the network compared to some baseline rate, or a rate given by processes unfolding on the network. Non-static networks have prompted the generalisation of existing static network measures and the development of new ones in order to characterise real-world temporal networks [146, 147]. In a similar way as to how the theory of static networks developed, the theory of temporal networks is growing fast with important developments including the statistical/descriptive characterisation of non-static networks [147] and the development of theoretical time-evolving networks without [147, 186, 251] or with dynamics [147, 260] on networks. This remains a highly active area with many new and exciting developments to follow.

Chapter 9
Non-Markovian epidemics

It has been long recognised that the assumption of exponentially distributed infectious periods with constant transmission rates is not realistic. However, these assumptions simplify simulation and are easier to translate into differential equations. In the spirit of focusing on mean-field or ODE models, we will show how our models can be extended to account for general transmission and recovery processes, with particular focus on approximating the time evolution of the epidemic.

Early studies of non-Markovian epidemics focused on SIR dynamics on fully connected networks, or homogeneously mixing populations, with the infection process being Markovian but with the infectious period taken from a general distribution [8, 278, 292, 293]. These approaches use probability theory arguments and typically focus on characterising the distribution of final epidemic sizes for finite populations, or on the average size in the infinite population limit. Similarly, the quasi-stationary distribution in a stochastic SIS model, again in a fully connected network, has been the subject of many studies [66, 230, 231]. More recently, it has been shown that one can readily apply results from queueing [19] or branching process [233] theory, or use martingales [65] to cast the same questions within a different framework and obtain results more readily.

Extending such results to networks that are not fully connected is non-trivial and we briefly summarise some of the recent results in this area. Work by Newman [234], later made more precise in [173] and extended in [211, 212], shows that the bond percolation approach, where the probability of transmission is effectively "smartly" averaged over transmission and recovery processes of a general form, leads to an implicit formula for the average final epidemic size, and this takes into account the degree distribution of the network. Much of this was discussed in Chapter 6. This provides an excellent approximation, as long as networks are not clustered and are created according to the Configuration Model.

Another promising approach in this direction is the message passing formalism discussed in [163, 327, 328]. This method uses a "cavity state" assumption, which

303
I.Z. Kiss et al., *Mathematics of Epidemics on Networks*, Interdisciplinary Applied Mathematics 46, DOI 10.1007/978-3-319-50806-1_9

is in practice equivalent to the "test node" assumption we have made previously. In [163], the authors show that this method can handle general transmission and recovery processes, and the resulting model is exact on tree networks. The reason for this exactness is intimately related to the fact that every vertex in a tree is a cut-vertex. Furthermore, for non-tree networks the same model provides a rigorous upper bound on the expected number of nodes infected by an epidemic.

The chapter is divided into three main parts with the first focusing on generalising the SIR pairwise model by introducing multiple stages of infection, thus allowing the infectious period to be gamma distributed. In the second part, the SIR pairwise model is extended to Markovian transmission but with a fixed infectious period followed by a generalisation to an arbitrary recovery process, all for homogenous networks. Finally, in the third part, we conclude with a general model for SIR epidemics on heterogeneous networks with arbitrary transmission and recovery processes.

9.1 Pairwise model with multiple stages of infection

In [199], an SIR model with gamma-distributed infectious periods was proposed. This model divides the I compartment into K successive compartments, $I_1, I_2, \ldots, I_K$, where infectious individuals behave in the same way, i.e. produce new infectious cases at the same rate β and move to the next compartment at rate $K\gamma$. This yields infectious periods which correspond to the sum of K independent and identically distributed exponential distributions, $Exp(K\gamma)$. This leads to a gamma distribution $\Gamma(K, K\gamma)$, with shape K and rate $K\gamma$, which can be conveniently tuned to a wide range of distributions.

9.1.1 Extended compartmental SIR model

To investigate such a model, we begin with the compartmental SIR model (1.2) as follows:

Compartmental SIR model

$$\dot{S} = -\beta SI/N, \tag{9.1a}$$
$$\dot{I} = \beta SI/N - \gamma I, \tag{9.1b}$$
$$\dot{R} = \gamma I, \tag{9.1c}$$

where the introduction of the additional infectious classes leads to the following:

Compartmental multi-stage SIR model

$$\dot{S} = -\beta SI/N, \tag{9.2a}$$

$$\dot{I}_1 = \beta SI/N - K\gamma I_1, \tag{9.2b}$$

$$\dot{I}_2 = K\gamma I_1 - K\gamma I_2, \tag{9.2c}$$

$$\vdots$$

$$\dot{I}_K = K\gamma I_{K-1} - K\gamma I_K, \tag{9.2d}$$

$$\dot{R} = K\gamma I_K, \tag{9.2e}$$

where $I = \sum_{k=1}^{K} I_k$.

Several important observations can be made. In both models:

1. the mean infectious period is $1/\gamma$,
2. the average number of infections caused before recovering is $R_0 = \beta/\gamma$,
3. the final epidemic size is given by solving

$$r_\infty = 1 - \exp(-R_0 r_\infty), \tag{9.3}$$

where $r_\infty = R(\infty)/N$, which is identical to (1.3) in the limit of a fully susceptible population at time $t = 0$ [204, 216].

However, differences between the two models manifest themselves in both the initial growth rate and the time evolution of the epidemic. Namely, in model (9.1), $I(t) \simeq I(0)\exp((\beta - \gamma)t)$, while in model (9.2), $I(t) \simeq I(0)\exp(\lambda t)$, where λ solves

$$R_0 = \frac{\lambda}{\gamma\left(1 - \left(\frac{\lambda}{K\gamma} + 1\right)^{-K}\right)}, \tag{9.4}$$

as shown in [319, 325]. This is illustrated in Fig. 9.1, where plots of the prevalence show clearly that higher values of K correspond to faster initial growth rate of the epidemic. The additional classes do not affect the basic reproductive ratio or final epidemic size, but result in epidemics which grow faster initially and have a shorter overall duration. Hence, some predictions remain universal, while others change when models are refined.

9.1.2 Pairwise model

We can translate this idea to the SIR pairwise model in order to investigate the impact of more realistic infectious period distributions on the epidemic threshold and final epidemic size [284]. The pairwise model with multiple infectious stages is as follows:

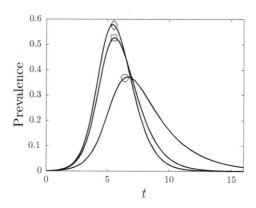

Fig. 9.1: A comparison of infection dynamics for a one- ($\bigcirc$), three- ($\square$) and five-stage ($\lozenge$) SI^KR models with influenza data $\beta = 1.66/\text{days}$ [168], and $\gamma^{-1} = 2.2\,\text{days}$ [325]. Each curve represents the sum of all I_i in model (9.2). Adding extra stages causes epidemics to occur earlier and result in a higher epidemic peak, although R_0 and the final epidemic size are identical for each curve.

Multi-stage SI^KR pairwise model for homogenous networks

$$[\dot{S}] = -\tau \sum_{i=1}^{K} [SI_i], \tag{9.5a}$$

$$[\dot{I_1}] = \tau \sum_{i=1}^{K} [SI_i] - K\gamma[I_1], \tag{9.5b}$$

$$[\dot{I_j}] = K\gamma[I_{j-1}] - K\gamma[I_j], \quad \text{for} \quad j = 2,3,\ldots,K, \tag{9.5c}$$

$$[\dot{R}] = K\gamma[I_K], \tag{9.5d}$$

$$[\dot{SS}] = -2\tau \sum_{i=1}^{K} [SSI_i], \tag{9.5e}$$

$$[\dot{SI_1}] = -(\tau + K\gamma)[SI_1] + \tau \left(\sum_{i=1}^{K} [SSI_i] - \sum_{i=1}^{K} [I_iSI_1] \right), \tag{9.5f}$$

$$[\dot{SI_j}] = -(\tau + K\gamma)[SI_j] + K\gamma[SI_{j-1}] - \tau \sum_{i=1}^{K} [I_iSI_j], \quad \text{for} \quad j = 2,3,\ldots,K, \tag{9.5g}$$

$$[\dot{SR}] = K\gamma[SI_K], \tag{9.5h}$$

where the closures in equations (4.6) and (4.7) lead to a closed self-consistent system. Figure 9.2 shows numerical tests for two network topologies and for two values of K, i.e. the trivial $K = 1$ case and $K = 3$, in order to illustrate the impact of multiple stages of infection. The agreement between output from the multi-stage pairwise model and results from explicit stochastic network simulation is good; in fact, in the large network limit, it will be exact for regular random networks.

This provides a heuristic validation of the new model. It is worth noting that for both networks, larger values of K result in epidemics that have a larger initial growth rate and a faster turnover. Typically, a large value of K leads to an epidemic which runs its course faster, as was the case for the simple compartmental model with multiple stages.

The probability that infection transmits across an isolated SI link is a fundamental quantity that appears in basic reproductive ratios and final epidemic size formulas. This is frequently called the *transmissibility*, and we denote it by $\tilde{\tau}$. Equally, the lifetime of an SI link (i.e. from the very moment that the I node becomes infected until either infection spreads or the infectious node recovers without infecting the susceptible node) plays a crucial role when considering network epidemics. Both will be covered in detail in the next section by considering a general recovery process.

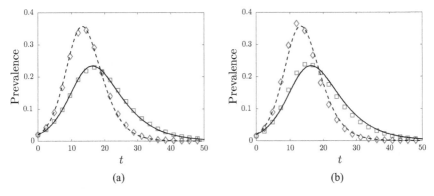

Fig. 9.2: Simulation of a SARS outbreak using $n\tau = 0.545$ [32, 261] and $\gamma^{-1} = 5.5$ days [325], with $n = 10$ and $N = 1000$. Lines correspond to the numerical solution of the pairwise model (9.5) ($K = 1$, solid lines; $K = 3$, dashed lines), while symbols represent the average of 250 outbreaks with ($\square$) $K = 1$ and ($\lozenge$) $K = 3$. We use (a) regular and (b) Erdős–Rényi random networks. Note that the fit is better for regular random networks.

The lifetime of an SI link and the probability of infection

As will become clear later, it is useful to consider the lifetime of an isolated SI link, assuming Markovian infection and a general recovery process. The isolation allows us to ignore transmissions to the S node from other sources. Let $f_I(t)$ be the density function of the infection process and T_I the time of transmission along the edge (ignoring the possibility of recovery first) so that

$$\xi_I(t) = \mathbb{P}(T_I > t) = \int_t^\infty f_I(u)du$$

is the associated survivor function. For a Markovian infection with rate τ, we get $f_I(t) = \tau \exp(-\tau t)$ and $\xi_I(t) = \exp(-\tau t)$.

Similarly, assume that the general recovery process is characterised by its density, $f_R(t)$, recovery time T_R and survivor function, ξ_R. Let T_{SI} be the random variable denoting the lifetime of the SI link. It is straightforward to define it as

$$\mathbb{P}(T_{SI} > t) = \mathbb{P}(T_I > t, T_R > t),$$

meaning that both recovery and infection happen after time t. However, $\mathbb{P}(T_{SI} > t) = \xi_{T_{SI}}$ is in fact the survivor function associated with T_{SI}. Thus,

$$\xi_{T_{SI}}(t) = \mathbb{P}(T_{SI} > t) = \mathbb{P}(T_I > t, T_R > t) = \mathbb{P}(T_I > t)\mathbb{P}(T_R > t) = \xi_I \xi_R,$$

and the density of the random variable T_{SI} is simply $-d\xi_{T_{SI}}/dt$, which leads to

$$f_{T_{SI}}(t) = f_I(t)\xi_R(t) + \xi_I(t)f_R(t).$$

Therefore, the expected lifetime of an isolated SI link is

$$\mathbb{E}(T_{SI}) = \int_0^\infty t f_{T_{SI}}(t)dt = \int_0^\infty t\left(f_I(t)\xi_R(t) + \xi_I(t)f_R(t)\right)dt.$$

By using $f_I(t) = \tau \exp(-\tau t)$ and $\xi_I(t) = \exp(-\tau t)$ in the above and integrating by parts, it can be shown that the average lifetime can be interpreted in terms of the Laplace transform of the density function corresponding to the recovery process at τ, $\mathcal{L}[f_R](\tau)$. Hence, in general, the following lemma holds.

Lemma 9.1 *The expected lifetime of an isolated SI link, with the general recovery process characterised by its density function $f_R(t)$ and a Markovian transmission process at rate τ, is*

$$\mathbb{E}(T_{SI}) = \frac{1}{\tau} - \frac{1}{\tau}\int_0^\infty \exp(-\tau t)f_R(t)dt = \frac{1 - \mathcal{L}[f_R](\tau)}{\tau}. \tag{9.6}$$

This result can be directly related to the probability of infection spreading across an isolated SI link. Let us first expand on this for the present setup, with the transmission process being Markovian and with a Γ-distributed infectious period. In this case, the following result for the per SI link probability of infection holds.

Lemma 9.2 *For the stochastic $SI^K R$ model, with the recovery times being gamma distributed,*

$$\Gamma(x; K, K\gamma) = \frac{1}{(K-1)!}(K\gamma)^K x^{K-1} e^{-K\gamma x}, \tag{9.7}$$

the probability of disease transmitting across an isolated SI link is given by

$$\tilde{\tau} = \int_0^\infty (1 - e^{-\tau t}) f_R(t) dt = 1 - \left(\frac{K\gamma}{\tau + K\gamma} \right)^K . \tag{9.8}$$

This is shown by integrating the probability of infection transmitting, $(1 - e^{-\tau t})$, over values of t given by the density $f_R(t)$. Alternatively, it can be shown that the probability of avoiding transmission in each stage of infection is $K\gamma/(\tau + K\gamma)$, so the probability of escaping infection in all K stages is $(K\gamma/(\tau + K\gamma))^K$. By rewriting expression (9.8) in the form

$$\tilde{\tau} = 1 - \left(\frac{K\gamma + \tau - \tau}{\tau + K\gamma} \right)^K = 1 - \left(1 - \frac{\tau}{\tau + K\gamma} \right)^K$$

and using the fact that $e^x = \lim_{n \to \infty} (1 + x/n)^n$, it follows that

$$\lim_{K \to \infty} \tilde{\tau}(K) = 1 - \exp\left(-\frac{\tau}{\gamma} \right). \tag{9.9}$$

In Table 9.1, we summarise the expected lifetime for a selection of possible recovery processes.

	$\mathbb{E}(T_{SI})$
Markovian	$\frac{1}{\tau + \gamma}$
Fixed length σ	$\frac{1}{\tau}(1 - e^{-\tau\sigma})$
$U(a,b)$	$\frac{1}{\tau}\left(1 - \frac{e^{-a\tau} - e^{-b\tau}}{\tau(b-a)}\right)$
$\Gamma(K, K\gamma)$	$\frac{1}{\tau}\left(1 - \left(\frac{K\gamma}{\tau + K\gamma}\right)^K\right)$
General	$\frac{1}{\tau}(1 - \mathcal{L}[f_R](\tau))$

Table 9.1: Expected lifetime of an isolated SI link for a range of recovery processes and Markovian transmission with rate τ.

Since the transmission process is Markovian with rate τ, the probability of infection transmitting across an SI link can also be obtained as $\tau \times \mathbb{E}(T_{SI})$, which for our current setup gives

$$\tau \times \frac{1}{\tau} \left(1 - \left(\frac{K\gamma}{\tau + K\gamma} \right)^K \right) = \tilde{\tau},$$

as expected, and this confirms our observation. This can be checked for all distributions shown in Table 9.1 by comparing directly to what one would get from evaluating $\int_0^\infty (1 - e^{-\tau t}) f_R(t) dt$.

Exercise 9.1. Check the result in Lemma 9.1 by evaluating

$$\mathbb{E}(T_{SI}) = \int_0^\infty t\left(f_I(t)\xi_R(t) + f_R(t)\xi_I(t)\right)dt.$$

Exercise 9.2. Check the result in Lemma 9.2 by evaluating

$$\tilde{\tau} = \int_0^\infty (1 - e^{-\tau t})\Gamma(t; K, K\gamma)dt.$$

The final epidemic size

The final epidemic size for the multi-stage pairwise model can be computed analytically, and this is summarised in the theorem below.

Theorem 9.3 *The final size of the epidemic* $r_\infty = [R](\infty)/N$ *in the pairwise model* (9.5) *with the classical closures, see Eqs.* (4.6) *and* (4.7), *and with a vanishingly small proportion of infected nodes at* $t = 0$ *is given by*

$$r_\infty = 1 - \theta^n, \quad \text{where } \theta = 1 - \tilde{\tau} + \tilde{\tau}\theta^{n-1}, \tag{9.10}$$

with $\tilde{\tau}$ *as defined in* (9.8). *Moreover, this is equivalent to*

$$s_\infty(=1 - r_\infty) = \left(1 - \tilde{\tau} + \tilde{\tau}s_\infty^{\frac{n-1}{n}}\right)^n, \tag{9.11}$$

which can be rearranged to give

$$\frac{s_\infty^{\frac{1}{n}} - 1}{\frac{1}{n-1}} = (n-1)\tilde{\tau}\left(s_\infty^{\frac{n-1}{n}} - 1\right). \tag{9.12}$$

In this chapter, we present three methods, corresponding to Theorems 9.3, 9.4 and 9.5 respectively, to prove this result. We begin with an early proof based on [166], with some of the omitted steps set in Exercise 9.3. In the next section we will present a different proof technique for a population in which all infections have exactly the same duration. The method presented there is perhaps more intuitive, and can be adapted to this model. Finally, we prove a general result using the edge-based compartmental model. Although we do not show it, the closure used here corresponds to the homogeneous network case for the edge-based compartmental model, that is, with $\psi(x) = x^n$. So with a few steps to demonstrate this equivalence (following Chapter 7), our final proof is probably the simplest.

Proof. We first introduce a parameter and some new variables

$$\zeta = \frac{n-1}{n}, \quad F = \frac{\sum_{i=1}^K [SI_i]}{[S]^\zeta}, \quad G = \frac{[SR]}{[S]^\zeta}, \quad M = \frac{[SS]}{\exp(n[S]^{1/n})[S]^{2\zeta}}, \quad L = \frac{[SS]}{[S]^\zeta},$$

and

$$P_i = \frac{[SI_i]}{[S]^\zeta} \quad \text{for } i = 1, 2, \dots, K. \tag{9.13}$$

Based on (9.5) and using that

$$[\dot{SR}] = -\tau \frac{[SR]\sum_{i=1}^{K}[SI_i]}{[S]} + K\gamma[SI_K],$$ (9.14)

the evolution equations for $F, G, L,$ and M are

$$\dot{F} = -\tau F - K\gamma P_K + \zeta\tau \frac{[SS]}{[S]}F,$$ (9.15a)

$$\dot{G} = K\gamma P_K,$$ (9.15b)

$$\dot{L} = -\zeta\tau \frac{[SS]}{[S]}F,$$ (9.15c)

$$\dot{M} = \tau M F.$$ (9.15d)

Given that $[I_i](0)=[I_i](\infty)=0$ for any $i=1,2,\ldots,K$, it implies that $F(0)=F(\infty)=0$. Integrating equation (9.15a) and using that $F = \frac{1}{\tau}\frac{\dot{M}}{M}$, $P_K = \frac{1}{K\gamma}\dot{G}$ and $\frac{[SS]}{[S]}F = -\frac{1}{\zeta\tau}\dot{L}$, leads to

$$0 = F(\infty) - F(0) = -\tau \int_0^\infty F(t)dt - K\gamma \int_0^\infty P_K(t)dt + \zeta\tau \int_0^\infty \frac{[SS](t)}{[S](t)}F(t)dt$$

$$= -[\ln(M(\infty)) - \ln(M(0))] - [G(\infty) - G(0)] - [L(\infty) - L(0)]$$

$$= -[\ln(M(\infty)) - \ln(M(0))] - \tilde{\tau}[L(\infty) - L(0)],$$

(9.16)

where in the last step we have used the fact that $G(0) = 0$ and the relation

$$G(\infty) = \frac{[SR]_\infty}{[S]_\infty^\zeta} = (\tilde{\tau} - 1)[L(\infty) - L(0)].$$ (9.17)

Furthermore, starting from system (9.5) it is possible to show that $[SS]_\infty$ can be written as

$$[SS]_\infty = \frac{n[S]_\infty^{2\zeta}}{N^{\zeta-1/n}}.$$ (9.18)

Substituting Eqs. (9.17) & (9.18) into Eq. (9.16) and using the fact that $[S](0) = N$ yields

$$0 = -nN^{1/n} + n[S]_\infty^{1/n} - \tilde{\tau}\left(\frac{n[S]_\infty^\zeta}{N^{\zeta-1/n}} - nN^{1/n}\right) \Leftrightarrow$$

$$0 = -1 + \left(\frac{[S]_\infty}{N}\right)^{1/n} - \tilde{\tau}\left[\left(\frac{[S]_\infty}{N}\right)^\zeta - 1\right].$$

Introducing the fraction of susceptible individuals as $s_\infty = [S]_\infty/N$, the above equation can be rewritten as

$$1 - s_\infty^{1/n} = \tilde{\tau}\left(1 - s_\infty^\zeta\right),$$

or alternatively, as another implicit equation for s_∞,

$$s_\infty = \left(1 - \tilde{\tau} + \tilde{\tau} s_\infty^\zeta\right)^n \Leftrightarrow s_\infty = \theta^n, \text{ where } \theta = \left(1 - \tilde{\tau} + \tilde{\tau} s_\infty^\zeta\right). \tag{9.19}$$

Since $[I]_i(\infty) = 0$, introducing $r_\infty = [R]_\infty / N$ yields the desired expression for the final size of an epidemic

$$r_\infty = 1 - s_\infty = 1 - \theta^n.$$

Using the fact that $\theta = s^{1/n}$ and that $\zeta = \frac{n-1}{n}$, equation (9.19) can be rewritten in the form

$$s_\infty^{1/n} = 1 - \tilde{\tau} + \tilde{\tau}\left(s_\infty^{1/n}\right)^{n-1} \Leftrightarrow \theta = 1 - \tilde{\tau} + \tilde{\tau}\theta^{n-1}.$$

This completes the proof.   □

In Fig. 9.3, the final epidemic size based on solving the implicit Eq. (9.10) shows clearly that the number of stages, and thus the precise shape of the distribution, despite their mean being the same, does affect the number of nodes that are eventually infected. This contrasts with the compartmental model, where different gamma distributions with the same mean led to the same final epidemic size.

This qualitative difference is because in a compartmental model, each transmission an individual makes goes to a new individual, and so the average number of transmissions corresponds directly to the average number of infections caused. In contrast, in a network model, transmissions may be repeated on the same edge and only the first has an effect. What matters is whether at least one transmission happens rather than how many transmissions occur. Thus, we expect $\tilde{\tau}$ to be more important than the mean number of transmissions. Indeed, as implied by Theorem 6.20, the final epidemic size we predict is exactly the same as long as we choose distributions such that the resulting $\tilde{\tau}$s are the same. This can be achieved even in the context of the multi-stage model by finding different (γ, K) pairs which lead to the same value of $\tilde{\tau}$.

An important connection between different modelling approaches can be made. In Chapter 6, Eq. (6.8) gives an implicit formula for the final epidemic size for the non-Markovian case derived using a percolation-based theory. Applying this to the current setup, Markovian transmission and gamma-distributed infectious periods lead to setting $p_i = \tilde{\tau}$, see Eq. (9.8), and the distribution of this is $\delta(p - \tilde{\tau})$. Plugging this into Eq. (6.8) leads to

$$\mathcal{A} = 1 - \theta^n, \tag{9.20a}$$

$$\theta = 1 - \tilde{\tau} + \tilde{\tau}\theta^{n-1}, \tag{9.20b}$$

(taking the smaller solution for θ if more than one exists) where we have used that $\psi(x) = \sum p_k x^k = x^n$, with $\psi'(x) = nx^{n-1}$, since the network is regular and we have replaced ω by θ. By making $\mathcal{A} = r_\infty$, it follows that the equations above are equivalent to (9.10). This illustrates the flexibility of modelling, where the network epidemic can be tackled as a percolation problem or a mean-field model such as pairwise. This can be exploited to select models based on specific questions. In this

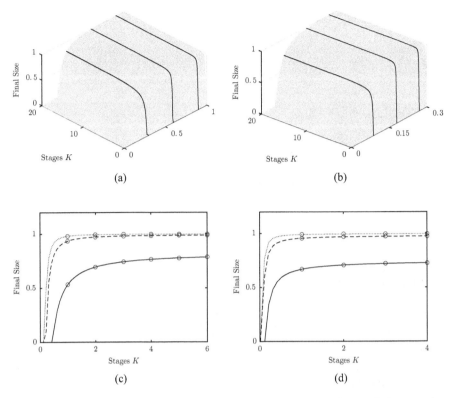

Fig. 9.3: Dependence of the final size of an epidemic (9.10) on the per-link transmission rate τ and the number of stages K in the pairwise model (9.5), with $\gamma = 0.4545$ for different average node degrees, and with $n = 4$ in (a, c) and $n = 10$ in (b, d). Panels (c) and (d) represent the cross sections of the surfaces along the black lines in (a) and (b), respectively. In (c), the following parameter values are considered: $\tau = 0.3$ (solid line), $\tau = 0.6$ (dashed line), $\tau = 0.9$ (dotted line), while in (d), we have $\tau = 0.09$ (solid line), $\tau = 0.18$ (dashed line), $\tau = 0.27$ (dotted). Circles correspond to integer values of K on each curve.

case, for example, the final epidemic size is arguably more straightforward to obtain via a percolation theory approach, while the time evolution of the epidemic needs a dynamical system-focused approach using the pairwise or edge-based compartmental models.

Perhaps equally importantly, the evolution equation for $[SI] = \sum_{j=1}^{K} [SI_j]$, which can be obtained from system (9.5) by summing the appropriate equations, shows that the single positive term that drives the creation of the SI links is

$$\tau \sum_{i=1}^{K} [SSI_i] \simeq \tau \frac{n-1}{n} \sum_{i=1}^{K} \frac{[SS][SS_i]}{[S]} = \tau \frac{n-1}{n} \frac{[SS]}{[S]} \sum_{i=1}^{K} [SI_i] = \tau \frac{n-1}{n} \frac{[SS]}{[S]} [SI],$$

which means that SI links are created at rate

$$\tau \frac{(n-1)}{n} \frac{[SS]}{[S]} \simeq \tau \frac{(n-1)}{n} \frac{n[S]\frac{[S]}{N}}{[S]} = \tau(n-1)\frac{[S]}{N}. \tag{9.21}$$

Combining this with the expected lifetime of an SI link leads to the basic reproductive ratio of SI links being

$$R_0^p = \frac{\tau(n-1)}{\tau+\gamma} \frac{[S]}{N} \simeq (n-1)\tilde{\tau}, \tag{9.22}$$

which allows us to reinterpret the final epidemic size relation (9.12) as

$$\frac{s_\infty^{\frac{1}{n}} - 1}{\frac{1}{n-1}} = R_0^p \left(s_\infty^{\frac{n-1}{n}} - 1 \right). \tag{9.23}$$

Hence, keeping the $(n-1)$ term explicitly in Eq. (9.12), and in some subsequent equations, allows us to show how R_0^p enters the implicit equation of the final epidemic size. We will show that the final epidemic size relation remains the same for Markovian transmission and general recovery processes. Moreover, we will discuss the limiting behaviour of Eq. (9.23) when $N \to \infty$ and $n \to \infty$.

Exercise 9.3. Using the equations of the pairwise model with multiple infectious stages (9.5) and Eq. (9.14), show that

a. the evolution equations for F, G, L and M given by system (9.15) are indeed correct;

b. the relation $G(\infty) = \frac{[SR]_\infty}{[S]_\infty^{\frac{\zeta}{\tau}}} = (\tilde{\tau} - 1)[L(\infty) - L(0)]$ holds, and that;

c. $[SS]_\infty$ can be expressed as $[SS]_\infty = \frac{n[S]_\infty^{2\zeta}}{N^{\zeta - 1/n}}$.

9.2 Pairwise model for epidemics with non-Markovian recovery times

In this section, we consider Markovian transmission and fixed infection duration, where an infected individual remains infected and infectious for precisely σ units of time, as considered in [182].

9.2.1 Derivation of the standard compartmental and pairwise model

Before we explain how to derive the pairwise model, we consider how the standard compartmental model can be extended to this case. This can be done by keeping in mind that all those who have become infected at time $(t - \sigma)$ will recover precisely at time t. This means that the recovery term, $-\gamma I(t)$, has to be replaced with $-\tau S(t - \sigma)I(t - \sigma)$, i.e. the rate at which new infectious individuals were created at time $(t - \sigma)$. This gives rise to the delay differential equation system below.

SIR compartmental model with constant infection duration

$$\dot{S}(t) = -\tau \frac{n}{N} S(t)I(t), \qquad (9.24a)$$

$$\dot{I}(t) = \tau \frac{n}{N} S(t)I(t) - \tau \frac{n}{N} S(t - \sigma)I(t - \sigma). \qquad (9.24b)$$

A similar argument is needed when generalising the pairwise model to this case. In particular, since infection is Markovian, the evolution equation for $[S]$ remains the same, namely

$$[\dot{S}] = -\tau[SI].$$

However, the evolution equation for $[I]$ changes as the $-\gamma[I]$ term is replaced by $-\tau[SI](t - \sigma)$, which represents the rate at which individuals got infected exactly σ time units ago. This leads to

$$[\dot{I}] = +\tau[SI] - \tau[SI](t - \sigma).$$

The evolution equation for $[SS]$ pairs remains unchanged as it only involves infection, and this is Markovian. Hence, we have

$$[\dot{SS}] = -2\tau[SSI].$$

Finally, the main challenge is posed by completing the evolution equation for $[SI]$ pairs. Terms involving transmission remain unchanged: (a) infection within the pair, $-\tau[SI]$; (b) infection from outside the pair, $-\tau[ISI]$; and (c) instantaneous creation of SI pairs, $+\tau[SSI]$. However, the challenge is to write down the recovery of SI pairs at time t. Naively, one would imagine that this is simply given by $-\tau[SSI](t - \sigma)$, which denotes the rate at which SI links were created σ times units ago. However, not all SI links created at time $(t - \sigma)$ will make it to time t. In particular, they are depleted by two different processes: (a) within-pair infection; and (b) infection from outside. Therefore, we are seeking a discount factor for the $-\tau[SSI](t - \sigma)$ term. This can be done by considering SI links as a cohort, say $c(t)$, and capture the decay of this variable from $(t - \sigma)$ to t. Hence, accounting for within pair infection and infection from outside, the evolution equation for $c(t)$ is

$$\dot{c}(t) = -\tau c(t) - \tau(n - 1)\frac{[SI]}{n[S]}c(t), \qquad (9.25)$$

where $(n-1)\frac{[SI]}{n[S]}$ is simply the expected number of infected neighbours of an S node with $(n-1)$ available stubs. Using separation of variables and integrating from $(t-\sigma)$ to t leads to

$$c(t) = c(t-\sigma)e^{-\tau \int_{t-\sigma}^{t}\left(1+(n-1)\frac{[SI](u)}{n[S](u)}\right)du}. \tag{9.26}$$

With the discount factor in mind, the evolution equation for $[SI]$ is

$$[\dot{SI}] = \tau([SSI] - [ISI] - [SI]) - \tau[SSI](t-\sigma)e^{-\tau\int_{t-\sigma}^{t}\left(1+(n-1)\frac{[SI](u)}{n[S](u)}\right)du}.$$

Therefore, summarising all the above and by including all the closures, the pairwise DDE with discrete and distributed delays for the non-Markovian case is

SIR pairwise model for homogenous networks with constant infection duration

$$[\dot{S}](t) = -\tau[SI](t), \tag{9.27a}$$

$$[\dot{I}](t) = \tau[SI](t) - \tau[SI](t-\sigma), \tag{9.27b}$$

$$[\dot{SS}](t) = -2\tau\zeta\frac{[SS](t)[SI](t)}{[S](t)}, \tag{9.27c}$$

$$[\dot{SI}](t) = -\tau[SI](t) - \tau\zeta\left(\frac{[SI](t)[SI](t)}{[S](t)} - \frac{[SS](t)[SI](t)}{[S](t)}\right)$$
$$\qquad - \tau\zeta\frac{[SS](t-\sigma)[SI](t-\sigma)}{[S](t-\sigma)}e^{-\int_{t-\sigma}^{t}\left([SI](u)\frac{\tau\zeta}{[S](u)}+\tau\right)du}, \tag{9.27d}$$

where $\zeta = (n-1)/n$.

9.2.2 Dynamics in time and final epidemic size

This pairwise system (9.27) can now be analysed both analytically and numerically. One of the first possible steps to assess the validity of such a model is to compare output from explicit stochastic simulation to the numerical solution of the pairwise model. Figure (9.4) shows that the pairwise model does well, especially for homogeneous networks, and that there are significant differences compared to the equivalent Markovian case, where the average time spent being infectious also equals σ. Similarly to the multi-stage model, the fixed infectious period leads to an epidemic with a larger initial growth rate when compared to the case of exponentially distributed infectious periods.

However, the new pairwise model can be used to analytically compute an implicit final epidemic size relation, which will shed light on how the non-Markovian or

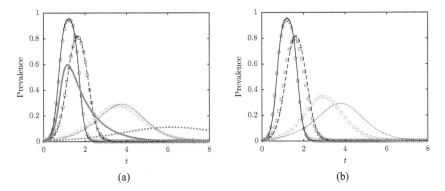

Fig. 9.4: Stochastic non-Markovian epidemics on networks with $N = 1000$ nodes. (a) Solid lines show the solution of (9.5) and circles/squares/diamonds correspond to simulation for homogeneous (random regular) graphs with $\langle k \rangle = 5$, 10 and 15, respectively. The dotted ($\langle k \rangle = 5$) and solid ($\langle k \rangle = 15$) grey lines correspond to purely Markovian epidemics. (b) The same as for panel (a) but for Erdős–Rényi random networks with $\langle k \rangle = 5$, 10 and 15. The transmission rate is $\tau = 0.55$ and the infectious period is fixed, $\sigma = 1$.

general recovery process impacts on the outcome of the epidemic. The result is summarised in the theorem below.

Theorem 9.4 *The final size of the epidemic,* $r_\infty = [R](\infty)/N = 1 - s_\infty = 1 - \frac{[S](\infty)}{[S](0)}$ *in the pairwise model (9.27) is given by*

$$\frac{s_\infty^{\frac{1}{n}} - 1}{\frac{1}{n-1}} = \frac{n-1}{N} \left(1 - e^{-\tau\sigma}\right) [S]_0 \left(s_\infty^{\frac{n-1}{n}} - 1\right). \tag{9.28}$$

Proof. Only the main steps of the proof are given. We start by reducing the four-dimensional system to just two equations. This can be done by noting that a first integral relating $[SS]$ and $[S]$ can be derived. As in the previous section, we set $\zeta = (n-1)/n$. By using the evolution equations for these two variables in system (9.27), we can write

$$\frac{d[SS]}{d[S]} = \frac{-2\tau\zeta \frac{[SS][SI]}{[S]}}{-\tau[SI]} = 2\zeta \frac{[SS]}{[S]}.$$

Solving this equation leads to $\frac{[SS]}{[S]^{2\zeta}} = \alpha$, where α is a constant. Thus, $\frac{[SS](t)}{[S]^{2\zeta}(t)}$ is an invariant quantity, and its value at $t = 0$ allows us to determine the constant α. This is given by

$$\alpha = \frac{[SS](0)}{[S]^{2\zeta}(0)} = \frac{[SS]_0}{[S]_0^{2\zeta}} = \frac{n[S]_0 \frac{[S]_0}{N}}{[S]_0^{2\zeta}} = \frac{n}{N} [S]_0^{\frac{2}{n}},$$

where we assumed that at time $t = 0$, nodes are placed at random and thus $[SS]_0 = n[S]_0 \frac{[S]_0}{N}$. This first integral allows us to reduce system (9.27) to just two equations:

$$[\dot{S}](t) = -\tau[SI](t), \tag{9.29a}$$

$$[\dot{SI}](t) = \tau\kappa[S]^{\frac{n-2}{n}}(t)[SI](t) - \tau[SI](t) - \tau\zeta\frac{[SI](t)}{[S](t)}[SI](t)$$

$$- \tau\kappa[S]^{\frac{n-2}{n}}(t-\sigma)[SI](t-\sigma)e^{-\int_{t-\sigma}^{t}\tau\zeta\frac{[SI](u)}{[S](u)}+\tau du}, \tag{9.29b}$$

where $\kappa = \frac{(n-1)}{N}[S]_0^{\frac{2}{n}}$. Noting that an equation of the form

$$\dot{w} = \text{in}(t) - \text{out}(t)w(t) - \text{in}(t-\sigma)e^{-\int_{t-\sigma}^{t}\text{out}(u)du}$$

can be solved to give

$$w(t) = \int_{t-\sigma}^{t} \text{in}(u)e^{-\int_u^t \text{out}(s)ds}du,$$

for $t \geq \sigma$, it follows that setting

$$\text{in}(t) = \tau\kappa[S]^{\frac{n-2}{n}}(t)[SI](t),$$

$$\text{out}(t) = \tau + \tau\zeta\frac{[SI](t)}{[S](t)},$$

the equation for $[SI](t)$, (9.29b), can be solved and yields

$$[SI](t) = \int_{t-\sigma}^{t} \tau\kappa[S]^{\frac{n-2}{n}}(u)[SI](u)e^{-\int_u^t \tau+\tau\zeta\frac{[SI](s)}{[S](s)}ds}du.$$

Substituting $[SI] = -[\dot{S}](t)/\tau$ in the integral above followed by some algebra leads to

$$[SI](t) = -\frac{[\dot{S}](t)}{\tau} = -\kappa[S]^{\zeta}(t)\int_{t-\sigma}^{t}[S]^{-\frac{1}{n}}(u)[S]'(u)e^{-\tau(t-u)}du. \tag{9.30}$$

By integrating the above from 0 to ∞ and after the explicit analytical evaluation of a double integral, we arrive at

$$\frac{[S]_\infty^{\frac{1}{n}} - [S]_0^{\frac{1}{n}}}{\frac{1}{n}} = \frac{n}{N}(1 - e^{-\tau\sigma})[S]_0^{\frac{n+1}{n}}\left(\frac{[S]_\infty^{\zeta}}{[S]_0^{\zeta}} - 1\right). \tag{9.31}$$

Setting $s_\infty = \frac{[S]_\infty}{[S]_0}$, the equation above can be written as

$$\frac{s_\infty^{\frac{1}{n}} - 1}{\frac{1}{n-1}} = \frac{n-1}{N}(1 - e^{-\tau\sigma})[S]_0\left(s_\infty^{\zeta} - 1\right),$$

which completes the proof.   □

From the invariance of $[SS]/[S]^{2\zeta} = [SS]/[S]^{2(n-1)/n}$, we see that if we define $\theta(t) = ([S](t)/S[0])^{1/n}$, then $[S](t) = S[0]\theta(t)^n$ and $[SS](t) = [SS](0)\theta(t)^{2n-2}$. We can think of θ as the probability that a random neighbour has not transmitted infection to an initially susceptible node: a single susceptible node has n neighbours which could transmit to it, while two connected susceptible nodes have $(2n-2)$ other neighbours which could transmit to them. In fact, the remainder of the proof is determining $\theta(\infty)$. This same invariance applies in the SI^KR model, and this proof can be adapted to that model. Several other important observations can be made:

1. From the equation of $[SI]$, see (9.29b), it is clear that the rate at which new SI links are generated close to $t = 0$ is

$$\tau\kappa[S]^{\frac{n-2}{n}}(t)\Big|_{[S](0)=[S]_0} = \tau\frac{(n-1)}{N}[S]_0^{\frac{2}{n}}[S]_0^{\frac{n-2}{n}} = \tau(n-1)\frac{[S]_0}{N}.$$

2. Now multiplying this by the average lifetime of an SI link, i.e. $\frac{1}{\tau}(1-e^{-\tau\sigma})$, allows us to re-introduce the pairwise basic reproductive ratio as

$$R_0^p = (n-1)\left(1-e^{-\tau\sigma}\right)\frac{[S]_0}{N}, \tag{9.32}$$

which in fact represents the average number of new SI edges created by an existing one.

3. Hence, the implicit equation for the final epidemic size can be written as

$$\frac{s_\infty^{\frac{1}{n}}-1}{\frac{1}{n-1}} = R_0^p\left(s_\infty^{\frac{n-1}{n}}-1\right). \tag{9.33}$$

The observations above provide a strong basis for extending the implicit final epidemic size formula for a general recovery process and lead to the conjecture that the following relation for the final epidemic size holds:

$$\frac{s_\infty^{\frac{1}{n}}-1}{\frac{1}{n-1}} = R_0^p\left(s_\infty^{\frac{n-1}{n}}-1\right) \text{ where } R_0^p = \tau(n-1)\mathbb{E}(T_{SI})\frac{[S]_0}{N}, \tag{9.34}$$

where, as before, $\mathbb{E}(T_{SI}) = \frac{1-\mathcal{L}[f_r](\tau)}{\tau}$ is the expected lifetime of an SI link, and it depends on the recovery process.

For the fully Markovian case, the final epidemic size formula reduces to

$$\frac{s_\infty^{\frac{1}{n}}-1}{\frac{1}{n-1}} = \frac{\tau(n-1)}{\tau+\gamma}\left(s_\infty^{\frac{n-1}{n}}-1\right),$$

which is identical to the result obtained in Chapter 4 in Eq. (4.17), after some appropriate algebraic manipulation. Similarly, for the multi-stage model, the conjectured formula gives

$$\frac{s_\infty^{\frac{1}{n}} - 1}{\frac{1}{n-1}} = \tau(n-1)\left(1 - \left(\frac{K\gamma}{\tau + K\gamma}\right)^K\right)\left(s_\infty^{\frac{n-1}{n}} - 1\right) = \tilde{\tau}(n-1)\left(s_\infty^{\frac{n-1}{n}} - 1\right),$$

which is identical to the previously derived final epidemic size formula in Eq. (9.12), upon assuming that $[S]_0/N = 1$.

Furthermore, it can be shown that in the limit of a fully connected network, $N \to \infty$ and $n = (N-1) \to \infty$, the left side of Eq. (9.34) leads to

$$\lim_{N,n\to\infty} \frac{s_\infty^{\frac{1}{n}} - 1}{\frac{1}{n-1}} = \ln(s_\infty).$$

Applying this limit in (9.34) and assuming an appropriate scaling between τ and n such that R_0^p has a well-defined limit, allows us to recover the final epidemic size formula corresponding to the simple compartmental model, namely

$$\ln(s_\infty) = R_0^p(s_\infty - 1) \Leftrightarrow r_\infty = 1 - e^{-R_0^p r_\infty}.$$

Our conjecture follows from Theorem 6.20, which we presented without proof. However, in the following section, we will prove this conjecture in a more general form by using the edge-based compartmental modelling technique. Thus, the only detail to complete the proof would be to show that our pairwise model here is equivalent to the corresponding edge-based compartmental model if the initial infection is randomly distributed and the nodes all have the same degree. Alternative rigorous proofs of this result can be found in [271, 328]. To confirm this result numerically, in Fig. 9.5 a series of final epidemic size curves are plotted for different choices of the infectious period distribution, but such that all have the same mean. Several important observations can be made. First, the agreement between simulation and analytical results is excellent. In Fig. 9.5a, it can be seen that for values of τ close to 0.083 there is huge variation in final epidemic size, despite the value of τ and mean recovery period being identical. Around this region, the final epidemic size can vary by as much 15%. This variation can be attributed solely to differences in the distribution of the infectious periods.

In Fig. 9.5b, we plot the final epidemic size against the values of R_0^p, and, as expected, the curves collapse onto a single line, and illustrate the universality of the newly defined pairwise basic reproductive ratio and implicit final epidemic size equation (9.34). The usefulness of the new threshold can be illustrated by considering a number of different recovery time distributions, e.g. $\Gamma(\frac{1}{2}, 2)$, $\text{Exp}(1)$, $\Gamma(2, \frac{1}{2})$ and Fixed(1), and by working out the explicit expression of R_0^p. Based on formula (9.34), the analytical expressions for R_0^p are

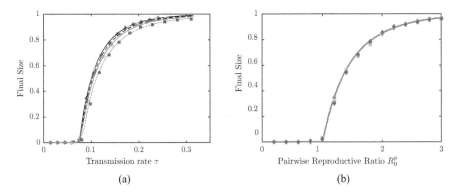

(a)   (b)

Fig. 9.5: Diamonds/circles/squares correspond to simulations using regular random networks with $\langle k \rangle = 15$ and using fixed and two different but gamma-distributed, $\Gamma(\alpha, \beta)$, infectious periods ($\bigcirc$, - shape $\alpha = 2$; scale $\beta = 1/2$; $\square$, - shape $\alpha = 1/2$; scale $\beta = 2$), respectively. (a) The solid lines correspond to the analytical final epidemic size for fixed (9.28) and general (9.34) infectious periods, with the grey solid line denoting the purely Markovian case. (b) Same as panel (a) but with the analytical and the simulated final epidemic sizes plotted against the pairwise basic reproductive ratio.

$$R^p_{0,\Gamma(\frac{1}{2},2)} = \mu \left(1 - \frac{1}{\sqrt{1+2\tau}} \right), \quad R^p_{0,\text{Exp}(1)} = \mu \left(\frac{\tau}{\tau+1} \right),$$

$$R^p_{0,\Gamma(2,\frac{1}{2})} = \mu \left(1 - \frac{4}{(2+\tau)^2} \right), \quad R^p_{0,\text{Fixed}(1)} = \mu \left(1 - e^{-\tau} \right),$$

where $\mu = \frac{(n-1)[S]_0}{N}$. For the expressions above, the following inequality holds:

$$R^p_{0,\Gamma(\frac{1}{2},2)} \leq R_{0,\text{Exp}(1)} \leq R^p_{0,\Gamma(2,\frac{1}{2})} \leq R^p_{0,\text{Fixed}(1)}.$$

It is important to note that (a) all infectious period distributions have the same mean 1 and (b) the variances satisfy the converse inequality, with higher variance in recovery time (i.e. 2, 1, 1/2 and 0) giving a smaller R^p_0 value, despite τ being fixed. Non-exponential infectious periods have a significant impact and highlight that is imperative to use the correct recovery process in order to avoid errors in fitting and estimating basic reproductive ratios and final epidemic sizes. It is worth noting that the analysis above can be made more rigorous by choosing a distribution with at least two parameters which allow for a better control over the mean and variance of the distribution. In this way, one can keep the mean fixed while changing the variance. This in turn will allow for a rigorous investigation of the impact of the variance in infectious periods on the pairwise basic reproductive ratio. We expect that, in general, higher variance leads to smaller R^p_0 values if the mean is kept fixed (see [270] for further analysis). There may be exceptions to this case. Lastly, we wish to point out that the growth-based basic reproductive ratio based on the pairwise model for purely Markovian epidemics in random regular networks is $\overline{R}_0 = \frac{\tau(n-2)}{\gamma}$, which, at the threshold, is equivalent to $R_0^p = 1$, namely

$$\overline{R}_0 = \frac{\tau(n-2)}{\gamma} = 1 \Leftrightarrow R_0^p = \frac{\tau(n-1)}{\tau+\gamma} = 1,$$

where we assumed that $[S]_0/N \simeq 1$ at $t = 0$ (see also Section 4.4).

9.2.3 Pairwise model for a general recovery process

The pairwise model with fixed infectious period (9.27) can be generalised further, as shown in [271], to account for a general recovery process, with probability density given by $f_I(t)$ and a corresponding survival function $\xi_I(t) = \int_0^t f_I(a)da$. The resulting pairwise model is as follows:

SIR pairwise model for homogenous networks with general recovery

$$[\dot{S}](t) = -\tau[SI](t), \tag{9.35a}$$

$$[\dot{SS}](t) = -2\tau\frac{n-1}{n}\frac{[SS](t)[SI](t)}{[S](t)}, \tag{9.35b}$$

$$[\dot{I}](t) = \tau[SI](t) - \int_0^t \tau[SI](t-a)f_I(a)da - \int_t^\infty i(0,a-t)\frac{f_I(a)}{\xi_I(a-t)}da, \tag{9.35c}$$

$$[\dot{SI}](t) = \tau\frac{n-1}{n}\frac{[SS](t)[SI](t)}{[S](t)} - \tau\frac{n-1}{n}\frac{[SI](t)}{[S](t)}[SI](t) - \tau[SI](t)$$
$$\quad - \int_0^t \tau\frac{n-1}{n}\frac{[SS](t-a)[SI](t-a)}{[S](t-a)}e^{-\int_{t-a}^t \tau\frac{n-1}{n}\frac{[SI](s)}{[S](s)}+\tau ds}f_I(a)da$$
$$\quad - \int_t^\infty \frac{n}{N}[S]_0 i(0,a-t)e^{-\int_0^t \tau\frac{n-1}{n}\frac{[SI](s)}{[S](s)}+\tau ds}\frac{f_I(a)}{\xi_I(a-t)}da, \tag{9.35d}$$

---

where $i(0,a)$ specifies the number of infected nodes of age a at time t. The full details of the technical derivation can be found in [271], as well the proof of the positivity of the solutions, the identification of a first integral of the system, which then allowed the authors to confirm the general final epidemic size relation Eq. (9.34) for this more general model. We note that a similar result is derived in [328], starting from the message passing model [163] but using a specific rather than general initial condition.

9.3 EBCM approach to non-Markovian dynamics

The complexity we have seen using pairwise approximations to study simple non-Markovian disease processes suggests that as the disease or network becomes more complicated, having a simpler modelling approach will be important. For SIR disease, we have such a framework, the EBCM approach, and we will use it to explore more complicated scenarios.

We consider a disease with non-Markovian transmission and recovery processes spreading through a static network with heterogeneous degree distribution. We define $\tau(a)$ and $\gamma(a)$ to be the instantaneous transmission and recovery rates for a node which has been infected for a units of time. That is, $\tau(a) = -\xi_I'(a)/\xi_I(a)$ and $\gamma(a) = -\xi_R'(a)/\xi_R(a)$. We take the network to be a Configuration Model network in which the probability of having degree k is $P(k)$.

We will assume that the disease is introduced by the instantaneous infection of a proportion ρ chosen uniformly at random from the population at time $t = 0$. Specifically, these individuals will all have an age of infection $a = 0$ at the initial time. We use the same variables as in our earlier EBCM equations, but use capital letters. We add $\phi_I(t,a)$ to denote the density for a neighbour of a test node to have been infected at time $(t-a)$, have not yet recovered and have not yet transmitted to the test node. Then, $\Phi_I(t) = \int_0^t \phi_I(t,a)\,da$. We similarly define $i(t,a)$ to be the density for a test node to have been infected at time $(t-a)$ and not yet recovered. Then, $I(t) = \int_0^t i(t,a)\,da$. Our variables are summarised in Table 9.2.

Variable	Definition
$\Theta(t)$	The probability that an initially susceptible test node has not received a transmission from a random neighbour by time t.
$\Phi_S(t)$	The probability that a random neighbour of a test node u is still susceptible.
$\Phi_I(t)$	The probability that a random neighbour of a test node u is infected, but has not transmitted to u.
$\phi_I(t,a)$	The density for a random neighbour of a test node u to be infected, have not transmitted to u by time t and be a units of time into its infection, $\Phi_I(t) = \int_0^t \phi_I(t,a)\,da$.
$\Phi_R(t)$	The probability that a random neighbour of a test node u has been infected and recovered without transmitting to u.
$S(t)$	The number of susceptible nodes.
$I(t)$	The number of infected nodes.
$i(t,a)$	The density of infected nodes that were infected at time $(t-a)$.
$R(t)$	The number of recovered nodes.
$\psi(x)$	The probability-generating function of the degree distribution: $\sum_{k=0}^{\infty} P(k)x^k$.

Table 9.2: The variables of the edge-based compartmental model.

By the arguments in Section 6.5, we know that $S(t) = N(1 - \rho)\psi(\Theta(t))$ and $\Phi_S(t) = (1 - \rho)\psi'(\Theta(t))/\langle K \rangle$, where $\psi(x) = \sum_k P(k)x^k$. The full system is given below.

EBCM model for non-Markovian transmission and recovery
Assuming that a fraction ρ of the population is randomly infected at time $t = 0$, the EBCM system is

$$\Phi_S(t) = (1 - \rho)\frac{\psi'(\Theta(t))}{\langle K \rangle}, \tag{9.36a}$$

$$\phi_I(t,0) = -\dot{\Phi}_S(t)$$

$$= \rho\delta(t) + (1 - \rho)\frac{\psi''(\Theta(t))}{\langle K \rangle}\int_0^t \tau(a)\phi_I(t,a)\,\mathrm{d}a, \tag{9.36b}$$

$$\left(\frac{\partial}{\partial t} + \frac{\partial}{\partial a}\right)\phi_I(t,a) = -(\tau(a) + \gamma(a))\phi_I(t,a) \qquad 0 \leq a \leq t, \tag{9.36c}$$

$$\Phi_I(t) = \int_0^t \phi_I(t,a)\,\mathrm{d}a, \tag{9.36d}$$

$$\frac{\mathrm{d}}{\mathrm{d}t}\Theta(t) = -\int_0^t \tau(a)\phi_I(t,a)\,\mathrm{d}a, \tag{9.36e}$$

$$\Phi_R(t) = \Theta - \Phi_S - \Phi_I, \tag{9.36f}$$

$$S(t) = N(1 - \rho)\psi(\Theta(t)), \tag{9.36g}$$

$$\left(\frac{\partial}{\partial t} + \frac{\partial}{\partial a}\right)i(t,a) = -\gamma(a)i(t,a) \qquad 0 \leq a \leq t, \tag{9.36h}$$

$$i(t,0) = -\dot{S}(t)$$

$$= N\rho\delta(t) + N(1 - \rho)\psi'(\Theta(t))\int_0^t \tau(a)\phi_I(t,a)\,\mathrm{d}a, \tag{9.36i}$$

$$I(t) = \int_0^t i(t,a)\,\mathrm{d}a, \tag{9.36j}$$

$$R(t) = N - S - I, \tag{9.36k}$$

where $\delta(t)$ is the Dirac delta function. We take the initial condition

$$\Theta(0) = 1. \tag{9.36l}$$

Theorem 9.5 *The final size of the epidemic* $r_\infty = R(\infty)/N$ *in the EBCM system* (9.36) *with a vanishingly small proportion of infected nodes at time* $t = 0$ *is*

$$r_\infty = 1 - \psi(\Theta(\infty)),$$

where $\Theta(\infty)$ is the smallest solution to

$$\Theta = 1 - \tilde{\tau} + \tilde{\tau}\frac{\psi'(\Theta)}{\langle K \rangle},$$

and $\tilde{\tau} = \int_0^\infty \tau(a)\exp(-\int_0^a \tau(\alpha) + \gamma(\alpha)\,d\alpha)\,da$ is the probability that infection spreads across an SI edge.

Proof. We have

$$\Theta(\infty) = 1 - \int_0^\infty \int_0^t \tau(a)\phi_I(t,a)\,da\,dt.$$

Interchanging the order of integration yields

$$\Theta(\infty) = 1 - \int_0^\infty \int_a^\infty \tau(a)\phi_I(t,a)\,dt\,da.$$

We set $u = (t-a)$ and note that $\phi_I(t,a) = \phi_I(u,0)\exp(-\int_0^a \tau(\alpha) + \gamma(\alpha)\,d\alpha)$. Then,

$$
\begin{aligned}
\Theta(\infty) &= 1 - \int_0^\infty \int_0^\infty \phi_I(u,0)\tau(a)e^{-\int_0^a \tau(\alpha)+\gamma(\alpha)\,d\alpha}\,du\,da \\
&= 1 - \left[\int_0^\infty \phi_I(u,0)\,du\right]\int_0^\infty \tau(a)e^{-\int_0^a \tau(\alpha)+\gamma(\alpha)\,d\alpha}\,da \\
&= 1 + \left[\int_0^\infty \dot{\Phi}_S(u)\,du\right]\int_0^\infty \tau(a)e^{-\int_0^a \tau(\alpha)+\gamma(\alpha)\,d\alpha}\,da \\
&= 1 + (\Phi_S(\infty) - \Phi_S(0))\tilde{\tau} \\
&= 1 + \tilde{\tau}\frac{\psi'(\Theta(\infty))}{\langle K \rangle} - \tilde{\tau}.
\end{aligned}
$$

All the calculations have been done for vanishingly small ρ.

One solution to this is always $\Theta = 1$, but above the epidemic threshold, a second, smaller solution appears. Repeating these calculations with small, but positive, ρ shows that this smaller solution is the correct limit. $\square$

If we take $\psi(x) = x^n$ for some fixed n and choose $\gamma(a)$ and $\tau(a)$ appropriately, then this corresponds to the assumptions of Theorems 9.3 and 9.4. Those theorems are a special case of this. To rigorously prove this, we would have to show that the pairwise equations reduce to these equations. This can be done following steps like those in Chapter 7. This would complete the proof of the final size relation conjectured in equation (9.34).

9.4 Conclusions and outlook

Further refinements for non-Markovian epidemics are provided in [61, 307, 311], where in [311], the authors investigate the impact of non-Markovian infectious process by considering Weibullean infection times and show that both the epidemic threshold and the steady state fraction of infecteds significantly change when

Weibull distributions with the same mean but with different start and tail points of the power law section are considered. In [61], the authors study an SIS epidemic where both infection and recovery are non-Markovian. By considering the system at the metastable state, a balance condition at equilibrium can be written down linking the average number of successful infection events that a node experiences or receives during a long time interval to the average number of infectious periods of the same node, though noting that the two should be equal. This approach leads to introducing the average number of times a node tries to infect its neighbour during its infectious period, and the balance condition works on the assumption that the process has been running for some time and by observing it over a long time window. This then leads to an equation that is similar to Eq. (3.9). It also turns out that there is a complex interaction between the performance of closures and whether the epidemic dynamics is non-Markovian. Results in [249] suggest that closures tend to perform better when the recovery period is fixed.

As we have shown, it is possible to systematically extend the pairwise and EBCM approaches to relax the assumption of Markovian transmission and recovery processes. Typically, such models lead to delay differential or integro-differential equations [163, 182, 271], whose analysis, be it analytical or numerical, is more challenging. Non-Markovian processes often require probabilistic approaches, such as queueing and renewal theory, and these may offer a strong alternative to mean-field models or can be used in conjunction to establish and extend the theory of studying the limiting behaviour of non-Markovian stochastic processes.

Although we have not discussed simulation techniques in this chapter, we end with some comments on simulation of non-Markovian epidemics. The algorithm in Fig. 6.18 will create a percolated network corresponding to a non-Markovian SIR disease, from which we can deduce properties of epidemics. For studying disease dynamics, the algorithms presented in Appendix A.1.2 are straightforward to adapt to non-Markovian processes for either SIS or SIR disease.

Chapter 10
PDE limits for large networks

In previous chapters, it was shown that dynamics on networks can be described by continuous-time Markov chains, where probabilities of states are determined by master equations. While limiting mean-field ODE models can provide a good approximation of the expected value of certain random variables, a PDE-based approach is needed if information on the probability distribution of these is desired. The aim of this chapter is to develop and study mathematical methods to estimate the accuracy of mean-field ODE and PDE approximations. This will be carried out for continuous-time Markov chains with state space $\{0, 1, \ldots, N\}$, for which transition from state k is possible only to states $k-1$ and $k+1$. These processes are called birth–death or one-step processes. Denoting by $p_k(t)$ the probability of state k at time t, the master equation of the process takes the form

$$\dot{p}_k = a_{k-1}p_{k-1} - (a_k + c_k)p_k + c_{k+1}p_{k+1}, \quad k = 0, \ldots, N. \tag{10.1}$$

The equation corresponding to $k = 0$ does not contain the first term on the right-hand side, while that corresponding to $k = N$ does not contain the third term, i.e. $a_{-1} = c_{N+1} = 0$. Moreover, the Markov chain requires that a_N and c_0 are set to zero. This will ensure that the sum of each column in the transition matrix is zero. The motivation for studying such a model is that an arbitrary binary network process, see Section 2.3.2, can be represented by a one-step process, as it will be shown later.

10.1 Model, methods and motivation

The infinite size limit, i.e. the case when $N \to \infty$, can be described by differential equations in the so-called density-dependent case, when the transition rates a_k and c_k can be given by non-negative, continuous functions $A, C : [0, 1] \to [0, +\infty)$ satisfying $A(1) = 0 = C(0)$ as follows:

© Springer International Publishing AG 2017
I.Z. Kiss et al., *Mathematics of Epidemics on Networks*, Interdisciplinary Applied Mathematics 46, DOI 10.1007/978-3-319-50806-1_10

$$\frac{a_k}{N} = A\left(\frac{k}{N}\right) \quad \text{and} \quad \frac{c_k}{N} = C\left(\frac{k}{N}\right). \tag{10.2}$$

We note that the conditions $A(1) = 0 = C(0)$ ensure $a_N = 0 = c_0$. The case when A and C are polynomials (referred to below as "the polynomial case") will play a crucial role in our investigation. The special case when these functions are linear or constant can be fully described mathematically, and will serve as motivation for studying the non-linear case. The large N limit of stochastic processes has been widely studied in the literature and several methods were developed, which will be dealt with now.

10.1.1 Aims and methods of investigation

The Markov chain described by the master equations (10.1) can be considered as a counting or birth–death process, for which k denotes the population size. Hence, once the system of master equations is solved for p_k, one can determine the (scaled) expected value (first moment) of the population size as

$$m_1(t) = \sum_{k=0}^{N} \frac{k}{N} p_k(t). \tag{10.3}$$

The aim of our investigation is to determine or approximate

- the expected value m_1 by deriving a mean-field ODE, and
- the distribution p_k by using a PDE.

This can be carried out by using the following methods:

- introduce mean-field approximations by deriving ODEs for the moments and using algebraic relations among them as closures,
- determine a PDE for the probability (or moment)-generating function, see Chapter 4 in [275], and solve it or approximate it with a solvable PDE, and
- derive the corresponding Fokker–Planck equations [263, 309] that can be considered as the continuous version of the master equation Eq. (10.1).

The aim of deriving mean-field type models is to find differential equations for m_1 (for more accurate approximations, including also higher order moments) that yield approximations for the expected values of random variables of interest without solving the large system corresponding to master equations (10.1). Moreover, estimating the accuracy of such approximation is also in the focus of the research. Such and similar questions were studied in the density-dependent case by several authors (see [35, 94, 189]). Differentiating (10.3), using the master equation Eq. (10.1) and the density-dependent case (10.2), one obtains $\dot{m}_1 = \sum_{k=0}^{N} \left[A\left(\frac{k}{N}\right) - C\left(\frac{k}{N}\right)\right] p_k$. If the application of function $F = A - C$ and the computation of the expected value commute as operations, the right-hand side yields $F(m_1)$ and the mean-field equation is obtained in the form $\dot{x} = A(x) - C(x)$. (Note that this is the case when the functions

A and *C* are linear; hence, x and m_1 coincide.) The accuracy of the mean-field approximation is then the difference $|x - m_1|$. The stochastic convergence of m_1 to x has been widely studied by using martingale theory [35, 94, 189]. In [29], uniform convergence was proved by using the approximation theory of operator semigroups and the authors showed that the difference is of order $1/N$. This operator semigroup approach enabled them to approximate not only the expected value but also the distribution p_k itself with a partial differential equation [28]. The approximation is based on introducing a two-variable function u for which $u(t, \frac{k}{N}) \approx p_k(t)$ and deriving the Fokker–Planck equation. Then, the master equation Eq. (10.1) can be considered to be the discretisation of the Fokker–Planck equation in an appropriate sense.

10.1.2 Motivation for the model: binary network dynamics

Several network dynamics can be described by this prototype model. For example, in the case of SIS epidemic dynamics on a complete graph, or on a Configuration Model random graph, $p_k(t)$ is the probability that there are k infected nodes. For a complete graph, $a_k = \tau k(N - k)$ and $c_k = \gamma k$, where $\tau = \beta/N$ is the transmission rate across an edge (with a given positive constant β) and γ is the rate of recovery of a node. (It is important here that the infection rate τ scales with $1/N$ because otherwise the infection pressure to a node would tend to infinity as the number of nodes, together with the degree of a node, tends to infinity.) For Configuration Model random graphs, there are many different states with k infected individuals, but we expect that the likely states all have similar properties, so that we can reasonably group all states with k infections together. For Configuration Model random graphs with different degree distributions, e.g. regular random graphs and power law graphs, the coefficient a_k was estimated numerically from simulations in [229]. In [180], the above master equation was used to describe the change in the number of edges in an adaptive network. In van Kampen's book [309], there are many physical and chemical examples of one-step processes.

Our more general motivation is a binary network dynamics for which each node of a network may have one of two statuses, Q and T. Recall first the formulation for a general binary dynamics from Chapter 2. The dynamic is determined by the $Q \to T$ and $T \to Q$ transition rates. We assume that if a node has status Q, then the transition rate of moving to status T is

$$f_{QT}(n_Q, n_T) = r_{QT}^0 + r_{QT}^Q n_Q + r_{QT}^T n_T, \tag{10.4}$$

where n_Q and n_T denote the number of neighbours of status Q and T, respectively (see (4.28a) in Section 4.6.2). This formula expresses that the transition rate depends on the number of neighbours of the two different types and there is a spontaneous transition (independently of the neighbours) characterised by the rate r_{QT}^0. Similarly, the rate of transition from status T to status Q can be given as $f_{TQ}(n_Q, n_T) = r_{TQ}^0 + r_{TQ}^Q n_Q + r_{TQ}^T n_T$. The following examples, also taken from Section 4.6.2, shed more light on the notation:

- SIS epidemic with infection rate τ and recovery rate γ:

$$f_{SI}(n_S, n_I) = \tau n_I, \quad f_{IS}(n_S, n_I) = \gamma.$$

- SIS epidemic with spontaneous infection at rate δ:

$$f_{SI}(n_S, n_I) = \delta + \tau n_I, \quad f_{IS}(n_S, n_I) = \gamma.$$

- Simplified voter model (each node has the same number of links) with two statuses Q and T:

$$f_{QT}(n_Q, n_T) = \alpha n_T, \quad f_{TQ}(n_Q, n_T) = \beta n_Q.$$

Consider now a general binary dynamic on a Configuration Model random graph, where the average node degree is n, and denote by $p_k(t)$ the probability that there are k nodes of status Q and $N - k$ nodes of status T. The rate of transition from k to $k+1$, i.e. a node of status T moves to status Q, can be given as $a_k = r_{TQ}^0(N-k) + r_{TQ}^Q e_{TQ}(k) + r_{TQ}^T e_{TT}(k)$, where $e_{TQ}(k)$ and $e_{TT}(k)$ denote the average number of QT and TT edges (doubly counted), respectively, when the number of Q nodes is k. Note that this involves an approximation, since the number of QT and TT edges depends not only on the number but also on the position of the Q nodes. A further approximation is needed when these edge numbers are expressed in terms of k. The simplest approximation is

$$e_{QT}(k) = kn\frac{N-k}{N}, \quad e_{TT}(k) = (N-k)n\frac{N-k}{N}, \quad e_{QQ}(k) = kn\frac{k}{N}.$$

This is based on the assumption that nodes are labelled at random with k pieces of Q labels and $N - k$ labels of type T. Hence, a proportion of $\frac{N-k}{N}$ out of kn stubs starting from a Q node will link to a node labelled with T. (The argument is similar for $e_{TT}(k)$ and for $e_{QQ}(k)$.) Substituting these formulas into the expression of a_k leads to

$$a_k = r_{TQ}^0(N-k) + r_{TQ}^Q kn\frac{N-k}{N} + r_{TQ}^T(N-k)n\frac{N-k}{N}.$$

Similar reasoning for the rate of transition from k to $k - 1$, when a node of status Q moves to status T, yields

$$c_k = r_{QT}^0 k + r_{QT}^Q kn\frac{k}{N} + r_{QT}^T kn\frac{N-k}{N}.$$

Observe that these are density-dependent coefficients satisfying (10.2) with the quadratic functions

$$A(z) = (1-z)(az+b) \quad \text{and} \quad C(z) = z(c+dz),$$

where

$$a = n(r_{TQ}^Q - r_{TQ}^T), \quad b = r_{TQ}^0 + n r_{TQ}^T, \quad c = r_{QT}^0 + n r_{QT}^T, \quad d = n(r_{QT}^Q - r_{QT}^T).$$

As an example, consider the SIS epidemic dynamics on an n-regular Configuration Model random graph (all nodes having degree n) with infection rate τ and recovery rate γ. Let $p_k(t)$ denote the probability of having k infected nodes at time t, i.e. $Q = I$ in the general context. Then $r_{TQ}^0 = 0$, because there is no spontaneous infection, $r_{TQ}^Q = \tau$ and $r_{TQ}^T = 0$. Concerning recovery, i.e. the transition from I to S, $r_{QT}^0 = \gamma$, because there is spontaneous recovery and $r_{QT}^Q = 0 = r_{QT}^T$. Thus, the coefficients in the functions A and C are $a = n\tau$, $b = 0$, $c = \gamma$ and $d = 0$, and the master equation is given by (10.1) with density-dependent coefficients (10.2).

Summarising, the master equation Eq. (10.1) with density-dependent coefficients given by quadratic functions A and C can describe an arbitrary binary dynamic, for which we assume linear transition functions[1] depending on the number of neighbours of the different statuses and the network as regular, i.e. the degree of each node is n.

> **Exercise 10.1.** Determine the coefficients in the functions A and C for the SIS epidemic propagation with spontaneous infection on an n-regular Configuration Model random graph with spontaneous infection rate δ, edge-wise infection rate τ and recovery rate γ.

> **Exercise 10.2.** Determine the coefficients in the functions A and C for the voter model on an n-regular Configuration Model random graph with transition rates α and β.

10.2 General results for one-step processes

Although the master equation Eq. (10.1) is a system of linear ODEs, its analytical description and, for large values of N, even its numerical treatment leads to mathematical difficulties. Fortunately, in many cases we do not need to know the probability of each state, but rather it suffices to know the expected number of nodes of a given status or higher order moments of the distribution. In this section, we derive differential equations for these objects/quantities. In later sections, we will investigate these equations in detail for various important cases.

---

[1] As before, we use "linear transition function" to mean all terms are zeroth or first order.

10.2.1 Differential equations for the moments and for the PGF

The nth moment of the probability distribution and its scaled version are

$$M_n(t) = \sum_{k=0}^{N} k^n p_k(t), \quad m_n(t) = \sum_{k=0}^{N} \frac{k^n}{N^n} p_k(t).$$

The probability-generating function (PGF) of the distribution is defined as

$$G(t,z) = \sum_{k=0}^{N} z^k p_k(t).$$

In order to derive ODEs for the moments and a PDE for the PGF, we will make use of the following lemma.

Lemma 10.1 *Let r_k $(k = 0,1,2,\ldots)$ be a sequence and let $r(t) = \sum_{k=0}^{N} r_k p_k(t)$, where $p_k(t)$ is given by (10.1). Then,*

$$\dot{r}(t) = \sum_{k=0}^{N} (a_k(r_{k+1} - r_k) + c_k(r_{k-1} - r_k)) p_k(t).$$

Proof. From (10.1), we obtain

$$\dot{r}(t) = \sum_{k=0}^{N} r_k \dot{p}_k(t) = \sum_{k=1}^{N} r_k a_{k-1} p_{k-1}(t) - \sum_{k=0}^{N} r_k (a_k + c_k) p_k(t) + \sum_{k=0}^{N-1} r_k c_{k+1} p_{k+1}(t) =$$

$$\sum_{k=0}^{N-1} r_{k+1} a_k p_k(t) - \sum_{k=0}^{N} r_k b_k p_k(t) + \sum_{k=1}^{N} r_{k-1} c_k p_k(t).$$

Using that $a_N = 0$, $c_0 = 0$ and $b_k = a_k + c_k$, we get

$$\dot{r}(t) = \sum_{k=0}^{N} (r_{k+1} a_k - r_k(a_k + c_k) + r_{k-1} c_k) p_k(t) = \sum_{k=0}^{N} (a_k(r_{k+1} - r_k) + c_k(r_{k-1} - r_k)) p_k(t).$$

Applying this Lemma to the nth moment, i.e. choosing $r_k = \frac{k^n}{N^n}$, leads to the following proposition.

Proposition 10.1 *The scaled nth moment, m_n, of the distribution p_k determined by the master equation Eq. (10.1) satisfies the differential equation*

$$\dot{m}_n = \frac{1}{N^n} \sum_{k=0}^{N} [a_k((k+1)^n - k^n) + c_k((k-1)^n - k^n)] p_k. \tag{10.5}$$

This proposition will enable us to study the moments. Applying it for $n = 1$ in the density-dependent case (10.2) leads to

$$\dot{m}_1 = \sum_{k=0}^{N} \left[A\left(\frac{k}{N}\right) - C\left(\frac{k}{N}\right) \right] p_k.$$

One way to get a closed system to approximate m_1 is simply to assume that the order of application of a non-linear function (such as A or C) and the expected value can be exchanged. For a linear function, this yields an exact relation, while for non-linear ones it is only an approximation. The quality of the approximation is best if the probability distribution is very narrow, so that the non-linear function varies little across the most likely states. This closure approximation implies

$$\dot{m}_1 \approx A\left(\sum_{k=0}^{N} \frac{k}{N} p_k \right) - C\left(\sum_{k=0}^{N} \frac{k}{N} p_k \right) = A(m_1) - C(m_1).$$

Introducing y_1 as the approximation of m_1, the approximating closed differential equation takes the form

$$\dot{y}_1 = A(y_1) - C(y_1). \tag{10.6}$$

In the polynomial case, the proposition makes it possible to derive an exact system of ODEs for the moments. This exact system can be used to derive closures and to investigate the performance of the closures analytically in two different ways. On the one hand, the difference between the exact and the closed system can be proved to decrease in a given order as N tends to infinity. On the other hand, lower and upper bounds can be derived for the exact value of the moments.

Applying Lemma 10.1 to the PGF, i.e. choosing $r_k = z^k$, yields the following proposition.

Proposition 10.2 *The probability-generating function G of the distribution p_k determined by the master equation Eq. (10.1) satisfies the differential equation*

$$\partial_t G(t,z) = \sum_{k=0}^{N} \left[a_k \left(z^{k+1} - z^k \right) + c_k \left(z^{k-1} - z^k \right) \right] p_k. \tag{10.7}$$

This proposition yields a partial differential equation for G in the polynomial case. This PDE can be solved analytically in the linear case, yielding an explicit formula for the evolution of the distribution p_k.

10.2.2 Fokker–Planck equation

The Fokker–Planck equation can be considered as the continuous version of the master equation Eq. (10.1). We wish to approximate the solution $p_k(t)$ by considering it as a discretisation of a continuous function $u(t,z)$ in the interval $[0,1]$, i.e.

$$u\left(t, \frac{k}{N} \right) = p_k(t) \tag{10.8}$$

for $0 \leq k \leq N$. Now, we derive an approximating PDE, called the Fokker–Planck equation, for the function $u(\cdot,\cdot)$ based on the ODE given by the master equation. This PDE is traditionally given in the form

$$\partial_t u(t,z) = \frac{1}{2}\partial_{zz}(g(z)u(t,z)) - \partial_z(h(z)u(t,z)), \tag{10.9}$$

see [263, 275, 309]. (We note that the factor $1/2$ is sometimes built in the coefficient function g.) The functions g and h will be determined in such a way that the finite difference discretisation of this PDE will yield the master equation Eq. (10.1). (In fact, any parabolic-type PDE with space-dependent coefficients could serve as the continuous version of the master equation.) The Fokker–Planck equation can be derived for more general (not only one-step) processes (see, e.g., [263, 275, 309]).

Derivation of the Fokker–Planck equation

We will assume an equation of the form (10.9) and derive the form that g and h must have. The following second order finite difference discretisation approximations will be used to relate the PDE and the master equation:

$$f(z-\Delta z) - 2f(z) + f(z+\Delta z) \approx \Delta z^2 f''(z), \quad f(z+\Delta z) - f(z-\Delta z) \approx 2\Delta z f'(z).$$

Applying these formulas with $z = k/N$ and $\Delta z = 1/N$ to the partial derivatives of the functions $g(z)u(t,z)$ and $h(z)u(t,z)$ with respect to z leads to

$$\partial_t u\left(t, \frac{k}{N}\right) = \frac{N^2}{2}(g_{k+1}x_{k+1} - 2g_k x_k + g_{k-1}x_{k-1}) - \frac{N}{2}(h_{k+1}x_{k+1} - h_{k-1}x_{k-1}), \tag{10.10}$$

where the notations $x_k = u\left(t, \frac{k}{N}\right)$, $g_k = g\left(\frac{k}{N}\right)$ and $h_k = h\left(\frac{k}{N}\right)$ are used. The above discretisation will be used also for $k=0$ and $k=N$; hence, two artificial mesh points are introduced at $z = -1/N$ and at $z = 1+1/N$. Differentiating (10.8) with respect to t and using the master equation Eq. (10.1) yields

$$\partial_t u\left(t, \frac{k}{N}\right) = \dot{p}_k = a_{k-1}p_{k-1} - (a_k + c_k)p_k + c_{k+1}p_{k+1}. \tag{10.11}$$

This equation can be considered as the discretisation of the Fokker–Planck equation. Upon substituting p_k by x_k for all k, the result must equal the right-hand side of (10.10). Matching coefficients of x_k for $0 < k < N$ leads to

$$a_k = \frac{N}{2}h_k + \frac{N^2}{2}g_k, \quad c_k = \frac{N^2}{2}g_k - \frac{N}{2}h_k. \tag{10.12}$$

Thus, we conclude that the desired functions g and h are defined in such a way that the relations

$$g\left(\frac{k}{N}\right) = g_k = \frac{1}{N^2}(a_k + c_k), \quad h\left(\frac{k}{N}\right) = h_k = \frac{1}{N}(a_k - c_k)$$

hold.

Derivation of the boundary conditions

Our PDE is second order in z, and so we require two boundary conditions. These will determine a_k and c_k for $k = 0$ and $k = N$. If we think of each point x_k as the centre of the interval $(x_k - 1/2N, x_k + 1/2N)$, then the integral of $u(t, z)$ from $-1/2N$ to $1 + 1/2N$ is approximated as a midpoint Riemann sum by $\sum_{k=0}^{N} p_k/N = 1/N$. That is

$$\int_{-\frac{1}{2N}}^{1+\frac{1}{2N}} u(t, z)\,dz = \frac{1}{N} \tag{10.13}$$

for all t. Integrating (10.9) on $[-\frac{1}{2N}, 1 + \frac{1}{2N}]$, we obtain

$$0 = \partial_t\left(\int_{-\frac{1}{2N}}^{1+\frac{1}{2N}} u(t, z)\,dz\right) = \frac{1}{2}\partial_z(gu)\left(1 + \frac{1}{2N}, t\right) - \frac{1}{2}\partial_z(gu)\left(-\frac{1}{2N}, t\right)$$
$$- (hu)\left(1 + \frac{1}{2N}, t\right) + (hu)\left(-\frac{1}{2N}, t\right),$$

which obviously holds if the boundary conditions

$$\frac{1}{2}\partial_z(gu)\left(-\frac{1}{2N}, t\right) - (hu)\left(-\frac{1}{2N}, t\right) = 0, \tag{10.14a}$$

$$\frac{1}{2}\partial_z(gu)\left(1 + \frac{1}{2N}, t\right) - (hu)\left(1 + \frac{1}{2N}, t\right) = 0 \tag{10.14b}$$

are satisfied.

It is useful to note that using these boundary conditions ensures that the conditions in (10.10) are satisfied for $k = 0$ and $k = N$. To see this for $k = 0$, we approximate the derivative of gu at $z = -1/2N$ with respect to z by central difference, and approximate the value of hu by the arithmetic mean at the two neighbouring values, x_{-1} and x_0. Then, Eq. (10.14a) yields

$$\frac{g_0 x_0 - g_{-1} x_{-1}}{2/N} - \frac{h_0 x_0 + h_{-1} x_{-1}}{2} = 0.$$

From this, we obtain

$$\frac{N^2}{2} g_0 x_0 - \frac{N}{2} h_0 x_0 = \frac{N^2}{2} g_{-1} x_{-1} + \frac{N}{2} h_{-1} x_{-1}.$$

Hence, for $k = 0$, Eq. (10.10) can be written as

$$\partial_t u(t,0) = -\frac{N}{2}(Ng_0 + h_0)x_0 + \frac{N}{2}(Ng_1 - h_1)x_1. \tag{10.15}$$

On the other hand, for $k = 0$, Eq. (10.11) takes the form $\partial_t u(t,0) = -a_0 p_0 + c_1 p_1$. Hence,

$$a_0 = \frac{N^2}{2}g_0 + \frac{N}{2}h_0, \quad c_1 = \frac{N^2}{2}g_1 - \frac{N}{2}h_1,$$

that is (10.12) holds also for $k = 0$. Similarly, it can be shown that the boundary condition (10.14b) implies that the relation holds also for $k = N$.

Fokker–Planck equation in the density-dependent case

In the density-dependent case, when the coefficients a_k and c_k are given by the functions A and C as $\frac{a_k}{N} = A\left(\frac{k}{N}\right)$ and $\frac{c_k}{N} = C\left(\frac{k}{N}\right)$, we obtain that g and h can be given as

$$g(z) = \frac{1}{N}(A(z) + C(z)), \quad h(z) = A(z) - C(z).$$

Summarising, the Fokker–Planck equation of the one-step process given by density dependent coefficients is as follows.

Fokker–Planck equation of the one-step process

$$\partial_t u(t,z) = \frac{1}{2N}\partial_{zz}((A(z) + C(z))u(t,z)) - \partial_z((A(z) - C(z))u(t,z)) \tag{10.16a}$$

subject to boundary conditions

$$\delta\partial_z((A + C)u)(-\delta, t) - ((A - C)u)(-\delta, t) = 0, \tag{10.16b}$$

$$\delta\partial_z((A + C)u)(1 + \delta, t) - ((A - C)u)(1 + \delta, t) = 0, \tag{10.16c}$$

where $\delta = 1/2N$.

10.3 One-step processes with linear coefficients

In the case when the coefficients a_k and c_k are linear in k, one can derive exact differential equations for the moments, and the probability distribution can be determined by solving the linear PDE for the probability-generating function (see Chapter 4 in [275] for the special case of the Poisson process). This is carried out in this section. If a_k and c_k are linear in k, then they can be given as

$$a_k = a(N - k) \text{ and } c_k = ck, \tag{10.17}$$

with some non-negative parameters a and c, since $a_N = 0 = c_0$.

10.3.1 Differential equation for the first moment

Substituting the coefficients (10.17) in (10.5) with $n = 1$ leads to

$$\dot{m}_1 = \sum_{k=0}^{N} \left[a \left(1 - \frac{k}{N} \right) - c \frac{k}{N} \right] p_k = a - (a + c)m_1. \tag{10.18}$$

Thus, the expected value $m_1(t)$ can be determined explicitly,

$$m_1(t) = \frac{a}{a + c} + Ke^{-(a+c)t},$$

where K is a constant determined by the initial condition. The steady state value is $m_1(\infty) = \frac{a}{a+c}$. The higher order moments can also be determined by solving simple differential equations.

10.3.2 Probability-generating function of the distribution

Applying (10.7) with the coefficients (10.17), one can derive a PDE for the probability-generating function. This is given by

$$\partial_t G(t, z) = \sum_{k=0}^{N} \left[a_k \left(z^{k+1} - z^k \right) + c_k \left(z^{k-1} - z^k \right) \right] p_k$$

$$= \sum_{k=0}^{N} \left[a(N - k) \left(z^{k+1} - z^k \right) + ck \left(z^{k-1} - z^k \right) \right] p_k.$$

The right-hand side can be expressed in terms of the partial derivatives of G with respect to z by using the identity

$$\sum_{k=0}^{N} kz^k p_k(t) = z \sum_{k=0}^{N} kz^{k-1} p_k = z\partial_z G(t, z).$$

Hence, $\partial_t G(t, z) = aN(z - 1)G(t, z) - (az + c)(z - 1)\partial_z G(t, z)$, i.e. the PGF satisfies the following first order PDE

$$\partial_t G(t, z) + (az + c)(z - 1)\partial_z G(t, z) = aN(z - 1)G(t, z). \tag{10.19}$$

This differential equation can be solved explicitly and then all information about the distribution can be obtained from the PGF. However, without solving the PDE the moments and the stationary distribution can be obtained directly from this partial differential equation. This will be carried out below.

Derivation of the ODE of the first moment from the PGF

This partial differential equation enables us to rederive the ODE (10.18) for the expected value as follows. Differentiating the PGF with respect to z yields $M_1(t) = \partial_z G(t, 1)$. Differentiating this equation with respect to t, we get $\dot{M}_1(t) = \partial_t \partial_z G(t, 1) = \partial_z \partial_t G(t, 1)$. Differentiating Eq. (10.19) with respect to z and then substituting $z = 1$ leads to $\dot{M}_1(t) = aNG(t, 1) - (a + c)\partial_z G(t, 1) = aN - (a + c)M_1(t)$. Dividing this by N, we recover Eq. (10.18) for the scaled expected value.

Stationary distribution

The stationary distribution is a solution of (10.19) which does not depend on time. Hence, it can be written as $G(t, z) = G^*(z)$ and $\partial_t G(t, z) = 0$. Therefore, Eq. (10.19) becomes an ODE for G^*:

$$(az + c)(z - 1)(G^*)'(z) = aN(z - 1)G^*(z).$$

Solving this equation, we obtain $\ln G^*(z) = N \ln(az + c) + K$, yielding

$$G^*(z) = \left(\frac{az + c}{a + c}\right)^N = (pz + 1 - p)^N,$$

where the initial condition $G^*(1) = 1$ was used and the parameter $p = \frac{a}{a+c}$ was introduced. This means that G^* is the generating function of the binomial distribution with parameter p. Thus, the stationary distribution of the master equation Eq. (10.1) with linear coefficients (10.17) is the binomial distribution $p_k = \binom{N}{k}p^k(1 - p)^{N-k}$.

10.3.3 Fokker–Planck equation

Let us now derive the Fokker–Planck equation when the coefficients are linear. Using Eq. (10.16a) with the choice $A(z) = a(1 - z)$ and $C(z) = cz$ leads to

$$\partial_t u(t, z) = \frac{1}{2N}\partial_{zz}(((c - a)z + a)u(t, z)) - \partial_z((a - (a + c)z)u(t, z)).$$

The solution in the steady state, i.e. when $\partial_t u(t, z) = 0$, can be determined analytically. The derivation is carried out first in the special case of $a = c$. Denoting the steady state solution by $U(z)$, it immediately follows that it satisfies the ODE

$$\frac{1}{2N}U''(z) = ((1-2z)U(z))'.$$

Integrating this equation leads to

$$\frac{1}{2N}U'(z) = (1-2z)U(z) + K.$$

The boundary condition at $z = -1/2N$ implies that $K = 0$. Then, the equation can be integrated again by separation of variables or integration by parts, and yields

$$U(z) = \text{const} \cdot \exp\left(-2N\left(z-\frac{1}{2}\right)^2\right).$$

The constant has to be chosen in such a way that the integral of U becomes $1/N$ (see (10.13)). This gives

$$U(z) = \frac{\sqrt{2}}{\sqrt{\pi N}}\exp\left(-2N\left(z-\frac{1}{2}\right)^2\right). \tag{10.20}$$

Note that this is an approximation of the binomial distribution. Namely, according to the Moivre–Laplace theorem the binomial distribution

$$B_k(N,p) = \binom{N}{k}p^k(1-p)^{N-k}$$

can be approximated by the normal distribution as

$$B_k(N,p) \approx \frac{1}{\sqrt{Np(1-p)}}\phi\left(\frac{k-Np}{\sqrt{Np(1-p)}}\right),$$

where

$$\phi(x) = \frac{1}{\sqrt{2\pi}}\exp\left(-\frac{x^2}{2}\right)$$

is the density function of the normal distribution. Applying this approximation for $p = 1/2$ yields

$$B_k(N,1/2) \approx U(k/N),$$

that is, the steady state of the Fokker–Planck equation can be considered as the continuous version of the binomial distribution, which is the steady state of the master equation Eq. (10.1). The accuracy of the Fokker–Planck equation is illustrated in the left panel of Fig. 10.1, where the exact steady state of the master equation is plotted together with function U. One can see that the agreement is excellent even for $N = 50$.

Returning to the general case, when $a = c$ is not assumed, the stationary solution U satisfies the differential equation

$$\frac{1}{2N}(((c-a)z+a)U(z))'' = (((a-(a+c)z)U(z))'.$$

Integrating this equation leads to

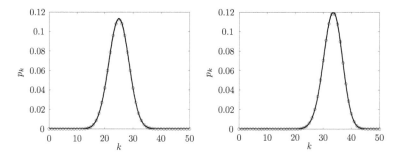

Fig. 10.1: The steady state of the distribution in the linear case, when $A(z) = a(1-z)$ and $C(z) = cz$ for $N = 50$. The binomial distribution as the exact solution of the master equation (circles) is shown together with U, the solution of the Fokker–Planck equation (solid curve). In the left panel, the case $a = c = 1$ is shown, when U is given by (10.20). In the right panel, the case $a = 2$, $c = 1$ is shown, when U is given by (10.21).

$$\frac{1}{2N}(((c-a)z+a)U(z))' = ((a-(a+c)z)U(z)$$

since the integrating constant is zero due to the boundary condition. This differential equation can be solved by separation of variables, yielding

$$U(z) = \frac{\text{const}}{(c-a)z+a}e^{H(z)}, \tag{10.21}$$

where the constant is determined in such a way that the integral of U is $1/N$, see (10.13), and

$$H(z) = \frac{2N}{(a-c)^2}\left[(a^2-c^2)z + 2ac\ln(a+(c-a)z)\right].$$

This steady state does not coincide with a normal distribution; however, in Section 10.5.3 it will be shown that it can be approximated by a normal distribution which is close to the corresponding binomial distribution. This is illustrated in the right panel of Fig. 10.1, where function U is plotted together with the binomial distribution.

10.3.4 Fokker-Planck equation for constant coefficients: random walk

It is worth briefly reviewing the simplest case of constant coefficients when the master equation Eq. (10.1) describes a random walk and the corresponding Fokker–Planck equation is the convection–diffusion (or advection–diffusion) equation with constant coefficients. Let a and c be positive numbers and consider the case of coefficients

$$a_k = a, \quad c_k = c \quad \text{for all } k \in \mathbb{Z}.$$

Here it is important to note that the boundary conditions $a_N = 0 = c_0$ are not satisfied; hence, an infinite state space, $\mathbb{Z}$, is considered.

Applying (10.5) with $n = 1$ and $n = 2$, the differential equations of the first two moments are

$$\dot{M}_1 = a - c \quad \text{and} \quad \dot{M}_2 = a + c + 2(a - c)M_1.$$

These can be solved yielding that the expected value is moving with constant velocity $a - c$ and the variance is increasing in time, which can be interpreted as a convection–diffusion process.

The analytical formula for the distribution can be obtained by using the PGF. Applying (10.7), the differential equation for the PGF is

$$\partial_t G(t,z) = \frac{1-z}{z}(c - az)G(t,z).$$

Hence,

$$G(t,z) = \exp\left(\frac{1-z}{z}(c - az)t\right).$$

The power series expansion in z yields a formula for the distribution. In the case $a = c = 1$, the distribution can be given as $p_k(t) = e^{-2t}I_{|k|}(2t)$, where I_j is the jth modified Bessel function (see [309] Chapter 6 (eq. 2.11) and [34] Section 2.1).

The case of constant coefficients can be considered to be density dependent with $A(z) = a/N$ and $C(z) = c/N$ according to (10.2). Then, using (10.16a) the Fokker–Planck equation takes the form

$$\partial_t u(t,z) = \frac{1}{2N^2}(a+c)\partial_{zz}u(t,z) - (a-c)\frac{1}{N}\partial_z u(t,z),$$

which is a convection–diffusion equation. The case when some boundary condition is imposed at $k = 0$ and at $k = N$ is studied in [34].

10.4 Mean-field equations for the moments

We have seen in the linear case that (10.5) gives rise to a differential equation for the expected value m_1 that can be solved. Hence, the expected value can be obtained without solving the large system of master equations (10.1). Using a simple non-linear case with quadratic coefficients we will show that the expected value cannot be determined in such a simple way in general. In this case, the differential equation derived from (10.5) will involve the second moment as well; hence, it will not be a self-contained differential equation for m_1.

10.4.1 The case of quadratic coefficients

Consider an example with quadratic coefficients

$$a_k = a\frac{k}{N}(N-k) \text{ and } c_k = ck \qquad (10.22)$$

corresponding to *SIS* epidemic propagation on a complete graph. Using (10.5) and (10.22), we obtain

$$\dot{m}_1(t) = \sum_{k=0}^N \left(a_k\left(\frac{k+1}{N}-\frac{k}{N}\right)+c_k\left(\frac{k-1}{N}-\frac{k}{N}\right)\right)p_k(t) = \sum_{k=0}^N \frac{1}{N}(a_k-c_k)p_k(t)$$

$$= \sum_{k=0}^N \frac{1}{N}\left(a\frac{k}{N}(N-k)-ck\right)p_k(t) = \sum_{k=0}^N \left(a\frac{k}{N}\left(1-\frac{k}{N}\right)-c\frac{k}{N}\right)p_k(t)$$

$$= (a-c)m_1(t) - am_2(t).$$

Hence, the differential equation of m_1 has the form

$$\dot{m}_1(t) = (a-c)m_1(t) - am_2(t). \qquad (10.23)$$

As we noted above, the differential equation for m_1 contains the second moment m_2; hence, we either need an algebraic approximation relating m_2 to m_1 in order to get an approximating differential equation for m_1 or need a further differential equation for m_2. The simplest closure approximation is

$$m_2 \approx m_1^2.$$

Introducing y_1 as the approximation of m_1, the approximating closed differential equation takes the form
$$\dot{y}_1 = (a-c)y_1 - ay_1^2.$$

In [29], the authors proved that in a bounded time interval the accuracy of the approximation can be estimated as

$$|y_1(t) - m_1(t)| \le \frac{K}{N},$$

where K is a constant depending on the length of the time interval.

In Section 10.6.2, we present the generalisation of the above estimate for the density-dependent case. In Section 10.4.2, we consider the case when the coefficients are polynomials of k and show that in this case a system of differential equations can be obtained for the moments. This system can be used to derive upper and lower estimates for the moments and enables us to get a simple proof, using only elementary ODE techniques, for the above statement. Finally, in Section 10.4.4 it is shown that closing the system at the level of the third moment produces a more accurate mean-field system.

10.4.2 Polynomial coefficients

Consider now the density-dependent case when the functions in (10.2) are polynomials of degree at most J given as

$$A(x) = \sum_{j=0}^{J} A_j x^j \quad \text{and} \quad C(x) = \sum_{j=0}^{J} C_j x^j \tag{10.24}$$

such that $A(1) = 0$ and $C(0) = 0$ hold and the values of A and C are non-negative in $[0, 1]$. Before deriving differential equations for the moments from (10.5), let us introduce

$$P_{k,n} = \frac{(k+1)^n - k^n - nk^{n-1}}{N^{n-1}}, \quad Q_{k,n} = \frac{(k-1)^n - k^n + nk^{n-1}}{N^{n-1}}$$

and

$$R_n(t) = \sum_{k=0}^{N} (a_k P_{k,n} + c_k Q_{k,n}) p_k(t). \tag{10.25}$$

Combining these with (10.5) leads to

$$\dot{m}_n(t) = \sum_{k=0}^{N} \left(\frac{a_k}{N} \left(n \frac{k^{n-1}}{N^{n-1}} + P_{k,n} \right) + \frac{c_k}{N} \left(-n \frac{k^{n-1}}{N^{n-1}} + Q_{k,n} \right) \right) p_k(t).$$

Hence,

$$\dot{m}_n(t) = n \sum_{k=0}^{N} \frac{a_k - c_k}{N} \left(\frac{k}{N} \right)^{n-1} p_k(t) + \frac{1}{N} R_n(t). \tag{10.26}$$

Now, use the facts that $a_k/N = A(k/N)$, $c_k/N = C(k/N)$ and the functions A and C are polynomials. Then, introducing $D_j = A_j - C_j$ leads to

$$\dot{m}_n(t) = n \sum_{k=0}^{N} \sum_{j=0}^{J} D_j \left(\frac{k}{N} \right)^{n+j-1} p_k(t) + \frac{1}{N} R_n(t),$$

which yields

$$\dot{m}_n(t) = n \sum_{j=0}^{J} D_j m_{n+j-1}(t) + \frac{1}{N} R_n(t). \tag{10.27}$$

For the large N limit, the following bound on the remainder term R_n is useful.

Proposition 10.3 *For any value of N, the inequalities $0 \leq R_n \leq \frac{n(n-1)}{2} B_c$ hold, where $B_c = \sum_{j=0}^{J}(|A_j| + |C_j|)$ is an upper bound for the coefficients of the polynomials.*

The statement can be proved by an elementary application of Taylor's theorem with Lagrangian remainder (see [29] for the detailed proof).

We note that using the binomial theorem, $P_{k,n}$ and $Q_{k,n}$ can be expressed in terms of the powers of k. Hence, R_n can be expressed as the linear combination of the moments. Therefore, system (10.27) is an infinite system of homogeneous linear differential equations for the moments m_n. This homogeneous linear system is not written in the usual matrix form because it is useful to separate the $O(\frac{1}{N})$ terms in order to handle the large N limit. In this limit, by neglecting the remainder term in (10.27), one arrives at the approximating system

$$\dot{y}_n(t) = n \sum_{j=0}^{J} D_j y_{n+j-1}(t), \tag{10.28}$$

where the new variable y_n is introduced to emphasise that the solutions of this system are only approximations of the original moments. Choosing appropriate initial conditions, this system leads to the mean-field equation as follows.

Consider the case of pure initial condition in (10.1), namely start the process from a given state $\ell \in \{0, 1, \ldots, N\}$, that is $p_\ell(0) = 1$ and $p_j(0) = 0$ for $j \neq \ell$. Then, the initial condition for the moments is $m_n(0) = (\ell/N)^n$. Consider the solution of (10.28) with this initial condition, i.e. with $y_n(0) = (\ell/N)^n$. The solution satisfying this initial condition can be written as $y_n(t) = y_1^n(t)$, where y_1 is the solution of the mean-field equation

$$\dot{y}_1 = \sum_{j=0}^{J} D_j y_1^j. \tag{10.29}$$

In order to check this, multiply this equation with $n y_1^{n-1}$. Then,

$$n y_1^{n-1} \dot{y}_1 = n \sum_{j=0}^{J} D_j y_1^{n+j-1},$$

which is exactly (10.28) once the substitution $y_k(t) = y_1^k(t)$ is carried out for all indices k.

Note that (10.29) is the same as (10.6) when the functions A and C are polynomials. This way, the infinite system for the moments led us to the mean-field equation. Note that we can arrive at this equation by using the closure relation as follows. Consider the case $n = 1$ in (10.27). Then, using that $R_1 = 0$, we get

$$\dot{m}_1 = \sum_{j=0}^{J} D_j m_j. \tag{10.30}$$

The closure approximation $m_n \approx m_1^n$ yields again the mean-field approximation (10.29). We emphasise repeatedly that the advantage of the mean-field approximation (10.29) with respect to the exact equation Eq. (10.30) is that the former can be solved independently of higher order moments, while the latter needs higher order moments, i.e. it is not a self-contained differential equation. Nevertheless, the exact equation Eq. (10.30) can be used in several ways:

- extended with the differential equations for all moments, i.e. considering (10.27) as an infinite system of ODEs, it can be used to prove that the difference of m_1 and y_1 is of order $1/N$ as N tends to infinity (see [176]),
- under certain assumptions regarding the sign of the coefficients D_j, it can be shown that y_1 is an upper bound on m_1, and introducing an appropriate system of differential equations for the newly defined functions $z_1, \ldots, z_d$, it can be proved that z_1 is a lower bound for m_1,
- higher order closures can be derived based on an *a priori* distribution for p_k.

Results obtained by using the above techniques will be discussed briefly in the next subsections.

10.4.3 Upper and lower bounds for the expected value

As a motivation for the ideas used during the course of the following derivation, consider the quadratic case studied in Section 10.4.1, when $a_k = a\frac{k}{N}(N-k)$ and $c_k = ck$. Then, according to (10.23) the differential equation of the first moment is $\dot{m}_1(t) = (a-c)m_1(t) - am_2(t)$. One of the main ideas is to apply Jensen's inequality to get an estimate of the second moment in terms of the first one. The classical Jensen's inequality [73] and the definition of the expected value yield the probabilistic version of Jensen's inequality.

Lemma 10.2 (Jensen's inequality) *If X is a random variable and $\varphi \colon \mathbb{R} \to \mathbb{R}$ is a convex function, then*

$$\varphi(\mathbb{E}[X]) \leq \mathbb{E}[\varphi(X)].$$

For concave φ, the reverse inequality holds.

Apply Jensen's inequality for the random variable X, for which $\mathbb{P}(X(t) = k) = p_k(t)$ with a fixed value of t and for $k \in \{0, 1, \ldots, N\}$. If φ is chosen as $\varphi(x) = x^n$, then due to Jensen's inequality,

$$m_1^n = (\mathbb{E}[X/N])^n \leq \mathbb{E}[(X/N)^n] = m_n \text{ for } n \geq 1. \tag{10.31}$$

Hence, the differential equation of the first moment yields

$$\dot{m}_1(t) = (a-c)m_1(t) - am_2(t) \leq (a-c)m_1(t) - am_1^2(t). \tag{10.32}$$

Recall that the mean-field equation is $\dot{y}_1(t) = (a-c)y_1(t) - ay_1^2(t)$, that is m_1 satisfies the inequality with the same right-hand side as that of the differential equation of y_1. Then, the following comparison result yields that $m_1(t) \le y_1(t)$.

Lemma 10.3 (Comparison) *Let $f : \mathbb{R}^2 \to \mathbb{R}$ be a continuous function. Assume that*

- *the initial value problem $x_2'(t) = f(t, x_2(t))$, $x_2(0) = x_0$ has a unique solution for $t \in [0, T]$;*
- *$x_1'(t) \le f(t, x_1(t))$ for $t \in [0, T]$; and $x_1(0) \le x_0$.*

Then, $x_1(t) \le x_2(t)$ for $t \in [0, T]$.

The result is standard in the theory of ODEs (see, e.g., [131]). We applied more general comparison results in Chapter 3.

This comparison of m_1 and y_1 can be extended to the general polynomial case when the differential equation of m_1 takes the form given in (10.30) and that of y_1 is of the type shown in (10.29). The inequality in (10.32) is based on the fact that the term containing the second moment has negative sign. Hence, in the generalisation we will need the sign condition $D_j \le 0$ for all $j \ge 2$. Then, applying Jensen's inequality in (10.30) we get

$$\dot{m}_1 = \sum_{j=0}^{J} D_j m_j \le \sum_{j=0}^{J} D_j m_1^j.$$

The right-hand side of the inequality is the same as that of (10.29); thus, the comparison lemma yields that $m_1(t) \le y_1(t)$ for all $t \ge 0$.

Let us turn now to the derivation of a lower bound for the expected value m_1. Applying again Jensen's inequality for the random variable X used previously and for $\varphi(x) = x^{(n+j-1)/n}$ with $j \ge 1$, yields

$$m_{n+j-1} = \mathbb{E}[(X/N)^{n+j-1}] \ge \mathbb{E}[(X/N)^n]^{\frac{n+j-1}{n}} = m_n^{\frac{n+j-1}{n}}. \tag{10.33}$$

In the case $j = 0$, the function $\varphi(x) = x^{(n-1)/n}$ is concave; hence, Jensen's inequality leads to

$$m_{n-1} = \mathbb{E}[(X/N)^{n-1}] \le \mathbb{E}[(X/N)^n]^{\frac{n-1}{n}} = m_n^{\frac{n-1}{n}}. \tag{10.34}$$

Using the differential equation of the nth moment (10.27), Proposition 10.3, applying (10.33) and (10.34), and assuming the sign condition $D_j \le 0$ for all $j \ge 2$ yield

$$m_n' \le n \sum_{j=0}^{J} D_j m_n^{\frac{n+j-1}{n}} + \frac{n(n-1)}{2N} B_c \tag{10.35}$$

for $n \ge 2$. The right-hand side of this inequality motivates the introduction of the z_n functions for $n = 1, 2, \ldots, J$. These solve the system of differential equations given below

$$\dot{z}_1 = \sum_{j=0}^{J} D_j z_j \tag{10.36a}$$

$$\dot{z}_n = n \sum_{j=0}^{J} D_j z_n^{\frac{n+j-1}{n}} + \frac{n(n-1)}{2N} B_c \quad (2 \le n \le J), \tag{10.36b}$$

and satisfy the initial condition $z_n(0) = m_1^n(0)$. By definition, $z_0(t) = 1$ for all t. Hence, the comparison lemma yields that $m_n \le z_n$ for $n \ge 2$. These inequalities enable us to compare z_1 to m_1. Namely, substituting $m_n \le z_n$ in (10.30) and exploiting the sign condition $D_j \ge 0$ for all $j \ge 2$ leads to

$$\dot{m}_1 \ge D_0 + D_1 y_1 + \sum_{j=2}^{J} D_j z_j. \tag{10.37}$$

Considering functions z_n $(n = 2,\dots,J)$ as fixed, then (10.37) and (10.36b) have the form $m_1'(t) \ge g_1(t, m_1(t))$ and $z_1'(t) = g_1(t, z_1(t))$, respectively, where $g_1(t,x) = D_0 + D_1 x + \sum_{j=2}^{J} D_j z_j(t)$. Thus, the comparison lemma implies $m_1 \ge z_1$. Summarising, the following theorem holds.

Theorem 10.4 *Consider the master equation Eq.* (10.1) *with density-dependent coefficients given in* (10.2) *and assume that the functions A and C are polynomials given in* (10.24). *Let* $D_j = A_j - C_j$ *be the difference of the coefficients of the polynomials. Let* m_1 *be the first moment, solving* (10.30); y_1 *be the mean-field approximation, solving* (10.29) *and satisfying the same initial condition as* m_1, *i.e.* $y_1(0) = m_1(0)$, *and* z_n *be the solution of* (10.36) *subject to the initial condition* $z_n(0) = m_1^n(0)$. *If the sign conditions* $D_j \le 0$ *hold for all* $j \ge 2$, *then*

$$z_1(t) \le m_1(t) \le y_1(t)$$

holds for all $t \ge 0$.

We note that it can be also proved that the difference of y_1 and z_1 is of order $1/N$ as N tends to infinity, implying that y_1 is an order $1/N$ approximation of the expected value m_1 (see in [5]). In that paper, the question of upper and lower bounds is dealt with in more detail and the necessity of the sign condition is discussed.

As an example, consider the general quadratic case when $A(z) = (1-z)(az+b)$ and $C(z) = z(c+dz)$. Then, the coefficient are $A_0 = b$, $A_1 = a-b$, $A_2 = -a$, $C_0 = 0$, $C_1 = c$ and $C_2 = d$; hence, $D_0 = b$, $D_1 = a-b-c$ and $D_2 = -a-d$. The sign condition of the above theorem holds if $a+d \ge 0$. The mean-field differential equation takes the form

$$\dot{y}_1 = D_0 + D_1 y_1 + D_2 y_1^2, \tag{10.38}$$

and the system for z_1 and z_2 is

$$\dot{z}_1 = D_0 + D_1 z_1 + D_2 z_2, \tag{10.39a}$$

$$\dot{z}_2 = 2D_0 z_2^{1/2} + 2D_1 z_2 + 2D_2 z_2^{3/2} + \frac{B_c}{N}, \tag{10.39b}$$

where $B_c = b + |a-b| + |a| + c + |d|$. The performance of the lower and upper bounds is illustrated in Fig. 10.2 for the case of SIS epidemic with spontaneous infection when $a = n\tau$, $b = \delta$, $c = \gamma$ and $d = 0$ (see Exercise 10.1). The mean-field equation Eq. (10.38) and the system (10.39) can be solved with an ODE solver subject to initial conditions $y_1(0) = \ell/N$, $z_1(0) = \ell/N$ and $z_2(0) = (\ell/N)^2$, where ℓ is the initial number of infected nodes. The solutions are shown in Fig. 10.2 for $N = 100$ together with the expected value $m_1(t) = \sum_{k=0}^{N} \frac{k}{N} p_k(t)$ obtained by solving the master equation Eq. (10.1) for p_k. One can see that the expected value is between the two bounds; in fact, it is hardly distinguishable from the solution of the mean-field equation. Note that the performance of the bounds is much better than in the case without spontaneous infection. Here, we get much closer bounds even for small values of N. The reader is asked to investigate the performance of the bounds for the case without spontaneous infection and for the voter model in the following two exercises.

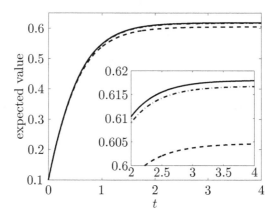

Fig. 10.2: Bounds for the expected value for SIS epidemic with spontaneous infection. The solution y_1 of the mean-field equation Eq. (10.38) (solid curve, upper bound), the first coordinate z_1 of the solution of system (10.39) (dashed curve, lower bound) and the expected value m_1 obtained from the solution of the master equation (dashed-dotted curve). The stationary part of the curves are enlarged in the inset. The parameter values are $N = 100$, $\gamma = 1$, $\tau = 0.05$, $n = 20$ and $\delta = 1$.

Exercise 10.3. Determine the upper and lower bounds in the case of SIS epidemic propagation on a homogeneous Configuration Model random graph with all nodes having degree n, infection rate τ and recovery rate γ by solving the differential equations (10.38) and (10.39). The coefficients are given by $a = n\tau$, $b = 0$, $c = \gamma$ and $d = 0$; use large values of N, such as $N = 10^6$ and $N = 10^7$. (For such large values of N, the exact master equation cannot be solved; hence, m_1 cannot be determined.)

Exercise 10.4. Determine the upper and lower bounds for the voter model on a homogeneous Configuration Model random graph with all nodes having degree n and transition rates α and β by solving the differential equations (10.38) and (10.39). The coefficients are given by $a = n\beta$, $b = 0$, $c = n\alpha$ and $d = -n\alpha$. In order to ensure the sign condition, take $\alpha < \beta$. Choose $N = 100$ and $N = 1000$ and solve the master equation Eq. (10.1) as well, and plot the exact value m_1 together with the upper and lower bounds.

10.4.4 Higher order closure based on an a priori distribution

The idea for deriving the simplest mean-field approximation (10.29), starting from the exact equation Eq. (10.30) was to use the approximation $m_n \approx m_1^n$. Now, to get a more accurate approximation, we keep the differential equation of the first moment exact and use the closure approximation in the differential equation of the second moment m_2. These kind of algebraic relations between the moments are referred to as higher order closures. The calculation can be carried out in the quadratic case when $J = 2$, that is A and C are quadratic polynomials in (10.24). In that case, the exact differential equations for the first two moments, using (10.27), take the form

$$\dot{m}_1 = D_0 + D_1 m_1 + D_2 m_2, \tag{10.40a}$$

$$\dot{m}_2 = 2(D_0 m_1 + D_1 m_2 + D_2 m_3) + \frac{1}{N}(E_0 + E_1 m_1 + E_2 m_2), \tag{10.40b}$$

where $E_j = A_j + C_j$ and R_2 was substituted for using (10.25). These equations are not closed or self-contained since the second moment depends on the third one and an equation for this is also needed.

The novel closure put forward here is based on the empirical observation that $p_k(t)$ is well approximated by a binomial or normal distribution. The justification of this approximation is discussed in Section 10.5.3. In the case of the binomial distribution $\mathcal{B}(n, p)$, the parameters n and p depend on time and will be specified in terms of the moments of the distribution. The first three moments of the binomial distribution can be specified in terms of the two parameters and are as follows:

$$Y_1 = np \tag{10.41a}$$

$$Y_2 = np + n(n-1)p^2 \tag{10.41b}$$

$$Y_3 = np + 3n(n-1)p^2 + n(n-1)(n-2)p^3. \tag{10.41c}$$

Using equations (10.41a) and (10.41b), n and p can be expressed in terms of Y_1 and Y_2 and lead to

$$p = 1 + Y_1 - \frac{Y_2}{Y_1}, \qquad n = \frac{Y_1^2}{Y_1 + Y_1^2 - Y_2}. \tag{10.42}$$

Plugging these expressions for p and n into (10.41c), the closure for the third moment is found to be

$$Y_3 = \frac{2Y_2^2}{Y_1} - Y_2 - Y_1(Y_2 - Y_1). \tag{10.43}$$

This relation defines the new triple closure and in terms of the density-dependent moments is equivalent to

$$y_3 = \frac{2y_2^2}{y_1} - y_1 y_2 + \frac{1}{N}(y_1^2 - y_2). \tag{10.44}$$

Using the equations for the first two moments (10.40a)–(10.40b) and the closure at the level of the third moment yields the new approximating system in the form

$$\dot{x}_1 = D_0 + D_1 x_1 + D_2 x_2, \tag{10.45a}$$

$$\dot{x}_2 = 2(D_0 x_1 + D_1 x_2 + D_2 x_3) + \frac{1}{N}(E_0 + E_1 x_1 + E_2 x_2), \tag{10.45b}$$

$$x_3 = \frac{2x_2^2}{x_1} - x_1 x_2 + \frac{1}{N}(x_1^2 - x_2). \tag{10.45c}$$

Summarising, the idea for deriving the higher order closure is to assume an *a priori* distribution for p_k that leads to an expression for the third moment in terms of the first two moments. As an alternative to binomial distribution, one can approximate the distribution p_k with a normal distribution. Then, a simple derivation yields that the third moment can be expressed as $y_3 = 3y_1 y_2 - 2y_1^3$. Hence an alternative closed system, based on the normal distribution approximation, can be written as

$$\dot{x}_1 = D_0 + D_1 x_1 + D_2 x_2,$$

$$\dot{x}_2 = 2(D_0 x_1 + D_1 x_2 + D_2 x_3) + \frac{1}{N}(E_0 + E_1 x_1 + E_2 x_2),$$

$$x_3 = 3x_1 x_2 - 2x_1^3.$$

Note that here the same variable names x_1, x_2 and x_3 are used, but strictly speaking these are different from those given in system (10.45).

These new closed systems were introduced for the case of SIS epidemic propagation in [176], where their performance was investigated in detail. Extensive numerical study showed that these closures give order $1/N^2$ approximation for the moments, in contrast to the $1/N$ accuracy of the usual mean-field approximation given by (10.29) and the pairwise approximation given by the widely used triple closure. We note that the $1/N$ accuracy of the usual mean-field approximation has already been proved rigorously (see [29] and Section 10.6.2). However, the order $1/N^2$ accuracy for the binomial and normal closures still awaits formal proof. The performance of the binomial closure for the SIS propagation is compared to that of the pairwise model in Chapter 4 (see Fig. 4.6).

Finally, we note that the well-known triple closures can be translated to a relation between the first three moments when the master equation Eq. (10.1) is based on the

SIS epidemic on a complete graph. In [176], it is shown that the closures of $[SSI]$ and $[ISI]$ triples lead to the relation

$$y_3 = -\frac{1}{N}y_1 + (1 + \frac{1}{N})y_2 - \frac{N-2}{N-1}\frac{(y_1 - y_2)^2}{1 - y_1}.$$

Using this closure instead of (10.44) leads to a system which is equivalent to the usual pairwise model. Recall that this closure yields only order $1/N$ accuracy compared to (10.45c) leading to order $1/N^2$ accuracy.

10.5 PDE approximation for the distribution

10.5.1 The PGF for polynomial coefficients

In this section, a parabolic partial differential differential equation is derived for the probability-generating function in the case when the coefficients of the master equation Eq. (10.1) are given by polynomials A and C according to (10.24). The derivation is based on the general result, Proposition 10.2, and uses the special form of the coefficients a_k and c_k. For simplicity, it is carried out in the case when A and C are quadratic polynomials. Since the polynomials have to satisfy the conditions $A(1) = 0$ and $C(0) = 0$, they can be written in the form

$$A(z) = (az + b)(1 - z), \qquad C(z) = z(c + dz). \tag{10.46}$$

Hence,

$$\frac{a_k}{N} = \left(a\frac{k}{N} + b\right)\left(1 - \frac{k}{N}\right), \qquad \frac{c_k}{N} = \frac{k}{N}\left(c + d\frac{k}{N}\right).$$

Substituting these expressions into (10.7) yields

$$\partial_t G(t,z) = N\sum_{k=0}^{N}\left[\left(a\frac{k}{N} + b\right)\left(1 - \frac{k}{N}\right)z^k(z-1) + \frac{k}{N}\left(c + d\frac{k}{N}\right)z^{k-1}(1-z)\right]p_k(t)$$

$$= N(1-z)\sum_{k=0}^{N}\left[\frac{a}{N^2}k^2z^k + \frac{d}{N^2}k^2z^{k-1} + \frac{b-a}{N}kz^k + \frac{c}{N}kz^{k-1} - bz^k\right]p_k(t).$$

The right-hand side can be expressed in terms of the partial derivatives of G with respect to z by using the following identities:

$$G(t,z) = \sum_{k=0}^{N}z^k p_k(t), \quad \partial_z G(t,z) = \sum_{k=0}^{N}kz^{k-1}p_k(t), \quad \partial_z^2 G(t,z) = \sum_{k=0}^{N}k(k-1)z^{k-2}p_k(t).$$

Using these relations, it is straightforward to verify

$$\sum_{k=0}^{N} kz^k p_k(t) = z\partial_z G(t,z) \quad \text{and} \quad \sum_{k=0}^{N} k^2 z^k p_k(t) = z^2 \partial_z^2 G(t,z) + z\partial_z G(t,z).$$

$$(10.47)$$

We note that further differentiation of G with respect to z enables us to determine $\sum_{k=0}^{N} k^j z^k p_k(t)$ in terms of the partial derivatives of G (at most of order j). Using (10.47) leads to a parabolic PDE for the probability-generating function G.

Proposition 10.4 *Consider the master equation Eq.* (10.1) *with density-dependent coefficients given in* (10.2) *and assume that the functions A and C are quadratic polynomials given in* (10.46). *Then, the probability-generating function $G(t,z) = \sum_{k=0}^{N} z^k p_k(t)$ satisfies the parabolic PDE*

$$\partial_t G = (1-z)\left[\frac{1}{N}(az^2 + dz)\partial_z^2 G + \left(az\frac{1-N}{N} + bz + c + \frac{d}{N} \right)\partial_z G - bNG \right].$$

We note that similar statement can be formulated in the case when A and C are polynomials of higher degree. The order of the partial differential equation is the same as the degree of the polynomials A and C. This partial differential equation can be used to derive differential equations (10.40a) and (10.40b) for the moments (see Exercise 10.5).

This PDE in Proposition 10.4 cannot be solved analytically in general. The approximation when the order $1/N$ terms are neglected is a first order PDE, which can be solved by using the method of characteristics. On the other hand, the partial differential equation can be used to determine the stationary distribution, of p_k describing the long time behaviour of the process. For the stationary distribution $\partial_t G = 0$ holds, i.e. G does not depend on t. Denoting the stationary solution now by $g(z) = G(t,z)$ leads to the following second order ordinary differential equation for g:

$$\frac{1}{N}(az^2 + dz)g''(z) + \left(az\frac{1-N}{N} + bz + c + \frac{d}{N} \right)g'(z) - bNg(z) = 0, \qquad (10.48)$$

subject to the initial condition $g(1) = 1$, since $G(t,1) = 1$ for any t. This is a second order linear ODE, for which an analytical formula is not available to express all solutions generally. However, for some special choices of the parameters a, b, c and d, the solution can be given in a simple form.

The simplest case is the linear one, when $a = d = 0$. Then, the differential equation is of first order and the solution can be obtained by integration as we saw in Section 10.3. Another case which is analytically tractable is $c = d = 0$. In that case, Eq. (10.48) is an Euler-type differential equation, the solution of which can be obtained in the form $g(z) = z^r$. The application of the Laplace transform may also help in finding the solution of (10.48). In the case $a = 0$, the Laplace transform reduces the differential equation to a first order one.

The case of SIS epidemic corresponds to $b = d = 0$. Then, the solution is the constant function $g(z) = 1$ describing the absorbing state of all nodes susceptible: $p_0 = 1$, $p_k = 0$ for $k \geq 1$. It would be desirable to determine the quasi-steady state from this differential equation; however, its derivation from this equation is not available. An alternative route to approximate the quasi-steady state is to start from the case of SIS epidemic with spontaneous infection when b is also positive and then let b tend to zero. In this case, an approximating solution can be determined by neglecting the order $1/N$ terms. Then, the first order differential equation can be written in the following form

$$(c + (b-a)z)g'(z) = bNg(z),$$

which can be solved by separation of variables. Using $g(1) = 1$, the solution is obtained as

$$g(z) = \left(\frac{1-p}{1-pz}\right)^r,$$

where $p = \frac{a-b}{c}$ and $r = \frac{bN}{a-b}$. If $p \in (0,1)$, i.e. $0 < a - b < c$, then this is the probability-generating function of the negative binomial distribution. Hence, for large N and for $0 < a - b < c$ the steady state distribution of p_k given by the master equation Eq. (10.1) can be approximated by a negative binomial distribution. It is important to note that the negative binomial distribution has an infinite range $k = 0, 1, 2, \ldots$. Our distribution, however, has a finite range $k = 0, 1, 2, \ldots, N$; hence, only the first $N + 1$ terms of the negative binomial distribution are used to approximate our distribution p_k. For a reasonable range of parameter values, the probabilities belonging to indices larger than N are negligible. The accuracy of this approximation is shown in Fig. 10.3 for $a = 0.2$, $b = 0.1$ and $c = 1$. We note that for larger values of a, the performance of the negative binomial distribution approximation is less convincing.

Summarising, the PDE for the PGF may yield the exact values for the distribution p_k; however, this is based on the analytical solution of the PDE, since its expansion according to the powers of z is also needed. Thus, explicit formulas for the distribution p_k are available only in special cases. As an alternative, the Fokker–Planck equation can be used, which is only an approximation of the original master equation but yields the approximating distribution directly. This will be considered in next subsections.

Exercise 10.5. Using that the moments can be expressed as $m_1(t) = \frac{1}{N}\partial_z G(t, 1)$ and $m_2(t) = \frac{1}{N^2}\partial_z^2 G(t, 1) + \frac{1}{N^2}\partial_z G(t, 1)$, derive the differential equation Eq. (10.40a) for m_1, starting from the PDE in Proposition 10.4.

10.5.2 First order PDE approximation

The Fokker–Planck equation Eq. (10.16a) is a parabolic partial differential equation with non-constant coefficients; hence, it cannot be solved analytically in general. In this section, we consider a solvable approximation, namely the following first order PDE that is obtained from (10.16a) by neglecting the order $1/N$ term:

$$\partial_t u(t,z) = \partial_z((C(z) - A(z))u(t,z)) \tag{10.49}$$

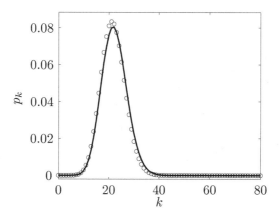

Fig. 10.3: The steady state of the distribution for SIS epidemics with spontaneous infection. The exact solution of the master equation at a sufficiently large time, $t = 20$, (circles) is shown together with the negative binomial distribution (solid curve) corresponding to the parameters $p = \frac{a-b}{c}$ and $r = \frac{bN}{a-b}$. The parameter values are $N = 200$, $a = n\tau$, $b = \delta$, $c = \gamma$ and $d = 0$, with $\gamma = 1$, $\tau = 0.02$, $n = 10$ and $\delta = 0.1$.

Solution of the first order PDE

This differential equation can be solved analytically by using the method of characteristics. First, it is useful to introduce the new unknown function v as $v(t,z) = (C(z) - A(z))u(t,z)$. The differential equation for v takes the form

$$\partial_t v(t,z) + (A(z) - C(z))\partial_z v(t,z) = 0.$$

The differential equations of the characteristics are

$$\frac{dt}{ds} = 1, \quad \text{and} \quad \frac{dz}{ds} = A(z) - C(z).$$

The first equation can be solved as $t = s + t_0$, the second one yields $D(z) = s + z_0$, where D is the integral of $1/(A - C)$, i.e. $D'(z) = \frac{1}{A(z) - C(z)}$, assuming that $A(z) \neq C(z)$. Therefore, $D(z) - t$ is constant; that is, all solutions of the PDE can be given in the form $v(t,z) = \varphi(D(z) - t)$, with an arbitrary differentiable function $\varphi : \mathbb{R} \to \mathbb{R}$. Then, the original unknown u can be given as

$$u(t,z) = \frac{\varphi(D(z)-t)}{C(z)-A(z)},$$

where the function φ is determined by the initial condition $u(0,z) = u_0(z)$. Namely, $\varphi(y) = u_0(z)(C(z) - A(z))$, where $y = D(z)$.

Solution in the quadratic case

If A and C are polynomials, then the integral of $1/(A-C)$ can be determined analytically; hence, the solution u can be given explicitly. This calculation is carried out below for the quadratic case when A and C are given as in (10.46). Then, $C(z) - A(z) = (a+d)z^2 + (b+c-a)z - b$, which can be written as $C(z) - A(z) = (a+d)(z-z_1)(z-z_2)$, where z_1 and z_2 are the roots of the quadratic polynomial. A simple calculation shows that

$$\frac{1}{A(z)-C(z)} = \frac{\mu}{z-z_1} - \frac{\mu}{z-z_2},$$

where $(a+d)\mu(z_2 - z_1) = 1$. Thus,

$$D(z) = \int \frac{1}{A(z)-C(z)} = \mu \ln(z-z_1) - \mu \ln(z-z_2) = \mu \ln\left(\frac{z-z_1}{z-z_2}\right).$$

The function $D(z) - t$ being constant is equivalent to

$$\frac{z-z_1}{z-z_2} e^{-\frac{t}{\mu}} = \text{const.}$$

Therefore, u can be written as

$$u(t,z) = \frac{1}{C(z)-A(z)} \varphi\left(\frac{z-z_1}{z-z_2} e^{-\frac{t}{\mu}}\right).$$

The function φ is determined by the initial condition as

$$u_0(z) = \frac{1}{C(z)-A(z)} \varphi\left(\frac{z-z_1}{z-z_2}\right).$$

Thus, $\varphi(y) = u_0(z)(C(z) - A(z))$, where $y = \frac{z-z_1}{z-z_2}$. Solving this equation for z leads to $z = \frac{yz_2 - z_1}{y-1}$. Therefore, u is given by

$$u(t,z) = u_0(h(t,z)) \frac{C(h(t,z)) - A(h(t,z))}{C(z)-A(z)}, \tag{10.50}$$

where

$$h(t,z) = \frac{\frac{z-z_1}{z-z_2}e^{-\frac{t}{\mu}}z_2 - z_1}{\frac{z-z_1}{z-z_2}e^{-\frac{t}{\mu}} - 1}.$$

This can be simplified to

$$h(t,z) = \frac{z_2(z-z_1) + z_1(z_2-z)e^{\frac{t}{\mu}}}{z - z_1 + (z_2-z)e^{\frac{t}{\mu}}}. \tag{10.51}$$

It is worth noting that $h(0,z) = z$ and that h satisfies the differential equation $\partial_t h = (C-A)\partial_z h$, which proves directly that the function u given by (10.50) is a solution of the PDE (10.49) and satisfies the initial condition $u(0,z) = u_0(z)$.

As an example, let us solve the differential equation corresponding to SIS dynamics when $b = d = 0$, $c = \gamma$ is the rate of recovery and $a = n\tau$, where τ is the transmission rate and n is the average degree of the network. Then, the roots of the polynomial $A - C$ are $z_1 = 1 - \frac{\gamma}{n\tau}$ and $z_2 = 0$, yielding $\mu = 1/(\gamma - n\tau)$. Then, (10.51) yields

$$h(t,z) = \frac{z(n\tau - \gamma)}{zn\tau - (\gamma - n\tau + zn\tau)e^{t(n\tau-\gamma)}},$$

and (10.50), after some algebra, leads to

$$u(t,z) = u_0(h(t,z))\frac{(\gamma - n\tau)^2 e^{t(n\tau-\gamma)}}{[zn\tau - (\gamma - n\tau + zn\tau)e^{t(n\tau-\gamma)}]^2}. \tag{10.52}$$

Consider the case of pure initial condition in (10.1), namely start the process from a given state $\ell \in \{0,1,\ldots,N\}$, that is $p_\ell(0) = 1$ and $p_j(0) = 0$ for $j \neq \ell$, meaning that there are ℓ infected individuals initially. Then, the continuous version of the initial condition can be given by the function $u_0 : [0,1] \to \mathbb{R}$, for which $u_0(z) = 1$ when $\ell/N - 1/2N \leq z \leq \ell/N + 1/2N$ and $u_0(z) = 0$ otherwise. The solution u can be determined explicitly and compared to the exact distribution p_k given by (10.1) for an arbitrary time t. Numerical comparison shows that for small values of time, the accuracy of the first order approximation u is good; however, as time increases, the performance of the approximation is poorer (see Exercise 10.6). This is explained by the fact that the diffusion term was neglected; hence, the step-function-like character of the initial condition is preserved for a long time. Thus, the first order PDE is not suitable for approximating the distribution p_k. However, interestingly, it can be used to approximate the expected value m_1 (see Section 10.6.1).

Exercise 10.6. Determine the approximation u given by (10.52) in the case of SIS epidemic when $A(z) = az(1-z)$ and $C(z) = cz$, with $a = n\tau$ and $c = \gamma$. Compare the approximation with the exact distribution obtained from the solution of the master equation Eq. (10.1) for different values of time.

10.5.3 Fokker–Planck equation

Consider now the Fokker–Planck equation in its general second order form (10.16a). Introducing the first and second order linear differential operators

$$Dv = \frac{1}{2}(gv)' - hv \quad \text{and} \quad Lv = (Dv)'$$

with

$$g(z) = \frac{1}{N}(A(z) + C(z)), \quad h(z) = A(z) - C(z),$$

the Fokker–Planck equation takes the form

$$\partial_t u = Lu. \tag{10.53}$$

According to (10.16b) and (10.16c), this is subject to boundary conditions

$$(Dv)(-\delta) = 0 = (Dv)(1 + \delta) \tag{10.54}$$

with $\delta = 1/2N$ and satisfies the initial condition

$$u(0, z) = u_0(z) \tag{10.55}$$

for $z \in [0, 1]$, where the initial function u_0 corresponds to the initial condition $p_k(0)$ in the sense that $u_0(z) = 1$ when $\ell/N - 1/2N \leq z \leq \ell/N + 1/2N$ and $u_0(z) = 0$ otherwise. Simple integration shows that for any time t

$$\int_{-\delta}^{1+\delta} u(t, z)dz = \int_{-\delta}^{1+\delta} u_0(z)dz = \frac{1}{N}.$$

Calculating the solution by using the Fourier method

The Fokker–Planck equation is a parabolic PDE with given initial and boundary conditions. Hence, its solution can be given by using the Fourier method. This starts from the eigenfunctions v_i and eigenvalues λ_i of the second order operator L satisfying

$$Lv_i = \lambda_i v_i, \quad (Dv_i)(-\delta) = 0 = (Dv_i)(1 + \delta).$$

Once it is shown that the eigenfunctions form a complete system in an appropriate function space, the solution of (10.53) subject to the boundary conditions (10.54) can be given as

$$u(t, z) = \sum_{i=0}^{\infty} c_i e^{\lambda_i t} v_i(z)$$

with coefficients c_i determined by the initial condition

$$u_0(z) = \sum_{i=0}^{\infty} c_i v_i(z).$$

The eigenvalue problem belonging to the Fokker–Planck equation can be transformed to a Sturm–Liouville problem; hence, it has countably many eigenvalues and its eigenfunctions form a complete system (see [263], p. 106). If $\lambda_0 = 0$, then there is a stationary solution v satisfying

$$Lv = 0, \qquad (Dv)(-\delta) = 0 = (Dv)(1+\delta).$$

Steady state

This stationary solution was investigated in detail when A and C are linear functions (see Section 10.3.3). The stationary solution can be obtained in the general case as follows. The definition of L yields that Dv is constant when v is the stationary solution. According to the boundary condition, this constant is zero, i.e. $Dv = 0$, yielding $\frac{1}{2}(gv)' = hv$. Introducing the function $w = gv$, this differential equation is equivalent to $w' = 2hw/g$. This can be integrated to yield $w(z) = K\exp(H(z))$, where K is a constant and H is the integral of $2h/g$, i.e. $H' = 2h/g$. Thus, the stationary solution is

$$v(z) = \frac{K}{g(z)}e^{H(z)}, \qquad \text{with} \quad H'(z) = 2\frac{h(z)}{g(z)},$$

where the constant is determined by $\int_{-\delta}^{1+\delta} v(z)\mathrm{d}z = \frac{1}{N}$.

In the density-dependent case, the integral of the function $\frac{A-C}{A+C}$ is needed. If A and C are polynomials, then its integral can be explicitly determined; however, the formulas become rather complicated even for low-degree polynomials. The case of first order polynomials, when $A(z) = a(1-z)$ and $C(z) = cz$, was solved in Section 10.3. Then, computing the integral of the function $\frac{a-(a+c)z}{a+(c-a)z}$ yields

$$H(z) = \frac{2N}{(a-c)^2}\left[(a^2 - c^2)z + 2ac\ln(a + (c-a)z)\right],$$

leading to a rather complicated formula for the stationary solution v.

Approximation of the steady state with normal distribution

A significantly simpler approximation, with normal distribution, can be derived by using the linear approximation of the coefficient functions $A - C$ and $A + C$. The approximation is based on the observation that the stationary distribution is concentrated around its expected value, which can be approximated by the steady state solution of the mean-field equation Eq. (10.6). The steady state is the solution $z^* \in [0,1]$ of the equation $A(z^*) = C(z^*)$, which exists because of the sign conditions $A(0) \geq 0$, $C(0) = 0$, $A(1) = 0$ and $C(1) \geq 0$. Here, for simplicity, it is assumed that the steady state is unique and $z^* \in (0,1)$. Then, the following zeroth and first order approximations are used: $A(z) + C(z) = A(z^*) + C(z^*)$ and $A(z) - C(z) = (A'(z^*) - C'(z^*))(z - z^*)$. Then,

$$\frac{A(z) - C(z)}{A(z) + C(z)} \approx \frac{(A'(z^*) - C'(z^*))(z - z^*)}{A(z^*) + C(z^*)},$$

the integral of which is a quadratic function. Hence, the following relation holds

$$H(z) \approx Nq(z - z^*)^2, \quad \text{with} \quad q = \frac{A'(z^*) - C'(z^*)}{A(z^*) + C(z^*)}.$$

Thus, the stationary distribution can be approximated by the normal distribution

$$w(z) = K e^{Nq(z-z^*)^2}, \tag{10.56}$$

and the constant is determined by the constraint that $\int_{-\delta}^{1+\delta} w(z) dz = \frac{1}{N}$. Note that the sign conditions on the functions A and C and the uniqueness of z^* ensure that $A'(z^*) - C'(z^*) < 0$, and hence, $q < 0$.

In the linear case, when $A(z) = a(1 - z)$ and $C(z) = cz$, the solution of the equation $A(z^*) = C(z^*)$ is $z^* = \frac{a}{a+c}$ and $q = -\frac{(a+c)^2}{2ac}$. Hence, the approximating formula takes the form

$$w(z) = K_1 \exp\left(-\frac{N}{2p(1-p)}(z - p)^2\right), \quad \text{with} \quad p = \frac{a}{a+c},$$

where K_1 is the normalisation constant. The exact formula for the steady state is

$$v(z) = \frac{K_2}{a + (c-a)z} \exp\left(\frac{2N}{(a-c)^2}\left[(a^2 - c^2)z + 2ac\ln(a + (c-a)z)\right]\right),$$

where K_2 is the normalisation constant. If a and c are of the same magnitude, then w yields an extremely accurate approximation of the exact solution v; in fact, they are visually indistinguishable if plotted on the same figure. In order to show the difference between them, they are plotted for $a = 10$ and $c = 1$ in Fig. 10.4 together with the steady state of the master equation, which is a binomial distribution with parameter $p = a/(a+c)$.

In the quadratic case of the SIS epidemic, when $A(z) = az(1 - z)$ and $C(z) = cz$ with $a = n\tau$ and $c = \gamma$, the solution of the equation $A(z^*) = C(z^*)$ is $z^* = \frac{a-c}{a}$ and $q = -\frac{a}{2c}$. Hence, the approximating formula takes the form

$$w(z) = K_1 \exp\left(-\frac{Na}{2c}\left(z - \frac{a-c}{a}\right)^2\right),$$

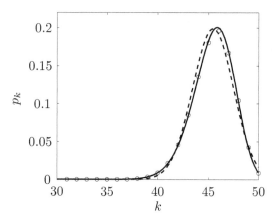

Fig. 10.4: The steady state of the distribution in the linear case, when $A(z) = a(1 - z)$ and $C(z) = cz$. The binomial distribution as the exact solution of the master equation (circles) is shown together with v, the solution of the Fokker–Planck equation (solid curve), and with w, the solution of the approximate Fokker–Planck equation (dashed curve). The parameter values are $N = 50$, $a = 10$ and $c = 1$.

where K_1 is the normalisation constant. The exact formula for the quasi-steady state cannot be given explicitly, because the stochastic model tends to the absorbing state. For comparison, the master equation Eq. (10.1) is solved on a long time interval and its final state is plotted in Fig. 10.5 together with w. We note that for smaller values of τ, when the solution of the master equation tends more quickly to the absorbing state, the performance of the approximation w is significantly poorer.

Exercise 10.7. Determine the approximation of the steady state using (10.56) in the case of the SIS epidemic with spontaneous infection, when $A(z) = (az + b)(1 - z)$ and $C(z) = cz$ with $a = n\tau$, $b = \delta$, and $c = \gamma$. Compare the approximation to the exact steady state obtained from the solution of the master equation Eq. (10.1) for a sufficiently large time for different values of τ, with $n = 10$, $\delta = 0.1$ and $\gamma = 1$.

10.6 The accuracy of the mean-field and Fokker–Planck approximations

10.6.1 Relation of the Fokker–Planck and mean-field equations

Returning to the general case, when A and C are polynomials, the formula for u enables us to derive an approximation for the moments. Recall that the nth moment is defined by

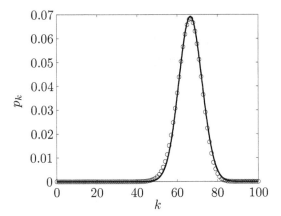

Fig. 10.5: The steady state of the distribution in the case of SIS epidemic propagation, when $A(z) = n\tau(1-z)$ and $C(z) = \gamma z$. The exact solution of the master equation at a sufficiently large time, $t = 10$, (circles) is shown together with w, the solution of the approximate Fokker–Planck equation (solid curve). The parameter values are $N = 100$, $n = 10$, $\tau = 0.3$ and $\gamma = 1$.

$$m_n(t) = \sum_{k=0}^{N} \frac{k^n}{N^n} p_k(t).$$

This sum can be approximated by an integral using the basic integral approximation

$$\int_{-\delta}^{1+\delta} f(z)dz \approx \sum_{k=0}^{N} \frac{1}{N} f\left(\frac{k}{N}\right), \quad \text{where} \quad \delta = \frac{1}{2N}.$$

This approximation is based on the division of the interval $[-\delta, 1+\delta]$ into $N+1$ parts, each of which has length $1/N$, and the midpoints of the subintervals are k/N for $k = 0, 1, \ldots, N$. Since $p_k(t)$ is approximated by $u(t, k/N)$, the integral approximation leads to

$$m_n(t) \approx N \int_{-\delta}^{1+\delta} z^n u(t,z)dz. \tag{10.57}$$

This approximation yields the time dependence of the moments explicitly once the function u is given. In the case of the first order PDE approximation (10.49), the function u can be calculated explicitly. For quadratic coefficients, it is given in (10.50). Now it will be shown how the integral approximation (10.57) yields a formula for the expected value using this solution u. For simplicity, the calculation is carried out in the case $b = 0$, when the roots of the quadratic polynomial $A - C$ are $z_1 = (a-c)/(a+d)$ and $z_2 = 0$. Therefore, (10.51) yields

$$h(t,z) = \frac{z_1 z T}{zT + z_1 - z}, \quad \text{with} \quad T = e^{t(c-a)}.$$

According to (10.57) and (10.50), the expected value can be approximated as

$$w_1(t) = N \int_{-\delta}^{1+\delta} z u_0(h(t,z)) \frac{C(h(t,z)) - A(h(t,z))}{C(z) - A(z)} dz.$$

The integral can be calculated by the substitution $y = h(t,z)$, which leads to

$$z = \frac{z_1 y}{z_1 T + (1 - T)y} =: g(y).$$

This enables us to reduce the interval of integration to $[q_1, q_2]$, where $q_1 = \ell/N - 1/2N$, $q_2 = \ell/N + 1/2N$ and ℓ is given by the initial condition $p_\ell(0) = 1$, since the initial function u_0 is zero outside this interval. Hence, using $C(z) - A(z) = (a + d)z(z - z_1)$ one obtains

$$w_1(t) = N \int_{q_1}^{q_2} g(y) \frac{(a+d)y(y - z_1)}{(a+d)g(y)(g(y) - z_1)} g'(y) dy,$$

which, after straightforward algebra, leads to

$$w_1(t) = \frac{N z_1}{1 - T} \left(q_2 - q_1 - B \ln \left(\frac{q_2 + B}{q_1 + B} \right) \right) \quad \text{with} \quad B = \frac{z_1 T}{1 - T}.$$

Observe that the term in the logarithm can be written as $(q_2 + B)/(q_1 + B) = 1 + (q_2 - q_1)/(q_1 + B)$ and that $q_2 - q_1 = 1/N$ is small. Therefore, the approximation $\ln(1 + x) \approx x$ can be used, and thus leads to

$$w_1(t) \approx \frac{N z_1}{1 - T} \left(\frac{1}{N} - \frac{B}{(q_1 + B)N} \right) = \frac{z_1 q_1}{q_1(1 - T) + z_1 T}.$$

A simple calculation shows that this is the solution of the mean-field equation $\dot{y}_1 = A(y_1) - C(y_1)$ subject to the initial condition $y_1(0) = q_1$, that is the approximation of the expected value obtained from the first order PDE is the same as the mean-field approximation. This observation can be generalised to any (not only quadratic) coefficients as follows.

The integral approximation of the moments enables us to derive a system of ODEs for the moments. Denoting the approximation in (10.57) by

$$w_n(t) = N \int_{-\delta}^{1+\delta} z^n u(t,z) dz,$$

simple differentiation, the use of the PDE (10.49) and integration by parts yield

$$\dot{w}_n(t) = N \int_{-\delta}^{1+\delta} z^n \partial_t u(t,z) dz = N \int_{-\delta}^{1+\delta} z^n \partial_z ((C(z) - A(z)) u(t,z)) dz$$

$$= -N \int_{-\delta}^{1+\delta} nz^{n-1}((C(z) - A(z))u(t,z)) = N \int_{-\delta}^{1+\delta} nz^{n-1} \sum_{j=0}^{J} D_j z^j u(t,z)$$

$$= n \sum_{j=0}^{J} D_j \int_{-\delta}^{1+\delta} z^{n-1+j} u(t,z) = n \sum_{j=0}^{J} D_j w_{n-1+j},$$

where it was assumed that A and C are polynomials given by (10.24), $D_j = A_j - C_j$ and the value of the function $(C - A)u$ is neglected at the boundaries. Thus, we derived the approximating system (10.28) for the moments, starting from the Fokker–Planck equation.

10.6.2 Mean-field equation

Consider now the general density-dependent case when the coefficients are given by (10.2). Using (10.5) with $n = 1$ leads to

$$\dot{m}_1 = \sum_{k=0}^{N} \left[A\left(\frac{k}{N}\right) - C\left(\frac{k}{N}\right) \right] p_k.$$

The closure relation can be obtained by assuming that the order of application of a non-linear function (such as A or C) and the expected value can be exchanged. (For a linear function, this yields an exact relation, while for non-linear ones it is only an approximation.) Thus, the closure approximation leads to

$$\sum_{k=0}^{N} \left[A\left(\frac{k}{N}\right) - C\left(\frac{k}{N}\right) \right] p_k \approx A\left(\sum_{k=0}^{N} \frac{k}{N} p_k \right) - C\left(\sum_{k=0}^{N} \frac{k}{N} p_k \right) = A(m_1) - C(m_1).$$

Introducing y_1 as the approximation of m_1, the approximating closed differential equation takes the form $\dot{y}_1 = A(y_1) - C(y_1)$ (see (10.6)).

The goal now is to prove that the approximation error, that is the difference between the mean-field approximation y_1 and the exact value m_1, is of order $1/N$. Thus, for a large system the mean-field equation gives a good approximation to the expected value. This question is studied in detail in [286] for the case of SIS epidemics and in [29] for general density-dependent processes, where the following theorem is proved.

Theorem 10.5 *Let the coefficients of* (10.1) *be given by* (10.2) *and let p_k be the solution of* (10.1) *satisfying the initial conditions $p_\ell(0) = 1$, $p_j(0) = 0$, $j \neq \ell$ with some $\ell \in \{0,1,2,\ldots,N\}$. Let m_1 be the expected value and y_1 be the solution of the mean-field equation Eq.* (10.6) *subject to the initial condition $y_1(0) = m_1(0) = \frac{\ell}{N}$. Then, for any $t_0 > 0$, there exists a constant K for which*

$$|y_1(t) - m_1(t)| \leq \frac{K}{N}, \, t \in [0,t_0].$$

In [29], it is shown that the statement holds also in the more general case of asymptotically density-dependent coefficients, which is defined as follows. The coefficients are given by the functions A_N and C_N as $a_k = A_N(k)$, $c_k = C_N(k)$, where the following limits

$$A(x) = \lim_{N \to \infty} \frac{A_N(Nx)}{N}, \quad C(x) = \lim_{N \to \infty} \frac{C_N(Nx)}{N},$$

exist such that A and C are continuous functions on the interval $[0,1]$. Moreover, there exists a constant L such that

$$\left| A(x) - \frac{A_N(Nx)}{N} \right| \le \frac{L}{N} \quad, \quad \left| C(x) - \frac{C_N(Nx)}{N} \right| \le \frac{L}{N} \tag{10.58}$$

hold for all $x \in [0,1]$ and for all $N \in \mathbb{N}$.

The theorem is proved in [29] by applying the variation of constant formula in the context of operator semigroups.

10.6.3 Fokker–Planck equation

The solution u of the Fokker–Planck equation was introduced as an approximation of the distribution p_k in the sense that $u(t, k/N) \approx p_k(t)$. In [28], Theorem 4.6, it was shown that this is an order $1/N^2$ approximation on finite time intervals. The rigorous statement can be formulated as follows.

Theorem 10.6 *Let the coefficients of* (10.1) *be given by* (10.2) *and let p_k be the solution of* (10.1) *satisfying the initial conditions $p_\ell(0) = 1$, $p_j(0) = 0$, $j \ne \ell$ with some $\ell \in \{0,1,2,\ldots,N\}$. Let u be the solution of the Fokker–Planck equation Eq.* (10.53) *subject to the boundary condition* (10.54) *and initial condition* (10.55). *Then, for any $t_0 > 0$, there exists a constant K for which*

$$\left| u\left(t, \frac{k}{N} \right) - p_k(t) \right| \le \frac{K}{N^2}, \quad t \in [0, t_0], \quad k = 0, 1, 2, \ldots, N.$$

We note that the proof is based on the application of the variation of constant formula in the context of operator semigroups. Comparing this theorem to Theorem 10.5, where only the expected value was approximated, it is important to note that here the order of approximation is $1/N^2$, while it was only order $1/N$ in Theorem 10.5. The reason for this difference is that the derivation of the Fokker–Planck equations used a second order approximation. However, the approximation in Theorem 10.5 corresponds to the case when the second order term in the Fokker–Planck equation, involving a factor of order $1/N$, is neglected. This further approximation leads to the loss of a $1/N$ order.

10.7 Conclusions and outlook

Binary processes on networks with N nodes can be modelled as one-step processes. The corresponding master equations form a system of linear differential equations with an $N \times N$ tridiagonal matrix structure. In the case of large networks, the solution of this linear system can be approximated by a parabolic linear partial differential equation, the so-called Fokker–Planck equation. The moments of the probability distribution determined by the master equations can be approximated by the solutions of mean-field equations.

The problem of rigorously linking exact stochastic models to mean-field approximations goes back to the early work of Kurtz [189]. Kurtz studied pure-jump density-dependent Markov processes, where, apart from providing a method for the derivation of the mean-field models, also used solid mathematical arguments to prove the stochastic convergence of the exact to the mean-field model. His earlier results, e.g. [189], relied on Trotter-type approximation theorems for operator semigroups. Later on, the results were embedded in a more general context of martingale theory [94]. These results were generalised in several directions, including non-Markovian processes [137].

The Fokker–Planck equation can be approximated by a first order PDE. This is the most intuitive approach since it is based on the idea that for large N the discrete distribution can be approximated by a continuous density function. The main steps of the proof that this approximation is of order $1/N$ can be found in the Appendix of [78]; the details of the rigorous mathematical proof are presented in [288].

The accuracy of mean-field approximations can be studied by ODE techniques. This is carried out in [286] for the case of SIS epidemics, and in [29] for general density-dependent processes. The accuracy of the Fokker–Planck approximation is investigated in [28] by using operator semigroup techniques.

Chapter 11
Disease spread in networks with large-scale structure

This book has developed analytic models of disease spread on networks. All of our tractable models require closure assumptions. The closure process assumes that we can explain the dynamics at the network scale by understanding the dynamics locally at the level of small units with random connections between these units. Unfortunately, for some networks these assumptions are not valid: there exists relevant structure at a scale larger than the scale of the closures. In such cases, if one wishes to characterise the spreading process, several options remain:

- use the original closures and accept errors which may be difficult to quantify;
- derive new closures at higher level (e.g., quadruples rather than triples); and
- use explicit stochastic network simulations to describe disease spread.

This chapter explores networks for which our closure assumptions break down. We will study a social network of Facebook friendships for the University of Oklahoma [303], the general class of Watts–Strogatz networks (including small-world networks) [324] and the Barabási–Albert preferential attachment networks [23].

A natural criticism of some of our models is that they do not account for clustering, and indeed the Facebook network and the Watts–Strogatz networks exhibit much more clustering than is found in random networks. For low levels of clustering or for networks with large typical degrees, we anticipate that the impact of clustering is relatively small, and so the models we have derived should perform well.

In most cases, simulation and predictions agree well, even though some assumptions of the closures break down for these networks. In general, reasonable agreement is found with a model which captures the degree distribution. Incorporating correlations between degrees improves the fit (but with many more equations).

11.1 A sample social network

We begin with a social network G collected from Facebook in September 2005 showing friendship relations in the University of Oklahoma [303]. This undirected

© Springer International Publishing AG 2017
I.Z. Kiss et al., *Mathematics of Epidemics on Networks*, Interdisciplinary Applied Mathematics 46, DOI 10.1007/978-3-319-50806-1_11

network has 17425 nodes and 892528 uniquely counted edges, giving $\langle k \rangle \simeq 103.50$. The degree distribution of this network is shown in Fig. 11.1. It is primarily an exponential distribution, but it has a few more high-degree nodes than would be predicted by an exponential distribution.

Looking at the higher order structure, we measure the number of triangles in the network using the *average clustering coefficient*, defined to be

$$C = \frac{1}{N} \sum_{u \in G} c_u,$$

where c_u is the probability that a pair of neighbours of u are connected, thus forming a triangle with u. The average clustering coefficient is 0.23. The assumption of our analytic models that there are no short cycles clearly does not hold.

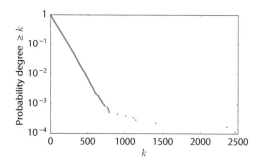

Fig. 11.1: The complementary cumulative degree distribution of the University of Oklahoma Facebook social network. The network appears to have a mostly exponential degree distribution, with a few high-degree nodes.

11.1.1 SIR epidemics

We consider the Markovian spread of an SIR disease through G. We can arbitrarily set $\gamma = 1$ and explore several values of τ. In [209], it was shown that the final size of discrete-time SIR epidemics in this network (and in other similar Facebook networks) is surprisingly well predicted by the implicit epidemic final size relation of system (6.6). This was unexpected because the prediction ignores clustering. Figure 11.2 shows that the agreement is good for continuous-time models as well. Thus, we have an accurate final epidemic size prediction using a theory which includes only the degree distribution and the rates τ and γ. Clustering appears unimportant.

We additionally find remarkably good agreement between the dynamic predictions of the EBCM model (which assumes no clustering) and simulated epidemics on G, shown in Fig. 11.3. The prediction shows more discrepancy than seen in the final epidemic size, but it is improved by including degree correlations. The specific

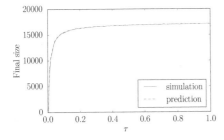
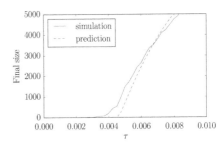

Fig. 11.2: Comparison of the final epidemic size of SIR epidemics (using the algorithm of Fig. 6.17) and the prediction of system (6.7) for the University of Oklahoma Facebook network. We set $\gamma = 1$ and look at different values of τ. There is a discrepancy near the epidemic threshold, but otherwise the models are in almost perfect agreement.

equations follow from Exercise 6.21. Our results suggest that degree correlations help the disease spread if the system is close to the threshold. This is because it provides the disease a high-degree region in which it can establish itself. Farther above the threshold, the degree correlations help some individuals escape infection because the resulting low-degree regions are more likely to escape infection.

We can offer an explanation of the agreement between simulation in a real-world network and a theory that ignores clustering following [214]. There is significant clustering in the network. This clustering should introduce correlations between the neighbours of a node u. We must explain why these correlations do not seem to affect the outcome.

To do this, we consider a triangle consisting of nodes u, v and w and assume that v is the first infected. If τ is small, then v is unlikely to transmit to w, and so the correlation between neighbours of u is generally small. On the other hand, if τ is large, then the disease is likely to spread quite well. In this network it is also likely that w has a degree near or above 100. Thus, if v is infected first and transmits to w, it is highly likely that w would have become infected quite soon regardless. Thus, although some correlation might be introduced by the clustering, it is small compared to what is expected even without clustering. Thus, we do not expect the clustering to play an important role in the spread of SIR disease in this network. Our arguments suggest that in more sparse networks, the behaviour may be different, and we can extend them to provide quantitative prediction of when clustering's impact is significant [214].

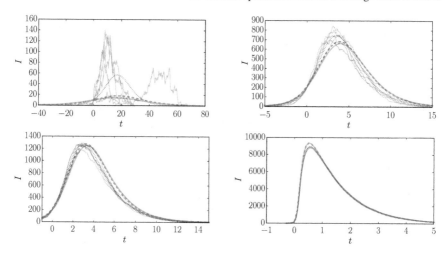

Fig. 11.3: Comparison of simulation and predictions for $I(t)$ from SIR epidemics in the University of Oklahoma Facebook network. For all cases, $\gamma = 1$. The values of τ are 0.005, 0.008, 0.01 and 0.1 in the top-left, top-right, bottom-left and bottom-right panel, respectively. Curves are shifted so that $t = 0$ when 200 nodes are infected, except for $\tau = 0.005$ for which $t = 0$ when 10 nodes are infected. Simulations (solid thin curves, using the algorithm of Fig. A.3) are well predicted by the EBCM model incorporating degree correlations (short-dashed from Exercise 6.21). The heterogeneous mean-field model (dashed, system (5.11)) and the EBCM model (solid, system (6.12)) also perform well.

11.1.2 SIS epidemics

We now consider the Markovian spread of SIS disease through the University of Oklahoma Facebook network G. We again take $\gamma = 1$ and vary τ. Figure 11.4 shows the spread of an SIS disease.

As in the SIR case, the system which incorporates degree correlations, the heterogeneous pairwise model of system (5.13), outperforms the homogeneous pairwise model, system (4.10), the heterogeneous mean-field model, system (5.10), and the homogeneous mean-field model, system (4.8). At small τ, the other models underestimate the equilibrium level, while at larger τ, they overestimate the equilibrium. The heterogeneous pairwise model provides a better approximation of both the early growth and the equilibrium level. The homogeneous mean-field model does quite badly, predicting no epidemic until larger values of τ.

We can understand this by noting that close to the threshold the degree correlations channel the disease into regions with many high-degree nodes in which it can persist. As the system moves farther above the threshold, the disease can persist through much of the network. Eventually, the high-degree nodes spend most of their time infected regardless of the degree distribution of their partners. After this, the degree correlations between low-degree nodes form a barrier, creating regions of low-degree nodes where the disease is often absent.

Unfortunately, the system of equations required for the heterogeneous pairwise model is quite large. In general, this is $\mathcal{O}(K^2)$, where K is the number of distinct degrees. In the Facebook network, the full system has nearly 690,000 equations. This can lead to numerical challenges.

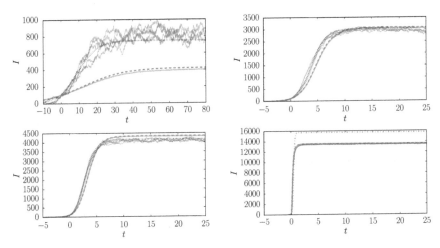

Fig. 11.4: Comparison of simulation and theory for $I(t)$ from SIS epidemics in the University of Oklahoma Facebook network. For all cases, $\gamma = 1$. The values of τ are 0.005, 0.008, 0.01 and 0.1 in the top-left, top-right, bottom-left and bottom-right panel, respectively. Curves are shifted so that $t = 0$ when 100 nodes are infected. Simulations (solid thin curves, using the algorithm of Fig. A.5) are well predicted by the heterogeneous pairwise model, which captures degree correlations (short-dashed, system (5.13)). The compact pairwise model (solid, system (5.18)) and the heterogeneous mean-field model (dashed, system (5.10)) perform reasonably well. The homogeneous pairwise model (dash-dot, system (4.10)) performs poorly, predicting no epidemic except for the largest τ.

11.2 Small-world networks

Small-world networks are often defined to be networks which have many short cycles, but with the typical path length between nodes being logarithmic in the population size N. Thus, each "[node] of the network is somehow 'close' to almost every other [node], even those that are perceived as likely to be far away" [323].

The canonical approach to generating a small-world network is through a Watts–Strogatz network [324]. We begin with nodes aligned in a ring, joining each node to the nearest k nodes (where k is even), with $k/2$ on either side. For the purposes of the construction, we think of each edge as having a "first" and a "second" node, where the edge connects the first node to the second node in a clockwise sense. We independently rewire each edge with a given probability p. If an edge is rewired, we disconnect the two nodes and join the "first" node to a randomly selected node in the network (which is not already its neighbour). Sample outputs of this algorithm are shown in Fig. 11.5.

Fig. 11.5: Sample small-world networks for $p = 0$, $p = 0.1$ and $p = 1$ with $k = 4$. Increasing values of p increases the amount of randomness. Note that even at $p = 1$, all nodes have degree at least $k/2$, so the $p = 1$ limit is not the same as an Erdős–Rényi network.

Figure 11.6 shows how clustering and the average shortest path length between nodes scale for Watts–Strogatz networks with different ps. For small p, the network is clustered and the typical path length between two nodes is of order N. For large p, the clustering is of order $1/N$ and the typical path length is of order $\log N$. The transitions of clustering and path length occur at values of p that are separated by orders of magnitude. The regime in which the typical path length is of order $\log N$, while the clustering is still large is the "small-world" regime.

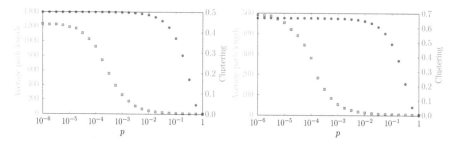

Fig. 11.6: Comparison of the average clustering coefficient and average shortest path length in small-world networks with $k = 4$ and $k = 10$ and $N = 10000$ nodes, based on 100 simulated small-world networks for each p. The regime of low shortest path and high clustering is the "small-world" regime. The average shortest path length is estimated by choosing 100 equally spaced nodes in the original ring and finding their average distance to all other nodes in the network.

The degrees in these networks are much more homogeneous than in the Facebook network. At $p = 0$, all nodes have degree k. As p increases, some heterogeneity occurs, but all nodes have at least $k/2$ edges. In the limit $N \to \infty$, a node's degree at $p = 1$ is $k/2$ plus a Poisson-distributed number with mean $k/2$.

11.2.1 SIR epidemics

We now explore SIR disease spread in Watts–Strogatz networks. We start with $k = 4$ and $N = 10000$ nodes. We take networks with $p = 0.2, 0.4, 0.7$ and 1. We compare our simulations with predictions in Fig. 11.7. The agreement between theory and simulation in these networks is not as good as observed in the Facebook network. In fact, referring to Fig. 11.6, we see that it is not until the networks are almost past the small-world regime before the fit becomes reasonable.

So it appears that the clustering of the Watts–Strogatz networks has more impact on disease transmission than the clustering of the Facebook network. For $k = 10$, we take $p = 0.01, 0.1, 0.5$ and 1. Compared to $k = 4$, the predictions begin to resemble simulations while closer to the highly clustered regime. In fact, revisiting our explanation for the Facebook network, we see that larger typical degrees should result in better predictions. So link density explains much of the apparent larger role of clustering in Watts–Strogatz networks.

We can explain more of the increased effect of clustering by noting that not only do the Watts–Strogatz networks have clustering, but they have structure at larger distances as well. The underlying ring structure results in more short cycles than can be explained just from triangles. So in the Facebook network, as the disease spreads out from an initial node, it quickly reaches nodes that are distributed throughout the network. After a short time, information about which node was initially infected is lost. In the Watts–Strogatz networks, however, the shortcuts are much rarer. Even after several generations, it is still possible to infer the region that contained the initial infection. This difference is why the models assuming that the neighbours are randomly selected do not perform as well in Watts–Strogatz networks.

11.2.2 SIS epidemics

We now consider SIS disease spread in Watts–Strogatz networks. We focus on $k = 4$ with the same values of p as in Fig. 11.7. We take $\gamma = 1$ and $\tau = 0.5$ (slightly smaller than the value in Fig. 11.7). Figure 11.8 shows simulations and theoretical predictions. The heterogeneous mean-field model significantly overestimates the equilibrium. The other models give reasonable predictions for the equilibrium, even when they do not capture the early growth well. Although the homogeneous mean-field model is a good match at $p = 0.4$, this is purely coincidental.

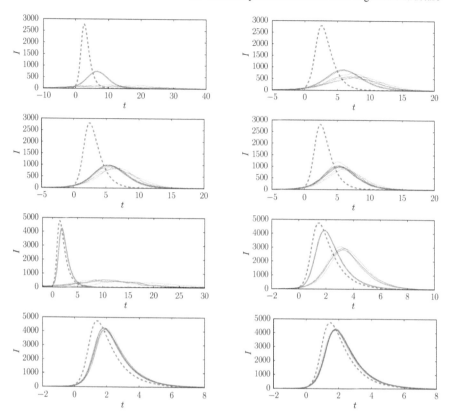

Fig. 11.7: Comparison of simulated and predicted dynamics for SIR disease spread in Watts–Strogatz networks of $N = 10000$ nodes. In all cases, $\gamma = 1$. In the top two rows, $k = 4$ and $\tau = 0.7$ with (top row) $p = 0.2$ and 0.4, (second row) $p = 0.7$ and 1. In the bottom two rows, $k = 10$ and $\tau = 0.5$, with (third row) $p = 0.01$ and 0.1, (bottom row) $p = 0.5$, and 1. We take $t = 0$ when 100 nodes are infected. If τ is large enough, simulations (solid thin curves, using the algorithm of Fig. A.3) are well predicted by the EBCM model (solid, system (6.12)) and the EBCM model incorporating degree correlations (short-dashed from Exercise 6.21). If τ is small, simulations are not well predicted. The heterogeneous mean-field model (dashed, system (5.11)), which ignores partnership duration, performs poorly in general.

11.3 Preferential attachment networks

We finally consider preferential attachment models. The canonical preferential attachment network of Barabási and Albert [23] begins with m nodes, all connected to each other. It sequentially adds nodes until there are N nodes. Each time a node joins the network, it connects to m existing nodes, selecting the neighbours with

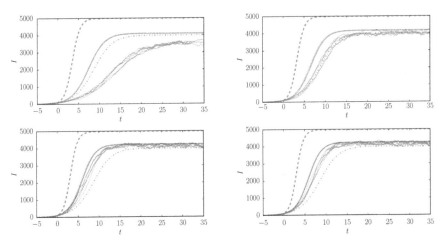

Fig. 11.8: Comparison of simulated and predicted dynamics for SIS disease spread in Watts–Strogatz networks of $N = 10000$ nodes. In all simulations, $\gamma = 1$, $k = 4$ and $\tau = 0.5$ with (top row) $p = 0.2$ and 0.4 and (bottom row) $p = 0.7$ and 1. We take $t = 0$ when 100 nodes are infected. The equilibrium value of simulations (solid thin curves, using the algorithm of Fig. A.5) is generally well predicted by the heterogeneous pairwise model (short-dashed, system (5.13)), the compact pairwise model (solid, system (5.18)) and the homogeneous pairwise model (dash-dot, system (4.10)). The transient growth is not. The heterogeneous mean-field model (dashed, system (5.10)) performs poorly in general.

probability proportional to their current degrees. The resulting network has a degree distribution with $P(k) \sim c_m k^{-\alpha_m}$ for some constants c_m and α_m. Figure 11.9 demonstrates this.

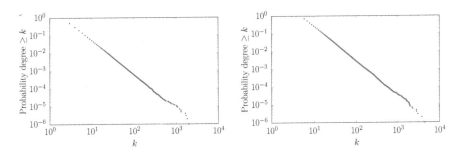

Fig. 11.9: The complementary cumulative degree distribution of Barabási–Albert networks with $N = 10^6$ nodes and (left) $m = 2$ and (right) $m = 5$. The average clustering coefficients of these networks are 0.00009 and 0.00017, respectively.

11.3.1 SIR epidemics

To investigate the spread of SIR disease, we compare simulations to predictions from analytic models. Agreement with the EBCM model is not as good as in some of the small-world networks. This is initially surprising because the clustering in the Barabási–Albert networks is so low. However, these networks contain degree correlations [101]. Figure 11.10 shows that almost all of the discrepancy is accounted for by incorporating correlations between degrees (as in Exercise 6.21).

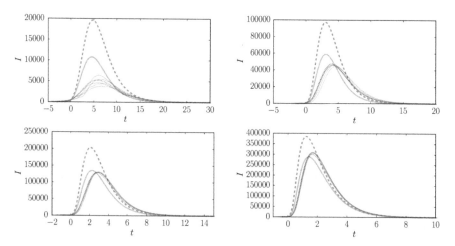

Fig. 11.10: Comparison of simulated and predicted SIR epidemics in a Barabási–Albert network with $N = 10^6$ and $m = 2$. The recovery rate is $\gamma = 1$ for all simulations, with (top row) $\tau = 0.15$ and 0.3, (bottom row) $\tau = 0.5$, and 1. We set $t = 0$ when 500 nodes are infected. Simulations (solid thin curves, using the algorithm of Fig. A.3) are generally well predicted by the EBCM model incorporating degree correlations (short-dashed from Exercise 6.21). The EBCM model without degree correlations (solid, system (6.12)) performs less well than in the Facebook network. The heterogeneous mean-field model (dashed, system (5.11)), which ignores partnership duration, performs poorly. Agreement is much better than for Watts–Strogatz networks.

Not all of Fig. 11.10 can be explained just from nearest-neighbour degree correlation. The additional discrepancy comes from longer-range correlations. As we might expect from the other networks we have looked at, if we take a network with higher degrees, the degree correlations become less important.

11.3.2 SIS epidemics

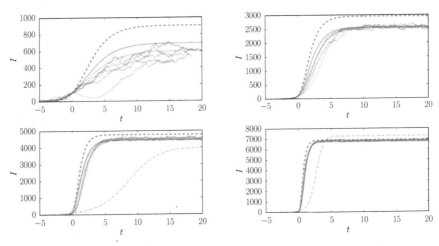

Fig. 11.11: Comparison of simulated and predicted SIS epidemics in a Barabási–Albert network with $N = 10^4$ and $m = 2$. The recovery rate is $\gamma = 1$ for all simulations, with (top row) $\tau = 0.15$ and 0.3, (bottom row) $\tau = 0.5$, and 1. We set $t = 0$ when 100 nodes are infected. Simulations (solid thin curves, using the algorithm of Fig. A.5) are well predicted by the heterogeneous pairwise model (short-dashed, system (5.13)). The compact pairwise model (solid, system (5.18)), which ignores degree correlations, provides reasonably good predictions as τ increases. The heterogeneous mean-field model (dashed, system (5.10)) performs less well. The homogeneous pairwise model (dash-dot, system (4.10)) performs poorly. In general, the theoretical predictions perform better than in the Watts–Strogatz networks.

We compare simulations with predictions for SIS epidemics in Fig. 11.11. The heterogeneous pairwise model performs well because it captures degree correlations. Close to the epidemic threshold, these correlations are particularly important, so the model performs significantly better than the others. Farther from the threshold, the compact pairwise performs almost as well. High above the threshold, the heterogeneous pairwise model becomes reasonable.

11.4 Conclusions and outlook

Throughout this book, we have derived analytic models describing the spread of epidemics on networks. These models typically involve assumptions about the underlying network structure. In this chapter, we have investigated what happens when these assumptions fail.

Our observations in this chapter show that models that incorporate heterogeneity in the network's degree distribution perform much better than models which assume that all nodes have the same degree. However, if the network has degree correlations, the models can perform poorly, particularly near the epidemic threshold. We can add degree correlations at the expense of a much larger system of equations, and the resulting models generally perform well.

The main property of real-world networks that is not captured by these models is clustering (or short cycles in general). The Watts–Strogatz model shows that short cycles can significantly alter the dynamics from what we would predict with our models. The fact that clustering does not play an important role in the behaviour of epidemics in the Facebook network of the University of Oklahoma, but does play an important role in Watts–Strogatz networks can be explained by the fact that (i) the high link density in the Facebook network may reduce the impact of clustering and (ii) the Watts–Strogatz networks have structure beyond just triangles. The latter can be explained by the fact that many nodes in the Watts–Strogatz networks may be reachable in a small number of steps from an initial node u, but a large fraction of these are only reachable because of a very small number of "shortcuts" in the network. Because the disease is likely to skip the shortcuts, its spread is much more restricted in such networks. Similar effects can be expected in spatially embedded networks, where again clustering couples with other properties that may invalidate the assumptions of even the most complex mean-field models.

In Table 11.1, we give a rough guide of how the choice of mean-field models depends on the network properties we want to capture. This is done with specific focus on SIR epidemics. Several caveats need to be taken into account. The table does not indicate how accurate the mean-field models are in predicting or agreeing with results based on simulations. Here, we simply specify which network properties are or can be included, but model accuracy will differ depending on network and model choice. For example, the individual-based model will always overestimate the true epidemic, and the pair-based model is exact on tree networks (see Chapter 3).

Table 11.1 allows us to identify the EBCM, see Chapter 6, as the most compact model with the fewest equations, but with further improvements needed when clustered networks are considered, although this remains a challenge for any other model which can capture clustering. Models where the number of equations scales with L, the number of distinct degrees in the network, are penalised by a high number of equations and limited analytical tractability.

We also note that models with multiple check marks ($\checkmark$), i.e., capable of accounting for many distinct network properties, may take on a slightly different form when all network features are included at once. For example, while the EBCM does well at individually capturing degree heterogeneity or even degree heterogeneity and mixing, it will need a specific form if clustered networks are to be modelled. In this case, the number of equations will grow and the degree distribution of the networks and mixing will be constrained by how the clustered network is built.

We have introduced a wide variety of models in this book, of varying complexity and detail. These models perform well for a range of networks, including clustered networks where their success may be surprising. Although these models will suffice for most static networks, there remain many directions for which more research is needed. Among these are dynamic networks, networks with multiple types or classes of nodes or edges, and diseases which violate the Markovian assumptions. Although we have touched on some of these in this book, much more research is ongoing.

Model \ Network property	Reg. deg.	Het. deg.	Mixing	Clustering	Number of equations
Individual-based (NIMFA)	✓	✓	✓	✗	$2N$
Pair-based	✓	✓	✓	✗	$\leq 2N + 4|E|$
Homogenous mean-field at single level	✓	✗	✗	✗	2
Homogenous pairwise	✓	✗	✗	Closure dependent	4
Heterogenous mean-field at single level	✓	✓	✓	✗	$\mathcal{O}(L)$
Heterogenous pairwise	✓	✓	✓	Closure dependent	$\mathcal{O}(L^2)$
Compact pairwise	✓	✓	✗	✗	$\mathcal{O}(L)$
Super-compact pairwise	✓	✓	✗	✗	4
Effective degree	✓	✓	✗	Closure dependent	$\mathcal{O}(M^2)$
Compact effective degree	✓	✓	✗	Closure dependent	$\mathcal{O}(M)$
EBCM	✓	✓	✓	✓ (for particular clustered networks)	$1 + 1$
Simulation	✓	✓	✓	✓	NA

Table 11.1: A summary of the relation between network properties and mean-field SIR models. The table captures the ability of mean-field models of capturing increasingly complex network properties, e.g., from degree and mixing to clustering. Regular or homogenous degree is abbreviated as Reg. deg, while Het. deg. denotes heterogeneous degree. The number of nodes in the network is N, the number of distinct degrees in the network is L and the maximum degree is M.

An important further step is to quantify the magnitude of the error made when we use different models. It is clear that models that include degree correlations perform better than those that do not, but how much better are they? It is clear that even the models that include degree correlations do not provide a perfect fit, but can we estimate how large the error is? More generally, we would like to be able to quickly measure a few properties of a network and identify the most appropriate model and estimate the error.

Finally, it is important that we better understand the true structure of empirical networks. This is a rapidly evolving area of study. Technological improvements give us access to much more data about contact structure. We are learning more about online social networks, the airline and road networks, animal movements between farms, and migration patterns within countries. All of these networks give us an insight into the constraints and opportunities that infectious diseases may face. Over the next decade, we expect to see empirical networks play a much larger role in infectious disease modelling.

Appendix A
Stochastic simulation of epidemics

In this appendix, we discuss some algorithms for performing stochastic simulations and comment on some issues related to interpreting the resulting output.

A.1 Efficient simulations

We begin with techniques for simulating disease spread in networks. We consider two general algorithms: The Gillespie algorithm and an event-driven algorithm. The Gillespie algorithm is a well-known algorithm for simulating Markovian processes where objects change status. The algorithm calculates the time to the next event and then separately calculates what that event will be. It then jumps to that time, updates statuses and repeats. This is particularly efficient for processes where all that matters is the number of objects of each status, such as a chemical reaction in a well-mixed vessel or disease spread in a well-mixed, mass-action population. For these, the information update at each event simply adjusts the counts. In a network however, knowing the exact node changing status is important. Choosing which node changes status can be a computationally slow process.

An alternative event-driven approach involves a priority queue. Rather than calculating at each time what the next event might be (chosen from all available events), when a node is infected, we calculate when it will transmit to others and when it will recover. These events are placed into the priority queue, and then the next event is removed from the queue and processed. The computational time required to add and remove events is roughly proportional to the log of the number of events already in the queue. This significantly reduces the amount of processing per event, producing a more efficient algorithm. An implementation of these methods is available at

https://springer-math.github.io/Mathematics-of-Epidemics-on-Networks/

Although the implementations of the SIS and SIR event-driven algorithms are quite different from the Gillespie algorithms, they simulate the same stochastic processes.

© Springer International Publishing AG 2017

I.Z. Kiss et al., *Mathematics of Epidemics on Networks*, Interdisciplinary Applied Mathematics 46, DOI 10.1007/978-3-319-50806-1

Thus, either could be used for the same purpose. In general, the event-driven algorithms are faster than the equivalent Gillespie algorithms and are more flexible in the case of non-Markovian transmission and recovery processes.

A.1.1 Gillespie algorithm

Input: Network G, per-edge transmission rate τ, recovery rate γ, set of index node(s) initial_infecteds, maximum time t_{max}.
Output: Lists times, S, I, and R giving number in each state at each time.

   function Gillespie_network_epidemic(G, τ, γ, initial_infections, t_{max})
      times, S, I, $R \leftarrow [0]$, $[|G|$-len(initial_infections)$]$, $[$len(initial_infections)$]$, $[0]$
      infected_nodes $\leftarrow$ initial_infections
      at_risk_nodes $\leftarrow$ uninfected nodes with infected neighbours
      for each node u in at_risk_nodes **do**
         infection_rate[u] = $\tau \times$ number of infected neighbours
      total_infection_rate $\leftarrow \sum_{u\in\text{at\_risk\_nodes}}$ infection_rate[u],
      total_recovery_rate $\leftarrow \gamma \times$ len(infected_nodes)
      total_rate $\leftarrow$ total_transmission_rate + total_recovery_rate
      time $\leftarrow$ exponential_variate(total_rate)
      while time$< t_{max}$ and total_rate> 0 **do**
         $r =$ uniform_random(0,total_rate)
         if $r <$total_recovery_rate **then**
            $u =$ random.choice(infected_nodes)
            remove u from infected_nodes
            reduce infection_rate[v] for u's susceptible neighbours v
         else
            choose u from at_risk_nodes with probability $\frac{\text{infection\_rate}[u]}{\text{total\_infection\_rate}}$.
            remove u from at_risk_nodes
            add u to infected_nodes
            for susceptible neighbours v of u **do**
               if v not in at_risk_nodes **then**
                  add v to at_risk_nodes
               update infection_rate[v]
         update times, S, I, and R
         update total_recovery_rate, total_infection_rate, and total_rate
         time $\leftarrow$ time + exponential_variate(total_rate)
      return times, S, I, R

Fig. A.1: Pseudocode for the Gillespie algorithm simulating an SIR epidemic in a network. Small changes are needed for simulating an SIS epidemic: when a node recovers, it needs to become susceptible and possibly needs to be placed into the at-risk node category.

Figure A.1 gives pseudocode adapting the well-known Gillespie algorithm [112, 113] (much of which was introduced earlier by Doob in [83]) to epidemics on networks. This algorithm uses iterative steps:

1. Find the rate of all possible events and compute the total rate of change occurring; this is done by finding the rate at which each infected node will recover and each at-risk susceptible node becomes infected;
2. Based on this rate of change, select the waiting time until the next event from an exponential distribution whose rate is the total rate of change;
3. Select which event occurs, with probability proportional to each event's rate; and
4. Update the rates based on which node changes status and repeat.

In more detail, these steps amount to determining the rate of infection of all susceptible nodes and the rate of recovery of all infectious nodes. Let these be denoted by r_u, where $u = 1, 2, 3, \ldots, N$ are the nodes. The infection rate of a susceptible node depends on how many infected neighbours it has, but the recovery rate of an infected node is independent of the network and status of neighbours. For example, Fig. A.2 illustrates the computation of some transition rates. Node 1 has one infected neighbour, so $r_1 = \tau$, and the two infected neighbours of node 3 yield $r_3 = 2\tau$. All infected nodes, e.g. nodes 5 and N have $r_5 = r_N = \gamma$. From this, the total rate of all transitions, denoted by T (total_rate in the pseudocode), is calculated from the current status of all individuals across the whole network. Therefore, $T = \sum_{u=1}^{N} r_u$, and the waiting time until the next event, t_{next}, is chosen from an exponential distribution with rate T. Because $r_u = \gamma$ if u is infected and 0 if u is susceptible and not in the at-risk nodes, we can write $T = \gamma[I] + \sum_{u \in \text{at\_risk\_nodes}} r_u$.

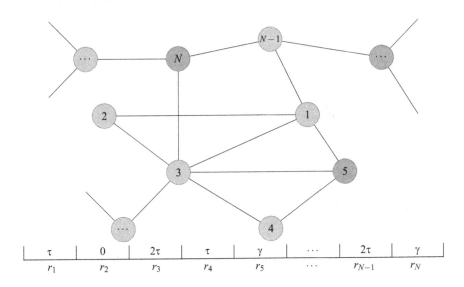

Fig. A.2: Illustrating the relation between the network, status of nodes and events rate vector used by the Gillespie algorithm. Susceptible nodes are denoted by (○) and infected nodes by (◉).

Next, a single event is chosen with probability proportional to its rate. Because all recoveries occur with the same rate, γ, the total recovery rate is $\gamma[I]$. So the probability the event is a recovery is $\gamma[I]/T$. If so, we select a random infected node. If a recovery does not occur, we select a random at-risk node with probability proportional to its rate. Because not all at-risk nodes have the same rate, this step of selecting the node is more involved than simply choosing a random node. It can be done in several ways:

- We can select a random number $r \in (0, T - \gamma[I])$. We iterate through the at-risk nodes, each time subtracting r_i until the result becomes negative. This is the node that recovers. This approach can be slow because, on average, half of the at-risk nodes will be processed in each iteration;
- We can *a priori* artificially inflate the transmission rates, treating all nodes as having the same infection rate $r^* \geq \max_{u \in \text{at\_risk\_nodes}} \{r_u\}$. This increases T to $\gamma[I] + r^*|\text{at\_risk\_nodes}|$. When a transmission is predicted to happen, we select a random at-risk node u whose true rate is $r_u \leq r^*$. We then correct for the fact that we have overestimated the risk of u by selecting an additional random number $p \in (0,1)$. If $p < r_u/r^*$, we "accept" the event: u is infected. Otherwise, we "reject" the event: no nodes change status, but the time t is increased. This can be slow if much processing time is spent on events in which nothing happens; and
- Alternately, we keep T as $T = \gamma[I] + \sum_{u \in \text{at\_risk\_nodes}} r_u$. We track the maximum transmission rate r^*. When a transmission event is going to occur, we choose a random node u from the at-risk nodes. We select an additional random number $p \in (0,1)$. If $p < r_u/r^*$, we infect u. If not, we choose a new random node u' from the at-risk nodes and a new p' and repeat. This can be slow if some of these events require repeated selections before finding the newly infected node.

These last two options use a technique known as rejection sampling. Whichever option is chosen, this is generally the slowest step of the algorithm. Once the time to next event and the event itself have been found, necessary rate updates are performed and the process begins again. The Gillespie algorithm allows us to exactly simulate the process.

A.1.2 Event-driven algorithm

We now consider another efficient algorithm for simulating SIR and SIS epidemics. This uses event-based simulation techniques to improve on Gillespie-style algorithms by avoiding the slow step of finding which node becomes infected. An additional advantage of this approach is that it can be easily generalised for non-Markovian processes.

A key observation is that when a node is infected, nothing that any of its neighbours does will affect when it recovers, who it transmits to or the timing of those transmissions. Thus, as soon as a node is infected, we can calculate when it will recover and when it will transmit to its neighbours. These events are inserted into a priority queue ordered by event time.[1] At each step of the simulation, the next event in the list is removed and processed. If it is a transmission, new events will be added to the queue corresponding to the recovery of the node and any transmissions from that node. This process iterates until a time specified by the user or no events remain in the queue (in which case no infection remains).

Although we assume here that the algorithm is performed using constant rates τ and γ, it is straightforward to adapt event-based simulations to other rules. The key requirement is that when a node becomes infected, we can calculate transmission and recovery times without considering events that have not happened yet. Thus, this approach can handle a non-Markovian epidemic process, as in Chapter 9, by some relatively simple adjustments using a different calculation of the duration of infection and the waiting time before transmission.

SIR epidemics

We begin by describing an event-based algorithm for continuous-time SIR disease transmission in an arbitrary unweighted, undirected network. The algorithm is based on the use of a priority queue Q. The underlying structure of Q allows for efficient addition of new events and removal of the earliest remaining event in the queue. Pseudocode for the algorithm is given in Figs. A.3 and A.4.

The algorithm repeatedly removes the earliest remaining event in Q. If the event is a recovery, the node recovers. If the event is a transmission and the recipient of the transmission has not yet been infected, the node is infected. Now that it is infected, the time of its recovery and the times at which it transmits to its neighbours will not be influenced by any external event. Thus, we can determine the future recovery time and times at which it may transmit to its neighbours immediately, without regard to any events that may happen between now and those events. We add these to Q, unless we can immediately rule them out of consideration. Since only the first transmission to a node v has any effect, we can rule out any transmissions for which the recipient has already been infected or has an earlier scheduled transmission. Similarly, we do not add events that would occur after the user-specified t_{max} to Q.

---

[1] A "priority queue" is a data structure which allows for insertion of new events in such a way that we can easily remove the next event to occur. When implemented efficiently, the computational time required to add or remove events is logarithmic in the number of events in the queue [289].

Input: Network G, per-edge transmission rate τ, recovery rate γ, set of index node(s) initial_infecteds, and maximum time t_{max}.
Output: Lists times, S, I, and R giving number in each state at each time.

```
function fast_SIR(G,τ, γ, initial_infecteds, tmax)
    times, S, I, R ← [0], [|G|], [0], [0]
    Q ← empty priority queue
    for u in G.nodes do
        u.status ← susceptible
        u.pred_inf_time ← ∞
    for u in initial_infecteds do
        Event ← {node: u, time: 0, action: transmit}
        u.pred_inf_time ← 0
        add Event to Q                                          ▷ ordered by time
    while Q is not empty do
        Event ← earliest remaining event in Q
        if Event.action is transmit then
            if Event.node.status is susceptible then
                process_trans_SIR(G, Event.node, Event.time, τ, γ, times, S, I, R, Q, tmax)
        else
            process_rec_SIR(Event.node, Event.time, times, S, I, R)
    return times, S, I, R
```

Fig. A.3: An efficient algorithm simulating continuous-time SIR epidemics in static networks with rates τ and γ. It relies on functions given in Fig. A.4. Events are stored in a priority queue Q, and the first is executed. If the event is a transmission and the recipient is susceptible, it becomes infected, its recovery is added to Q and transmissions to its neighbours may be added. If the event is a recovery, the node recovers.

SIS epidemics

With some modest alterations, we can adapt the event-driven SIR disease algorithm to SIS disease. Figures A.5 and A.6 show the pseudocode. The major change is that we must consider the possibility of successful transmissions to a node after its first infection because it may recover to a susceptible state in between. This involves some changes to the algorithm to find the next transmission, and whenever we encounter a transmission event, after checking its effect, we have to consider the possibility that the original source may transmit again.

```
function process_trans_SIR(G, u, t, τ, γ, times, S, I, R, Q, tmax)
    append times, S, I, and R with t, S.last−1, I.last+1, and R.last
    u.status ← infected
    u.rec_time ← t+exponential_variate(γ)
    if u.rec_time< tmax then
        newEvent ← {node: u, time: u.rec_time, action: recover}
        add newEvent to Q
    for v in G.neighbours(u) do
        find_trans_SIR(Q, t, τ, u, v, tmax)
function find_trans_SIR(Q, t, τ, source, target, tmax)
    if target.status is susceptible then
        inf_time ← t+exponential_variate(τ)
        if inf_time < minimum(source.rec_time, target.pred_inf_time, tmax) then
            newEvent ← {node: target, time: inf_time, action: transmit}
            add newEvent to Q
            target.pred_inf_time ← inf_time
function process_rec_SIR(u, t, times, S, I, R)
    append times, S, I, and R with t, S.last, I.last−1, and R.last+1
    u.status ← recovered
```

Fig. A.4: Auxiliary functions for fast_SIR in Fig. A.3. process_rec_SIR handles a recovering node. process_trans_SIR takes a newly infected node u and creates a recovery event for u if the recovery happens before t_{max}. For each neighbour v of u, it calls find_trans_SIR to determine when u would transmit to v, adding a transmission event to Q if the event might occur (u still infected and v not known to be already infected by then).

A.2 Time shifting of simulation results

We now consider a technical issue related to comparing deterministic models with stochastic simulations. When a stochastic simulation begins with a small number of infections, it may die out, it may have unusually fast early growth or it may take a long time before successfully becoming established and growing. The deterministic models do not have this variability.

When we use a simulation to study epidemics, we must account for these stochastic delays and the fact that some realisations die out. If our goal is to predict the future spread of an epidemic that has already established itself, we should obviously discard the outbreaks that died out stochastically.

It is natural to consider the average of the simulations that do not die out. However, there is an additional issue. The stochastic effects that may accelerate or delay the onset of the epidemic mean that simulated epidemics become large at different times. In Fig. A.7, the epidemics are effectively time translations of one another. Thus, the peaks occur with some stochastic delay. If the population model and the

Input: Network G, transmission rate per edge τ, recovery rate γ, set of index node(s)
  initial_infecteds, and maximum time t_{max}.
Output: t: list of times and I: list containing number infected at each time.

> **function** fast_SIS(G,τ, γ, initial_infecteds, t_{max})
>   initialise Q, node statuses and return variables as in fast_SIR, but include a source for
> infections with recovery time 0.
>   **while** Q is not empty **do**
>   Event $\leftarrow$ earliest remaining event in Q
>   **if** Event.action is transmit **then**
>   **if** Event.node.status is susceptible **then**
>   process_trans_SIS(G, Event.node, Event.time, τ, γ, times, S, I, Q, t_{max})
>   find_next_trans_SIS(Q, Event.source, Event.node, t)   ▷ needed for SIS model
>   **else**
>   process_rec_SIS(Event.node, t, S, I)
>   **return** times, S, I

Fig. A.5: An efficient event-based algorithm simulating continuous-time SIS epidemics in static networks. It relies on functions given in Fig. A.6. It is similar to the algorithm in Fig. A.3. When nodes recover, they become susceptible again. An important change compared to the SIR version is that when a transmission event occurs, we check the possibility that the source of that transmission might cause another transmission later. This requires that we include the source in every transmission event.

disease model accurately represent the real-world epidemic, we would expect the real epidemic to have the same shape, with some stochastic delay. If we try to capture this by taking the average of many simulations, the result will have a different shape. By averaging many peaks that do not align, we get a lower, wider peak which does not resemble any single realisation.

Thus, we have two options to align trajectories:

- We can take many simulations and shift time so that $t = 0$ corresponds to the prevalence crossing some threshold size; or
- We can initialise our simulations with a large number of infections such that the stochastic noise is negligible.

Both of these assume that the epidemic is in a large enough population so that it is reasonable to talk about the deterministic phase of an epidemic.

```
function process_trans_SIS(G, u, t, τ, γ, times, S, I, Q, tmax)
    append times, S, and I with t, S.last−1, and I.last+1
    u.status ← infected
    u.rec_time ← t+exponential_variate(γ)
    if u.rec_time < tmax then
        newEvent ← {node: u, time: u.rec_time, action: recover}
        add newEvent to Q
    for v in G.neighbours(u) do
        find_next_trans_SIS(Q, t, τ, u, v, tmax)
function find_next_trans_SIS(Q, t, τ, source, target, tmax)
    if target.rec_time < source.rec_time then
        transmission_time = max(t, target.rec_time)+exponential_variate(τ)
        if transmission_time < source.rec_time then
            newEvent ← {node: target, time: transmission_time, action: transmit, source: source}
            push(Q, newEvent)
function process_rec_SIS(u, times, S, I)
    append times, S, and I with t, S.last+1, and I.last−1
    u.status ← susceptible
```

Fig. A.6: Auxiliary functions for event-based SIS simulation. Recovery is handled as in the SIR case except that the node becomes susceptible again. Each transmission is handled similarly, but when finding the next transmission event, there are fewer events we can exclude from adding to Q. We find the first transmission from the source to the recipient and add it to Q unless the recipient is infected and will not recover prior to the transmission or the source will recover prior to the transmission. Since multiple transmissions from a source to a target may cause infection, after each transmission event, fast_SIS also calls find_next_trans.

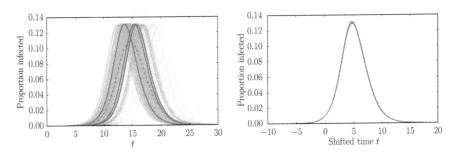

Fig. A.7: (Left) SIR simulations in a 10^6 node Configuration Model network with $P(2) = P(3) = P(4) = P(5) = 1/4$ and $τ = γ = 1$. Simulations were run until 200 epidemics occurred, using the algorithm in Fig. A.3. These epidemics are shown: three are highlighted, and the average of all 200 is dashed. The average does not resemble individual realisations. (Right) Shifting time so that $t = 0$ when 1% of the population is infected aligns individual realisations and the average.

References

1. Allard, A., Hébert-Dufresne, L., Young, J.G., Dubé, L.J.: General and exact approach to percolation on random graphs. Phys. Rev. E **92**(6), 062807 (2015)
2. Allard, A., Noël, P.A., Dubé, L.J., Pourbohloul, B.: Heterogeneous bond percolation on multitype networks with an application to epidemic dynamics. Phys. Rev. E **79**(3), 036113 (2009)
3. Andersson, H., Britton, T.: Stochastic Epidemic Models and Their Statistical Analysis, vol. 151. Springer Science & Business Media, New York (2012)
4. Anderson, R.M., May, R.M.: Infectious Diseases of Humans: Dynamics and Control, vol. 1. Oxford University Press, Oxford (1991)
5. Armbruster, B., Besenyei, A., Simon, P.L.: Bounds for the expected value of one-step processes. arXiv preprint arXiv:1505.00898 (2015)
6. Artalejo, J.R.: On the time to extinction from quasi-stationarity: a unified approach. Phys. A: Stat. Mech. Appl. **391**(19), 4483–4486 (2012)
7. Artalejo, J.R., Economou, A., Lopez-Herrero, M.J.: Stochastic epidemic models with random environment: quasi-stationarity, extinction and final size. J. Math. Biol. **67**(4), 799–831 (2013)
8. Ball, F.: A unified approach to the distribution of total size and total area under the trajectory of infectives in epidemic models. Adv. Appl. Probab. **18**(2), 289–310 (1986)
9. Ball, F., Lyne, O.D.: Stochastic multi-type SIR epidemics among a population partitioned into households. Adv. Appl. Probab. **33**(1), 99–123 (2001)
10. Ball, F.G., Lyne, O.D.: Epidemics among a population of households. In: Mathematical Approaches for Emerging and Reemerging Infectious Diseases: Models, Methods, and Theory, pp. 115–142. Springer, New York (2002)
11. Ball, F.G., Lyne, O.D.: Optimal vaccination policies for stochastic epidemics among a population of households. Math. Biosci. **177**, 333–354 (2002)
12. Ball, F., Neal, P.: A general model for stochastic SIR epidemics with two levels of mixing. Math. Biosci. **180**(1), 73–102 (2002)
13. Ball, F., Neal, P.: Network epidemic models with two levels of mixing. Math. Biol. **212**(1), 69–87 (2008)
14. Ball, F., Sirl, D.: An SIR epidemic model on a population with random network and household structure, and several types of individuals. Adv. Appl. Probab. **44**(1), 63–86 (2012)
15. Ball, F., Sirl, D.: Acquaintance vaccination in an epidemic on a random graph with specified degree distribution. J. Appl. Probab. **50**(4), 1147–1168 (2013)
16. Ball, F., Sirl, D., Trapman, P.: Threshold behaviour and final outcome of an epidemic on a random network with household structure. Adv. Appl. Probab. **41**(3), 765–796 (2009)
17. Ball, F., Sirl, D., Trapman, P.: Analysis of a stochastic SIR epidemic on a random network incorporating household structure. Math. Biosci. **224**(2), 53–73 (2010)

18. Ball, F., Britton, T., Sirl, D.: A network with tunable clustering, degree correlation and degree distribution, and an epidemic thereon. J. Math. Biol. **66**(4–5), 979–1019 (2013)

19. Ball, F., Britton, T., Neal, P.: On expected durations of birth-death processes, with applications to branching processes and SIS epidemics. J. Appl. Probab. **53**(1), 203–215 (2016)

20. Ball, F., Britton, T., House, T., Isham, V., Mollison, D., Pellis, L., Tomba, G.S.: Seven challenges for metapopulation models of epidemics, including households models. Epidemics **10**, 63–67 (2015)

21. Bansal, S., Khandelwal, S., Meyers, L.A.: Exploring biological network structure with clustered random networks. BMC Bioinformatics **10**(1), 405 (2009)

22. Bansal, S., Read, J., Pourbohloul, B., Meyers, L.A.: The dynamic nature of contact networks in infectious disease epidemiology. J. Biol. Dyn. **4**(5), 478–489 (2010)

23. Barabási, A.L., Albert, R.: Emergence of scaling in random networks. Science **286**(5439), 509–512 (1999)

24. Barbour, A.D., Reinert, G.: Approximating the epidemic curve. Electron. J. Probab **18**(54), 1–30 (2013)

25. Barr, D.R., Thomas, M.U.: An eigenvector condition for Markov chain lumpability. Oper. Res. **25**, 1028–1031 (1977)

26. Barrat, A., Barthélemy, M., Vespignani, A.: Dynamical processes on complex networks. Cambridge University Press, New York (2008)

27. Barthélemy, M., Barrat, A., Pastor-Satorras, R., Vespignani, A.: Dynamical patterns of epidemic outbreaks in complex heterogeneous networks. J. Theor. Biol. **235**(2), 275–288 (2005)

28. Bátkai, A., Havasi, Á., Horváth, R., Kunszenti-Kovács, D., Simon, P.L.: PDE approximation of large systems of differential equations. Oper. Matrices **9**(1), 147–163 (2015)

29. Bátkai, A., Kiss, I.Z., Sikolya, E., Simon, P.L.: Differential equation approximations of stochastic network processes: an operator semigroup approach. Netw. Heterog. Media **7**(1), 43–58 (2012)

30. Battiston, F., Nicosia, V., Latora, V.: Structural measures for multiplex networks. Phys. Rev. E **89**(3), 032804 (2014)

31. Bauch, C.T.: The spread of infectious diseases in spatially structured populations: an invasory pair approximation. Math. Biosci. **198**(2), 217–237 (2005)

32. Bauch, C.T., Lloyd-Smith, J.O., Coffee, M.P., Galvani, A.P.: Dynamically modeling SARS and other newly emerging respiratory illnesses: past, present, and future. Epidemiology **16**(6), 791–801 (2005)

33. Bearman, P.S., Moody, J., Stovel, K.: Chains of affection: the structure of adolescent romantic and sexual networks. Am. J. Sociol. **110**(1), 44–91 (2004)

34. ben Avraham, D., Bollt, E.M., Tamon, C.: One-dimensional continuous-time quantum walks. Quant. Inf. Process. **3**(1–5), 295–308 (2004)

35. Bobrowski, A.: Functional Analysis for Probability and Stochastic Processes: An Introduction. Cambridge University Press, Cambridge (2005)

36. Boccaletti, S., Latora, V., Moreno, Y., Chavez, M., Hwang, D.U.: Complex networks: structure and dynamics. Phys. Rep. **424**(4), 175–308 (2006)
37. Boccaletti, S., Bianconi, G., Criado, R., Del Genio, C.I., Gómez-Gardeñes, J., Romance, M., Sendiña-Nadal, I., Wang, Z., Zanin, M.: The structure and dynamics of multilayer networks. Phys. Rep. **544**(1), 1–122 (2014)
38. Bodó, Á., Katona, G.Y., Simon, P.L.: SIS epidemic propagation on hypergraphs. Bull. Math. Biol. **78**(4), 713–735 (2016)
39. Boguñá, M., Pastor-Satorras, R.: Epidemic spreading in correlated complex networks. Phys. Rev. E **66**(4), 047104 (2002)
40. Boguñá, M., Pastor-Satorras, R., Vespignani, A.: Absence of epidemic threshold in scale-free networks with degree correlations. Phys. Rev. Lett. **90**(2), 028701 (2003)
41. Boguñá, M., Castellano, C., Pastor-Satorras, R.: Nature of the epidemic threshold for the susceptible-infected-susceptible dynamics in networks. Phys. Rev. Lett. **111**(6), 068701 (2013)
42. Boguná, M., Lafuerza, L.F., Toral, R., Serrano, M.A.: Simulating non-Markovian stochastic processes. Phys. Rev. E **90**(4), 042108 (2014)
43. Bollobás, B.: Random Graphs, 2nd edn. Cambridge University Press, Cambridge (2001)
44. Bollobás, B., Kozma, R., Miklos, D.: Handbook of Large-Scale Random Networks, vol. 18. Springer Science & Business Media, Berlin, Heidelberg (2010)
45. Bornholdt, S., Schuster, H.G.: Handbook of Graphs and Networks: From the Genome to the Internet. Wiley, Berlin (2006)
46. Box, G.E.P., Hunter, J.S., Hunter, W.G.: Statistics for experimenters: design, innovation, and discovery. AMC **10**, 12 (2005)
47. Brauer, F., Castillo-Chavez, C.: Mathematical Models in Population Biology and Epidemiology, 2nd edn. Springer, New York (2012)
48. Brauer, F., van den Driessche, P., Wu, J.: Mathematical Epidemiology, vol. 1945. Springer, Berlin, Heidelberg (2008)
49. Britton, T.: Stochastic epidemic models: a survey. Math. Biosci. **225**(1), 24–35 (2010)
50. Britton, T., Trapman, P.: Inferring global network properties from egocentric data with applications to epidemics. Math. Med. Biol. **32**(1), 101–114 (2015)
51. Britton, T., Deijfen, M., Martin-Löf, A.: Generating simple random graphs with prescribed degree distribution. J. Stat. Phys. **124**(6), 1377–1397 (2006)
52. Broder, A., Kumar, R., Maghoul, F., Raghavan, P., Rajagopalan, S., Stata, R., Tomkins, A., Wiener, J.: Graph structure in the web. Comput. Netw. **33**(1), 309–320 (2000)
53. Brown, R.F.: A Topological Introduction to Nonlinear Analysis. Springer, Berlin (1993)
54. Butler, B.K., Siegl, P.H.: Sharp bounds on the spectral radius of nonnegative matrices and digraphs. Linear Algebra Appl. **439**(5), 1468–1478 (2013)
55. Caldarelli, G.: Scale-Free Networks: Complex Webs in Nature and Technology. Oxford University Press, Oxford (2007)

56. Cardy, J.L., Grassberger, P.: Epidemic models and percolation. J. Phys. A: Math. Gen. **18**(6), L267–L271 (1985)
57. Castellano, C., Pastor-Satorras, R.: Thresholds for epidemic spreading in networks. Phys. Rev. Lett. **105**(21), 218701 (2010)
58. Castellano, C., Fortunato, S., Loreto, V.: Statistical physics of social dynamics. Rev. Mod. Phys. **81**(591), 591–646 (2009)
59. Castillo-Chavez, C., Song, B.: Dynamical models of tuberculosis and their applications. Math. Biosci. Eng. **1**(2), 361–404 (2004)
60. Cator, E., Van Mieghem, P.: Nodal infection in Markovian susceptible-infected-susceptible and susceptible-infected-removed epidemics on networks are non-negatively correlated. Phys. Rev. E **89**(5), 052802 (2014)
61. Cator, E., van de Bovenkamp, R., Van Mieghem, P.: Susceptible-infected-susceptible epidemics on networks with general infection and cure times. Phys. Rev. E **87**(6), 062816 (2013)
62. Centers for Disease Control and Prevention: Severe acute respiratory syndrome — Singapore, 2003. Morb. Mortal. Wkly. Rep. **52**(18), 405–411 (2003)
63. Chatterjee, S., Durrett, R.: Contact processes on random graphs with power law degree distributions have critical value 0. Ann. Probab. **37**(6), 2332–2356 (2009)
64. Christakis, N.A., Fowler, J.H.: Social network sensors for early detection of contagious outbreaks. PLoS ONE **5**(9), e12948 (2010)
65. Clancy, D.: SIR epidemic models with general infectious period distribution. Stat. Probab. Lett. **85**, 1–5 (2014)
66. Clancy, D., Mendy, S.T.: Approximating the quasi-stationary distribution of the SIS model for endemic infection. Methodol. Comput. Appl. Probab. **13**(3), 603–618 (2011)
67. Clarke, J., White, K.A.J., Turner, K.: Approximating optimal controls for networks when there are combinations of population-level and targeted measures available: chlamydia infection as a case-study. Bull. Math. Biol. **75**(10), 1747–1777 (2013)
68. Cohen, R., Havlin, S.: Complex Networks: Structure, Robustness and Function. Cambridge University Press, Cambridge (2010)
69. Courtesy of Salathé Lab, Penn State University: Flu Outbreaks Modeled by new Study of Classroom Schedules. http://science.psu.edu/news-and-events/2013-news/Salathe2-2013. Accessed 31 July 2016
70. Cowan, J.D.: Proceedings of the 1990 Conference on Advances in Neural Information Processing Systems 3: Stochastic Neurodynamics. Morgan Kaufmann Publishers Inc., San Francisco, CA (1990)
71. Cowan, N.J., Chastain, E.J., Vilhena, D.A., Freudenberg, J.S., Bergstrom, C.T.: Nodal dynamics, not degree distributions, determine the structural controllability of complex networks. PLoS ONE **7**(6), e38398 (2012)
72. Cozzo, E., Banos, R.A., Meloni, S., Moreno, Y.: Contact-based social contagion in multiplex networks. Phys. Rev. E **88**(5), 050801 (2013)

73. Cvetkovski, Z.: Inequalities: Theorems, Techniques and Selected Problems. Springer Science & Business Media, Berlin/Heidelberg (2012)

74. Daley, D.J., Gani, J.: Epidemic Modelling: An Introduction, vol. 15. Cambridge University Press, Cambridge (2001)

75. Danon, L., Ford, A.P., House, T., Jewell, C.P., Keeling, M.J., Roberts, G.O., Ross, J.V., Vernon, M.C.: Networks and the epidemiology of infectious disease. Interdisc. Perspect. Infect. Dis. **2011**, Article ID 284909, 1–28. (2011).

76. Decreusefond, L., Dhersin, J.S., Moyal, P., Tran, V.C.: Large graph limit for an SIR process in random network with heterogeneous connectivity. Ann. Appl. Probab. **22**(2), 541–575 (2012)

77. de Oliveira, M.J., Mendes, J.F.F., Santos, M.A.: Nonequilibrium spin models with Ising universal behaviour. J. Phys. A: Math. Gen. **26**, 2317 (1993)

78. Diekmann, O., Heesterbeek, J.A.P.: Mathematical Epidemiology of Infectious Diseases: Model Building, Analysis and Interpretation. Wiley, New York (2000)

79. Diekmann, O., Heesterbeek, J.A.P., Metz, J.A.J.: On the definition and the computation of the basic reproduction ratio $\mathcal{R}_0$ in models for infectious diseases in heterogeneous populations. J. Math. Biol. **28**(4), 365–382 (1990)

80. Diekmann, O., De Jong, M.C.M., Metz, J.A.J.: A deterministic epidemic model taking account of repeated contacts between the same individuals. J. Appl. Probab. **35**(2), 448–462 (1998)

81. Diekmann, O., Heesterbeek, H., Britton, T.: Mathematical Tools for Understanding Infectious Disease Dynamics. Princeton University Press, Princeton (2012)

82. Diestel, R.: Graph Theory. Springer, Heidelberg, New York (2005)

83. Doob, J.L.: Markoff chains–denumerable case. Trans. Am. Math. Soc. **58**(3), 455–473 (1945)

84. Dorogovtsev, S.N., Mendes, J.F.F., Samukhin, A.N.: Giant strongly connected component of directed networks. Phys. Rev. E **64**(2), 025101 (2001)

85. Dorogovtsev, S.N., Goltsev, A.V., Mendes, J.F.F.: Ising model on networks with an arbitrary distribution of connections. Phys. Rev. E **66**(1), 016104-1–016104-5 (2002)

86. Draief, M., Massoulié, L.: Epidemics and Rumours in Complex Networks. Cambridge University Press, New York (2010)

87. Durrett, R.: Random Graph Dynamics. Cambridge University Press, Cambridge (2007)

88. Durrett, R.: Some features of the spread of epidemics and information on a random graph. Proc. Natl. Acad. Sci. **107**(10), 4491–4498 (2010)

89. Eames, K.T.D.: Modelling disease spread through random and regular contacts in clustered populations. Theor. Popul. Biol. **73**(1), 104–111 (2008)

90. Eames, K.T.D., Keeling, M.J.: Modeling dynamic and network heterogeneities in the spread of sexually transmitted diseases. Proc. Natl. Acad. Sci. **99**(20), 13330–13335 (2002)

91. Eames, K.T.D., Keeling, M.J.: Contact tracing and disease control. Proc. R. Soc. Lond. B: Biol. Sci. **270**(1533), 2565–2571 (2003)

92. Easley, D., Kleinberg, J.: Networks, Crowds, and Markets: Reasoning About a Highly Connected World. Cambridge University Press, Cambridge (2010)

93. Estrada, E.: The Structure of Complex Networks: Theory and Applications. Oxford University Press, Oxford (2011)

94. Ethier, S.N., Kurtz, T.G.: Markov Processes: Characterization and Convergence, vol. 282. Wiley, Hoboken/New Jersey (2009)

95. Euler, L.: Solutio problematis ad geometriam situs pertinentis. Commentarii academiae scientiarum Petropolitanae **8**, 128–140 (1741)

96. Feld, S.L.: Why your friends have more friends than you do. Am. J. Sociol. **96**(6), 1464–1477 (1991)

97. Fennell, P.G., Melnik, S., Gleeson, J.P.: Limitations of discrete-time approaches to continuous-time contagion dynamics. Phys. Rev. E **94**(5), 052125 (2016)

98. Ferreira, S.C., Castellano, C., Pastor-Satorras, R.: Epidemic thresholds of the susceptible-infected-susceptible model on networks: a comparison of numerical and theoretical results. Phys. Rev. E **86**(4), 041125 (2012)

99. Fiedler, M.: Special Matrices and Their Applications in Numerical Mathematics. Dover Publications, New York (2008)

100. Filliger, R., Hongler, M.O.: Lumping complex networks. In: Lectures and Gallery of Madeira Math Encounters XXXV (2008). http://ccm.uma.pt/mme35/

101. Fotouhi, B., Rabbat, M.G.: Degree correlation in scale-free graphs. Eur. Phys. J. B **86**(12), 1–19 (2013)

102. Frasca, M., Sharkey, K.J.: Discrete-time moment closure models for epidemic spreading in populations of interacting individuals. J. Theor. Biol. **399**, 13–21 (2016)

103. Frieze, A., Karoński, M.: Introduction to Random Graphs. Cambridge University Press, Cambridge (2015)

104. Fu, X., Small, M., Chen, G.: Propagation Dynamics on Complex Networks: Models, Methods and Stability Analysis. Wiley, Chichester/UK (2013)

105. Funk, S., Jansen, V.A.A.: Interacting epidemics on overlay networks. Phys. Rev. E **81**(3), 036118 (2010)

106. Funk, S., Gilad, E., Watkins, C., Jansen, V.A.A.: The spread of awareness and its impact on epidemic outbreaks. Proc. Natl. Acad. Sci. **106**(16), 6872–6877 (2009)

107. Funk, S., Salathé, M., Jansen, V.A.A.: Modelling the influence of human behaviour on the spread of infectious diseases: a review. J. R. Soc. Interface **7**(50), 1247–1256 (2010)

108. Gantmacher, F.R.: The Theory of Matrices. Taylor & Francis, Providence/Rhode Island (1964)

109. Gershgorin, S.A.: Uber die abgrenzung der eigenwerte einer matrix. Izvestiya Rossiiskoi Akademii Nauk, Seriya Matematicheskaya **6**, 749–754 (1931)

110. Gertsbakh, I.B.: Epidemic process on a random graph: some preliminary results. J. Appl. Probab. **14**(03), 427–438 (1977)

111. Ghoshal, G., Zlatic, V., Caldarelli, G., Newman, M.E.J.: Random hypergraphs and their applications. Phys. Rev. E **79**, 066118 (2009)
112. Gillespie, D.T.: A general method for numerically simulating the stochastic time evolution of coupled chemical reactions. J. Comput. Phys. **22**(4), 403–434 (1976)
113. Gillespie, D.T.: Exact stochastic simulation of coupled chemical reactions. J. Phys. Chem. **81**(25), 2340–2361 (1977)
114. Gleeson, J.P.: Bond percolation on a class of clustered random networks. Phys. Rev. E **80**(3), 036107 (2009)
115. Gleeson, J.P.: High-accuracy approximation of binary-state dynamics on networks. Phys. Rev. Lett. **107**(6), 068701 (2011)
116. Gleeson, J.P.: Binary-state dynamics on complex networks: pair approximation and beyond. Phys. Rev. X **3**(2), 021004 (2013)
117. Gleeson, J.P., Melnik, S., Hackett, A.: How clustering affects the bond percolation threshold in complex networks. Phys. Rev. E **81**(6), 066114 (2010)
118. Goldstein, E., Paur, K., Fraser, C., Kenah, E., Wallinga, J., Lipsitch, M.: Reproductive numbers, epidemic spread and control in a community of households. Math. Biosci. **221**(1), 11–25 (2009)
119. Golub, G.H., Van Loan, C.F.: Matrix Computations, vol. 3. JHU Press, Baltimore/Maryland (2012)
120. Granell, C., Gómez, S., Arenas, A.: Dynamical interplay between awareness and epidemic spreading in multiplex networks. Phys. Rev. Lett. **111**(12), 128701 (2013)
121. Grassberger, P.: On the critical behavior of the general epidemic process and dynamical percolation. Math. Biosci. **63**, 157–172 (1983)
122. Green, D.M., Kiss, I.Z.: Large-scale properties of clustered networks: implications for disease dynamics. J. Biol. Dyn. **4**(5), 431–445 (2010)
123. Griffeath, D.: Additive and Cancellative Interacting Particle Systems. Springer, Berlin (1979)
124. Grimmett, G., Stirzaker, D.: Probability and Random Processes, 3rd edn. Oxford University Press, Oxford (2001)
125. Grindrod, P., Higham, D.J.: Evolving graphs: dynamical models, inverse problems and propagation. Proc. R. Soc. Lond. A: Math. Phys. Eng. Sci. **466**(2115), 753–770 (2010)
126. Gross, T., Blasius, B.: Adaptive coevolutionary networks: a review. J. R. Soc. Interface **5**(20), 259–271 (2008)
127. Gross, T., Kevrekidis, I.G.: Robust oscillations in SIS epidemics on adaptive networks: coarse graining by automated moment closure. EPL (Europhys. Lett.) **82**(3), 38004 (2008)
128. Gross, T., D'Lima, C.J.D., Blasius, B.: Epidemic dynamics on an adaptive network. Phys. Rev. Lett. **96**(20), 208701 (2006)
129. Hadjichrysanthou, C., Sharkey, K.J.: Epidemic control analysis: designing targeted intervention strategies against epidemics propagated on contact networks. J. Theor. Biol. **365**, 84–95 (2015)

130. Hagberg, A.A., Schult, D.A., Swart, P.J.: Exploring network structure, dynamics, and function using NetworkX. In: Proceedings of the 7th Python in Science Conferences (SciPy 2008), vol. 2008, pp. 11–16 (2008)

131. Hale, J.K.: Ordinary Differential Equations. Dover Books on Mathematics Series. Dover Publications, New York (2009)

132. Halliday, D.M., Rosenberg, J.R., Amjad, A.M., Breeze, P., Conway, B.A., Farmer, S.F.: A framework for the analysis of mixed time series/point process data—theory and application to the study of physiological tremor, single motor unit discharges and electromyograms. Progr. Biophys. Mol. Biol. **64**(2), 237–278 (1995)

133. Harada, Y., Ezoe, H., Iwasa, Y., Matsuda, H., Sato, K.: Population persistence and spatially limited social interaction. Theor. Popul. Biol. **48**(1), 65–91 (1995)

134. Harris, T.E.: On a class of set-valued Markov processes. Ann. Probab. **4**(2), 175–194 (1976)

135. Hastings, M.B.: Systematic series expansions for processes on networks. Phys. Rev. Lett. **96**(14), 148701 (2006)

136. Hatzopoulos, V., Taylor, M., Simon, P.L., Kiss, I.Z.: Multiple sources and routes of information transmission: implications for epidemic dynamics. Math. Biosci. **231**(2), 197–209 (2011)

137. Hayden, R.A., Horváth, I., Telek, M.: Mean field for performance models with generally-distributed timed transitions. In: Quantitative Evaluation of Systems, pp. 90–105. Springer, Heidelberg/New York (2014)

138. Hébert-Dufresne, L., Patterson-Lomba, O., Goerg, G.M., Althouse, B.M.: Pathogen mutation modeled by competition between site and bond percolation. Phys. Rev. Lett. **110**, 108103 (2013)

139. Heesterbeek, H., Anderson, R.M., Andreasen, V., Bansal, S., De Angelis, D., Dye, C., Eames, K.T.D., Edmunds, W.J., Frost, S.D.W., Funk, S., Hollingsworth, T.D., House, T., Isham, V., Klepac, P., Lessler, J., Lloyd-Smith, J.O., Metcalf, C.J.E., Mollison, D., Pellis, L., Pulliam, J.R.C., Roberts, M.G., Viboud, C.: Modeling infectious disease dynamics in the complex landscape of global health. Science **347**(6227), aaa4339 (2015)

140. Hethcote, H.W., Yorke, J.A.: Gonorrhea Transmission Dynamics and Control. Lecture Notes in Biomathematics, vol. 56. Springer, Berlin/Heidelberg (1984)

141. Hethcote, H.W., Yorke, J.A., Nold, A.: Gonorrhea modeling: a comparison of control methods. Math. Biosci. **58**(1), 93–109 (1982)

142. Hirsch, M.W., Smith, H.: Monotone dynamical systems. In: A. Cañada, P. Drábek, A. Fonda (eds.) Handbook of Differential Equations: Ordinary Differential Equations, vol. 2, pp. 239–357. Elsevier BV, Amsterdam (2005)

143. Hladish, T., Melamud, E., Barrera, L.A., Galvani, A., Meyers, L.A.: Epifire: an open source C++ library and application for contact network epidemiology. BMC Bioinformatics **13**(1), 1 (2012)

144. Hoen, A.G., Hladish, T.J., Eggo, R.M. Lenczner, M., Brownstein, J.S., Meyers, L.A.: Epidemic wave dynamics attributable to urban community struc-

ture: a theoretical characterization of disease transmission in a large network. J. Med. Internet Res. **17**(7) (2015).

145. Holley, R.A., Liggett, T.M.: Ergodic theorems for weakly interacting infinite systems and the voter model. Ann. Probab. **3**(4), 643–663 (1975)

146. Holme, P., Saramäki, J.: Temporal networks. Phys. Rep. **519**(3), 97–125 (2012)

147. Holme, P., Saramäki, J.: Temporal Networks. Springer, Berlin (2013)

148. House, T., Keeling, M.J.: Deterministic epidemic models with explicit household structure. Math. Biosci. **213**(1), 29–39 (2008)

149. House, T., Keeling, M.J.: The impact of contact tracing in clustered populations. PLoS Computat. Biol. **6**(3), e1000721 (2010)

150. House, T., Keeling, M.J.: Epidemic prediction and control in clustered populations. J. Theor. Biol. **272**(1), 1–7 (2011)

151. House, T., Keeling, M.: Insights from unifying modern approximations to infections on networks. J. R. Soc. Interface **8**(54), 67–73 (2011)

152. House, T., Davies, G., Danon, L., Keeling, M.J.: A motif-based approach to network epidemics. Bull. Math. Biol. **71**(7), 1693–1706 (2009)

153. Isham, V., Medley, G.: Models for Infectious Human Diseases: Their Structure and Relation to Data, vol. 6. Cambridge University Press, Cambridge (1996)

154. Jackson, M.O.: Social and Economic Networks, vol. 3. Princeton University Press, Princeton (2008)

155. Jacobi, M.N., Görnerup, O.: A spectral method for aggregating variables in linear dynamical systems with application to cellular automata renormalization. Adv. Complex Syst. **12**(1–25) (2009)

156. Janson, S., Luczak, M., Windridge, P.: Law of large numbers for the SIR epidemic on a random graph with given degrees. Rand. Struct. Alg. **45**(4), 726–763 (2014)

157. Jones, P.W., Smith, P.: Stochastic Processes: An Introduction, 2nd edn. CRC Press, Boca Raton/FL (2012)

158. Juher, D., Ripoll, J., Saldaña, J.: Outbreak analysis of an SIS epidemic model with rewiring. J. Math. Biol. **67**(2), 411–432 (2013)

159. Juher, D., Kiss, I.Z., Saldaña, J.: Analysis of an epidemic model with awareness decay on regular random networks. J. Theor. Biol. **365**, 457–468 (2015)

160. Kamke, E.: Zur theorie der systeme gewöhnlicher differentialgleichungen. II. Acta Math. **58**(1), 57–85 (1932)

161. Karlin, S., Taylor, H.M.: A First Course in Stochastic Processes, 2nd edn. Academic Press, New York (1975)

162. Karlin, S., Taylor, H.M.: A Second Course in Stochastic Processes. Academic Press, New York (1981)

163. Karrer, B., Newman, M.E.J.: Message passing approach for general epidemic models. Phys. Rev. E **82**(1), 016101 (2010)

164. Karrer, B., Newman, M.E.J.: Random graphs containing arbitrary distributions of subgraphs. Phys. Rev. E **82**(6), 066118 (2010)

165. Keeling, M.J.: The ecology and evolution of spatial host-parasite systems. Ph.D. thesis, University of Warwick (1995)

166. Keeling, M.J.: The effects of local spatial structure on epidemiological invasions. Proc. R. Soc. Lond. Ser. B: Biol. Sci. **266**(1421), 859–867 (1999)

167. Keeling, M.J., Eames, K.T.D.: Networks and epidemic models. J. R. Soc. Interface **2**(4), 295–307 (2005)

168. Keeling, M.J., Rohani, P.: Modeling Infectious Diseases in Humans and Animals. Princeton University Press, Princeton (2008)

169. Keeling, M.J., Ross, J.V.: On methods for studying stochastic disease dynamics. J. R. Soc. Interface **5**(19), 171–181 (2008)

170. Keeling, M.J., Rand, D.A., Morris, A.J.: Correlation models for childhood epidemics. Proc. R. Soc. Lond. Ser. B: Biol. Sci. **264**(1385), 1149–1156 (1997)

171. Kemeny, J.G., Snell, J.L.: Finite Markov Chains, 2nd edn. Springer, New York (1976)

172. Kenah, E., Miller, J.C.: Epidemic percolation networks, epidemic outcomes, and interventions. Interdiscip. Perspect. Infect. Dis. (2011)

173. Kenah, E., Robins, J.M.: Second look at the spread of epidemics on networks. Phys. Rev. E **76**(3), 036113 (2007)

174. Kermack, W.O., McKendrick, A.G.: A contribution to the mathematical theory of epidemics. R. Soc. Lond. Proc. Ser. A **115**, 700–721 (1927)

175. Kirkwood, J.D.: Statistical mechanics of fluid mixtures. J. Chem. Phys. **3**, 300–313 (1935)

176. Kiss, I.Z., Simon, P.L.: New moment closures based on *a priori* distributions with applications to epidemic dynamics. Bull. Math. Biol. **74**(7), 1501–1515 (2012)

177. Kiss, I.Z., Green, D.M., Kao, R.R.: The effect of contact heterogeneity and multiple routes of transmission on final epidemic size. Math. Biosci. **203**(1), 124–136 (2006)

178. Kiss, I.Z., Green, D.M., Kao, R.R.: The network of sheep movements within Great Britain: network properties and their implications for infectious disease spread. J. R. Soc. Interface **3**(10), 669–677 (2006)

179. Kiss, I.Z., Simon, P.L., Kao, R.R.: A contact-network-based formulation of a preferential mixing model. Bull. Math. Biol. **71**(4), 888–905 (2009)

180. Kiss, I.Z., Berthouze, L., Taylor, T.J., Simon, P.L.: Modelling approaches for simple dynamic networks and applications to disease transmission models. Proc. R. Soc. Lond. A: Math. Phys. Eng. Sci. **468**(2141), 1332–1355 (2012)

181. Kiss, I.Z., et al.: Exact deterministic representation of Markovian SIR epidemics on networks with and without loops. J. Math. Biol. **70**(3), 437–464 (2015)

182. Kiss, I.Z., Röst, G., Vizi, Z.: Generalization of pairwise models to non-Markovian epidemics on networks. Phys. Rev. Lett. **115**(7), 078701 (2015)

183. Kivelä, M., Arenas, A., Barthélemy, M., Gleeson, J.P., Moreno, Y., Porter, M.A.: Multilayer networks. J. Complex Netw. **2**(3), 203–271 (2014)

184. Koch, D., Illner, R., Ma, J.: Edge removal in random contact networks and the basic reproduction number. J. Math. Biol. **67**(2), 217–238 (2013)

185. Kolmogorov, A.N.: The local structure of turbulence in incompressible viscous fluid for very large Reynolds numbers. Dokl. Akad. Nauk SSSR **30**, 299–303 (1941)

186. Krings, G., Karsai, M., Bernhardsson, S., Blondel, V.D., Saramäki, J.: Effects of time window size and placement on the structure of an aggregated communication network. EPJ Data Sci. **1**(4), 1–16 (2012)

187. Kuehn, C.: Moment closure–a brief review: control of self-organizing nonlinear systems. Springer, 253–271 (2016)

188. Kurant, M., Thiran, P.: Layered complex networks. Phys. Rev. Lett. **96**(13), 138701 (2006)

189. Kurtz, T.G.: Solutions of ordinary differential equations as limits of pure jump Markov processes. J. Appl. Probab. **7**(1), 49–58 (1970)

190. Kuulasmaa, K.: The spatial general epidemic and locally dependent random graphs. J. Appl. Probab. **19**(4), 745–758 (1982)

191. Kuulasmaa, K., Zachary, S.: On spatial general epidemics and bond percolation processes. J. Appl. Probab. **21**(4), 911–914 (1984)

192. Lajmanovich, A., Yorke, J.A.: A deterministic model for Gonorrhea in a non-homogeneous population. Math. Biosci. **28**(3), 221–236 (1976)

193. Lanchier, N., Neufer, J.: Stochastic dynamics on hypergraphs and the spatial majority rule model. J. Stat. Phys. 21–45 (2013)

194. Leone, M., Vazquez, A., Vespignani, A., Zecchina, R.: Ferromagnetic ordering in graphs with arbitrary degree distribution. Eur. Phys. J. B **28**, 191–197 (2002)

195. Leung, K.Y.: Dangerous connections: the spread of infectious diseases on dynamic networks. Ph.D. thesis, Utrecht University (2016)

196. Leung, K.Y., Kretzschmar, M., Diekmann, O.: *SI* infection on a dynamic partnership network: characterization of R_0. J. Math. Biol. **71**(1), 1–56 (2015)

197. Lindquist, J., Ma, J., van den Driessche, P., Willeboordse, F.H.: Effective degree network disease models. J. Math. Biol. **62**(2), 143–164 (2011)

198. Liu, Y.Y., Slotine, J.J., Barabási, A.L.: Controllability of complex networks. Nature **473**(7346), 167–173 (2011)

199. Lloyd, A.: Realistic distributions of infectious periods in epidemic models: changing patterns of persistence and dynamics. Theor. Popul. Biol. **60**, 59–71 (2001)

200. Lokhov, A.Y., Mézard, M., Ohta, H., Zdeborová, L.: Inferring the origin of an epidemic with a dynamic message-passing algorithm. Phys. Rev. E **90**(1), 012801 (2014)

201. Lovász, L.: Large Networks and Graph Limits, vol. 60. American Mathematical Society, Providence/Rhode Island (2012)

202. Ludwig, D.: Final size distributions for epidemics. Math. Biosci. **23**, 33–46 (1975)

203. Lusher, D., Koskinen, J., Robins, G.: Exponential Random Graph Models for Social Networks: Theory, Methods, and Applications. Cambridge University Press, Cambridge (2012)

204. Ma, J., Earn, D.J.D.: Generality of the final size formula for an epidemic of a newly invading infectious disease. Bull. Math. Biol. **68**(3), 679–702 (2006)

205. Marceau, V., Noël, P.A., Hébert-Dufresne, L., Allard, A., Dubé, L.J.: Adaptive networks: coevolution of disease and topology. Phys. Rev. E **82**(3), 036116 (2010)
206. Marceau, V., Noël, P.A., Hébert-Dufresne, L., Allard, A., Dubé, L.J.: Modeling the dynamical interaction between epidemics on overlay networks. Phys. Rev. E **84**(2), 026105 (2011)
207. Matsuda, H., Ogita, N., Sasaki, A., Satō, K.: Statistical mechanics of population the Lattice Lotka-Volterra Model. Progr. Theor. Phys. **88**(6), 1035–1049 (1992)
208. May, R.M., Anderson, R.M.: Transmission dynamics of HIV infection. Nature **326**, 137 (1987)
209. Melnik, S., Hackett, A., Porter, M.A., Mucha, P.J., Gleeson, J.P.: The unreasonable effectiveness of tree-based theory for networks with clustering. Phys. Rev. E **83**(3), 036112 (2011)
210. Meyers, L.A., Pourbohloul, B., Newman, M.E., Skowronski, D.M., Brunham, R.C.: Network theory and SARS: predicting outbreak diversity. J. Theor. Biol. **232**(1), 71–81 (2005)
211. Miller, J.C.: Epidemic size and probability in populations with heterogeneous infectivity and susceptibility. Phys. Rev. E **76**(1), 010101(R) (2007)
212. Miller, J.C.: Bounding the size and probability of epidemics on networks. J. Appl. Probab. **45**, 498–512 (2008)
213. Miller, J.C.: Percolation and epidemics in random clustered networks. Phys. Rev. E **80**(2), 020901(R) (2009)
214. Miller, J.C.: Spread of infectious disease through clustered populations. J. R. Soc. Interface **6**(41), 1121–1134 (2009)
215. Miller, J.C.: A note on a paper by Erik Volz: SIR dynamics in random networks. J. Math. Biol. **62**(3), 349–358 (2011)
216. Miller, J.C.: A note on the derivation of epidemic final sizes. Bull. Math. Biol. **74**(9), 2125–2141 (2012)
217. Miller, J.C.: Epidemics on networks with large initial conditions or changing structure. PLoS ONE **9**(7), e101421 (2014)
218. Miller, J.C.: Complex contagions and hybrid phase transitions. J. Complex Netw. **4**(2), 201–223 (2016)
219. Miller, J.C., Kiss, I.Z.: Epidemic spread in networks: Existing methods and current challenges. Math. Modell. Nat. Phenom. **9**(2), 4 (2014)
220. Miller, J.C., Volz, E.M.: Incorporating disease and population structure into models of SIR disease in contact networks. PLoS ONE **8**(8), e69162 (2013)
221. Miller, J.C., Volz, E.M.: Model hierarchies in edge-based compartmental modeling for infectious disease spread. J. Math. Biol. **67**(4), 869–899 (2013)
222. Miller, J.C., Slim, A.C., Volz, E.M.: Edge-based compartmental modelling for infectious disease spread. J. R. Soc. Interface **9**(70), 890–906 (2012)
223. Mollison, D.: Epidemic Models: Their Structure and Relation to Data, vol. 5. Cambridge University Press, Cambridge (1995)
224. Molloy, M., Reed, B.: A critical point for random graphs with a given degree sequence. Rand. Struct. Alg. **6**(2), 161–179 (1995)

225. Molloy, M., Reed, B.: The size of the giant component of a random graph with a given degree sequence. Comb. Probab. Comput. **7**(03), 295–305 (1998)
226. Moreno, Y., Pastor-Satorras, R., Vespignani, A.: Epidemic outbreaks in complex heterogeneous networks. Eur. Phys. J. B: Condens. Matter Complex Syst. **26**(4), 521–529 (2002)
227. Morris, A.J.: Representing spatial interactions in simple ecological models. Ph.D. thesis, University of Warwick (1997)
228. Müller, M.: Über das fundamentaltheorem in der theorie der gewöhnlichen differentialgleichungen. Mathematische Zeitschrift **26**(1), 619–645 (1927)
229. Nagy, N., Kiss, I.Z., Simon, P.L.: Approximate master equations for dynamical processes on graphs. Math. Modell. Nat. Phenom. **9**(02), 43–57 (2014)
230. Nåsell, I.: The quasi-stationary distribution of the closed endemic SIS model. Adv. Appl. Probab. **28**(03), 895–932 (1996)
231. Nåsell, I.: On the quasi-stationary distribution of the stochastic logistic epidemic. Math. Biosci. **156**(1), 21–40 (1999)
232. Nåsell, I.: Extinction and quasi-stationarity in the Verhulst logistic model. J. Theor. Biol. **211**(1), 11–27 (2001)
233. Neal, P.: Endemic behaviour of SIS epidemics with general infectious period distributions. Adv. Appl. Probab. **46**(1), 241–255 (2014)
234. Newman, M.E.J.: Spread of epidemic disease on networks. Phys. Rev. E **66**(1), 016128 (2002)
235. Newman, M.E.J.: Component sizes in networks with arbitrary degree distributions. Phys. Rev. E **76**(4), 045101 (2007)
236. Newman, M.E.J.: Networks: An Introduction. Oxford University Press, Oxford (2009)
237. Newman, M.E.J.: Random graphs with clustering. Phys. Rev. Lett. **103**(5), 058701 (2009)
238. Newman, M., Barabasi, A.L., Watts, D.J.: The Structure and Dynamics of Networks. Princeton University Press, Princeton (2006)
239. Noël, P.A., Allard, A., Hébert-Dufresne, L., Marceau, V., Dubé, L.J.: Propagation on networks: an exact alternative perspective. Phys. Rev. E **85**(3), 031118 (2012)
240. Noël, P.A., Davoudi, B., Brunham, R.C., Dubé, L.J., Pourbohloul, B.: Time evolution of disease spread on finite and infinite networks. Phys. Rev. E **79**(2), 026101 (2009)
241. Nold, A.: Heterogeneity in disease-transmission modeling. Math. Biosci. **52**(3), 227–240 (1980)
242. Nowzari, C., Preciado, V.M., Pappas, G.J.: Analysis and control of epidemics: a survey of spreading processes on complex networks. IEEE Control Syst. **36**(1), 26–46 (2016)
243. Pastor-Satorras, R., Vespignani, A.: Epidemic dynamics and endemic states in complex networks. Phys. Rev. E **63**(6), 066117 (2001)
244. Pastor-Satorras, R., Vespignani, A.: Epidemic spreading in scale-free networks. Phys. Rev. Lett. **86**, 3200–3203 (2001)

245. Pastor-Satorras, R., Vespignani, A.: Epidemic dynamics in finite size scale-free networks. Phys. Rev. E **65**(3), 035108 (2002)

246. Pastor-Satorras, R., Rubi, M., Diaz-Guilera, A.: Statistical Mechanics of Complex Networks, vol. 625. Springer Science & Business Media, Berlin/Heidelberg/New York (2003)

247. Pastor-Satorras, R., Castellano, C., Van Mieghem, P., Vespignani, A.: Epidemic processes in complex networks. Rev. Mod. Phys. **87**, 925 (2015)

248. Pellis, L., Ball, F., Bansal, S., Eames, K., House, T., Isham, V., Trapman, P.: Eight challenges for network epidemic models. Epidemics **10**, 58–62 (2015)

249. Pellis, L., House, T., Keeling, M.J.: Exact and approximate moment closures for non-Markovian network epidemics. J. Theor. Biol. **382**, 160–177 (2015)

250. Perko, L.: Differential Equations and Dynamical Systems, vol. 7. Springer Science & Business Media, New York (2001)

251. Perra, N., Gonçalves, B., Pastor-Satorras, R., Vespignani, A.: Activity driven modeling of time varying networks. Sci. Rep. **2**, Article No. 469, 1–7 (2012)

252. Picard, P.: Sur les modèles stochastique logistiques en démographie. Ann. Inst. Henri Poincaré B **II**, 151–172 (1965)

253. Porter, M.A., Gleeson, J.P.: Dynamical Systems on Networks: A Tutorial. Springer International Publishing, Heidelberg/New York (2016)

254. Pourbohloul, B., Brunham, R.C.: Network models and transmission of sexually transmitted diseases. Sex. Transm. Dis. **31**(6), 388–390 (2004)

255. Prakash, B.A., Chakrabarti, D., Valler, N.C., Faloutsos, M., Faloutsos, C.: Threshold conditions for arbitrary cascade models on arbitrary networks. Knowl. Inf. Syst. **33**(3), 549–575 (2012)

256. Rand, D.A.: Advanced ecological theory: principles and applications. In: Correlation Equations and Pair Approximations for Spatial Ecologies, pp. 100–142. Blackwell Science, Oxford (1999)

257. Rattana, P., Blyuss, K.B., Eames, K.T.D., Kiss, I.Z.: A class of pairwise models for epidemic dynamics on weighted networks. Bull. Math. Biol. **75**(3), 466–490 (2013)

258. Rattana, P., Miller, J.C., Kiss, I.Z.: Pairwise and edge-based models of epidemic dynamics on correlated weighted networks. Math. Modell. Nat. Phenom. **9**(02), 58–81 (2014)

259. Renshaw, E.: Modelling Biological Populations in Space and Time. Cambridge University Press, Cambridge (1991)

260. Ribeiro, B., Perra, N., Baronchelli, A.: Quantifying the effect of temporal resolution on time-varying networks. Sci. Rep. **3**, Article No. 3006, 1–5 (2013)

261. Riley, S., Fraser, C., Donnelly, C.A., Ghani, A.C., Abu-Raddad, L.J., Hedley, A.J., Leung, G.M., Ho, L.M., Lam, T.H., Thach, T.Q., Chau, P., Chan, K.P., Lo, S.V., Leung, P.Y., Tsang, T., Ho, W., Lee, K.H., Lau, E.M.C., Ferguson, N.M., Anderson, R.M.: Transmission dynamics of the etiological agent of SARS in Hong Kong: impact of public health interventions. Science **300**(5627), 1961–1966 (2003)

262. Risau-Gusmán, S., Zanette, D.H.: Contact switching as a control strategy for epidemic outbreaks. J. Theor. Biol. **257**(1), 52–60 (2009)

263. Risken, H.: The Fokker-Planck Equation: Methods of Solution and Applications. Springer Series in Synergetics. Springer, Berlin, Heidelberg (2012)
264. Ritchie, M., Berthouze, L., House, T., Kiss, I.Z.: Higher-order structure and epidemic dynamics in clustered networks. J. Theor. Biol. **348**, 21–32 (2014)
265. Ritchie, M., Berthouze, L., Kiss, I.Z.: Beyond clustering: mean-field dynamics on networks with arbitrary subgraph composition. J. Math. Biol. **72**(1–2), 255–281 (2016)
266. Ritchie, M., Berthouze, L., Kiss, I.Z.: Generation and analysis of networks with a prescribed degree sequence and subgraph family: higher-order structure matters. J. Complex Netw. (2016)
267. Rock, K., Brand, S., Moir, J., Keeling, M.J.: Dynamics of infectious diseases. Rep. Progr. Phys. **77**(2), 026602 (2014)
268. Rogers, L.C.G., Pitman, J.W.: Markov functions. Ann. Probab. **9**, 537–711 (1981)
269. Rogers, T., Clifford-Brown, W., Mills, C., Galla, T.: Stochastic oscillations of adaptive networks: application to epidemic modelling. J. Stat. Mech.: Theor. Exp. **2012**(08), P08018 (2012)
270. Röst, G., Vizi, Z., Kiss, I.Z.: Impact of non-Markovian recovery on network epidemics. In: Biomat 2015: Proceedings of the International Symposium on Mathematical and Computational Biology. World Scientific, New York (2015)
271. Röst, G., Vizi, Z., Kiss, I.Z.: Pairwise approximation for SIR type network epidemics with non-Markovian recovery. arXiv preprint arXiv:1605.02933 (2016)
272. Salinelli, E., Tomarelli, F.: Discrete Dynamical Models, vol. 76. Springer, Heidelberg/New York (2014)
273. Saramäki, J., Kaski, K.: Modelling development of epidemics with dynamic small-world networks. J. Theor. Biol. **234**(3), 413–421 (2005)
274. Saumell-Mendiola, A., Serrano, M.Á.., Boguñá, M.: Epidemic spreading on interconnected networks. Phys. Rev. E **86**(2), 026106 (2012)
275. Scott, M.: Applied stochastic processes in science and engineering. University of Waterloo, eBook (2013)
276. Sedgewick, R.: Algorithms in C, Part 5: Graph Algorithms. Addison-Wesley (2002)
277. Sélley, F., Besenyei, Á., Kiss, I.Z., Simon, P.L.: Dynamic control of modern, network-based epidemic models. SIAM J. Appl. Dyn. Syst. **14**(1), 168–187 (2015)
278. Sellke, T.: On the asymptotic distribution of the size of a stochastic epidemic. J. Appl. Probab. **20**(02), 390–394 (1983)
279. Sharkey, K.J.: Deterministic epidemiological models at the individual level. J. Math. Biol. **57**, 311–331 (2008)
280. Sharkey, K.J.: Deterministic epidemic models on contact networks: correlations and unbiological terms. Theor. Popul. Biol. **79**, 115–129 (2011)
281. Sharkey, K.J., Wilkinson, R.R.: Complete hierarchies of SIR models on arbitrary networks with exact and approximate moment closure. Math. Biosci. **264**, 74–85 (2015)

282. Sharkey, K.J., Fernandez, C., Morgan, K.L., Peeler, E., Thrush, M., Turnbull, J.F., Bowers, R.G.: Pair-level approximations to the spatio-temporal dynamics of epidemics on asymmetric contact networks. J. Math. Biol. **53**(1), 61–85 (2006)

283. Sharkey, K.J., et al.: Exact equations for SIR epidemics on tree graphs. Bull. Math. Biol. **77**(4), 614–645 (2015)

284. Sherborne, N., Blyuss, K.B., Kiss, I.Z.: Dynamics of multi-stage infections on networks. Bull. Math. Biol. **77**(10), 1909–1933 (2015)

285. Shkarayev, M.S., Tunc, I., Shaw, L.B.: Epidemics with temporary link deactivation in scale-free networks. J. Phys. A: Math. Theor. **47**(45), 455006 (2014)

286. Simon, P.L., Kiss, I.Z.: From exact stochastic to mean-field ODE models: a new approach to prove convergence results. IMA J. Appl. Math. **78**(5), 945–964 (2013)

287. Simon, P.L., Kiss, I.Z.: Super compact pairwise model for SIS epidemic on heterogeneous networks. J. Complex Netw. **4**(2), 187–200 (2016)

288. Simon, P.L., Taylor, M., Kiss, I.Z.: Exact epidemic models on graphs using graph-automorphism driven lumping. J. Math. Biol. **62**(4), 479–508 (2011)

289. Skiena, S.S.: The Algorithm Design Manual, 2nd edn. Springer Science & Business Media, New York (2009)

290. Smith, H.L.: Monotone Dynamical Systems: An Introduction to the Theory of Competitive and Cooperative Systems, vol. 41. American Mathematical Society, Providence/Rhode Island (2008)

291. Sood, V., Redner, S.: Voter model on heterogeneous graphs. Phys. Rev. Lett. **94**(17), 178701-1–178701-4 (2005)

292. Startsev, A.N.: On the distribution of the size of an epidemic in a non-Markovian model. Theor. Probab. Appl. **41**(4), 730–740 (1997)

293. Startsev, A.N.: Asymptotic analysis of the general stochastic epidemic with variable infectious periods. J. Appl. Probab. **38**(01), 18–35 (2001)

294. Szabó, A., Simon, P.L., Kiss, I.Z.: Detailed study of bifurcations in an epidemic model on a dynamic network. Differ. Equ. Appl. **4**, 277–296 (2012)

295. Szabó-Solticzky, A., et al.: Oscillating epidemics in a dynamic network model: stochastic and mean-field analysis. J. Math. Biol. **72**(5), 1153–1176 (2016)

296. Szarski, J.: Differential inequalities. Instytut Matematyczny Polskiej Akademi Nauk (Warszawa) (1965)

297. Tarjan, R.: Depth-first search and linear graph algorithms. SIAM J. Comput. **1**(2), 146–160 (1972)

298. Taylor, T.J., Kiss, I.Z.: Interdependency and hierarchy of exact and approximate epidemic models on networks. J. Math. Biol. **69**(1), 183–211 (2014)

299. Taylor, M., Simon, P.L., Green, D.M., House, T., Kiss, I.Z.: From Markovian to pairwise epidemic models and the performance of moment closure approximations. J. Math. Biol. **64**(6), 1021–1042 (2012)

300. Taylor., M., Taylor, T.J., Kiss, I.Z.: Epidemic threshold and control in a dynamic network. Phys. Rev. E **85**, 016103 (2012)

301. Trapman, P.: On analytical approaches to epidemics on networks. Theor. Popul. Biol. **71**(2), 160–173 (2007)

302. Trapman, P.: Reproduction numbers for epidemics on networks using pair approximation. Math. Biosci. **210**(2), 464–489 (2007)
303. Traud, A.L., Kelsic, E.D., Mucha, P.J., Porter, M.A.: Comparing community structure to characteristics in online collegiate social networks. SIAM Rev. **53**(3), 526–543 (2011)
304. Tunc, I., Shkarayev, M.S., Shaw, L.B.: Epidemics in adaptive social networks with temporary link deactivation. J. Stat. Phys. **151**(1–2), 355–366 (2013)
305. Valdez, L.D., Macri, P.A., Braunstein, L.A.: Temporal percolation of the susceptible network in an epidemic spreading. PLoS ONE **7**(9), e44188 (2012)
306. van Baalen, M.: Pair approximations for different spatial geometries. In: The Geometry of Ecological Interactions: Simplifying Spatial Complexity, pp. 359–387. Cambridge University Press, Cambridge (2000)
307. van de Bovenkamp, R., Van Mieghem, P.: Survival time of the susceptible-infected-susceptible infection process on a graph. Phys. Rev. E **92**(3), 032806 (2015)
308. van den Driessche, P., Watmough, J.: Reproduction numbers and sub-threshold endemic equilibria for compartmental models of disease transmission. Math. Biosci. **180**(1), 29–48 (2002)
309. van Kampen, N.G.: Stochastic Processes in Physics and Chemistry, vol. 1. Elsevier, Amsterdam (1992)
310. Van Mieghem, P.: The n-intertwined SIS epidemic network model. Computing **93**(2–4), 147–169 (2011)
311. Van Mieghem, P., van de Bovenkamp, R.: Non-Markovian infection spread dramatically alters the susceptible-infected-susceptible epidemic threshold in networks. Phys. Rev. Lett. **110**(10), 108701 (2013)
312. Van Mieghem, P., Omic, J., Kooij, R.: Virus spread in networks. IEEE/ACM Trans. Netw. **17**(1), 1–14 (2009)
313. Van Mieghem, P., Sahneh, F.D., Scoglio, C.: An upper bound for the epidemic threshold in exact Markovian SIR and SIS epidemics on networks. In: 2014 IEEE 53rd Annual Conference on Decision and Control (CDC), pp. 6228–6233. IEEE (2014)
314. Vazquez, F., Eguíluz, V.M.: Analytical solution of the voter model on uncorrelated networks. New J. Phys. **10**(6), 063011, 1–19 (2008)
315. Volz, E.M.: Random networks with tunable degree distribution and clustering. Phys. Rev. E **70**(5), 056115 (2004)
316. Volz, E.M.: SIR dynamics in random networks with heterogeneous connectivity. J. Math. Biol. **56**(3), 293–310 (2008)
317. Volz, E.M., Meyers, L.A.: Susceptible–infected–recovered epidemics in dynamic contact networks. Proc. R. Soc. Lond. B: Biol. Sci. **274**(1628), 2925–2934 (2007)
318. Volz, E.M., Miller, J.C., Galvani, A., Meyers, L.A.: Effects of heterogeneous and clustered contact patterns on infectious disease dynamics. PLoS Comput. Biol. **7**(6), e1002042 (2011)

319. Wallinga, J., Lipsitch, M.: How generation intervals shape the relationship between growth rates and reproductive numbers. Proc. R. Soc. Lond. B: Biol. Sci. **274**(1609), 599–604 (2007)
320. Wang, H., Li, Q., D'Agostino, G., Havlin, S., Stanley, H.E., Van Mieghem, P.: Effect of the interconnected network structure on the epidemic threshold. Phys. Rev. E **88**(2), 022801 (2013)
321. Wang, W., Tang, M., Zhang, H.F., Gao, H., Do, Y., Liu, Z.H.: Epidemic spreading on complex networks with general degree and weight distributions. Phys. Rev. E **90**(4), 042803 (2014)
322. Wang, W., Liu, Q.H., Zhong, L.F., Tang, M., Gao, H., Stanley, H.E.: Predicting the epidemic threshold of the susceptible-infected-recovered model. Sci. Rep. **6**, 24676, 1–12 (2016)
323. Watts, D.J.: Networks, dynamics, and the small-world phenomenon 1. Am. J. Soc. **105**(2), 493–527 (1999)
324. Watts, D.J., Strogatz, S.H.: Collective dynamics of 'small-world' networks. Nature **393**(6684), 440–442 (1998)
325. Wearing, H.J., Rohani, P., Keeling, M.J.: Appropriate models for the management of infectious diseases. PLoS Med. **2**(7), 621 (2005)
326. Wilkinson, R.R., Sharkey, K.J.: An exact relationship between invasion probability and endemic prevalence for Markovian SIS dynamics on networks. PLoS ONE **8**(7), e69028 (2013)
327. Wilkinson, R.R., Sharkey, K.J.: Message passing and moment closure for susceptible-infected-recovered epidemics on finite networks. Phys. Rev. E **89**(2), 022808-1–022808-6 (2014)
328. Wilkinson, R.R., Ball, F.G., Sharkey, K.J.: The relationships between message passing, pairwise, Kermack-McKendrick and stochastic SIR epidemic models. arXiv preprint arXiv:1605.03555 (2016)
329. Yan, G., Tsekenis, G., Barzel, B., Slotine, J.J., Liu, Y.Y., Barabási, A.L.: Spectrum of controlling and observing complex networks. Nat. Phys. **11**(9), 779–786 (2015)
330. Yap, H.P.: Some Topics in Graph Theory. London Mathematical Society, Lecture Notes, vol. 108. Cambridge University Press, Cambridge (1986)
331. Youssef, M., Scoglio, C.: Mitigation of epidemics in contact networks through optimal contact adaptation. Math. Biosci. Eng. **10**(4), 1227–1251 (2013)

Index

© Springer International Publishing AG 2017
I.Z. Kiss et al., *Mathematics of Epidemics on Networks*, Interdisciplinary Applied
Mathematics 46, DOI 10.1007/978-3-319-50806-1

CPSIA information can be obtained
at www.ICGtesting.com
Printed in the USA
LVHW081737300821
696471LV00001B/1